Electromagnetic Field Theory for Engineers and Physicists

T0178380

Günther Lehner

Electromagnetic Field Theory for Engineers and Physicists

Translated by Matt Horrer

 Springer

Prof. Dr.rer.nat. Günther Lehner (em.)
Universität Stuttgart
Fak.05 Informatik, Elektrotechnik
und Informationstechnik
Pfaffenwaldring 47
70569 Stuttgart

Translator
Matt Horrer
Raleigh
North Carolina
USA
mhorrer@ieee.org

ISBN 978-3-642-42529-5   ISBN 978-3-540-76306-2 (eBook)
DOI 10.1007/978-3-540-76306-2
Springer Heidelberg Dordrecht London New York

Cover design: eStudio Calamar S.L.

Printed on acid-free paper

Springer is part of Springer Science+Business Media (www.springer.com)

The Author's Preface

This book deals with the fundamental principles of electrodynamics, i .e. the theory of electromagnetic fields as given by Maxwell's equations. It is an outgrowth from the lectures, which the author has been giving to the students of electrical engineering at the University of Stuttgart, Germany, for approximately a quarter of a century. For the textbook, the contents of the lectures have been supplemented by a chapter on numerical methods for the solution of boundary and initial value problems, which provides a rough first survey over the methods available only, without going into details. Furthermore, there are several appendices devoted to some more special topics, as among others to the problem of the possibility of an extremely small but nonzero restmass of the photon, which would lead to Proca's equations, a modified version of Maxwell's equations; to the important question of eventually existing magnetic monopoles; to the deeper meaning of the electromagnetic potentials in view of quantum mechanics and the Bohm-Aharonov-effects. The last appendix covers a brief survey of special relativity, because this, in principle, is an essential part of electrodynamics, which is inevitably needed for its real understanding.

The treatment is based on Maxwell's equations from the beginning. They are described and explained in Chapter 1. The following chapters are devoted to electrostatics; to the important mathematical tools of electromagnetic field theory (method of separation of variables using cartesian coordinates, cylindrical coordinates, and spherical coordinates; conformal mapping for plane problems); to stationary current density fields; to magnetostatics; to quasi stationary time dependent problems as field-diffusion, skin effect etc.; and finally electromagnetic waves and dipole radiation. Everything in these chapters is derived from Maxwell's equations, except the additionally necessary assumptions characterizing various media, their conductivity, polarizability, and magnetizability.

The basic concepts of vector analysis are also developed from the beginning together with Maxwell's equations. The divergence (div) is defined as the small volume limit of the surface integral (flux) of a vector field and the rotation (curl) as small surface limits of three line integrals (circulations) of a vector field. These definitions immediately clarify the plausible meaning of both of these operators of vector analysis. The divergence being the volume density of sources or sinks, the rotation being the three dimensional surface density of circulation. The integral theorems of Gauss and Stokes are immediately plausible consequences of these definitions also. This procedure provides an easy and well comprehensible access to the realm of vector analysis. It also very clearly demonstrates the physical meaning of Maxwell's equations. Helmholtz's theorem (presented in one of the appendices) teaches us that each vector field is completely defined by its divergence and its rotation. So it is obvious that we need four equations to describe electric and magnetic fields, two for their sources and sinks and two for their circulations. Thus, Maxwell's equations, often considered to be almost incomprehensible, are hoped to become really plausible.

The different chapters contain a variety of analytical solutions of boundary and initial value problems. It is quite often claimed that this is no longer of interest and that such problems nowadays are usually treated numerically by computers. The author cannot share this opinion. People trying to solve electrodynamical boundary and initial value problems numerically, without having studied and understood the theoretical background and not having seen examples of a variety of fields, often if not mostly obtain faulty results. Having little or no feeling for the matter, they may believe in the correctness of their results. It is not at all easy to test if the solutions are really correct. The availability of many analytical solutions is a very valuable and even indispensable tool for testing numerical programs. It is always advisable to solve similar problems analytically when doing numerical work and to test the numerical methods by comparing the results.

Initially conceived for students of engineering and physics, the textbook turned out to be useful for professionally working engineers and physicists also. That is why six editions of the original German version of the book have appeared already. The author together with the translator hopes that the present English translation will be as useful for its readers.

Stuttgart, 2009   Günther Lehner

Table of Contents

viii   Table of Contents

List of Symbols

General Symbols

$\sim$	(*For example* $\tilde{f}$) refers to a function f which is derived via an integral transform (Fourier, Hankel, Laplace transform)
$\sim$	Proportionality sign (*for example* $F \sim m$)
*	(*For example* z*, w*) refers to the complex conjugate of a quantity (*e.g.* for z, w), or to a dual quantity (*e.g.* A* to A, φ* to φ)
$n, \perp$	A perpendicular component if used as an index
$t, \|\|$	A tangential component if used as an index
$\oint$	The circle indicates that the integral is to be taken over a closed contour (line integral) or over a closed surface (surface integral).
∇ , $\nabla \bullet$	Nabla symbol "del" and "del dot" operator (del in Cartesian coordinates: $\langle \frac{\partial}{\partial x}, \frac{\partial}{\partial y}, \frac{\partial}{\partial z} \rangle$)
$\nabla \bullet \mathbf{a}$	Divergence of the vector $\mathbf{a}$
$\nabla \times a$	Circulation of the vector $\mathbf{a}$ (curl)
$\nabla^2 \equiv \Delta$	Laplace operator
$\wedge$	Quantum mechanical operators, *e.g.* $\hat{H}$.
$\cdot$	Used to indicate multiplication of scalar quantities
$\bullet$	Scalar multiplication of vectors, scalar product, dot product
$\times$	Vector product of two vector quantities
$\mathbf{a\,b}$	Dyadic product. It is an operator whose result is a tensor matrix with elements: $(\mathbf{a\,b})_{ik} = (a_i b_k)$

Latin Letters

$\mathbf{a}, a_x, a_y, a_z$	A vector and its cartesian components a_x, a_y, a_z
$a, \mathbf{a}$	(Surface) area, magnitude and vector (where the are caould be confused with the vector potential)
$A, \mathbf{A}$	(Surface) area magnitude and vector
$\mathbf{A}$	Magnetic vector potential
$\mathbf{A}*$	Electric vector potential (in analogy to the magnetic vector potential).
$arg(z)$	Argument of a complex number (phase angle)
$ber(), bei()$	Kelvin's function
$\mathbf{B}$	Magnetic flux density, magnetic induction, B-field
$\mathbf{B}_0$	Amplitude of the magnetic field of an electromagnetic wave
B_n	Normal component of $\mathbf{B}$
B_t	Tangential component of $\mathbf{B}$
c	Speed of light in vacuum
c_G	Group velocity of light
c_{Ph}	Phase velocity of light
c_{ik}	Influence coefficients
C_{ik}	Capacitance coefficient
C	Capacitance
C'	Capacitance per unit length
$cos\ (), cosh()$	Cosine and hyperbolic cosine function
$\boldsymbol{D}$	Displacement field (electric)
D_n	Normal component of $\mathbf{D}$

D_t	Tangential component of **D**
$d\mathbf{A}, d\mathbf{a}$	Differential of a surface element (vector quantity)
dA, da	Magnitude of the vector of the surface element
$\frac{\partial}{\partial t}, \frac{\partial}{\partial x}, \ldots$	Partial derivative with respect to t, x, ...
$\frac{\partial \phi}{\partial n}$	Normal component of the gradient of the function ϕ
dt	Differential of time
$d\mathbf{s}$	Differential of a line element (vector)
ds	The magnitude of the differential of a line element
$d\tau$	Differential of a volume element
$d\Omega$	Differential of a solid angle
$d\alpha$	Differential of an angle
d	Distance, thickness of a layer, Skin depth
$\nabla \bullet \mathbf{a}$	Divergence of the vector **a**
E	Electric Field vector, **E**-field
E	Magnitude of **E** or the complex field $E = E_x + iE_y$
E_n	Normal component of **E**
E_t	Tangential component of **E**
E_0	Amplitude of the electric field of an electromagnetic wave
$\mathbf{E}_{i0}, \mathbf{E}_{r0}, \mathbf{E}_{t0}$	Amplitude of the incident, reflected, transmitted wave
$\mathbf{E}_i$	Incident electromagnetic wave
$\mathbf{E}_e$	Driving electric field

$E\left(\frac{\pi}{2}, k\right)$	Total elliptic integral of the 2nd kind
$\mathbf{e}_u$	Unity vector in the direction of the u coordinate
$\mathbf{e}_u\mathbf{e}_u$	Dyadic product
e	Electron charge
e	The number e (=2.718...), the basis of the natural logarithm
$\exp(x) = e^x$	Exponential function
$erf()$	Error function
$erfc()$	Complement to the Error function [1 - erf()]
$f()$	A function
f	Frequency
$\mathbf{F}$	Force vector
F	Magnitude of the force
F	Potential, besides φ and ϕ
G	Conductance
G'	Conductance per unit length
$G(\mathbf{r}, \mathbf{r}_0)$	Green's function
$G_D(\mathbf{r}, \mathbf{r}_0)$	Green's function for Dirichlet's boundary value problem
$G_N(\mathbf{r}, \mathbf{r}_0)$	Green's function for Neumann's boundary value problem
$\mathbf{g}, \mathbf{g}_e$	Electric current density, volume current density
$\mathbf{g}_m$	Magnetic current density (duality to the electric current density)
$\mathbf{g}_{mag}$	Magnetization current density (volume bound)
∇f	Gradient of the function f (grad f)

h	Planck's Constant
$\hbar$	h bar, Planck's constant divided by 2π $(h/2\pi)$
$\mathbf{H}$	Magnetic field, H-field
$\mathbf{H}_0$	Amplitude of the H-field of an electromagnetic wave
H_n	Normal component of $\mathbf{H}$
H_t	Tangential component of $\mathbf{H}$
$H(x\text{-}x_0)$	Heaviside's step function (generalized function)
H	Hamiltonian
$\hat{H}$	Hamilton operator
I	Current
$I_m(\ )$	Modified Bessel function of the 1st kind of order m
i	Imaginary number symbol $i = \sqrt{-1}$
$J_m(\ )$	Bessel function of order m
$K_m(\ )$	Modified Bessel function of the 2nd kind of order m
$K\left(\dfrac{\pi}{2}, k\right)$	Total elliptical integral of the 1st kind
$\mathbf{k}$	Surface current density
$\mathbf{k}_{mag}$	Surface bound current density
$\mathbf{k}$	Wave vector = vector of wave number
k	Magnitude of the wave vector (wave number)
L, l	Length
l	Propagation number, wave number
L_{ik}	Coefficient of inductance

L, L_{ii}	Self inductance
L'	Self inductance per unit length
$\ln()$	Natural logarithm
$L = (L_{ik})$	Lorentz transform
L^{-1}	Inverse Lorentz transform
L^{-T}	Transposed Lorentz transform
$\mathcal{L}\{f(t)\}$	Laplace transform of the function f
$\mathcal{L}^{-1}\{\tilde{f}(p)\}$	Inverse Laplace transform of the function f
m	Whole number, integer
m	Mass
m_0	Rest mass
$\mathbf{m}$	Magnetic dipole moment
m	Magnitude of the magnetic dipole moment
$\mathbf{M}$	Magnetization (spatial density of $\mathbf{m}$)
n	Refractive index
n	Whole number, integer
n	Number of turns per unit length
$\mathbf{n}$	Vector normal to a surface
N	Total number of turns
N	Abbreviation for the frequently occurring quantity associated with wave guides: $N = \varepsilon\mu\omega^2 - \mu\kappa\, i\omega - k_z^2$
$N_m()$	Neumann's function of index m
P	Point in space
P	Power

$\mathbf{P}$	Polarization (spatial density of $\mathbf{p}$)
$\mathbf{p}$	Electric dipole moment
p	Magnitude of the electric dipole moment
$\mathbf{p}$	Momentum vector
p	Magnitude of the momentum vector
$\hat{\mathbf{p}}$	The momentum vector operator
p_k	A component of the canonical momentum
p	Complex number (particularly used with Laplace transformation)
$P_n^m(\ )$	Associated Legendre function
$P_n(\ ) = P_n^0$	Legendre function
p_{ik}	Coefficients of the potential
Q, Q_e	Electric charge
q	Electric line charge density
Q_m	Magnetic charge
q_k	Canonical spatial coordinate
R	Resistance
R'	Resistance per unit length
R_{mag}	Magnetic resistance
R	Reflectance
R_S	Impedance of radiation
$\mathbf{r}$	Position vector
$\dot{\mathbf{r}}$	Velocity (time derivative of the position vector)
$\ddot{\mathbf{r}}$	Acceleration (2nd time derivative of the position vector)

$\mathbf{r, r'}$	Position vector in the reference frames Σ and Σ', respectively
r, R	Radius in spherical coordinates (along with θ, φ)
r	Radius in cylindrical coordinates (along with φ, z)
$\nabla \times \mathbf{a}$	curl of a vector (circulation)
$\mathbf{S}$	Poynting vector
$sin\,()$, $sinh()$	Sine and hyperbolic sine function
$tan\,()$	Tangent function
t	Time
t_0	Diffusion time
t_r	Relaxation time
T	Transmittance
U	Potential energy
$\mathbf{u, u'}$	Velocity vector in the reference frames Σ and Σ', respectively
u	Power density
u	Real part of a complex function $u + i\,v$
u_1, u_2, u_3	Coordinates in general
V	Volume
$\mathbf{v}$	Velocity vector
v	Absolute value of the velocity
v_G	Group velocity
v_{Ph}	Phase velocity
V	Voltage
V_{21}	Voltage or potential difference between the two points 1 and 2
V_i	Induced Voltage or electromotive force (EMF)

W	Energy, work
w	Energy density
w	Complex function, complex potential
x	Cartesian coordinate
y	Cartesian coordinate
Y_n^m	Spherical harmonic function
z	Cartesian coordinate
z	Complex number x + i y
z^*	Conjugate complex number (to the number z) x - i y
Z	Characteristic Impedance
Z_0	Characteristic Impedance for vacuum
Z_m	Cylindrical function for index m

Greek Letters

α	Angle
α	Damping factor (negative imaginary part of the complex wave number $k = \beta - i\alpha$)
α	Sommerfeld's fine structure constant $\alpha = \dfrac{e^2}{2h}\sqrt{\dfrac{\mu_0}{\varepsilon_0}} \approx \dfrac{1}{137}$
β	Angle
β	Phase angle of a complex number
β	Real part of the complex wave number $k = \beta - i\alpha$
β	Common abbreviation used in the theory of relativity for v/c,
δ_{ik}	Kronecker symbol

(δ_{ik})	Identity operator, unit operator
$\delta(x - x_0)$	One-dimensional Dirac delta function
$\delta(\mathbf{r} - \mathbf{r}_0)$	Three-dimensional Dirac delta function
Δ	Difference
Δ	Determinant (of a matrix)
Δ	Laplace operator (*e.g.* $\Delta = \dfrac{\partial^2}{\partial x^2} + \dfrac{\partial^2}{\partial y^2} + \dfrac{\partial^2}{\partial z^2} = \nabla^2$)
Δ_2	Laplace operator in a plane (two dimensional), *e.g.* $\Delta_2 = \dfrac{\partial^2}{\partial x^2} + \dfrac{\partial^2}{\partial y^2}$
ε	Permittivity
ε_0	Absolute Permittivity, dielectric constant of free space
ε_r	Relative permittivity
$\boldsymbol{\epsilon}$	Permittivity tensor
ε_{ik} , ε_{xy}	Components of the permittivity tensor ($\boldsymbol{\epsilon}$)
ζ	Dimensionless Cartesian coordinate: z/l
η	Dimensionless Cartesian coordinate: y/l
η	Real part of the angular velocity $\omega = \eta + i\sigma$
ϑ	Angle
θ	Polar angle (2nd coordinate of spherical coordinates)
κ	Electric conductivity
κ	Compton wave number $\kappa = \dfrac{m_0 c}{\hbar} = \dfrac{2\pi}{\lambda_c}$

λ	Wavelength
λ_c	Cutoff wavelength
λ_c	Compton wavelength
λ_{mn}	n^{th} zero of $J_m(x)$
μ	Permeability
μ_0	Permeability of vacuum
μ_r	Relative permeability
$\boldsymbol{\mu}$	Permeability tensor
μ_{ik}, μ_{xy}	Components of the Permeability tensor
μ_{mn}	n^{th} zero of the derivative $J'_m(x)$
ξ	Dimensionless coordinate (x/l)
π	pi = 3.1415...
$\boldsymbol{\Pi}_e$	The electric Hertz vector
$\boldsymbol{\Pi}_m$	The magnetic Hertz vector
ρ	Radius
ρ, ρ_e	Electric volume charge density,
ρ_m	Magnetic volume charge density
ρ_{mag}	Fictitious magnetic volume charge density
σ	Electric surface charge density
σ_{mag}	Fictitious magnetic surface charge density
σ	Imaginary part of the angular frequency $\omega = \eta + i\sigma$

$\displaystyle\sum_{i=1}^{n}$	Sum, indexed by i running from i = 1 to i = n
Σ, Σ'	Reference frames, inertial frames
τ	Dimensionless time
τ	Surface density of the electric dipole moment
τ	Volume, and $d\tau$: volume element
φ	Azimuthal angle (cylindrical and spherical coordinates)
φ	Phase angle
φ	Scalar potential
Φ	Scalar potential
Φ	Magnetic flux
χ	Electric susceptibility
χ_m	Magnetic susceptibility
ψ	Flux function
ψ	Scalar magnetic potential
ψ	Wave function in quantum mechanics
ω	Angular frequency = $2\pi f$
ω_c	Cutoff frequency
ω_{nmp}	Resonant frequency, Eigenfrequency of a cavity (m, n, p are integer numbers)
Ω	Electric flux
Ω	Solid angle
Ω	Dimensionless angular frequency

1   Maxwell's Equations

1.1   Introduction

In this book, we describe the principles which govern electric and magnetic or electromagnetic *fields* and *waves*. This area of knowledge, frequently referred to as *Electromagnetism*, has a long history and is associated with many famous names among which Maxwell has a prominent place. Maxwell was the one who, in the nineteenth century, gave electromagnetism its final form, by fixing an inconsistency and summarizing the then voluminous material into few equations, through which everything else can be derived. These equations are called Maxwell's equations. They form the foundation of the so-called *Classical Electromagnetism*. The first chapter of this book shall serve to introduce these equations.

We have to emphasize, however, that Classical Electromagnetism, which is mostly expressed through Maxwell's equations is not really complete. The 20th century brought insights that have caused extensions in two different directions. The first is related to Albert Einstein and leads to the *Theory of Relativity*. Application of this fundamental idea is intimately related, but not limited to electromagnetism. One could even go as far as stating that Classical Electromagnetism can only be understood, and its full importance recognized, through the perspective of the Theory of Relativity. Later we will discuss, that electromagnetic fields propagate in the form of waves. The thereby created *electromagnetic waves* manifest themselves in manifold ways: as radio waves, heat radiation, visible light, x-rays, gamma rays, etc. In vacuum the velocity of this propagation is the speed of *light in vacuum* ($c \approx 3 \cdot 10^8 m/s$). The Theory of Relativity elevates the speed of light to a quantity that is fundamental for the structure of space and time and thus making it a fundamental constant of nature. Besides this, electromagnetic waves have also brought another important knowledge. Light consists, as we have known since Planck, of individual particles called *photons*. Together with other fundamental discoveries, which we do not want to discuss here, this has lead to *Quantum Electrodynamics*. This theory treats electromagnetic fields as what they, according to the current state of knowledge, really are: namely waves and particles simultaneously. That is to say, it describes how they are created, destroyed, how they interact with other matter, etc.

Of these three closely related theories – Classical Electromagnetism, Special Relativity, and quantum-electrodynamics – we will only deal with classical electrodynamics. Nevertheless, occasionally it will be necessary to mention facts that go beyond it, and to clarify a situation may require use of elements from other theories, *for example* the Theory of Relativity. This restriction is purely of didactical nature and certainly not based on the idea that only classical electrodynamics is of practical value. The opposite would be true. To mention just a few examples: the characteristics and behavior of electrons in metals (band

G. Lehner, *Electromagnetic Field Theory for Engineers and Physicists*,
DOI 10.1007/978-3-540-76306-2_1, © Springer-Verlag Berlin Heidelberg 2010

model), behavior of semiconductors and consequently that of transistors, the processes in photoelectric cells, the achievements of laser technology, effects – equally as strange as important – such as superconductivity, etc., can only be discussed and understood by means of quantum theory.

1.2   Charge and Coulomb's Law

In the following section we will derive Maxwell's equations. We will do this in an abbreviated form, but roughly following the historical derivation. We will begin by considering an historically old experience, which most of us have experienced many times. If certain objects are rubbed and then separated, they exert a force on each other. Rubbing changes these objects. They are transformed into a state which we will call *electric* or *electrically charged* – whatever that may mean. To learn about those forces, we conduct the following thought experiment.

We start by choosing three different objects (A, B, C), which were electrically charged by rubbing them. There are now the following possibilities:

1.   A and B attract each other
2.   A and C attract each other

What would be the force between B and C? Is the answer to this question trivial? Can we make a prediction? In any case, the experiment provides us with the answer:

3.   B and C repel each other

Is this surprising? Is it by chance? No, it is not chance, but a law of nature. We can repeat this experiment infinitely often and always get the same result: If A attracts both B and C, then B and C repel each other. There are other possibilities:

1.   A and B repel each other
2.   A and C repel each other
3.   B and C repel each other

also:

1.   A and B attract each other
2.   A and C repel each other
3.   B and C attract each other

This result may be so familiar to us that we take it for granted and it may appear trivial, but this is not so. Were we to instead deal with gravitational forces or nuclear forces, our experiment would exhibit different results. Strictly speaking, our result is correct only under the implicit assumption that the electric force is greater than any other kind of potentially superimposed force, like gravitational or nuclear forces. This restriction is very important in natural behavior. The nucleus of an atom consists partly of particles that repel each other. The nucleus would burst apart if there were not attracting forces that more than compensate the repelling electric force. Gravitation, while an attracting force, is too weak to prevent destruction of the nucleus. One needs to remain conscious of this fact when, in the following, we make statements about electric forces.

We can summarize the experience with electrically charged objects in the following way:

1. There are two sorts of electrical charge, which we term positive and negative charges.

2. Like charges repel, while opposite charges attract each other

These qualitative statements are, however, insufficient. At the end of the day, we want to formulate physical laws quantitatively. We will utilize the following experimental result: To begin, one can measure the force between charged objects, *for example*, by utilizing springs. We measure the force of A on B, and A on C. Next, we combine B and C, and then measure the force that A exerts on the combined object B+C. We will find that this force is the sum of the individual forces of the previous experiment.

This is a principal realization, whose consequence is far reaching. For now, we simply want to justify the right to expand on our qualitative statements on charge into a more quantitative one based on the magnitude of charge. We will call the charge quantity Q. The question of units with which to express Q shall be left for later. For now, assume we have already defined the unit and found a method to measure the charge quantity Q in this unit. This allows us to measure charges Q_1 and Q_2 and so the force between the two charges, which in turn, enables us to formulate *Coulomb's law*:

1. The force between two charges Q_1 and Q_2 is proportional to both Q_1 and Q_2 and also inversely proportional to the square of the distance r_{12}^2 between them

$$F \sim \frac{Q_1 Q_2}{r_{12}^2} \tag{1.1}$$

2. The axis of the force lies on the direct line between the charges; it is repelling for like charges, and attractive for opposite charges.

The fact that F_{12} being proportional to $1/r_{12}^2$ is of great significance *i.e.* it is an *inverse square law.*. We will come back to discuss the consequences of this law later. This property is shared between electric and gravitational forces.

Forces are vector quantities. An arbitrary force **F** is therefore determined by three components, *for example* in a Cartesian coordinate system:

$$\mathbf{F} = \langle F_x, F_y, F_z \rangle \tag{1.2}$$

Suppose there is a charge Q_1 at point $\mathbf{r}_1$

$$\mathbf{r}_1 = \langle x_1, y_1, z_1 \rangle, \tag{1.3}$$

and a charge Q_2 at point $\mathbf{r}_2$

$$\mathbf{r}_2 = \langle x_2, y_2, z_2 \rangle \tag{1.4}$$

then we can write Coulomb's law, combining all those statements, in the following way:

$$\mathbf{F}_{12} = \frac{Q_1 Q_2}{4\pi\varepsilon_0} \frac{\mathbf{r}_2 - \mathbf{r}_1}{|\mathbf{r}_2 - \mathbf{r}_1|^3} \qquad (1.5)$$

where $\mathbf{F}_{12}$ is the force that Q_1 exerts on Q_2. Conversely, the force that Q_2 exerts on Q_1 is:

$$\mathbf{F}_{21} = \frac{Q_1 Q_2}{4\pi\varepsilon_0} \frac{\mathbf{r}_1 - \mathbf{r}_2}{|\mathbf{r}_1 - \mathbf{r}_2|^3} \qquad (1.6)$$

It follows from these relations that

$$\mathbf{F}_{12} + \mathbf{F}_{21} = 0 \qquad (1.7)$$

Here, $4\pi\varepsilon_0$ is just an arbitrary proportionality constant. It is arbitrary because we still have to select the units for force, charge, and length. We will later make a selection, which in turn uniquely defines the physical constant ε_0, the so-called *permittivity* or *dielectric constant in free space*. Currently we have created a fairly simple world, which consists only of charges in an otherwise empty space (vacuum).

1.3   Electric Field Strength E and Displacement Field D

A single charge in the otherwise empty space causes that space to change. A second charge brought into this space experiences a force at every point of that space. This force is expressed by Coulomb's law and varies from point to point. At this stage it will be beneficial to introduce the concept of the *electric field*. It is the quintessence of all possible effects by such forces at the different locations of this space, which become obvious only after we place a charge at a particular point.

The term field, more generally, refers to a quantity of any kind that is a function of space (and possibly of time). This book will also deal with various kinds of fields.

The electric field strength is described by a vector quantity represented by the symbol $\mathbf{E}$. It is defined as the force in the field per unit charge.

$$\mathbf{E} = \frac{\mathbf{F}}{Q} \qquad (1.8)$$

This definition makes sense because the force, according to Coulomb's law, is proportional to Q and thus E is independent on the (test) charge.

Furthermore, Coulomb's law states that a charge Q_1 at location $\mathbf{r}_1$, at an arbitrary field point $\mathbf{r}$ produces the following electric field:

$$\mathbf{E}(\mathbf{r}) = \frac{Q_1}{4\pi\varepsilon_0} \frac{\mathbf{r} - \mathbf{r}_1}{|\mathbf{r} - \mathbf{r}_1|^3} \qquad (1.9)$$

For reasons which we will be able to understand only later, we will now define only for vacuum, the vector quantity of the *electric displacement Field* **D** in the following way:

$$\boxed{\mathbf{D} = \varepsilon_0 \mathbf{E}} \tag{1.10}$$

1.4 Electric Flux

By means of **D** and Fig. 1.1 we define the *electric flux*:

$$\boxed{\Omega = \int_A \mathbf{D} \bullet d\mathbf{A} = \int_A D_n dA} \tag{1.11}$$

D_n is the component of **D** that is perpendicular (normal) to the surface element $d\mathbf{A}$. The dot indicates a scalar or dot product of two vectors. The vector $d\mathbf{A}$ is always perpendicular to the surface element and its magnitude $|d\mathbf{A}|$ equals the value of its surface area. That is:

$$|d\mathbf{A}| = dA \tag{1.12}$$

The term electric flux is based on the analogy to a moving fluid, where the velocity is:

$$\mathbf{v}(\mathbf{r}, t)$$

If the fluid is incompressible, then the amount of fluid that moves through a surface A per unit of time can be expressed as

$$\int_A \mathbf{v} \bullet d\mathbf{A}$$

This is called flux through the surface. This analogy is frequently used for the definition of all kinds of fluxes. We now want to ponder about the question, how much electric flux passes through an arbitrary closed surface if there are charges somewhere, that is to say, charges can be inside or outside the space enclosed by our surface.

The answer to this question is rather easy if we limit ourselves to the area of a sphere (radius r_0), with a charge Q_1 is at its center. (For a closed surface let $d\mathbf{A}$ be always oriented outwardly, see Fig. 1.2.)

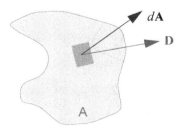

Fig. 1.1

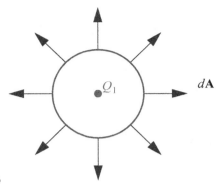

Fig. 1.2

$$\Omega = \oint_A D_n dA = \frac{Q_1}{4\pi r_0^2} \oint_A dA = \frac{Q_1}{4\pi r_0^2} 4\pi r_0^2 = Q_1 \qquad (1.13)$$

Use was made of the fact that for symmetry reasons $D_n = D = |\mathbf{D}|$. So, in this case, one finds that the flux is the charge itself. How would this result change if we changed the spherical surface into an arbitrary one? To formally solve the integral of the flux (1.11) could become very difficult. A trick, however, allows to reduce the new problem into the already solved one. We surround the charge simultaneously by an arbitrarily large surface of a sphere A_1, centered at the location of the charge Q_1, and an arbitrary surface A_2 (see Fig. 1.3). As a result, for every small cone we find the following relation:

$$\mathbf{D}_1 \bullet d\mathbf{A}_1 = \mathbf{D}_2 \bullet d\mathbf{A}_2$$

This is a consequence of:

$$\mathbf{D} \bullet d\mathbf{A} = D \, dA_t, \qquad (1.14)$$

where dA_t is the component of $d\mathbf{A}$ parallel to $\mathbf{D}$ and the fact that although D decreases with $1/r^2$ on one hand, on the other, dA_t, increases with r^2, given that r

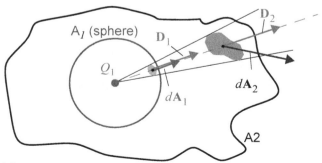

Fig. 1.3

is the distance from the charge. This means, that independently of the shape of A_2, the same flux passes through it as passes through the surface of the sphere A_1.

Let us next study the flux through a closed surface where the charge is outside of it. The same arguments as before shall serve to analyze the situation in Fig. 1.4, and to show that the flux through this closed surface will vanish entirely. Every flux entering the surface will exit it as well. In summary one writes:

$$\Omega = \left\{ \begin{array}{c} Q_1 \\ 0 \end{array} \right\} \text{ if } Q_1 \text{ is } \left\{ \begin{array}{c} \text{inside} \\ \text{outside} \end{array} \right\} \text{ of the closed surface} \qquad (1.15)$$

What happens if there are multiple charges in our space? We start with the statement that in order to determine the total force caused by all charges simultaneously, it is permissible to add the forces exerted by those charges. As forces are vectors, this addition is a vectorial addition. Addition is also permitted for the electric field. This only seemingly trivial fact has received its own term:

Superposition principle, which applies to the electric field

We have made use of this principle before, when we introduced charge. We must emphasize: The superposition principle does not state that it is allowed to add forces as vectors. This fact is a basic principle of mechanics and is the reason for the usefulness of vectors altogether. The crucial point is that the force between charges is independent of the existence of other charges in its vicinity, *i.e.*, is not changed by those other charges. This, however, is highly nontrivial and perhaps not even true under all circumstances (it may not apply, *for instance,* when we deal with very strong fields).

The superposition principle allows us to write an expression for n charges Q_i at the locations $\mathbf{r}_i$

$$\mathbf{E}(\mathbf{r}) = \sum_{i=1}^{n} \mathbf{E}_i = \sum_{i=1}^{n} \frac{Q_i}{4\pi\varepsilon_0} \frac{\mathbf{r} - \mathbf{r}_i}{|\mathbf{r} - \mathbf{r}_i|^3} . \qquad (1.16)$$

The flux Ω through any arbitrary closed surface is thus

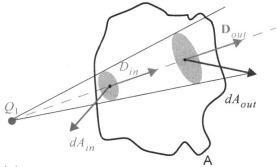

Fig. 1.4

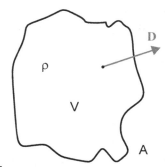

Fig. 1.5

$$\Omega = \oint \mathbf{D} \bullet d\mathbf{A} = \oint \sum \mathbf{D}_i \bullet d\mathbf{A} = \sum \oint \mathbf{D}_i \bullet d\mathbf{A} = \sum_{enclosed} Q_i$$

finally

$$\boxed{\Omega = \sum_{enclosed} Q_i} . \tag{1.17}$$

Ω equals the sum of all charges inside this closed surface. Instead of point charges, we are able to study continuously distributed charges in some space. This requires the definition of the volume charge density $\rho(\mathbf{r}, t)$. It is defined as the differential quotient

$$\boxed{\rho = \lim_{d\tau \to 0} \frac{dQ}{d\tau}} , \tag{1.18}$$

with dQ being the charge contained in the volume element $d\tau$. The total charge in a volume V is thus

$$Q = \int_V \rho \, d\tau . \tag{1.19}$$

This, on the other hand, equals the electric flux that passes through the surface of this volume, and enables us to write for any volume (see Fig. 1.5)

$$\boxed{\oint_A \mathbf{D} \bullet d\mathbf{A} = \int_V \rho \, d\tau} . \tag{1.20}$$

With this, we found a fundamental relation. It is the integral form of one of the (in total four) Maxwell's equations. Before discussing it in more detail, we will need to introduce several other terms and concepts.

1.5   Divergence of a Vector Field and Gauss' Integral Theorem

Equation (1.20) is applicable for any volume, in particular for an infinitesimally small one. This allows one to rewrite

$$\int_V \rho \, d\tau = \rho V = \oint_A \mathbf{D} \bullet d\mathbf{A}$$

or

$$\rho \;=\; \lim_{V \to 0} \frac{\oint_A \mathbf{D} \bullet d\mathbf{A}}{V} \tag{1.21}$$

This is a very important relation in vector analysis. The divergence $div\ \mathbf{a}$ or $\nabla \bullet \mathbf{a}$ for an arbitrary vector field $\mathbf{a(r)}$ is defined as the limit:

$$\boxed{\nabla \bullet \mathbf{a} \;=\; \lim_{V \to 0} \frac{\oint_A \mathbf{a} \bullet d\mathbf{A}}{V}} \; . \tag{1.22}$$

Comparison of (1.21) with (1.22) reveals

$$\boxed{\nabla \bullet \mathbf{D} \;=\; \rho} \; . \tag{1.23}$$

This is the equivalent of (1.20), the differential form of Maxwell's equation. We will verify that it is in fact a differential equation.

The way we derived this equation also illustrates its significance. We use our previous example of the incompressible fluid where $\oint_A \mathbf{v} \bullet d\mathbf{A}$. Therefore, $\nabla \bullet \mathbf{v}$ can only be non-zero, if fluid flows out of the volume element (source), or flows into it (sink). To apply this to our field lines $\mathbf{E}$ or $\mathbf{D}$, we can say that they can only originate at locations where electric charges are (Fig. 1.6).

Electric charges are sources or sinks of the electric field

Divergence is a mathematical term suited for this fact and is a measure of the strength of the source or sink.

At this point one should be alerted to what our conclusions are based on. They are a consequence of Coulomb's law, or more precisely of the $1/r^2$ dependency in it. Would this dependency be any different, the relation between D, A, and Q would not hold: $\oint \mathbf{D} \bullet d\mathbf{A} \neq Q$ and $\nabla \bullet \mathbf{D} \neq \rho$. In view of the streaming fluid and the $1/r^2$ dependency, however, we find our results to be rather trivial. A water fountain idealized as a point source pours water evenly in all directions and produces a purely radial flux field, with $v_r \propto (1/r^2)$. The flux $\oint_A \mathbf{v} \cdot d\mathbf{A}$ which does not enclose a source has to be zero. On the other hand, we have to note that any, even the slightest deviation from Coulomb's law, would be significant and would result

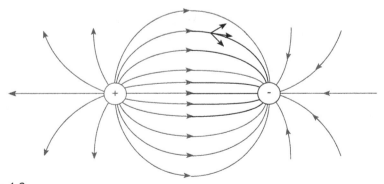

Fig. 1.6

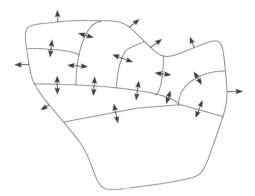

Fig. 1.7

in a quantitatively different electrodynamics. For that reason, it was interesting and necessary to verify by measurements, whether any deviation could be found. Up to now, not even the most precise measurements have found any deviation. It cannot be excluded, however, that such deviations might be found in the future when even more precise measurements become available. In such a case, this will require that this theory be modified at least in parts. These are areas of concern, which reach far into the domain of Quantum Mechanics and Relativity. They are related to the question whether the rest mass of photons is actually zero or not. Appendix A.1 will deal with this topic in more detail.

The above definition of the divergence leads to a for us very important theorem. We want to integrate $\nabla \bullet \mathbf{a}$, the divergence of a vector field $\mathbf{a}$ over the volume shown in Fig. 1.7. We use the following fact:

$$\int_V \nabla \bullet \mathbf{a} \, d\tau \ = \ \sum_i \left[\lim_{V_i \to 0} \frac{\oint \mathbf{a} \bullet d\mathbf{A}}{V_i} \right] V_i$$

This means to separate the macroscopic volume into many microscopic volume elements and then calculate the divergence for each such micro element by taking the limit of V approaching zero. In this case, all the surface integrals inside cancel because each surface occurs twice, each time with a different sign, as the normal vector for each of those two surface elements has the opposite direction. What remains is the surface integral over the outer surface of the macroscopic volume, *i.e.*:

$$\boxed{\int_V \nabla \bullet \mathbf{a} \, d\tau \ = \ \oint_A \mathbf{a} \bullet d\mathbf{A}}$$ (1.24)

This is *Gauss' Integral Theorem*.

This equation formally establishes the relation between (1.20) and (1.23). Using (1.24) in (1.20) gives:

$$\oint_A \mathbf{D} \bullet d\mathbf{A} \ = \ \int_V \rho \, d\tau \ = \ \int_V \nabla \bullet \mathbf{D} \, d\tau \ .$$

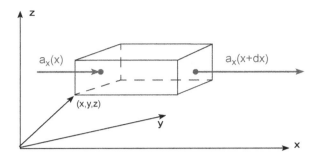

Fig. 1.8

Because this has to be true for an arbitrary volume, the integrands have to be equal to each other, *i.e.*

$$\nabla \bullet \mathbf{D} = \rho \ .$$

This means that (1.23) follows from (1.20). Conversely, from (1.23) follows

$$\int_V \nabla \bullet \mathbf{D} d\tau = \int_V \rho \, d\tau = \oint_A \mathbf{D} \bullet d\mathbf{A}$$

and hence (1.20). In conclusion, we realize that Gauss' integral theorem provides the rigorous formal proof to our previous plausibility arguments

The definition of the divergence in (1.22) is didactically advantageous, but impractical for actual computations. Therefore, we will calculate $\nabla \bullet \mathbf{a}$ in the Cartesian components of $\mathbf{a}$:

$$\mathbf{a} = \langle a_x(x, y, z), a_y(x, y, z), a_z(x, y, z) \rangle \tag{1.25}$$

We write the related surface integral and take the limit of its volume as it goes to zero (see Fig. 1.8),

$$\nabla \bullet \mathbf{a} = \lim_{V \to 0} \frac{\oint \mathbf{a} \bullet d\mathbf{A}}{V}$$

$$= \lim_{dx, dy, dz \to 0} \frac{1}{dx\,dy\,dz} \Big\{ [a_x(x + dx) - a_x(x)]dy\,dz + [a_y(y + dy) - a_y(y)]dx\,dz$$

$$+ [a_z(z + dz) - a_z(z)]dx\,dy \Big\}$$

$$= \lim_{dx, dy, dz \to 0} \frac{\left[a_x(x) + \dfrac{\partial a_x}{\partial x}dx - a_x(x)\right]dy\,dz + [\ldots + \ldots - \ldots]dx\,dz + [-\ldots]dx\,dy}{dx\,dy\,dz}$$

$$= \lim_{dx, dy, dz \to 0} \frac{\dfrac{\partial a_x}{\partial x}dx\,dy\,dz + \dfrac{\partial a_y}{\partial y}dx\,dy\,dz + \dfrac{\partial a_z}{\partial z}dx\,dy\,dz}{dx\,dy\,dz}$$

$$= \frac{\partial a_x}{\partial x} + \frac{\partial a_y}{\partial y} + \frac{\partial a_z}{\partial z}$$

that is

$$\nabla \bullet \mathbf{a} = \frac{\partial a_x}{\partial x} + \frac{\partial a_y}{\partial y} + \frac{\partial a_z}{\partial z} \qquad (1.26)$$

The divergence is a scalar quantity, formally expressed as the scalar product (dot product) of the vector operator ∇ (Nabla or "del"), for Cartesian coordinates with the vector $\mathbf{a}$, and using the Cartesian unity vectors $\mathbf{e}_x, \mathbf{e}_y, \mathbf{e}_z$ as

$$\nabla = \left\langle \frac{\partial}{\partial x}, \frac{\partial}{\partial y}, \frac{\partial}{\partial z} \right\rangle = \mathbf{e}_x\frac{\partial}{\partial x} + \mathbf{e}_y\frac{\partial}{\partial y} + \mathbf{e}_z\frac{\partial}{\partial z} . \qquad (1.27)$$

1.6   Work and the Electric Field

A charge Q within the reach of an electric field experiences the force $Q\mathbf{E}$ and moves, if not held fixed in place. The field performs work on the charge. Conversely, to move the charge against the field requires one to do work.

If we move the charge from the starting point P_A along the contour C_1 to an endpoint P_E (Fig. 1.9), then the total work we have to do is given by

$$W_1 = -\int_{C_1} \mathbf{F} \bullet d\mathbf{s} = -Q\int_{C_1} \mathbf{E} \bullet d\mathbf{s} . \qquad (1.28)$$

This is because dW for the path element $d\mathbf{s}$ is

$$dW_1 = -\mathbf{F} \bullet d\mathbf{s} . \qquad (1.29)$$

We could have moved the charge along path C_2 with the result:

$$W_2 = -\int_{C_2} \mathbf{F} \bullet d\mathbf{s} = -Q\int_{C_2} \mathbf{E} \bullet d\mathbf{s}. \qquad (1.30)$$

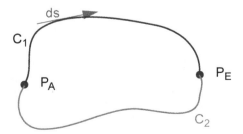

Fig. 1.9

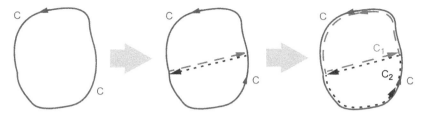

Fig. 1.10

Initially, we will only deal with time-independent fields. Suppose the results W_1 and W_2 would be different. We then could take advantage of this to build a perpetuum mobile (1st kind). Suppose, for instance, $W_2 > W_1$, then it would be possible to move the charge from P_A to P_B via path C_1 and then back to P_A via path C_2. We would need to invest the work W_1, but gain work W_2 on the way back. Overall, the work of the closed loop would be $W_2 - W_1 > 0$. Repeating this process would manifest itself as a perpetuum mobile. Of course, we have reasons to assume that this is impossible. The theorem of conservation of energy requires to state that:

$$W_1 = W_2 . \tag{1.31}$$

or

$$\int_{C_1} \mathbf{E} \bullet d\mathbf{s} - \int_{C_2} \mathbf{E} \bullet d\mathbf{s} = 0 . \tag{1.32}$$

Consequently, the work over any closed contour is

$$\boxed{\oint \mathbf{E} \bullet d\mathbf{s} = 0} . \tag{1.33}$$

This important relation was derived without using the knowledge about electric fields we have gained so far. We need to verify that the electric fields, in fact, meet this requirement. Again, we are currently studying time-independent fields only. Time dependency will come in later, and we will find that (1.33) is not applicable in such a case. Nevertheless, if (1.33) applies to a single point charge, it also applies to an arbitrary distribution of charges at rest. The reason for this is the superposition principle. It is therefore sufficient to prove (1.33) for a point charge.

Before starting our proof, we will investigate some simple properties of line integrals over closed curves. Fig. 1.10 shows a closed curve C, which is separated into two closed curves C_1 and C_2 by inserting a line. We get:

$$\oint_C \mathbf{a} \bullet d\mathbf{s} = \int_{C_1} \mathbf{a} \bullet d\mathbf{s} + \int_{C_2} \mathbf{a} \bullet d\mathbf{s} .$$

Notice that the two newly added path elements identically cancel. This kind of subdivision can be repeated by individually subdividing C_1 and C_2, respectively, and so on. If we now study the field of a point charge, we can start with any closed contour. We can reduce the integral $\oint \mathbf{E} \bullet d\mathbf{s}$ to integrals over the kind shown in Fig. 1.11. For this case we write :

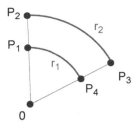

Fig. 1.11

$$\oint \mathbf{E} \bullet d\mathbf{s} = \left(\int_{P_1}^{P_2} + \int_{P_2}^{P_3} + \int_{P_3}^{P_4} + \int_{P_4}^{P_1} \right) \mathbf{E} \bullet d\mathbf{s} = 0 \, .$$

On the paths (arcs) from P_2 to P_3 and P_4 to P_1, $\mathbf{E}$ is perpendicular to $d\mathbf{s}$

$$\int_{P_2}^{P_3} \mathbf{E} \bullet d\mathbf{s} + \int_{P_4}^{P_1} \mathbf{E} \bullet d\mathbf{s} = 0 \, .$$

On the other two paths from P_1 to P_2 and P_3 to P_4, on the other hand, $\mathbf{E}$ and $d\mathbf{s}$ are parallel and anti-parallel, respectively. This means that

$$\int_{P_1}^{P_2} \mathbf{E} \bullet d\mathbf{s} = \int_{P_1}^{P_2} E \, ds = -\int_{P_3}^{P_4} E \, ds = \int_{P_3}^{P_4} -\mathbf{E} \bullet d\mathbf{s} \, .$$

This finalizes the proof. Unfortunately the perpetuum mobile is not a feasible option. Relation (1.33) will prove to be far-reaching. First, we need to introduce some terms, which will be done in the next section.

1.7   The Rotation of a Vector Field and Stoke's Integral Theorem

Consider an arbitrary vector field $\mathbf{a}(\mathbf{r})$. For any closed curve, we can write the line integral $\oint \mathbf{a} \bullet d\mathbf{s}$. We may also look at arbitrarily small area elements, and the line integrals corresponding to their boundary. Reducing the size of the area elements more and more will make the line integrals smaller and smaller, such that they will vanish in that limit. However, the ratio of the line integral over its related area element will converge towards a limit. We define a new vector field which we call the circulation or curl of $\mathbf{a}$ ($curl(\mathbf{a})$ or $\nabla \times \mathbf{a}$) as follows:

We choose three perpendicular, but otherwise arbitrary area elements dA_1, dA_2, dA_3, which share a common center in space. Together they form a right handed system. With this we write the limit.

$$\lim_{dA_i \to 0} \frac{\oint \mathbf{a} \bullet d\mathbf{s}}{dA_i} = r_i \qquad (i = 1, 2, 3) \tag{1.34}$$

Note that the line integral in the numerator is an infinitesimal loop extending over the boundary of the area element and the orientation is such that the line integral forms a right handed system with the vector dA_1 (Fig. 1.12).

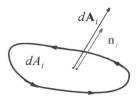

Fig. 1.12

The limits r_i are the components of the vector that represents $\nabla \times \mathbf{a}$, the circulation of the vector field $\mathbf{a}$ in the coordinate system defined by the three area elements. If

$$\mathbf{n}_i = \frac{d\mathbf{A}_i}{A_i} \qquad where \quad \mathbf{n}_i \bullet \mathbf{n}_k = \delta_{ik} = \begin{cases} 1 & for \quad i = k \\ 0 & for \quad i \neq k \end{cases}$$

represents the unit vector perpendicular to the area element $d\mathbf{A}_i$, then

$$\nabla \times \mathbf{a} = r_1 \mathbf{n}_1 + r_2 \mathbf{n}_2 + r_3 \mathbf{n}_3 = \langle r_1, r_2, r_3 \rangle$$

or

$$\mathbf{n}_i \bullet (\nabla \times \mathbf{a}) = r_i = (\nabla \times \mathbf{a})_i$$

It is not trivial that one can regard the components r_i as those of a vector. We still need to prove that those components transform like a vector when transforming the coordinate system (of course into another orthogonal one). We will not carry out this proof here, but leave it to Vector Analysis. Using the definition of *curl* provides *Stoke's Integral Theorem* almost immediately. One looks at an arbitrary area and calculate the related curl. One separates this area into many arbitrarily small sub-areas and then apply the definition of curl onto those. The result is a sum of line integrals where all internal parts cancel, and only one line integral over the outer boundary remains – as was shown in section 1.6 (using Fig. 1.10)

$$\boxed{\int_A \nabla \times \mathbf{a} \bullet d\mathbf{A} = \oint_C \mathbf{a} \bullet d\mathbf{s}} \ . \tag{1.35}$$

As before, orientation of the surface area and direction of the contour integral have to form a right handed system. Applying eq. (1.35) to the electric field and using (1.33) we get

$$\oint \mathbf{E} \bullet d\mathbf{s} = \int \nabla \times \mathbf{E} \bullet d\mathbf{A} = 0 \ .$$

This has to be true for an arbitrary curve or its surface area which it surrounds. Consequently it must be true that

$$\boxed{\nabla \times \mathbf{E} = 0} \ . \tag{1.36}$$

Conversely, eq. (1.33) follows from (1.35) and (1.36). This is because of

$$\oint \mathbf{E} \bullet d\mathbf{s} = \int \nabla \times \mathbf{E} \bullet d\mathbf{A} = 0$$

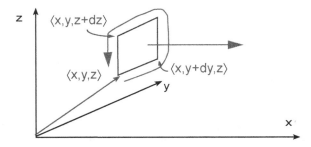

Fig. 1.13

This means that eqs. (1.33) and (1.36) are equivalent. Each follows form the other.

It is time to summarize our previous results. We found two important integral relations, namely (1.20) and (1.33);

$$\oint_A \mathbf{D} \bullet d\mathbf{A} = \int_V \rho \, d\tau \qquad\qquad \oint \mathbf{E} \bullet d\mathbf{s} = 0$$

These have two equivalent differential relations, (1.23) and (1.36)

$$\nabla \bullet \mathbf{D} = \rho \qquad\qquad \nabla \times \mathbf{E} = 0$$

The relation between them is established via the integral theorems (1.24) and (1.35)

$$\int_V \nabla \bullet \mathbf{a} \, d\tau = \oint_A \mathbf{a} \bullet d\mathbf{A} \qquad\qquad \int_A \nabla \times \mathbf{a} \bullet d\mathbf{A} = \oint_C \mathbf{a} \bullet d\mathbf{s}$$

(Gauss)   (Stokes)

This pair of relations were derived for the electrostatic case, *i.e.* charges at rest. We will need to modify one of the relations later, when we study time-dependent systems. We will then explain why $\nabla \bullet \mathbf{D} = \rho$ or its respective integral formulation can be applied without change, while the other ($\nabla \times \mathbf{E} = 0$) requires modification.

It turns out that above definition is rather impractical, should one actually try to calculate the curl of a vector field. Therefore, we will write the curl of a given field **a** in its Cartesian components (Fig. 1.13).

It is sufficient to just calculate the x component and then generalize this result, which can be done rather easily. Based on definition (1.34) and the additional constraint that the orientation of the path forms a right-handed system with $\nabla \times \mathbf{a}$, we write:

$$(\nabla \times \mathbf{a})_x = \lim_{dy, dz \to 0} \frac{a_y(z)dy + a_z(y + dy)dz - a_y(z + dz)dy - a_z(y)dz}{dy \, dz}$$

$$= \lim_{dy,\, dz \to 0} \frac{a_y(z)\,dy + a_z(y)\,dz + \dfrac{\partial a_z}{\partial y}\,dy\,dz - a_y(z)\,dy - \dfrac{\partial a_y}{\partial z}\,dy\,dz - a_z(y)\,dz}{dy\,dz}$$

$$= \lim_{dy,\, dz \to 0} \frac{\left(\dfrac{\partial a_z}{\partial y} - \dfrac{\partial a_y}{\partial z}\right)dy\,dz}{dy\,dz} = \frac{\partial a_z}{\partial y} - \frac{\partial a_y}{\partial z}.$$

Therefore formally expressed as the vector product of the Nabla operator with the vector **a**:

$$\boxed{\nabla \times \mathbf{a} = \left\langle \frac{\partial a_z}{\partial y} - \frac{\partial a_y}{\partial z}, \frac{\partial a_x}{\partial z} - \frac{\partial a_z}{\partial x}, \frac{\partial a_y}{\partial x} - \frac{\partial a_x}{\partial y} \right\rangle} \qquad (1.37)$$

Or written in the form of a determinant:

$$\nabla \times \mathbf{a} = \begin{vmatrix} \mathbf{e}_x & \mathbf{e}_y & \mathbf{e}_z \\ \dfrac{\partial}{\partial x} & \dfrac{\partial}{\partial y} & \dfrac{\partial}{\partial z} \\ a_x & a_y & a_z \end{vmatrix} \qquad (1.38)$$

The determinant is the usual form used in Vector Calculus to express the vector product. The vectors $\mathbf{e}_x, \mathbf{e}_y, \mathbf{e}_z$ are the unit vectors in x-, y-, and z-direction, respectively

$$\begin{aligned} \mathbf{e}_x &= \langle 1, 0, 0 \rangle \\ \mathbf{e}_y &= \langle 0, 1, 0 \rangle \\ \mathbf{e}_z &= \langle 0, 0, 1 \rangle \end{aligned} \qquad (1.39)$$

It is appropriate to deal with the *curl* in more detail, but we will abstain from this here and leave it to Vector Calculus to fill in the intricate details. Nevertheless, it shall be noted that the reader should not conclude from the notation *curl* **a** = $\nabla \times \mathbf{a}$ that *curl* **a** is perpendicular to ∇ or to **a**. In contrast to a real vector, it is not possible to assign a direction to the del (∇) operator. The vector resulting from $\nabla \times \mathbf{a}$ can point in any direction relative to **a**. $\nabla \times \mathbf{a}$ can be perpendicular to **a**, it can also be parallel to it. The reader should convince himself of this by studying a few examples.

Because of Gauss's theorem, for an arbitrary closed surface one has

$$\oint_A (\nabla \times \mathbf{a}) \bullet d\mathbf{A} = \int_V \nabla \bullet (\nabla \times \mathbf{a})\, d\tau$$

And because of Stoke's theorem

$$\oint_A (\nabla \times \mathbf{a}) \bullet d\mathbf{A} = 0$$

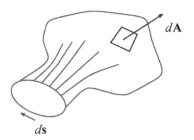

Fig. 1.14

The right side is zero because the line integral that is initially there, decreases and finally vanishes when transitioning from an open surface to a closed one (Fig. 1.14). Therefore, for an arbitrary volume

$$\oint_A \nabla \bullet (\nabla \times \mathbf{a}) d\tau = 0$$

and because this is true for any volume, the integrand has to vanish

$$\boxed{\nabla \bullet (\nabla \times \mathbf{a}) = 0} \qquad (1.40)$$

Eq. (1.40) is an important relation. It says that the curl of an arbitrary vector field has no sources. Proof of this relation can also be shown by directly applying equations (1.26) and (1.37):

$$\nabla \bullet (\nabla \times \mathbf{a}) = \frac{\partial}{\partial x}\left(\frac{\partial a_z}{\partial y} - \frac{\partial a_y}{\partial z}\right) + \frac{\partial}{\partial y}\left(\frac{\partial a_x}{\partial z} - \frac{\partial a_z}{\partial x}\right) + \frac{\partial}{\partial z}\left(\frac{\partial a_y}{\partial x} - \frac{\partial a_x}{\partial y}\right)$$

$$= 0$$

The divergence of a vector field depends rather plausibly on the sources and sinks of that field. The curl also has a plausible meaning. Let us, for instance, analyze a rotating rigid body (Fig. 1.15). Its angular velocity is ω. Thus, a point at distance r from the center axis has the velocity

$$v = |\mathbf{v}| = \omega r .$$

The angular velocity is oftentimes regarded as a vector whose magnitude is ω and its direction points along the rotational axis of a right handed system. The curl of $\mathbf{v}$ also has the direction of the axis, thus proportional to ω.

One finds that (see also Fig. 1.16)

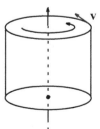

Fig. 1.15

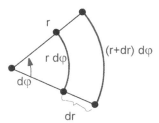

Fig. 1.16

$$|\nabla \times \mathbf{v}| = \lim_{d\varphi,\, dr \to 0} \frac{\omega(r+dr)(r+dr)d\varphi - \omega rr d\varphi}{r d\varphi dr}$$

$$= \lim_{d\varphi,\, dr \to 0} \frac{2\omega r d\varphi dr - \omega dr^2 d\varphi}{r d\varphi dr} = 2\omega \; .$$

that is

$$\nabla \times \mathbf{v} = 2\omega \tag{1.41}$$

In the area of hydrodynamics, flows whose rotation does not vanish are referred to as eddies. This refers to the circulation or rotation. Generalizing, we call those fields curl free or irrotational whose circulation vanishes, while those whose circulation is non-zero are termed rotational. Consequently, for electrostatics, one has time-independent fields which have sources, but no circulation.

1.8   Potential and Voltage

An electrostatic field may be described by different, but equivalent terms:

- •   It is irrotational
- •   The integral $\oint \mathbf{E} \bullet d\mathbf{s}$ vanishes
- •   The integral $\int_{P_A}^{P_E} \mathbf{E} \bullet d\mathbf{s}$ solely depends on the points P_A and P_E

  but not on the particular path taken from P_A to P_E.

This allows to express the field in a unique way by a scalar function, which is closely related to the previously outlined line integral describing the work. In section 1.6 we found:

$$\int_{C_1} \mathbf{E} \bullet d\mathbf{s} = \int_{C_2} \mathbf{E} \bullet d\mathbf{s} \; .$$

This is true for any path C_1 or C_2 between the points P_A to P_E. (see Fig. 1.9). Therefore, the potential function (or simply potential) can be defined by

$$\boxed{\varphi(\mathbf{r}) = \varphi_0 - \int_{r_0}^{r} \mathbf{E} \bullet d\mathbf{s}} \; . \tag{1.42}$$

The choice of the starting point $\mathbf{r}_0$, for which the potential may assume the value φ_0 is arbitrary. This is rather insignificant as just the potential difference (Voltage) is the important property. Accordingly, the voltage between two points $P_1(\mathbf{r}_1)$ and $P_2(\mathbf{r}_2)$ is given by

$$V_{21} = \varphi_2 - \varphi_1 = \varphi_0 - \int_{r_0}^{r_2} \mathbf{E} \bullet d\mathbf{s} - \varphi_0 + \int_{r_0}^{r_1} \mathbf{E} \bullet d\mathbf{s} \tag{1.43}$$

$$= \int_{r_2}^{r_1} \mathbf{E} \cdot d\mathbf{s}$$

or more compactly

$$\boxed{V_{21} = \varphi_2 - \varphi_1 = \int_{2}^{1} \mathbf{E} \bullet d\mathbf{s}} . \tag{1.44}$$

V_{21} is the work which can be gained per unit of charge when it is moved from P_2 to P_1. Consequently, the dimension for V_{21} is given as energy over charge. For two closely spaced neighboring points, the infinitesimal potential difference is

$$d\varphi = -\mathbf{E} \bullet d\mathbf{s} = - (E_x dx + E_y dy + E_z dz)$$

$$= \left(\frac{\partial \varphi}{\partial x}dx + \frac{\partial \varphi}{\partial y}dy + \frac{\partial \varphi}{\partial z}dz\right) \tag{1.45}$$

$$= (\nabla \varphi) \bullet d\mathbf{s}$$

This is the gradient which, in general, is obtained from a function f in the following way:

$$\boxed{\nabla f(\mathbf{r}) = \langle \frac{\partial f}{\partial x}, \frac{\partial f}{\partial y}, \frac{\partial f}{\partial z}\rangle = \operatorname{grad} f} \tag{1.46}$$

From eq. (1.45) follows

$$\boxed{\mathbf{E} = -\nabla \varphi} . \tag{1.47}$$

The scalar function φ describes the field completely because the gradient provides all the components of the field. The function φ is the favored way to describe the field, as this requires one to deal with a simple function instead of three functions (one for each coordinate component).

Of course we have also

$$\boxed{\nabla \times (\nabla \varphi) = 0} . \tag{1.48}$$

This is true for any function φ. The fact that the curl of the field vanishes, is the prerequisite that allows for the definition of a unique potential function. There exists a vector field for every potential, while the converse does not hold. Namely there is no unique potential for every vector field (nevertheless, in rotational fields, it is possible, and is sometimes done, to define not unique potentials). In general, it is possible to prove the generality of (1.48) by using (1.46) and (1.37). For instance, the relation for the x component is:

$$\frac{\partial}{\partial y}\frac{\partial \varphi}{\partial z} - \frac{\partial}{\partial z}\frac{\partial \varphi}{\partial y} = 0$$

The potential expresses the ability to do work, work of a particle at a particular spatial location within the field. Is there a charged particle at point $\mathbf{r}$, then the equation of motion applies

$$m\ddot{\mathbf{r}} = Q\mathbf{E} \tag{1.49}$$

If we multiply this equation with $\dot{\mathbf{r}}$

$$m\ddot{\mathbf{r}} \bullet \dot{\mathbf{r}} = Q\mathbf{E} \bullet \dot{\mathbf{r}}$$

we get

$$\frac{\mathrm{d}}{\mathrm{d}t}\left[\frac{1}{2}m\dot{\mathbf{r}}^2\right] = -Q\frac{\mathrm{d}\varphi}{\mathrm{d}t}$$

or

$$\frac{\mathrm{d}}{\mathrm{d}t}\left[\frac{1}{2}m\dot{\mathbf{r}}^2 + Q\varphi\right] = 0$$

and

$$\boxed{\frac{1}{2}m\dot{\mathbf{r}}^2 + Q\varphi = const} \tag{1.50}$$

This is the law on conservation of energy. It states that the sum of the kinetic and potential energy of a particle is constant. If one lets, for instance, a particle start at location $\mathbf{r}_0$ with velocity $\dot{\mathbf{r}} = \mathbf{v} = 0$, where the potential is φ_0, one finds

$$\frac{1}{2}mv^2 + Q\varphi = Q\varphi_0$$

or

$$\boxed{v = \sqrt{\frac{2Q}{m}(\varphi_0 - \varphi)} = \sqrt{\frac{2Q}{m}V}} \tag{1.51}$$

Here, V is the voltage through which the particle has "fallen". This relation between velocity and potential difference has many applications (for instance X-rays, electronic optics, etc.)

The locations of a potential field at which φ is constant is defined as an *equipotential surface*. For a displacement $d\mathbf{s}$ along such an equipotential surface one has

$$d\varphi = -\mathbf{E} \bullet d\mathbf{s} = 0. \tag{1.52}$$

Consequently $\mathbf{E}$ is perpendicular to $d\mathbf{s}$, *i.e.* $\mathbf{E}$ is perpendicular to the equipotential surface. Equipotential surfaces and field lines are important to illustrate fields (Fig. 1.17). Oftentimes many field lines are combined to form flux pipes (Fig. 1.18). There are no sources in the charge-free space.

$$\nabla \bullet \mathbf{D} = 0$$

or

$$\oint \mathbf{D} \bullet d\mathbf{A} = 0.$$

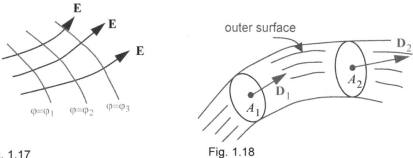

Fig. 1.17   Fig. 1.18

If one applies this onto a piece of the flux pipe, one obtains:

$$\oint \mathbf{D} \bullet d\mathbf{A} = \oint_{A_1} \mathbf{D} \bullet d\mathbf{A} + \oint_{\substack{outer \\ surface}} \mathbf{D} \bullet d\mathbf{A} + \oint_{A_2} \mathbf{D} \bullet d\mathbf{A} = 0 \, .$$

For the outer surface (skin) the relation is

$$\mathbf{D} \bullet d\mathbf{A} = 0 \, .$$

One finds thus

$$\oint_{A_1} \mathbf{D} \bullet d\mathbf{A} + \oint_{A_2} \mathbf{D} \bullet d\mathbf{A} = 0 \, .$$

This means for an infinitesimally small cross-section, if the surface elements are perpendicular to the fields, that:

$$- D_1 dA_1 + D_2 dA_2 = 0 \, .$$

or

$$\frac{D_1}{D_2} = \frac{A_2}{A_1} \, .$$

If the field components depend on the location

$$E_x = E_x(x, y, z)$$
$$E_y = E_y(x, y, z)$$
$$E_z = E_z(x, y, z) \quad ,$$

then the equations for the field lines can be obtained from the differential equations

$$E_x : E_y : E_z = dx : dy : dz \quad .$$

1.9   The Electric Current and Ampere's Law

The discovery of electric forces between electrically charged bodies has led to the previously discussed electrostatic concepts. Besides those, another type of force has been known for a long time, the so-called *magnetic force*, whose close relationship to the electric forces is a rather late discovery.

The earth, for instance, is surrounded or penetrated by a strange field, which expresses itself by exerting forces on specific materials. This field or those forces, respectively, have peculiar characteristics. For instance, they exhibit a force on a magnetic needle by trying to align it into a specific direction, while they exert no or only a minor net force on the needle as a whole. The primary effect is a torque and to a lesser degree *net forces*, which may even vanish entirely.

Historically, these phenomena were explained in terms of "magnetic charges", which were thought to be located in the magnetic poles of a magnet. This linguistic use is more confusing than helpful, and we will not introduce these concepts here in this way. Magnetic forces are – as much as we know today – of a different kind, as electrostatic ones which we have dealt with so far. We will refrain from using this only seemingly apparent analogy that suggests magnetic fields as the result of magnetic charges. *Based on our current knowledge, there are no magnetic charges. The cause of magnetic fields is rather an electric current, i.e. moving electric charges.* By experiment, one finds that a current carrying wire in the vicinity of a magnetic needle exhibits a magnetic field that influences the needle. Before we study this in more detail, we have to define the electric current and electric current density. Observe an infinitesimal area element $d\mathbf{A}$ that is perpendicular to the flow of the charge and through which in the time interval dt flows the charge d^2Q. Then the vector of the current density is defined as

$$\mathbf{g} = \frac{d^2Q}{dtdA}\frac{d\mathbf{A}}{dA} \tag{1.53}$$

The flux of $\mathbf{g}$ through a surface A is the electric current I.

$$I = \int_A \mathbf{g}\cdot d\mathbf{A} = \int_A \frac{d^2Q}{dt} = \frac{d}{dt}\int_A dQ = \frac{dQ}{dt} \tag{1.54}$$

This means that I is the total charge that flows through the surface per unit of time.

There are materials, within which charges can move freely, the so-called *conductors*. This is in distinction to *insulators*, where this is not possible under normal conditions (or only to a very limited extend). Thus, a current may flow in a conductor. It is then surrounded by a magnetic field. The simplest case is for a straight and infinitely long wire. In this case, one finds that the force on a needle of a compass is inversely proportional to its distance to the wire (that is, with increasing distance it decreases by $1/r$) and that the magnetic needle orients itself in a tangential way along concentric circles that surround the wire (Fig. 1.19).

To describe this situation one introduces the so-called magnetic field intensity H. The accompanying field surrounds the infinitely long, straight wire in the shape of closed loops. We will calculate the integral $\oint \mathbf{H}\bullet d\mathbf{s}$ for any such closed loop. First one determines the case when the loop does not enclose the wire, *i.e.* the current I. As we found already before, it is possible to reduce those integrals to ones of the form as shown in Fig. 1.20.

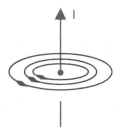

Fig. 1.19

In this case one has

$$\oint \mathbf{H} \bullet d\mathbf{s} = \left(\int_1^2 + \int_2^3 + \int_3^4 + \int_4^1 \right) \mathbf{H} \bullet d\mathbf{s}$$

where

$$\int_1^2 \mathbf{H} \bullet d\mathbf{s} = \int_3^4 \mathbf{H} \bullet d\mathbf{s} = 0$$

because of $\mathbf{H} \perp d\mathbf{s}$, and

$$\int_2^3 \mathbf{H} \bullet d\mathbf{s} = -\int_4^1 \mathbf{H} \bullet d\mathbf{s}$$

because of

$$\int_2^3 \mathbf{H} \bullet d\mathbf{s} = -\frac{C}{r_2} r_2 \varphi = -C\varphi$$

and

$$\int_4^1 \mathbf{H} \bullet d\mathbf{s} = +\frac{C}{r_1} r_1 \varphi = C\varphi.$$

C is a constant, which we will leave undetermined for now. The conclusion is that for a path that does not enclose any current one has:

$$\oint \mathbf{H} \bullet d\mathbf{s} = 0.$$

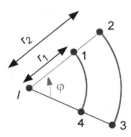

Fig. 1.20

Next we study a path C that encloses the wire. If we add a circle that also encloses the wire, which we connect to the given path, in the way specified by Fig. 1.21, then we obtain an overall path that does not include the wire and consequently the integral $\oint \mathbf{H} \bullet d\mathbf{s}$ vanishes. Because of the fact that the two integrals over the link between the circle and the initial path identically cancel, one finds that the overall path including the original path C and the line integral over the circle, which is oriented in the negative sense (clockwise), also vanish.

$$\oint_C \mathbf{H} \bullet d\mathbf{s} + \oint_K \mathbf{H} \bullet d\mathbf{s} = 0 .$$

or

$$\oint_C \mathbf{H} \bullet d\mathbf{s} = \oint_K \mathbf{H} \bullet d\mathbf{s} = \frac{C}{r_0} 2\pi r_0 = 2\pi C .$$

All such line integrals give or result in the same, non-zero value. Furthermore, one can experimentally confirm that all forces, and thus also the fields are proportional to the current. The introduced constant C is therefore also proportional to the current. If there are several currents, one only needs to add those to obtain the overall current, where only this sum is relevant. We summarize that for an arbitrary path the integral is

$$\oint \mathbf{H} \bullet d\mathbf{s} \sim I ,$$

where I is the sum of all currents that are enclosed by a particular path. One can choose the proportionality constant as one wishes, as long as one realizes that this impacts the units for current I and field $\mathbf{H}$ and length $d\mathbf{s}$, which are not yet defined. So, one may simply choose

$$\boxed{\oint \mathbf{H} \bullet d\mathbf{s} = I} , \tag{1.55}$$

This is known as *Ampere's Law*. It is more general, and its validity goes beyond the example of a straight wire, *i.e.* it applies to any distribution of currents, which can easily be verified by experiment. It contains everything relevant for the relation between time-independent currents and magnetic fields. We will need to make modifications for the time-dependent case. One can rewrite the previous equations a little:

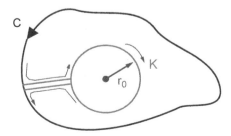

Fig. 1.21

Fig. 1.22

$$\oint \mathbf{H} \bullet d\mathbf{s} = \int_A \nabla \times \mathbf{H} \bullet d\mathbf{A} = I = \int_A \mathbf{g} \bullet d\mathbf{A} \ .$$

This is true for any surface. Therefore the integrands of the two surface integrals have to be equal and it must hold that

$$\boxed{\nabla \times \mathbf{H} = \mathbf{g}} \ . \tag{1.56}$$

Equations (1.55) and (1.56) are equivalent. One follows from the other. Both state that currents are the cause of magnetic fields. Their significance is similar to equations (1.20) and (1.23), which described the relation between the electric field and charges. There is, however, a big difference: charges cause sources of the electric field; currents cause a circulation of the magnetic field. Electrostatic fields are always free of circulation, while we will find that magnetic fields are always free of sources (to be more specific, that the magnetic induction or **B** field, which we will need to introduce, never has sources).

If we look at the results (1.55) and (1.56) more closely, one finds that there are some difficulties and even contradictions, which indicate that their current form can not be correct for time-dependent systems. Imaging two charged bodies with charges +Q and -Q, respectively, as shown in Fig. 1.22. Those bodies exert a force on each other. If we connect the two bodies by means of a conducting wire, then the charges have a path to follow the electric field. The result is an electric current, originating at the positively charged body, leading towards the negatively charged one. If we attempt to use, *for instance,*. eq. (1.55) in this situation, we experience some difficulties. Since the wire is neither closed nor extends towards infinity, we have trouble deciding if a particular path encloses the wire or not. This difficulty is even more apparent if we use (1.56). It gives

$$\nabla \bullet (\nabla \times \mathbf{H}) = \nabla \bullet \mathbf{g} = 0 , \tag{1.57}$$

This implies that the current density is source free. Obviously this is not so, as the current density originates at the charged body. During this process the charge changes as some of it is transported by the current to the other body. To enable us to discuss this in more detail, we will study the principle of conservation of charge in section 1.10.

1.10 The Principle of Charge Conservation and Maxwell's First Equation

Consider an arbitrary volume. Charges contained in it may flow in, or out of it. This is the only way for the overall charge in the volume to change. The only other possibility would be that charges spontaneously appear or disappear. Our experience teaches us that this is not the case. This is the *Principle of Conservation of Electric Charge.*

Expressed in a more general way, we find that the overall charge in the universe is constant (probably zero). Although there exist processes where new charges are created, this does not change the overall charge balance, because always the same number of positive as negative charges are created. Our experience up to now is that naturally occurring charges always come in multiples of an elementary charge. For instance, in the negative charge of an electron or in the positive charge of its counterpart, the proton. It is possible that a photon creates a pair of particles with opposite charges (*for example* a particle, antiparticle pair; electron, positron or proton, antiproton). We need to mention particles with charges that are one third or two thirds of the elementary charge (quarks) which, however, do not change the principle of charge conservation.

This principle is mathematically formulated as follows:

$$\oint \mathbf{g} \bullet d\mathbf{A} = -\frac{\partial}{\partial t}\int \rho \cdot d\tau = \int \nabla \bullet \mathbf{g} \cdot d\tau .$$

Consequently

$$\boxed{\nabla \bullet \mathbf{g} + \frac{\partial \rho}{\partial t} = 0} . \tag{1.58}$$

This is the continuity equation. It is an expression of charge conservation. On the other hand

$$\rho = \nabla \bullet \mathbf{D}$$

and therefore

$$\nabla \bullet \left(\mathbf{g} + \frac{\partial \mathbf{D}}{\partial t}\right) = 0 . \tag{1.59}$$

This means that the vector sum $\mathbf{g} + \partial \mathbf{D}/\partial t$ is source free. Therefore, it is possible to express it as the curl of a suitable vector field, as according to (1.40) the divergence of any curl vanishes:

$$\nabla \times \mathbf{a} = \mathbf{g} + \frac{\partial \mathbf{D}}{\partial t} . \tag{1.60}$$

At this point it is plausible to identify the vector $\mathbf{a}$ with the magnetic field intensity $\mathbf{H}$. In the time-independent case this would correctly lead to Ampere's law (1.55). It was Maxwell who recognized that this is incorrect in the general case. One obtains Maxwell's first equation as the correct generalization of Ampere's law for time-dependent processes.

$$\nabla\times\mathbf{H} = \mathbf{g} + \frac{\partial\mathbf{D}}{\partial t}.$$

(1.61)

The term $\mathbf{g} + \partial\mathbf{D}/\partial t$ is the total current density. It consists of two parts the free current density ($\mathbf{g}_f$) and the displacement current density $\mathbf{g}_D = \partial\mathbf{D}/\partial t$

Maxwell's first equation fixes the inconsistencies we experienced at the end of the previous section. There exists a field between the charged bodies. The electric field changes as the current flows and causes the displacement current which closes the circuit For every closed path we have

$$\oint \mathbf{H} \bullet d\mathbf{s} = \int_A \nabla\times\mathbf{H} \bullet d\mathbf{A} = \int_A \left(\mathbf{g} + \frac{\partial\mathbf{D}}{\partial t}\right) \bullet d\mathbf{A}.$$

(1.62)

The result of the integration becomes unique. That is, for a given path, it does not depend of the chosen surface area. If this were not the case, there would be no Stokes integral theorem. Let it be also noted that this can also be shown using Gauss' law and the relation $\nabla\bullet(\nabla\times\mathbf{A}) = 0$.

To derive eq. (1.59), we have used the relation $\nabla\bullet\mathbf{D} = \rho$ and thereby made a generalization, which is not quite natural and should not be made without qualifications. We have derived the equation $\nabla\bullet\mathbf{D} = \rho$ from Coulomb's law for charges at rest. Notice that the reverse conclusion may not be made. It is not necessarily possible to derive Coulomb's law from $\nabla\bullet\mathbf{D} = \rho$. Field lines originating at a charge could be organized in an unsymmetrical way, for which Coulomb's law does not apply anymore, still leaving the total flux equivalent to the charge. We do not need to assume this kind of field for charges at rest, since due to symmetry considerations no particular direction is favored. That is the reason why Coulomb's law applies to charges at rest. In order to find it, one needs to apply the symmetry argument to $\nabla\bullet\mathbf{D} = \rho$. For moving charges the situation is more complicated. The symmetry consideration is not valid anymore because the field of a moving charge is actually not spherically symmetric and Coulomb's law is not valid in this case. Still, $\nabla\bullet\mathbf{D} = \rho$ is applicable or $\oint\mathbf{D} \bullet d\mathbf{A} = Q$, respectively. Although our starting point was Coulomb's law as some basic fact, one now finds that the relation $\nabla\bullet\mathbf{D} = \rho$ is more basic and more generally applicable. It could even be seen as the real definition of charge, because for every charge, moving or at rest, belongs a corresponding flux and there is no flux without charge. Fig. 1.23 gives a qualitative picture of a charge at rest and one that moves with a uniform velocity. The field of the uniformly moving charge can be derived from that of the charge at rest by facilitating the Lorentz contraction. The distortion of the field can only be understood in the context of the Theory of Relativity. Nevertheless, this distortion is correctly described in Classical Electrodynamics. The magnetic forces caused by moving charges are exactly the consequences of the distortion of the electric field. The magnetic forces are thus also of electrical nature. They are based on the changes of the electric field due to motion. The distortion of the field of a moving charge is a relativistic effect, that is, it is noticeable at very high speeds, *i.e.* close to the speed of light and it would disappear if the speed of light were not finite. In this case there would be no magnetism. Because Classical

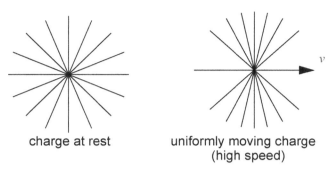

charge at rest   uniformly moving charge
                      (high speed)

Fig. 1.23

Electrodynamics already contains magnetism, which is actually a relativistic effect. It could survive the changes brought into physics by the Theory of Relativity without changing it. Let us also note here that the field of a moving charge is not irrotational. (A very worth reading discussion of those issues can be found in [1]).

Besides the vector $\mathbf{H}$, we introduce the vector $\mathbf{B}$, the magnetic flux density or magnetic induction. For vacuum one has

$$\boxed{\mathbf{B} = \mu_0 \mathbf{H}} \,. \tag{1.63}$$

The forces exerted by magnetic fields are actually based on $\mathbf{B}$. It would be most appropriate to call $\mathbf{B}$ the magnetic field strength, what some authors actually do. μ_0 is the *permeability of free space*. Key to comprehension of the so far described magnetic forces is the recognition that they only apply to moving charges.

$$\boxed{\mathbf{F} = Q\mathbf{v} \times \mathbf{B}} \,. \tag{1.64}$$

This is the Lorentz force. Is there also an electric field, then one gets the overall

$$\mathbf{F} = Q(\mathbf{E} + \mathbf{v} \times \mathbf{B}) \,. \tag{1.65}$$

The Lorentz force is perpendicular to $\mathbf{v}$ and $\mathbf{B}$. This results in strange effects. For instance parallel currents attract each other (Fig. 1.24). The current I_2 (I_1) causes a field $\mathbf{B}_1$ ($\mathbf{B}_2$) at the location of the current I_2 (I_1) and this field induces the Lorentz force $\mathbf{F}_{12}$ ($\mathbf{F}_{21}$). It is interesting to study the force of a current carrying wire loop that is placed in the field of another current (Fig. 1.25). The current I causes the field $\mathbf{B}$ at the location of the loop S that carries the current I_s. The Lorentz force acts only on currents that are perpendicular to $\mathbf{B}$, and causes a torque, in much the same way as we have described for the compass needle. As far as we know today, all magnetic materials are characterized by currents that circulate within them (Ampere's molecular currents). This is apart from phenomenons related to the spin–a basic property of elementary particles. The spin of those particles causes them to act as if they carried circulating currents. We conclude that there are no magnetic charges and all magnetic forces are ultimately Lorentz forces (disregarding again the effects of the spin of the elementary particles).

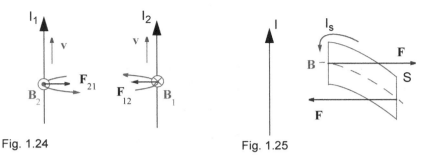

Fig. 1.24   Fig. 1.25

1.11 Faraday's Law of Induction

In addition to the above described experience, there is one more basic property. Faraday's law of induction, usually referred to as the Law of Induction, formulates this phenomenon in the following way:

If a magnetic flux through the surface of a closed loop changes over time, then there will be an induced voltage in that loop, which is proportional to the change of the magnetic flux.

By magnetic flux we mean the flux of the magnetic induction

$$\phi = \int \mathbf{B} \bullet d\mathbf{A}$$ (1.66)

The voltage that we find in a closed loop is the electromotive force (EMF) and is given by the integral

$$\oint \mathbf{E} \bullet d\mathbf{s} ,$$

which vanishes in the electrostatic case. Here, this is not the case and one has now

$$\oint \mathbf{E} \bullet d\mathbf{s} = - \frac{\partial}{\partial t}\phi ,$$ (1.67)

where we have already decided that a possible proportionality constant is dimensionless and equal to 1. One may write as well

$$\oint \mathbf{E} \bullet d\mathbf{s} = \int_{A} (\nabla \times \mathbf{E}) \bullet d\mathbf{A} = - \frac{\partial}{\partial t}\int_{A} \mathbf{B} \bullet d\mathbf{A} = -\int_{A} \frac{\partial}{\partial t}\mathbf{B} \bullet d\mathbf{A} .$$

Since this is true for every surface A, one may also write

$$\nabla \times \mathbf{E} = - \frac{\partial}{\partial t}\mathbf{B} .$$ (1.68)

The two equivalent relations (1.67) and (1.68) represent Maxwell's 2nd equation, in integral and differential from, respectively.

One takes the divergence of both sides in eq. (1.68) to find

$$\nabla \bullet (\nabla \times \mathbf{E}) = -\nabla \bullet \frac{\partial}{\partial t}\mathbf{B} = - \frac{\partial}{\partial t}\nabla \bullet \mathbf{B} = 0 .$$

Note that $\nabla\bullet$ and $\partial/\partial t$ commute. Consequently, the divergence of **B** may only be a time-independent spatial function:

$$\nabla\bullet\mathbf{B} = f(\mathbf{r}) \ . \tag{1.69}$$

From our experience, we conclude that

$$f(\mathbf{r}) \equiv 0 \ , \tag{1.70}$$

and therefore

$$\boxed{\nabla\bullet\mathbf{B} = 0} \ . \tag{1.71}$$

The field lines of B are thus free of sources and sinks. As previously determined, this means that there are no magnetic charges at which field lines could begin or end. A frequent conclusion is that this means magnetic field lines have to either close on themselves, or extend into infinity. This conclusion is, however, incorrect. There are examples for fields whose lines do neither (a more detailed discussion on this is found in Chapter 5, Section 5.11.2, which deals with magnetostatics).

1.12 Maxwell's Equations

In the previous Sections we have introduced all of Maxwell's equations. There is a differential and an integral form for each of the those four equations. We summarize them here, side by side.

differential form	integral form
$\nabla\times\mathbf{H} = \mathbf{g} + \dfrac{\partial\mathbf{D}}{\partial t}$	$\oint_C \mathbf{H}\bullet d\mathbf{s} = \int_A \left(\mathbf{g} + \dfrac{\partial\mathbf{D}}{\partial t}\right)\bullet d\mathbf{A}$
$\nabla\times\mathbf{E} = -\dfrac{\partial\mathbf{B}}{\partial t}$	$\oint_C \mathbf{E}\bullet d\mathbf{s} = -\dfrac{\partial}{\partial t}\int_A \mathbf{B}\bullet d\mathbf{A}$
$\nabla\bullet\mathbf{B} = 0$	$\oint_A \mathbf{B}\bullet d\mathbf{A} = 0$
$\nabla\bullet\mathbf{D} = \rho$	$\oint_A \mathbf{D}\bullet d\mathbf{A} = \int_V \rho\, dt$

$$(1.72)$$

These are two vector and two scalar equations for five vector quantities (**E**, **D**, **H**, **B**, **g**) and one scalar quantity (ρ). Obviously, since every vector equation is equivalent to three scalar ones, there are more unknowns (5 times $3 + 1 = 16$) than equations (2 times $3 + 2 = 8$). If we consider, as we have seen in the previous Section, that the relation $\nabla\bullet\mathbf{B} = 0$ follows from Maxwell's second equation, or more precisely, $\nabla\bullet\mathbf{B} = 0$ serves as an initial condition within the system of Maxwell's equations, then the discrepancy grows even larger. Now we have 7 equations for 16 unknowns. That means, to supplement Maxwell's equations we need 9 more. At least for vacuum, we have met some of those equations already:

$$\left.\begin{array}{l} \mathbf{D} = \varepsilon_0 \mathbf{E} \\ \mathbf{B} = \mu_0 \mathbf{H} \end{array}\right\} . \tag{1.73}$$

If we deal with matter in general, then we need to describe $\mathbf{D}$ in some form as a function of $\mathbf{E}$ and $\mathbf{B}$ as a function of $\mathbf{H}$ (which we will discuss later in more detail). These relations are called the *constitutive relations*:

$$\left.\begin{array}{l} \mathbf{D} = \mathbf{D}(\mathbf{E}) \\ \mathbf{B} = \mathbf{B}(\mathbf{H}) \end{array}\right\} . \tag{1.74}$$

We gain another equation from the fact that electric currents in conductors are caused by electric fields and thus somehow depend on the electric field

$$\mathbf{g} = \mathbf{g}(\mathbf{E}) . \tag{1.75}$$

In the simplest case, and usually the most important one finds that the volume current density $\mathbf{g}$ is proportional to $\mathbf{E}$ (this is Ohm's law)

$$\mathbf{g} = \kappa \mathbf{E} . \tag{1.76}$$

The coefficient κ is the *specific electric conductivity*. Summarizing, one can say that *(1.74)* and *(1.75)*, *supplement Maxwell's equations (1.72), making it a complete system of equations.*

Maxwell's equations are linear. This is a consequence of the superposition principle for both the electric and the magnetic fields (for the magnetic field, it is contained in the reflections that led to (1.55)). Linearity is the formal expression for the superposition principle. Linearity is also important for practical applications, that is, to solve specific problems. Linear equations are much easier to solve than non-linear ones. Linearity is lost when the supplementing "material equations" (1.74) and (1.75) become non-linear, which is a possible scenario.

Maxwell's equations exhibit a high degree of symmetry, which gives them kind of an aesthetic charm. This symmetry is particularly obvious in the case of vacuum with no charges or currents present. Here we get

$$\left| \begin{array}{c} \nabla \times \mathbf{H} = \dfrac{\partial \mathbf{D}}{\partial t} \\[2ex] \nabla \times \mathbf{E} = -\dfrac{\partial \mathbf{B}}{\partial t} \end{array} \right| \begin{array}{c} \nabla \bullet \mathbf{B} = 0 \\[2ex] \nabla \bullet \mathbf{D} = \rho \end{array} \left| \begin{array}{c} \mathbf{B} = \mu_0 \mathbf{H} \\[2ex] \mathbf{D} = \varepsilon_0 \mathbf{E} \end{array} \right| \tag{1.77}$$

We will see that this symmetry results in important consequences. A changing electric field $(\partial \mathbf{D}/\partial t)$ causes a magnetic circulation $(\nabla \times \mathbf{H})$, which is also varying in time and causes an electric circulation $(\nabla \times \mathbf{E})$, etc. This describes the mechanism of generation and propagation of electromagnetic waves (Fig. 1.26), to which radio waves, light, heat radiation, etc. owe their existence.

This symmetry is somewhat lost when we introduce charges and currents. This result is somewhat unsatisfactorily. At least as far as we know today, this asymmetry is based on the fact that there are no magnetic charges which would serve as sources of the magnetic field. There are a number of scientists who do not believe that this is the last word on that matter. In fact, it is conceivable that such magnetic charges, though not yet discovered, do exist. Thus the search for such charges continues. If they exist, this would require one to make modifications to Maxwell's equations. It is a useful exercise to determine how this would need to be done. Besides the spatial density of electric charges (ρ_e) there would be magnetic charges (ρ_m). Both could cause currents ($\mathbf{g}_e$, $\mathbf{g}_m$). In addition to the principle of conservation of electric charges

$$\nabla \bullet \mathbf{g}_e + \frac{\partial \rho_e}{\partial t} = 0, \tag{1.78}$$

we would require the conservation of magnetic charges

$$\nabla \bullet \mathbf{g}_m + \frac{\partial \rho_m}{\partial t} = 0. \tag{1.79}$$

Then

$$\nabla \bullet \mathbf{D} = \rho_e \tag{1.80}$$

and

$$\nabla \bullet \mathbf{B} = \rho_m. \tag{1.81}$$

This gives us

$$\nabla \bullet \left(\mathbf{g}_e + \frac{\partial \mathbf{D}}{\partial t} \right) = 0.$$

$$\nabla \bullet \left(\mathbf{g}_m + \frac{\partial \mathbf{B}}{\partial t} \right) = 0.$$

Those equations can be satisfied with the Ansatz

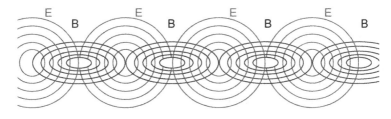

Fig. 1.26

$$\nabla \times \mathbf{a} = \mathbf{g}_m + \frac{\partial \mathbf{B}}{\partial t} \,,$$

$$\nabla \times \mathbf{b} = \mathbf{g}_e + \frac{\partial \mathbf{D}}{\partial t} \,.$$

We know that **b** represents **H**. In a similar way, to arrive correctly at Maxwell's 2nd equation from $\mathbf{g}_m = 0$, we identify **a** as representing **-E**. The overall result would then be:

$$\nabla \times \mathbf{H} = \frac{\partial \mathbf{D}}{\partial t} + \mathbf{g}_e \qquad \qquad \nabla \bullet \mathbf{B} = \rho_m$$

$$\nabla \times \mathbf{E} = -\frac{\partial \mathbf{B}}{\partial t} - \mathbf{g}_m \qquad \qquad \nabla \bullet \mathbf{D} = \rho_e \qquad \qquad (1.82)$$

Should magnetic charges be discovered some day, Maxwell's equations were to modify in the indicated way. Some further remarks on magnetic charges are treated in Appendix A.2.

We now return to Maxwell's equations without magnetic charges. They describe a vast abundance of phenomena, with which we will involve ourselves in the following. This is usually done step by step and we will proceed in the same manner. That is to say, we will not attempt to solve the full system of Maxwell's equations right way, but start with fields that do not exhibit any time dependency. Then we get

$$\nabla \times \mathbf{H} = \mathbf{g}$$

$$\nabla \times \mathbf{E} = 0$$

$$\nabla \bullet \mathbf{B} = 0$$

$$\nabla \bullet \mathbf{D} = \rho \,.$$

Two of those equations depend solely on electrostatic quantities and we already know these equations

$$\nabla \times \mathbf{E} = 0$$

$$\nabla \bullet \mathbf{D} = \rho \,.$$

They define the *electrostatics*, which will be our first area of study. Of course do we need to include the relation $\mathbf{D} = \mathbf{D}(\mathbf{E})$. The other two equations

$$\nabla \times \mathbf{H} = \mathbf{g}$$

$$\nabla \bullet \mathbf{B} = 0$$

define the *magnetostatic* case, if we add the constitutive relation between **B** and **H**. This topic will occupy the second principle part of this book. Only in the last, the third principle part will we deal with the complete set of Maxwell's equations where we cover topics like skin effect, wave propagation, radiation, antennas, etc.

We base all this on our current state of knowledge, which is not necessarily final. It is always possible that new discoveries some day will mandate us to modify our understanding and our theories. We already came across several questions, for which there are currently no final answers, such as whether or not Coulomb's law is exact, or whether magnetic charges actually do exist, etc. This lies within the nature of science. Of course even though our current answers may only be preliminary, they are nevertheless interesting and important enough to study.

1.13 System of Units

Initially, we left open the practical question of which units one should employ for the various quantities That is, which system of units one should introduce. We will now remedy this situation.

There are quite a number of different systems of units in use and there are many discussions on which one would be best for whatever reason. Those discussions are not profitable and we will refrain from doing it here. This book will use a single system consistently, namely the MKSA system, which is used internationally, and which in some countries is mandated by law.

Every system of units is based on basic units from which other quantities are derived. The MKSA system got its name form the fact that meter, kilogram, second, and Ampere were chosen as its basis. Naturally, every basic unit needs a firm definition, *i.e.* it has to be defined through a reference or a "normal". The term needs a clear definition and experimentally, the quantity needs to be readily reproducible. The normal could be of a physical *prototype* or a *natural phenomenon*. For the MKSA system the four basic units are defined in the following way:

1. 1 Meter (m): Since 1983, one meter is defined by the propagation time of light. Specifically, the distance that light in vacuum travels during

$$\frac{1}{299792458} s \, .$$

Previously (1889 - 1960) the definition of the meter was in terms of a prototype bar that was kept in Paris (France), which consisted of 90% platinum and 10 % iridium. This prototype of the meter was supposed to be exactly one ten-millionth part of the distance from an Earth pole to the equator (but it was not accurate). Between 1963 and 1983 the definition was based on spectroscopy, *i.e.* the measure of the spectral line of a particular wave length of Krypton-86.

2. 1 Second (s): Recently, the definition of time is also based on spectroscopy, namely the time span of

   9192631770 periods

   of a particular radiation of caesium. Before that, 1 second was defined as the 86400th fraction of a mean solar day of the year 1900.

3.   1 Kilogram (kg): It was, and still is defined as the mass equivalent to that of a prototype consisting of platinum and iridium which is kept in Paris (France).

4.   1 Ampere (A): The definition used to be that of a current, exactly constant in time, which in one second deposits $1.118 \text{ mg} = 1.118 \cdot 10^{-6}\text{kg}$ of silver from an aqueous solution of silver-nitrate. The unit defined in this way is now called the international Ampere and differs slightly from the so-called absolute Ampere, which is the one that is used today as the basic unit for current. To understand this definition, we need to remember the attracting force between two parallel, current carrying, conducting wires, described in Section 1.10 (see also Fig. 1.24). If we take two parallel wires of infinite length at a distance r from each other that carry the currents I_1 and I_2 then the magnetic field B_1 that I_1 produces at the location of the current I_2 is

$$B_1 = \mu_0 H_1 = \mu_0 \frac{I_1}{2\pi r} . \tag{1.83}$$

This results from the definition of **B** by (1.63) and from equations (1.55) or (1.56), exploding the symmetry of the problem. If we choose for instance (1.55) we get

$$\oint \mathbf{H} \bullet d\mathbf{s} = 2\pi r H_1 = I_1 ,$$

that is

$$H_1 = \frac{I_1}{2\pi r} .$$

The force exerted on the second wire follows from (1.83), together with (1.64). The current in a wire consists of moving charges. The force on a single charge in the second wire that moves with velocity v is

$$F = Qv\frac{\mu_0 I_1}{2\pi r} ,$$

and the overall force on the whole wire is

$$F_t = \sum_i Q_i v_i \frac{\mu_0 I_1}{2\pi r} , \tag{1.84}$$

where the sum extends over all charges in wire 2. This is an infinite sum as the wire is infinitely long and thus contains an infinite number of charges. However, the force per unit length remains constant.

$$\frac{F_t}{L} = \frac{\mu_0 I_1}{2\pi r} \frac{\sum_i Q_i v_i}{L} . \tag{1.85}$$

The expression $(\sum_i Q_i v_i)/L$ is nothing more than the current I_2.

$$\frac{\sum_i Q_i v_i}{L} = I_2 \ . \tag{1.86}$$

Because of the definition of the current as the charge per unit of time that crosses a particular cross section we find the overall relation

$$\boxed{\frac{F}{L} = \frac{\mu_0 I_1 I_2}{2\pi r}} \ . \tag{1.87}$$

Next, we consider two infinitely long, infinitesimally narrow, straight wires at a distance of $1m$ apart. We also require the wires to be parallel and carry the same current $I = I_1 = I_2$. If each one exerts the force of $2 \cdot 10^{-7}$ Newton per meter of its length, then the related current amounts to $I = I_1 = I_2 = 1A$ (Ampere). Here, Newton is the unit of force in the MKS-system.

$$1N = 1 \text{ Newton} = 1\frac{mkg}{s^2}$$

This results in

$$2 \cdot 10^{-7} \ \frac{N}{m} = \frac{\mu_0 A^2}{2\pi m} \ .$$

With this definition of the basic unit Ampere, we have also defined μ_0 :

$$\boxed{\mu_0 = 4\pi \cdot 10^{-7} \ \frac{N}{A^2}} \ . \tag{1.88}$$

We have now introduced those four basic units. We will derive the other units from these. There are the purely mechanical units. The unit for force was already used:

$$1 \text{ Newton} = 1N = 1\frac{mkg}{s^2} \ ,$$

the unit for energy

$$1 \text{ Joule} = 1J = 1Nm = 1\frac{m^2 kg}{s^2} \ ,$$

and the unit for power

$$1 \text{ Watt} = 1W = 1\frac{Nm}{s} = 1\frac{J}{s} = 1\frac{m^2 kg}{s^3} \ .$$

This leads us to the electrical units. From the definition of current,

$$I = \frac{dQ}{dt}$$

we can derive the unit for charge

$$1 \text{ Coulomb} = 1C = 1As \ .$$

Note the interesting fact that the charge is just a deduced quantity, although it is of fundamental importance and was the actual starting point of our discussion. Charges occur in nature only in multiples of the so-called elementary charge (disregarding quarks). This charge is very small, and just a tiny fraction of one Coulomb, namely

$$e \approx 1.6 \cdot 10^{-19} \ C \ .$$ (1.89)

By defining e in this way, we set the charge of an electron to $-e$ and that of a proton or positron to $+e$. Because of

$$\mathbf{F} = Q\mathbf{E} \ ,$$

The unit of the electric field is $1 N/C$, and thus the unit of the potential becomes $1 \ Nm/C$ ($\mathbf{E} = -\nabla\varphi$). It is called Volt:

$$1 \ \text{Volt} = 1V = 1\frac{Nm}{C} = 1\frac{J}{C} = 1\frac{W}{A} \ .$$

Therefore

$$1V \cdot 1A = 1W$$
$$1V \cdot 1C = 1J$$

From Coulomb's law

$$|\mathbf{F}| = \frac{Q_1 Q_2}{4\pi\varepsilon_0 r^2} \ ,$$

one derives the dimension for ε_0, $[\varepsilon_0]$

$$[\varepsilon_0] = \frac{C^2}{Nm^2} = \frac{C^2}{Jm} = \frac{C^2}{CVm} = \frac{As}{Vm} \ ,$$

The numerical value can be obtained by measurement. It depends on the chosen system of units. In our system of units

$$\varepsilon_0 = 8.855 \cdot 10^{-12} \frac{As}{Vm} \ .$$ (1.90)

The previously mentioned unit for the electric field (1 N/C) may also be expressed as 1 V/m. This settles the units for the electric displacement (D) to

$$1\frac{As}{Vm}\frac{V}{m} = 1\frac{As}{m^2} = 1\frac{C}{m^2} \ ,$$

which is obvious from the relation

$$\oint \mathbf{D} \bullet d\mathbf{A} = Q \ .$$

This definition allows us to write μ_0 in a different form

$$[\mu_0] = \frac{N}{A^2} = \frac{VC}{mA^2} = \frac{VAs}{A^2 m} = \frac{Vs}{Am} \ ,$$

and finally

$$\mu_0 = 1.2566 \cdot 10^{-6} \frac{Vs}{Am} \ .$$ (1.91)

When comparing the definitions in (1.90) and (1.91) it strikes us that the product of μ_0 and ε_0 has a purely mechanical dimension.

$$[\mu_0 \varepsilon_0] = \frac{Vs}{Am} \frac{As}{Vm} = \left(\frac{s}{m}\right)^2 ,$$

The numerical value for this gives

$$\boxed{\frac{1}{\mu_0 \varepsilon_0} = 9 \cdot 10^{16} \left(\frac{m}{s}\right)^2 = c^2} , \qquad\qquad (1.92)$$

that is the square of the speed of light. This is no coincidence. Historically, this was the first indication that light is of an electromagnetic nature, which will occupy us in later sections.

The unit for current density is $1 A/m^2$. The unit for the magnetic field intensity H results from Maxwell's first equation (1.61) to $1 A/m$. Because of the relation $B = \mu_0 H$, this determines the units for B to $1 Vs/Am \cdot 1 A/m.= 1 Vs/m^2$. This unit is called Tesla

$$1 \text{ Tesla} = 1 T = 1 \frac{Vs}{m^2} .$$

The unit for the magnetic flux is expressed in *Weber* and results to

$$1 \frac{Vs}{m^2} 1 m^2 = 1 Vs = 1 \text{ Weber} = 1 Wb .$$

Another derived unit is that for Resistance

$$1 \text{ Ohm} = 1 \Omega = 1 \frac{V}{A} ,$$

for Capacitance

$$1 \text{ Farad} = 1 F = 1 \frac{C}{V} = 1 \frac{As}{V} = 1 \frac{s}{\Omega} ,$$

and for Inductance

$$1 \text{ Henry} = 1 H = 1 \frac{Vs}{A} = 1 \Omega s .$$

Those quantities have not been introduced yet and we will need to make up for this later. The definitions for 1 Henry and 1 Farad are also used to express μ_0 in Henry per meter (H/m) and ε_0 in Farad per meter (F/m).

Every physical unit has to be understood as the product of a numerical value and a unit:

quantity = numerical value • unit

Examples for this are found in equations (1.88), (1.89), (1.90), and (1.92) of this section. The usual rules for calculations apply for such products, which should be clear from the way we derived the relations.

Finally, to conclude this section, we will state some useful conversion factors towards other frequently used units of measurement.

1 Tesla = 1 T = 10^4 Gauss

1 Maxwell = 1M = 10^{-8} Weber

1 electronvolt = 1 eV = $1.6 \; 10^{-19}$ Joule .

2 Basics of Electrostatics

2.1 Fundamental Relations

The fundamental relations of electrostatics were introduced in Chapter 1. Before discussing electrostatics in more detail, we summarize the basic results.

The force between two charges Q_1 and Q_2 is given by Coulomb's law:

$$\mathbf{F}_{12} = \frac{Q_1 Q_2}{4\pi\varepsilon_0} \cdot \frac{\mathbf{r}_2 - \mathbf{r}_1}{|(\mathbf{r}_2 - \mathbf{r}_1)|^3} \ . \tag{2.1}$$

This, in turn, determines the electric field that a charge Q_1 at location $\mathbf{r}_1$ produces at location $\mathbf{r}$ in free space

$$\mathbf{E}(\mathbf{r}) = \frac{Q_1}{4\pi\varepsilon_0} \cdot \frac{\mathbf{r} - \mathbf{r}_1}{|(\mathbf{r} - \mathbf{r}_1)|^3} \ , \tag{2.2}$$

while the electric displacement is defined to be

$$\mathbf{D}(\mathbf{r}) = \varepsilon_0 \mathbf{E} = \frac{Q_1}{4\pi} \cdot \frac{\mathbf{r} - \mathbf{r}_1}{|(\mathbf{r} - \mathbf{r}_1)|^3} \ . \tag{2.3}$$

For an arbitrary charge distribution it follows that

$$\oint \mathbf{D} \bullet d\mathbf{A} = Q = \int_V \rho \, d\tau \ . \tag{2.4}$$

or

$$\nabla \bullet \mathbf{D} = \rho \ . \tag{2.5}$$

For charges at rest (this is what is discussed in this chapter) one has

$$\oint \mathbf{E} \bullet d\mathbf{s} = 0 \tag{2.6}$$

or

$$\nabla \times \mathbf{E} = 0 \ . \tag{2.7}$$

This is the basis for defining a potential function

$$\varphi(\mathbf{r}) = \varphi_0 - \int_{\mathbf{r}_0}^{\mathbf{r}} \mathbf{E} \bullet d\mathbf{s} \ . \tag{2.8}$$

Conversely, the electric field is given by

$$\mathbf{E} = -\nabla\varphi \ . \tag{2.9}$$

Because of (2.5), it is also true that

$$\nabla \bullet \mathbf{E} = \frac{\rho}{\varepsilon_0} \ . \tag{2.10}$$

Using (2.9) one obtains

$$\nabla \bullet (-\nabla\varphi) = \frac{\rho}{\varepsilon_0}$$

G. Lehner, *Electromagnetic Field Theory for Engineers and Physicists*,
DOI 10.1007/978-3-540-76306-2_2, © Springer-Verlag Berlin Heidelberg 2010

or

$$\nabla\bullet(\nabla\varphi) \ = \ \nabla^2\varphi \ = \ \Delta\varphi \ = \ -\frac{\rho}{\varepsilon_0} \ . \tag{2.11}$$

This is Poisson's equation, which is going to play a central role . For the specific case of $\rho = 0$, one obtains Laplace's equation

$$\nabla^2\varphi \ = \ 0 \ . \tag{2.12}$$

Expressed in Cartesian coordinates

$$\nabla^2\varphi \ = \ \frac{\partial}{\partial x}\frac{\partial\varphi}{\partial x} + \frac{\partial}{\partial y}\frac{\partial\varphi}{\partial y} + \frac{\partial}{\partial z}\frac{\partial\varphi}{\partial z} \ = \ \frac{\partial^2}{\partial x^2}\varphi + \frac{\partial^2}{\partial y^2}\varphi + \frac{\partial^2}{\partial z^2}\varphi \ ,$$

that is

$$\nabla^2 \ = \ \frac{\partial^2}{\partial x^2} + \frac{\partial^2}{\partial y^2} + \frac{\partial^2}{\partial z^2} \ . \tag{2.13}$$

The symbols ∇^2 or Δ represent the *Laplace operator* also referred to as *Laplacian*.

2.2   Field Intensity and Potential for a given Charge Distribution

If one places a point charge Q_1 at location $\mathbf{r}_1$, one finds the electric field intensity to be

$$\mathbf{E}(\mathbf{r}) \ = \ \frac{Q_1}{4\pi\varepsilon_0} \cdot \frac{\mathbf{r} - \mathbf{r}_1}{|\mathbf{r} - \mathbf{r}_1|^3} \tag{2.14}$$

or written in terms of its components:

$$\left.\begin{array}{l}
E_x \ = \ \dfrac{Q_1}{4\pi\varepsilon_0} \cdot \dfrac{x - x_1}{\sqrt{(x-x_1)^2 + (y-y_1)^2 + (z-z_1)^2}^3} \\[3ex]
E_y \ = \ \dfrac{Q_1}{4\pi\varepsilon_0} \cdot \dfrac{y - y_1}{\sqrt{(x-x_1)^2 + (y-y_1)^2 + (z-z_1)^2}^3} \\[3ex]
E_z \ = \ \dfrac{Q_1}{4\pi\varepsilon_0} \cdot \dfrac{z - z_1}{\sqrt{(x-x_1)^2 + (y-y_1)^2 + (z-z_1)^2}^3}
\end{array}\right\} \ . \tag{2.15}$$

To calculate the potential we will start from the general definition:

$$\varphi \ = \ \varphi_B - \int_{\mathbf{r}_B}^{\mathbf{r}} \mathbf{E} \bullet d\mathbf{s} \ , \tag{2.16}$$

where φ_B is an arbitrarily chosen reference potential evaluated at point $\mathbf{r}_B$ (reference point). To calculate φ, one needs to evaluate the line integral along some path from $\mathbf{r}_B$ to $\mathbf{r}$. Since the integral is independent of the chosen path, one

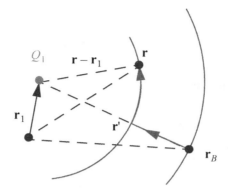

Fig. 2.1

has the freedom to choose any convenient path as proven in the previous chapter (Section 1.6 - 1.8).

We will take advantage of this freedom and simplify an otherwise difficult task. We choose the path indicated in Fig. 2.1. Starting at our reference point $\mathbf{r}_B$, we head towards the charge at $\mathbf{r}_1$ until we reach the concentric sphere around Q_1 on which our field point $\mathbf{r}$ lies. This is at point $\mathbf{r}'$, where

$$|\mathbf{r} - \mathbf{r}_1| = |\mathbf{r}' - \mathbf{r}_1| \ .$$

Then we continue on the sphere, centered at the location of Q_1, towards the field point $\mathbf{r}$ until we reach it. We write the integral

$$\varphi(\mathbf{r}) = \varphi_B - \int_{\mathbf{r}_B}^{\mathbf{r}'} \mathbf{E} \bullet d\mathbf{s} - \int_{r'}^{\mathbf{r}} \mathbf{E} \bullet d\mathbf{s}$$

$$= \varphi_B - \int_{\mathbf{r}_B}^{\mathbf{r}'} \mathbf{E} \bullet d\mathbf{s} = \varphi_B - \int_{|\mathbf{r}_B - \mathbf{r}_1|}^{|\mathbf{r}' - \mathbf{r}_1|} \frac{Q_1}{4\pi\varepsilon_0 x^2} \cdot dx$$

$$= \varphi_B + \frac{Q_1}{4\pi\varepsilon_0 |\mathbf{r}' - \mathbf{r}_1|} - \frac{Q_1}{4\pi\varepsilon_0 |\mathbf{r}_B - \mathbf{r}_1|} \ ,$$

$$\varphi(\mathbf{r}) = \varphi_B + \frac{Q_1}{4\pi\varepsilon_0 |\mathbf{r} - \mathbf{r}_1|} - \frac{Q_1}{4\pi\varepsilon_0 |\mathbf{r}_B - \mathbf{r}_1|} \ . \tag{2.17}$$

If we decide to pick $\varphi_B = 0$ for a reference point at infinity, then

$$\boxed{\varphi(\mathbf{r}) = \frac{Q_1}{4\pi\varepsilon_0 |\mathbf{r} - \mathbf{r}_1|} = \frac{Q_1}{4\pi\varepsilon_0 \sqrt{(x - x_1)^2 + (y - y_1)^2 + (z - z_1)^2}}} \ . \tag{2.18}$$

We use this to calculate the electric field

$$\mathbf{E} = -\nabla\varphi \ ,$$

and find the field components according to (2.15).

If one has many point charges $Q_1, Q_2, ..., Q_i, ...$ at the locations $\mathbf{r}_1, \mathbf{r}_2, ..., \mathbf{r}_i, ...$, then, because of the superposition principle (which applies not only to the field, but also to the potential), one uses

$$\varphi = \sum_i \frac{Q_i}{4\pi\varepsilon_0 |\mathbf{r} - \mathbf{r}_i|} . \tag{2.19}$$

In general, the charge distribution may be continuous. When the charge density is given as a function of the location $\mathbf{r}$, then

$$\boxed{\varphi(\mathbf{r}) = \frac{1}{4\pi\varepsilon_0} \int \frac{dQ'}{|\mathbf{r} - \mathbf{r}'|} = \frac{1}{4\pi\varepsilon_0} \int \frac{\rho(\mathbf{r}')d\tau'}{|\mathbf{r} - \mathbf{r}'|}} , \tag{2.20}$$

where $d\tau'$ is the volume element in the space of the vector $\mathbf{r}'$, *i.e.*

$$d\tau' = dx'dy'dz' . \tag{2.21}$$

The corresponding electric field is

$$\mathbf{E} = -\nabla\varphi(\mathbf{r}) = -\frac{1}{4\pi\varepsilon_0} \int \nabla_r \frac{\rho(\mathbf{r}')d\tau'}{|\mathbf{r} - \mathbf{r}'|} . \tag{2.22}$$

Notice that the gradient operator operates on $\mathbf{r}$ only and not on $\mathbf{r}'$. To highlight this, the del operator is marked with the index r. Now

$$\nabla_r \frac{1}{|\mathbf{r} - \mathbf{r}'|} = \nabla_{(x, y, z)} \frac{1}{\sqrt{(x - x')^2 + (y - y')^2 + (z - z')^2}}$$

$$= -\frac{2(\mathbf{r} - \mathbf{r}')}{2\sqrt{(x - x')^2 + (y - y')^2 + (z - z')^2}^3} = -\frac{\mathbf{r} - \mathbf{r}'}{|\mathbf{r} - \mathbf{r}'|^3} , \tag{2.23}$$

and finally

$$\boxed{\mathbf{E}(\mathbf{r}) = \frac{1}{4\pi\varepsilon_0} \int \frac{\rho(\mathbf{r}')(\mathbf{r} - \mathbf{r}')}{|\mathbf{r} - \mathbf{r}'|^3} d\tau'} . \tag{2.24}$$

Sometimes one deals with situations where there are charges distributed on surfaces or line elements (*surface charge, line charge*). One then defines the surface charge density σ as the charge per unit area,

$$\sigma = \frac{dQ}{dA} . \tag{2.25}$$

The associated potential is then given by

$$\varphi(\mathbf{r}) = \frac{1}{4\pi\varepsilon_0} \int \frac{\sigma(\mathbf{r}')}{|\mathbf{r} - \mathbf{r}'|} dA' , \tag{2.26}$$

and the electric field

$$\mathbf{E}(\mathbf{r}) = \frac{1}{4\pi\varepsilon_0} \int \frac{\sigma(\mathbf{r}')(\mathbf{r} - \mathbf{r}')}{|\mathbf{r} - \mathbf{r}'|^3} dA' . \tag{2.27}$$

Similarly, the line charge density or linear density q is defined as the charge per unit length,

$$q = \frac{dQ}{dl} \, , \tag{2.28}$$

with an associated potential

$$\varphi(\mathbf{r}) = \frac{1}{4\pi\varepsilon_0}\int\frac{q(\mathbf{r}')}{|\mathbf{r}-\mathbf{r}'|}dl' \, , \tag{2.29}$$

and the electric field

$$\mathbf{E}(\mathbf{r}) = \frac{1}{4\pi\varepsilon_0}\int\frac{q(\mathbf{r}')(\mathbf{r}-\mathbf{r}')}{|\mathbf{r}-\mathbf{r}'|^3}dl' \, . \tag{2.30}$$

In principle, these formulas allow for the calculation of the potential and electric field for an arbitrary distribution of point charge, line-, surface-, and volume charge densities, as well as any combination thereof. However, carrying out the integrals is not always easy for real-world problems, and the mathematical difficulties may be appreciable. Nevertheless, it is frequently possible to simplify the task by taking advantage of symmetries. This is illustrated in the next section via some specific examples.

2.3   Specific Charge Distributions

2.3.1   One-dimensional, Planar Charge Distributions

In this case, ρ is a function of only one Cartesian coordinate (*e.g.* x):

$$\rho = \rho(x) \, .$$

Here, it is better not to use the general integrals of the previous section, but start with considering the symmetry. $\mathbf{E}$ and $\mathbf{D}$ have to depend on x only and can only have an x-component. For the same reasons, the potential can only have an x-dependency. This enables us to start with the relations

$$\nabla\bullet\mathbf{D} = \frac{\partial D_x}{\partial x} = \rho(x) \tag{2.31}$$

and

$$\nabla^2\varphi = \frac{\partial^2}{\partial x^2}\varphi = -\frac{\rho(x)}{\varepsilon_0} \, , \tag{2.32}$$

which allows one to calculate D_x and φ by integrating once and twice, respectively. Thus

$$D_x(x) = D_x(a) + \int_a^x\rho(x')dx' = \frac{1}{2}\int_{-\infty}^x\rho(x')dx' - \frac{1}{2}\int_x^\infty\rho(x')dx' \tag{2.33}$$

It is left as an exercise for the reader to determine how the constant of integration $D_x(a)$ needs to be chosen, in order to arrive at the result.

2.3.2 Spherically Symmetric Distributions

If the charge distribution depends solely on the distance r to a center

$$r = \sqrt{x^2 + y^2 + z^2} \; , \tag{2.34}$$

then it is spherically symmetric and

$$\rho = \rho(r) \; .$$

It would be very difficult to calculate φ and $\mathbf{E}$ by using the general integrals. Exploiting the symmetry simplifies the problem dramatically, however. One may assume that $\mathbf{E}$ and $\mathbf{D}$ only have components that point towards or away from the center (radial components E_r and D_r), and that these depend on r only. One surrounds the center of symmetry with a concentric sphere, which allows one to apply relation (1.20), and immediately solve this problem:

$$\boxed{\oint_A \mathbf{D} \bullet d\mathbf{A} = \int_V \rho d\tau}$$

that is

$$\oint_A D_r(r)dA = D_r 4\pi r^2 = \int_0^r \rho(r')4\pi r'^2 dr'$$

or

$$D_r(r) = \frac{1}{r^2}\int_0^r \rho(r')r'^2 dr' \tag{2.35}$$

and

$$E_r(r) = \frac{1}{\varepsilon_0 r^2}\int_0^r \rho(r')r'^2 dr' \; . \tag{2.36}$$

Finally, if we also choose $\varphi = 0$ for the limit of $r \to \infty$, then

$$\varphi(r) = -\int_\infty^r \frac{1}{\varepsilon_0 r'^2}\left(\int_0^{r'} \rho(r'')r''^2 dr''\right)dr' \; . \tag{2.37}$$

On the other hand we have

$$\frac{\partial}{\partial r}\varphi(r) = -\frac{1}{\varepsilon_0 r^2}\int_0^r \rho(r'')r''^2 dr''$$

$$r^2\frac{\partial}{\partial r}\varphi(r) = -\frac{1}{\varepsilon_0}\int_0^r \rho(r'')r''^2 dr''$$

$$\frac{\partial}{\partial r}\left(r^2\frac{\partial}{\partial r}\varphi(r)\right) = -\frac{1}{\varepsilon_0}\rho(r)r^2$$

and thus

$$\boxed{\frac{1}{r^2}\frac{\partial}{\partial r}\left(r^2\frac{\partial}{\partial r}\varphi(r)\right) = -\frac{\rho(r)}{\varepsilon_0}} \; . \tag{2.38}$$

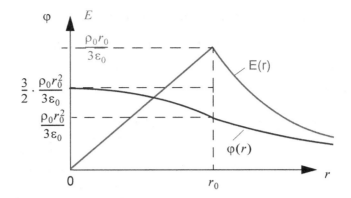

Fig. 2.2

This is nothing other than Poisson's differential equation for the specific case of spherical symmetry. In a later section, we will find that $(1/r^2)(\partial/\partial r)(r^2)(\partial/\partial r)$ is nothing else than the radial part of Laplace's operator ∇^2.

To illustrate this, we use a simple **example**. Suppose a sphere with radius r_0 is filled with the uniform volume charge density ρ_0. There shall be no other charges. The electric field then results in

for $r \leq r_0$: $E_r(r) = \dfrac{1}{\varepsilon_0 r^2} \int_0^r \rho_0 r'^2 dr' = \dfrac{1}{\varepsilon_0 r^2} \rho_0 \dfrac{r^3}{3} = \dfrac{\rho_0}{3\varepsilon_0} r$

for $r \geq r_0$: $E_r(r) = \dfrac{1}{\varepsilon_0 r^2} \int_0^{r_0} \rho_0 r'^2 dr' = \dfrac{1}{\varepsilon_0 r^2} \rho_0 \dfrac{r_0^3}{3} = \dfrac{\rho_0 r_0^3}{3\varepsilon_0} \dfrac{1}{r^2}$

For the potential one finds

for $r \leq r_0$: $\varphi(r) = -\int_\infty^r E_r(r') dr' = -\int_\infty^{r_0} \dfrac{\rho_0 r_0^3}{3\varepsilon_0} \dfrac{1}{r'^2} dr' - \int_{r_0}^r \dfrac{\rho_0}{3\varepsilon_0} r' dr'$

$$= \dfrac{\rho_0 r_0^3}{3\varepsilon_0} \dfrac{1}{r_0} - \dfrac{\rho_0}{3\varepsilon_0} \dfrac{(r^2 - r_0^2)}{2} = -\dfrac{\rho_0}{3\varepsilon_0} \dfrac{3r_0^2 - r^2}{2}$$

for $r \geq r_0$: $\varphi(r) = \dfrac{\rho_0 r_0^3}{3\varepsilon_0} \dfrac{1}{r}$

A plot of these relations is given in Fig. 2.2.

Conversely, it is also possible to find the charge density for a given potential. One might ask, what charge density gives the spherically symmetric potential $(Q_0/4\pi\varepsilon_0 r)$? If one wants to proceed formally, then one may use eq. (2.38) to calculate $\rho(r)$.

$$\frac{Q_0}{4\pi\varepsilon_0 r^2}\frac{\partial}{\partial r}r^2\frac{\partial}{\partial r}\left(\frac{1}{r}\right) = \frac{Q_0}{4\pi\varepsilon_0}\frac{1}{r^2}\frac{\partial}{\partial r}r^2\left(-\frac{1}{r^2}\right) = -\frac{Q_0}{4\pi\varepsilon_0}\frac{1}{r^2}\frac{\partial}{\partial r}1 = 0 \ .$$

The result is thus that $\rho = 0$. This not entirely correct. In order to differentiate, one needs to exclude the origin. However, this is where the charge $Q = Q_0$ is located, and it is precisely this charge that creates the given potential. This example illustrates that one needs to be mathematically very careful when dealing with point charges. To remedy this, we will introduce the Dirac δ function in a later section. It enables one to formally treat point charges in the same way as other distributions. The point charge could also be somewhat hidden and thus less obvious than in this trivial example. Take, for instance, the potential

$$\varphi = \frac{Q_0}{4\pi\varepsilon_0 r}\exp\left(-\frac{r}{r_D}\right) \ .$$

This is the so-called *shielded Coulomb potential* (in contrast to the ordinary Coulomb potential ($Q_0/4\pi\varepsilon_0 r$)). It is relevant for the theory of electrolytes and plasmas, which will not be covered here. The volume charge density for this potential is:

$$\rho(r) = -\varepsilon_0\frac{1}{r^2}\frac{\partial}{\partial r}r^2\frac{\partial}{\partial r}\left[\frac{Q_0}{4\pi\varepsilon_0 r}\exp\left(-\frac{r}{r_D}\right)\right]$$

$$= -\frac{Q_0}{4\pi}\frac{1}{r^2}\frac{\partial}{\partial r}r^2\left[-\frac{1}{r^2}\exp\left(-\frac{r}{r_D}\right) - \frac{1}{rr_D}\exp\left(-\frac{r}{r_D}\right)\right]$$

$$= -\frac{Q_0}{4\pi}\frac{1}{rr_D^2}\exp\left(-\frac{r}{r_D}\right) \ .$$

This allows one to calculate *for example*, the charge within a sphere of radius r.

$$\int_0^r \rho_0(r')4\pi r'^2 dr' = -\int_0^r\frac{Q_0}{4\pi}\frac{1}{r'r_D^2}\exp\left(-\frac{r'}{r_D}\right)4\pi r'^2 dr'$$

$$= Q_0\left(1+\frac{r}{r_D}\right)\exp\left(-\frac{r}{r_D}\right) - Q_0$$

One finds the electric field intensity

$$E_r = -\frac{\partial}{\partial r}\varphi(r) = \frac{Q_0}{4\pi\varepsilon_0}\left(\frac{1}{r^2}+\frac{1}{rr_D}\right)\exp\left(-\frac{r}{r_D}\right) \ .$$

From this we obtain the charge inside a sphere of radius r:

$$4\pi\varepsilon_0 r^2 E_r = Q_0\left(1+\frac{r}{r_D}\right)\exp\left(-\frac{r}{r_D}\right) \ .$$

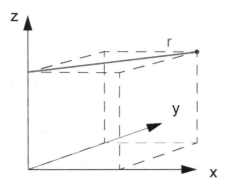

Fig. 2.3

There appears to be a contradiction. The integration over the charge density led to a charge that is smaller by Q_0. The puzzle is resolved if one looks closely at E_r for a very small radius:

$$E_r = \frac{Q_0\left(1 + \dfrac{r}{r_D}\right)\exp\left(-\dfrac{r}{r_D}\right)}{4\pi\varepsilon_0 r^2} \Rightarrow \frac{Q_0}{4\pi\varepsilon_0 r^2} \; .$$

This is the field generated by a point charge located the origin (or the potential $Q_0/4\pi\varepsilon_0 r$ of a point charge at the origin) for very small radii. This point charge is not included in our expression for ρ and in the integral over it. This clears up the apparent contradiction. Again, it appears that is necessary to be very careful. The total charge outside of the origin is just $-Q_0$, the overall charge is then zero. The charge outside cancels the point charge, and shields it, which is why we refer to this as a shielded Coulomb potential.

2.3.3   Cylindrically Symmetric Distributions

If the charge density depends solely on the distance r from an axis, then this distribution is deemed to be cylindrically symmetric (Fig. 2.3)

$$\rho = \rho(r) \tag{2.39}$$

with

$$r = \sqrt{x^2 + y^2} \; . \tag{2.40}$$

If we replace the concentric sphere of Section 2.3.2 by a coaxial cylinder, then we may proceed in much the same way as before. Starting from

$$\oint_A \mathbf{D} \bullet d\mathbf{A} = \int_V \rho \, d\tau \, ,$$

now, on a per unit length basis, we obtain:

$$2\pi r D_r = \int_0^r \rho(r') 2\pi r' dr'$$

or

$$D_r = \frac{1}{r} \int_0^r \rho(r') r' dr',$$ (2.41)

where D_r represents the component of $\mathbf{D}$ which points away radially from the axis. This is the only component of $\mathbf{D}$, which results from the symmetry of the problem. From this we get

$$E_r = \frac{1}{\varepsilon_0 r} \int_0^r \rho(r') r' dr',$$ (2.42)

and

$$\varphi = -\frac{1}{\varepsilon_0} \int_{r_B}^r \frac{1}{r'} \left(\int_0^{r'} \rho(r'') r'' dr'' \right) dr',$$ (2.43)

if the potential $\varphi = 0$ for $r = r_B$. Then

$$\frac{\partial \varphi}{\partial r} = -\frac{1}{\varepsilon_0 r} \int_0^r \rho(r'') r'' dr''$$

$$r \frac{\partial \varphi}{\partial r} = -\frac{1}{\varepsilon_0} \int_0^r \rho(r'') r'' dr''$$

$$\frac{\partial}{\partial r} \left(r \frac{\partial \varphi}{\partial r} \right) = -\frac{\rho(r) r}{\varepsilon_0}$$

that is

$$\boxed{\frac{1}{r} \frac{\partial}{\partial r} \left(r \frac{\partial \varphi}{\partial r} \right) = -\frac{\rho}{\varepsilon_0}}.$$ (2.44)

This is again Poisson's equation, now for this specific case of cylindrical symmetry. As an **example**, consider a cylinder of radius r_0 with uniform charge density ρ_0. There shall be no other charges. Then the electric field is

for $r \le r_0$: $E_r(r) = \frac{1}{\varepsilon_0 r} \int_0^r \rho_0 r' dr' = \frac{\rho_0}{2\varepsilon_0} r$

for $r \ge r_0$: $E_r(r) = \frac{1}{\varepsilon_0 r} \int_0^{r_0} \rho_0 r' dr' = \frac{\rho_0 r_0^2}{2\varepsilon_0} \cdot \frac{1}{r}$

For the potential, under the assumption that $r_B > r_0$, one finds

for $r \le r_0$: $\varphi = -\int_{r_B}^r E_r(r') dr' = -\int_{r_B}^{r_0} E_r(r') dr' - \int_{r_0}^r E_r(r') dr'$

$$= -\frac{\rho_0 r_0^2}{2\varepsilon_0} \ln \frac{r_0}{r_B} - \frac{\rho_0}{2\varepsilon_0} \left(\frac{r^2 - r_0^2}{2} \right)$$

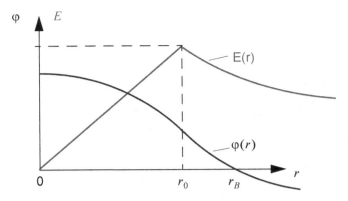

Fig. 2.4

$$\varphi \;=\; -\frac{\rho_0 r_0^2}{2\varepsilon_0}\!\left(\ln\frac{r_0}{r_B} + \frac{\dfrac{r^2}{r_0^2}-1}{2} \right)$$

for $r \geq r_0$: $\varphi = -\int_{r_B}^{r} E_r(r')dr' = -\frac{\rho_0 r_0^2}{2\varepsilon_0}\ln\frac{r}{r_B}\,.$

These relations are shown in Fig. 2.4.

An interesting limit is obtained in the case of a line charge at the axis. In this case, r_0 approaches zero, but it does this in a way that the product $\rho_0 r_0^2 \pi = q$ remains finite. The ρ_0, therefore, needs to become infinite. In this case the field becomes

$$E_r = \frac{q}{2\pi\varepsilon_0 r}\,, \tag{2.45}$$

and

$$\varphi = -\frac{q}{2\pi\varepsilon_0}\ln\frac{r}{r_B}\,, \tag{2.46}$$

where $r_B (0 < r_B < \infty)$ is the radius where φ vanishes. Here, φ is called the *logarithmic potential*, and is typical of the straight and uniform line charge. The potential shown in Fig. 2.4 is not a true logarithmic potential because it is not logarithmic for every radius r.

2.4   The Field Generated by two Point Charges

The field generated by two point charges as a special case from the potential of eq. (2.19)

$$\varphi = \frac{1}{4\pi\varepsilon_0}\left[\frac{Q_1}{\sqrt{(x-x_1)^2+(y-y_1)^2+(z-z_1)^2}}\right.$$
$$\left. + \frac{Q_2}{\sqrt{(x-x_2)^2+(y-y_2)^2+(z-z_2)^2}}\right] . \tag{2.47}$$

Taking the gradient gives in column vector notation:

$$\mathbf{E} = -\nabla\varphi = \begin{bmatrix} E_x \\ E_y \\ E_z \end{bmatrix} = \begin{bmatrix} \frac{1}{4\pi\varepsilon_0}\left[\frac{Q_1(x-x_1)}{\sqrt{(x-x_1)^2+(y-y_1)^2+(z-z_1)^2}^3}\right. \\ \left. + \frac{Q_2(x-x_2)}{\sqrt{(x-x_2)^2+(y-y_2)^2+(z-z_2)^2}^3}\right] \\ \frac{1}{4\pi\varepsilon_0}\left[\frac{Q_1(y-y_1)}{\sqrt{(x-x_1)^2+(y-y_1)^2+(z-z_1)^2}^3}\right. \\ \left. + \frac{Q_2(y-y_2)}{\sqrt{(x-x_2)^2+(y-y_2)^2+(z-z_2)^2}^3}\right] \\ \frac{1}{4\pi\varepsilon_0}\left[\frac{Q_1(z-z_1)}{\sqrt{(x-x_1)^2+(y-y_1)^2+(z-z_1)^2}^3}\right. \\ \left. + \frac{Q_2(z-z_2)}{\sqrt{(x-x_2)^2+(y-y_2)^2+(z-z_2)^2}^3}\right] \end{bmatrix} \tag{2.48}$$

We will use a coordinate system according to Fig. 2.5 and then simplify above expression somewhat.

We have

$$\left. \begin{aligned} \mathbf{r}_1 &= \langle -\tfrac{d}{2}, 0, 0 \rangle \\ \mathbf{r}_2 &= \langle +\tfrac{d}{2}, 0, 0 \rangle \end{aligned} \right\} \tag{2.49}$$

and

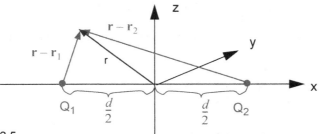

Fig. 2.5

$$E_x = \frac{1}{4\pi\varepsilon_0}\left[\frac{Q_1\left(x+\frac{d}{2}\right)}{\sqrt{\left(x+\frac{d}{2}\right)^2+y^2+z^2}^{\,3}} + \frac{Q_2\left(x-\frac{d}{2}\right)}{\sqrt{\left(x-\frac{d}{2}\right)^2+y^2+z^2}^{\,3}}\right]$$

$$E_y = \frac{1}{4\pi\varepsilon_0}\left[\frac{Q_1 y}{\sqrt{\left(x+\frac{d}{2}\right)^2+y^2+z^2}^{\,3}} + \frac{Q_2 y}{\sqrt{\left(x-\frac{d}{2}\right)^2+y^2+z^2}^{\,3}}\right] \qquad (2.50)$$

$$E_z = \frac{1}{4\pi\varepsilon_0}\left[\frac{Q_1 z}{\sqrt{\left(x+\frac{d}{2}\right)^2+y^2+z^2}^{\,3}} + \frac{Q_2 z}{\sqrt{\left(x-\frac{d}{2}\right)^2+y^2+z^2}^{\,3}}\right]$$

A remarkable fact is that there exists a point where the field vanishes. This is a special point and, because of the frequently mentioned analogy to flux problems, is called the *stagnation point*. To calculate its coordinates x_s, y_s, z_s, one sets all three components of **E** in eq. (2.50) equal to zero and then solves this equation for $x = x_s, y = y_s, z = z_s$. We skip this simple calculation and just give the result:

$$x_s = \begin{cases} \dfrac{d}{2}\dfrac{\sqrt{|Q_1|}+\sqrt{|Q_2|}}{\sqrt{|Q_1|}-\sqrt{|Q_2|}} & \text{for charges of opposite signs} \\[4mm] \dfrac{d}{2}\dfrac{\sqrt{|Q_1|}-\sqrt{|Q_2|}}{\sqrt{|Q_1|}+\sqrt{|Q_2|}} & \text{for charges of same sign} \end{cases} \qquad (2.51)$$

$$y_s = 0\ ,$$

$$z_s = 0\ .$$

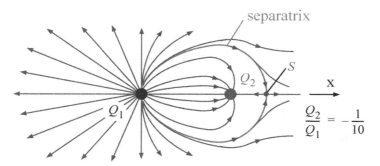

separatrix

Q_2

S

Q_1

x

$$\frac{Q_2}{Q_1} = -\frac{1}{10}$$

Fig. 2.6

The stagnation point always lies on the straight line connecting the two charges. For charges of the same sign, it lies between the two charges and closer to the one with the smaller magnitude. For charges of opposite signs, it lies outside closer to the charge with the smaller magnitude.

The stagnation point exhibits a strange property, namely that force lines cut one another here, which is possible only because the field vanishes at this specific point.

Knowledge of the location of the stagnation point is rather useful in being able to generate a qualitative picture of the field. Let us investigate the case of opposite charges as shown in Fig. 2.6, where *for example*, $Q_1 > 0$, $Q_2 < 0$, $|Q_1| > |Q_2|$. Some of the force lines which originate at Q_1 end at Q_2. Since it was given that $|Q_1| > |Q_2|$, not all can end at Q_2. Those which can not end at the other charge, extend to infinity. This is plausible, as from a great distance the configuration has to appear as that of a point charge of value $Q_1 + Q_2$. There are, therefore, two kinds of force lines: those that end at Q_2, and those extending to infinity. They can be found in the different regions of Fig. 2.6, which would provide the full 3D picture if rotated around the x-axis. The border of the two regions is made up of force lines that run through the stagnation point and from that point on, they can no longer be uniquely traced. Those lines of force are sometimes referred to as *separatrices*, *i.e.* as lines that separate different regions. Another interesting task is to analyze the equipotential surfaces (see Fig. 2.7).

Again, the equipotential surface which passes through the stagnation point plays a prominent role. It is also called separatrix. It separates the space in three different regions. The first region encloses just one charge, the second just the other, and the third region encloses both charges.

For charges of the same sign, we show the electric force lines and the equipotential surfaces in Fig. 2.8 and Fig. 2.9. The separatrices are highlighted

It is possible to show that the angle between those equipotential surfaces which pass through the stagnation point and the x-axis is the same in both cases and for all charges. One finds

$$\tan\alpha = \sqrt{2}, \qquad \alpha = 55° .$$

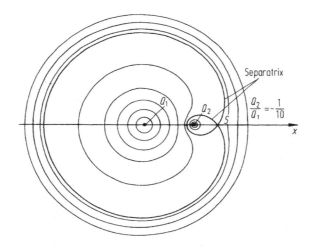

Fig. 2.7

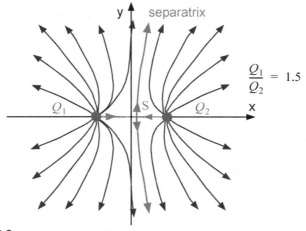

Fig. 2.8

This is even true for rotationally symmetric charge distributions of all kinds, not only for the case of the two point charges discussed here.

If there are more than two charges, the configurations can become rather complicated. Then, knowing the location of the stagnation points, is a particularly useful means to understand the structure of such a field.

For future purposes, we investigate here the equipotential surface $\varphi = 0$ for the special case of two opposite charges, where

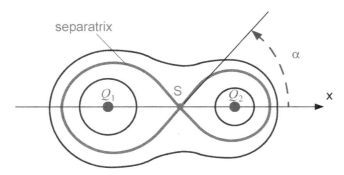

Fig. 2.9

$$0 = \frac{1}{4\pi\varepsilon_0}\left[\frac{Q_1}{\sqrt{\left(x+\frac{d}{2}\right)^2+y^2+z^2}} + \frac{Q_2}{\sqrt{\left(x-\frac{d}{2}\right)^2+y^2+z^2}}\right]$$

or

$$\frac{|Q_1|}{\sqrt{\left(x+\frac{d}{2}\right)^2+y^2+z^2}} = \frac{|Q_2|}{\sqrt{\left(x-\frac{d}{2}\right)^2+y^2+z^2}}$$

Taking the square gives

$$Q_1^2\left(x^2-dx+\frac{d^2}{4}+y^2+z^2\right) = Q_2^2\left(x^2+dx+\frac{d^2}{4}+y^2+z^2\right)$$

or

$$\left(x-\left(\frac{d}{2}\right)\frac{Q_1^2+Q_2^2}{Q_1^2-Q_2^2}\right)^2+y^2+z^2 = \frac{d^2Q_1^2Q_2^2}{(Q_1^2-Q_2^2)^2}$$

which is the equation for a sphere. Its characteristics are illustrated in Fig. 2.10. As before, we assumed $|Q_1| > |Q_2|$. The distances of the charge from the center of the sphere are

$$r_1 = x_s+\frac{d}{2} = d\frac{Q_1^2}{Q_1^2-Q_2^2} \qquad (2.52)$$

$$r_2 = x_s-\frac{d}{2} = d\frac{Q_2^2}{Q_1^2-Q_2^2} \qquad (2.53)$$

From this, we find

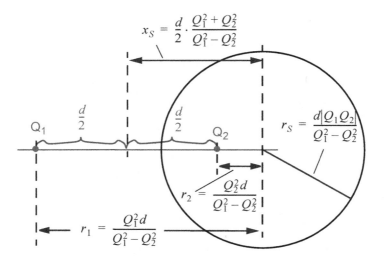

Fig. 2.10

$$\frac{r_1}{r_2} = \frac{Q_1^2}{Q_2^2} \tag{2.54}$$

and

$$r_1 r_2 = \frac{d^2 Q_1^2 Q_2^2}{(Q_1^2 - Q_2^2)^2} = r_S^2 , \tag{2.55}$$

that is, the product of the two distances equals the square of the radius of the sphere. We will use this result when we discuss image charges.

Also interesting is the case of opposite charges having the same magnitude.

$$|Q_1| = |Q_2| = Q$$

that is

$$Q_2 = -Q_1$$

Based on eq. (2.51), the stagnation point has now moved to infinity. All force lines that originate at Q_1 (if Q_1 is positive) end at Q_2. This results in a field as illustrated in Fig. 2.11, and is called a *dipole field*. One can assign a dipole moment to these charges (Fig. 2.12). The dipole moment is a vector quantity which points from the positive to the negative charge, whose magnitude is

$$|Q||d| = |Q||\mathbf{r}_+ - \mathbf{r}_-| ,$$

where

$$\mathbf{p} = |Q|(\mathbf{r}_+ - \mathbf{r}_-) \tag{2.56}$$

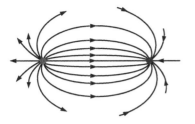

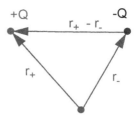

Fig. 2.11

Fig. 2.12

$$p = |\mathbf{p}| = |Q|d \qquad (2.57)$$

$$d = |\mathbf{r}_+ - \mathbf{r}_-| . \qquad (2.58)$$

If we let Q approach infinity and d approach zero in such a manner that p remains finite, then we have created an *ideal dipole*. It will be discussed in detail in the next section.

2.5   The Ideal Dipole

2.5.1   The Ideal Dipole and its Potential

Consider a charge $-Q$ be at location $\mathbf{r}_1$ and a charge $+Q$ located at $\mathbf{r}_1 + d\mathbf{r}_1$. The dipole moment is (see Fig. 2.12)

$$\mathbf{p} = Qd\mathbf{r}_1 .$$

Imagine increasing Q and decreasing $d\mathbf{r}_1$ at the same time, in such a way as to keep $\mathbf{p}$ fixed. The related potential for the charges is

$$\varphi = \frac{Q}{4\pi\varepsilon_0}\left[\frac{1}{|\mathbf{r} - (\mathbf{r}_1 + d\mathbf{r}_1)|} - \frac{1}{|\mathbf{r} - \mathbf{r}_1|}\right] .$$

In terms of the Taylor expansion:

$$\frac{1}{|\mathbf{r} - (\mathbf{r}_1 + d\mathbf{r}_1)|} = \frac{1}{|\mathbf{r} - \mathbf{r}_1|} + dx_1\frac{\partial}{\partial x_1}\frac{1}{|\mathbf{r} - \mathbf{r}_1|} +$$

$$+ dy_1\frac{\partial}{\partial y_1}\frac{1}{|\mathbf{r} - \mathbf{r}_1|} + dz_1\frac{\partial}{\partial z_1}\frac{1}{|\mathbf{r} - \mathbf{r}_1|} + \dots$$

$$= \frac{1}{|\mathbf{r} - \mathbf{r}_1|} + d\mathbf{r}_1 \bullet \nabla_{r_1}\frac{1}{|\mathbf{r} - \mathbf{r}_1|} + \dots .$$

The gradient operator is marked with the index $\mathbf{r}_1$ to express the fact that the derivatives are with respect to the components of $\mathbf{r}_1$, The potential is now

$$\varphi = \frac{Q}{4\pi\varepsilon_0} d\mathbf{r}_1 \bullet \nabla_{\mathbf{r}_1} \frac{1}{|\mathbf{r}-\mathbf{r}_1|}$$

$$= -\frac{Q}{4\pi\varepsilon_0} d\mathbf{r}_1 \bullet \nabla_{\mathbf{r}} \frac{1}{|\mathbf{r}-\mathbf{r}_1|} \quad ,$$

having used the fact that

$$\nabla_{\mathbf{r}_1} \frac{1}{|\mathbf{r}-\mathbf{r}_1|} = -\nabla_{\mathbf{r}} \frac{1}{|\mathbf{r}-\mathbf{r}_1|} \quad .$$

Finally we get for the potential

$$\varphi = -\frac{\mathbf{p} \bullet \nabla_{\mathbf{r}} \dfrac{1}{|\mathbf{r}-\mathbf{r}_1|}}{4\pi\varepsilon_0} = \frac{\mathbf{p} \bullet (\mathbf{r}-\mathbf{r}_1)}{4\pi\varepsilon_0 |\mathbf{r}-\mathbf{r}_1|^3} \quad , \tag{2.59}$$

where $\mathbf{r}$ is the observation point and $\mathbf{r}_1$ the location of the dipole $\mathbf{p}$. Using the angle θ between $\mathbf{p}$ and $\mathbf{r}-\mathbf{r}_1$ as illustrated in Fig. 2.13, we write

$$\varphi = \frac{p \cos\theta}{4\pi\varepsilon_0 |\mathbf{r}-\mathbf{r}_1|^2} \tag{2.60}$$

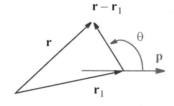

Fig. 2.13   Fig. 2.14

The dipole field shall be discussed in more detail. It is rotationally symmetric around the axis parallel to the orientation of $\mathbf{p}$. We choose that to be the z-axis of a Cartesian coordinate system (Fig. 2.14). For this situation one obtains

$$\varphi = \frac{p \cos\theta}{4\pi\varepsilon_0 r^2} = \frac{pz}{4\pi\varepsilon_0 (x^2+y^2+z^2)^{3/2}} \quad ,$$

with

$$\cos\theta = \frac{z}{\sqrt{x^2+y^2+z^2}} \quad .$$

Consequently

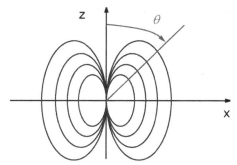

Fig. 2.15

$$E_x = -\frac{\partial \varphi}{\partial x} = \frac{3pxz}{4\pi\varepsilon_0(x^2+y^2+z^2)^{5/2}}$$

$$E_y = -\frac{\partial \varphi}{\partial y} = \frac{3pyz}{4\pi\varepsilon_0(x^2+y^2+z^2)^{5/2}}$$ (2.61)

$$E_z = -\frac{\partial \varphi}{\partial z} = \frac{p}{4\pi\varepsilon_0(x^2+y^2+z^2)^{3/2}}\left(3\frac{z^2}{x^2+y^2+z^2}-1\right)$$

Because of the rotational symmetry, it is sufficient to calculate the field in a plane, e.g., for the x-z-plane (y=0) (see Fig. 2.15).

$$E_x = \frac{3pxz}{4\pi\varepsilon_0(x^2+z^2)^{5/2}} = \frac{3p\cos\theta\sin\theta}{4\pi\varepsilon_0 r^3}$$

$$E_y = 0$$ (2.62)

$$E_z = \frac{p(3\cos^2\theta-1)}{4\pi\varepsilon_0 r^3}$$

If we transform to spherical coordinates (r, θ, φ) then the azimuthal component vanishes. The remaining components are:

$$E_r = E_x\sin\theta + E_z\cos\theta = \frac{2p\cos\theta}{4\pi\varepsilon_0 r^3}$$

$$E_\theta = E_x\cos\theta - E_z\sin\theta = \frac{p\sin\theta}{4\pi\varepsilon_0 r^3}$$ (2.63)

All lines of force pass through the origin. This may initially seem surprising, but is quite plausible if we consider Fig. 2.15 as having emerged as the limit from Fig. 2.11.

Oftentimes, one deals not with individual dipoles, but rather a collection of dipoles distributed over a volume, surface, or a line, more or less densely filled with dipoles. Like for potentials for volume, surface, or line charges, where one employs the superposition principle to integrate over the potentials of point charges, one similarly makes use of the superposition of the potentials (2.59) of the "point dipole".

2.5.2   Volume Distribution of Dipoles

If the dipoles are distributed within a volume, the resulting volume density is defined by the quantity

$$\mathbf{P} = \frac{d\mathbf{p}}{d\tau} \ .$$

This is the *polarization*, which turns out to be an important quantity. This distribution generates a potential

$$\left. \begin{aligned} \varphi &= -\int \frac{\mathbf{P}(\mathbf{r}') \bullet \nabla_{\mathbf{r}} \frac{1}{|\mathbf{r}-\mathbf{r}'|} d\tau'}{4\pi\varepsilon_0} \\ &= +\int \frac{\mathbf{P}(\mathbf{r}') \bullet \nabla_{\mathbf{r}'} \frac{1}{|\mathbf{r}-\mathbf{r}'|} d\tau'}{4\pi\varepsilon_0} \end{aligned} \right\} \tag{2.64}$$

This expression leads to interesting consequences. We start by considering the following integral

$$\frac{1}{4\pi\varepsilon_0} \int \nabla_{r'} \bullet \left[\mathbf{P}(\mathbf{r}') \frac{1}{|\mathbf{r}-\mathbf{r}'|} \right] d\tau'$$

$$= \frac{1}{4\pi\varepsilon_0} \int \frac{\nabla_{r'} \bullet \mathbf{P}(\mathbf{r}')}{|\mathbf{r}-\mathbf{r}'|} d\tau' + \frac{1}{4\pi\varepsilon_0} \int \mathbf{P}(\mathbf{r}') \bullet \nabla_{r'} \frac{1}{|\mathbf{r}-\mathbf{r}'|} d\tau'$$

$$= \frac{1}{4\pi\varepsilon_0} \oint \frac{\mathbf{P}(\mathbf{r}') \bullet d\mathbf{A}'}{|\mathbf{r}-\mathbf{r}'|} \ ,$$

where we have used the vector formula
$$\nabla \bullet (f\mathbf{a}) = f\nabla \bullet \mathbf{a} + \mathbf{a} \bullet \nabla f \ .$$

Thus

$$\varphi = -\frac{1}{4\pi\varepsilon_0} \int \frac{\nabla_{r'} \bullet \mathbf{P}(\mathbf{r}')}{|\mathbf{r}-\mathbf{r}'|} d\tau' + \frac{1}{4\pi\varepsilon_0} \oint \frac{\mathbf{P}(\mathbf{r}') \bullet d\mathbf{A}'}{|\mathbf{r}-\mathbf{r}'|} \ . \tag{2.65}$$

Comparison of this equation with equations (2.20) and (2.26) reveals that it is possible to think of a volume distribution of dipoles as the result of superposition of a volume charge distribution and a surface charge distribution, namely by

$$\boxed{\rho(\mathbf{r}') = -\nabla \bullet \mathbf{P}(\mathbf{r}')} \tag{2.66}$$

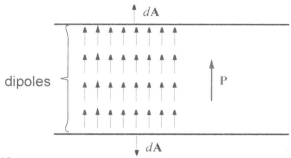

Fig. 2.16

and

$$\sigma(\mathbf{r}') = \frac{\mathbf{P}(\mathbf{r}') \bullet d\mathbf{A}'}{dA'} = \mathbf{P}(\mathbf{r}') \bullet \mathbf{n} \qquad (2.67)$$

This important result can also be made plausible, by considering this example.
Consider the disk shown in Fig. 2.16, filled with uniform polarization **P**.
The charges of the dipoles inside the volume cancel each other. Only at the surface,
there will be a net, bound charge. The net charge at the top is a positive surface
charge and the one at the bottom is negative. One may think of this as the result of
two disks of uniform volume charge which are slightly displaced against each other
(Fig. 2.17). If the volume charges are $+\rho$ and $-\rho$, and the displacement is d, then
we find for the polarization $P = \rho d$ and the surface charges are $\pm\rho d = \pm P$. The
reason is that **P** is perpendicular to the surface of the disk.

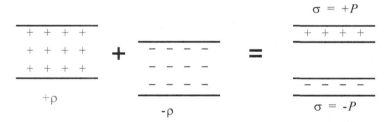

Fig. 2.17

The general case is illustrated in Fig. 2.18, where the surface charge is given by

$$\sigma(\mathbf{r}') = \frac{\rho d \cdot dA \cdot \cos\gamma}{dA} = \frac{\mathbf{P}(\mathbf{r}') \bullet d\mathbf{A}}{dA} \ ,$$

which is also the result previously obtained in a more formal manner. If the
polarization is not uniform, then the charges inside do not cancel entirely and there
remains a net volume charge. Fig. 2.19 illustrates this case. It shows a volume with

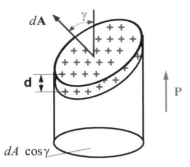

Fig. 2.18

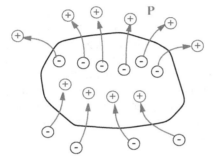

Fig. 2.19

dipoles inside. At the end of each dipole vector, there is a positive and at its beginning a negative charge. We have

$$\oint \mathbf{P} \bullet d\mathbf{A} = -Q = -\int_V \rho \, d\tau \ ,$$

i.e., the overall flux of the polarization $\mathbf{P}$ over the surface is equivalent to the negative of the charge in the volume (a vector $\mathbf{P}$ that points outward represents a negative charge inside). Conversely,

$$\oint \mathbf{P} \bullet d\mathbf{A} = \int_V \nabla \bullet \mathbf{P} \, d\tau \ ,$$

and comparison mandates that

$$\rho = -\nabla \bullet \mathbf{P} \ .$$

2.5.3   Surface Distributions of Dipoles (Dipole Layers)

If we place dipoles on a surface, one obtains a so-called *double layer* or *dipole layer*. The name stems from the fact that it is equivalent to two layers of opposite charges. As shown in Fig. 2.20, let **p** point in the direction of $d\mathbf{A'}$. We define the surface density of the dipole moment

$$\tau = \frac{dp}{dA'} \tag{2.68}$$

Using (2.60), we find for the potential

$$\varphi = \int_A \frac{\tau(\mathbf{r'})\cos\theta}{4\pi\varepsilon_0|\mathbf{r} - \mathbf{r'}|^2} \ dA' \ . \tag{2.69}$$

With the solid angle element

$$d\Omega = \frac{\cos\theta \ dA'}{|\mathbf{r} - \mathbf{r'}|^2} \tag{2.70}$$

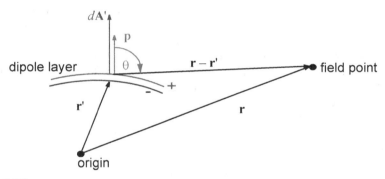

Fig. 2.20

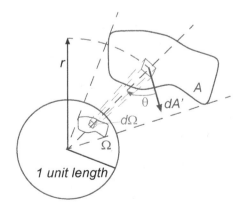

Fig. 2.21

by which the surface element dA' is seen from the field point. The element of the solid angle $d\Omega$ is the projection of the surface element onto a unit sphere centered about the field point, as illustrated in Fig. 2.21. It can be calculated by eq. (2.70). Consequently, $d\Omega$ and Ω are dimensionless quantities. This definition is in analogy to that of the "plane" angle (see Section 2.5.4, on line dipoles and in particular Fig. 2.29, which is the equivalent of Fig. 2.21 for that case). The result is

$$\boxed{\varphi = \frac{1}{4\pi\varepsilon_0}\int_A \tau\, d\Omega}\,.\qquad(2.71)$$

Specifically, for a surface with uniform surface density of the dipole moment we find

$$\boxed{\varphi = \frac{\tau}{4\pi\varepsilon_0}\Omega}\,,\qquad(2.72)$$

where Ω is the solid angle under which the uniform dipole layer appears when looking from the field point. Confusion with the electric flux (for which Ω was also used) should not be an issue.

As an **example**, let us consider a sphere whose surface is uniformly covered with outwardly facing dipoles. We may picture this uniform dipole layer as consisting of two concentric spheres with opposite charges, where the charge is very large and the difference of their radius is very small. For all points inside the sphere we have $\Omega = -4\pi$ (the negative sign is a result of the definition of θ in Fig. 2.20). Conversely, for all points outside we have $\Omega = 0$. Therefore (for outwardly oriented dipoles) we get

$$\boxed{\varphi = \begin{cases} -\dfrac{\tau}{\varepsilon_0} & inside \\[2mm] 0 & outside \end{cases}}\,.\qquad(2.73)$$

When passing through the dipole layer from inside to outside, the potential experiences a discontinuity by τ/ε_0.

This result can be generalized. It applies to a dipole layer of any shape and is independent of whether τ is uniform or not. Passing through a dipole layer in the direction of the dipole increases the potential by τ/ε_0, where τ is now a function of the location. The potential difference depends on how one passes through the dipole layer. One proves this generalized claim by beginning with a surface that is covered with electric charges. Let the surface charge density at a particular point be σ and the electric displacement just above that point be $\mathbf{D}_1$, and the one underneath $\mathbf{D}_2$. One can split $\mathbf{D}_1$ and $\mathbf{D}_2$ into their parallel (tangential) D_t and perpendicular (normal) D_n components with respect to the surface (Fig. 2.22). Now, one applies eq. (2.4) to the small cylinder shown in Fig. 2.22 whose extent perpendicular to the surface shall be so small that the contribution of the sides of the cylinder vanishes. The remaining contribution is

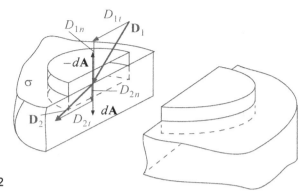

Fig. 2.22

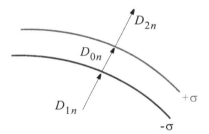

Fig. 2.23

$$(D_{2n} - D_{1n})dA = \sigma dA$$

or

$$\boxed{D_{2n} - D_{1n} = \sigma} .$$

(2.74)

Notice that we did not make any statement about the tangential components, here, which will be covered in a later section. Now consider the case of two parallel surfaces in close vicinity with surface charges of opposite sign (Fig. 2.23), for which one finds:

$$D_{0n} - D_{1n} = -\sigma,$$

$$D_{2n} - D_{0n} = +\sigma$$

From these two equations it follows:

$$D_{2n} = D_{1n} = D_n$$

and

$$D_{0n} = D_n - \sigma$$

Therefore, the normal component of **D** remains unchanged by the dipole layer. The normal component of **D** within the layer is decreased by the value of σ. The voltage when passing through the layer in perpendicular, positive direction is:

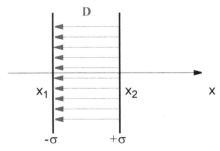

Fig. 2.24

$$\delta\varphi = -E_{0n}d = -\frac{D_{0n}}{\varepsilon_0}d = -\frac{D_n - \sigma}{\varepsilon_0}d \;. \tag{2.75}$$

As before, the positive direction is the direction of the dipole moment. D_n is finite, while d is arbitrarily small and σ is large enough to make σd finite, *i.e.*, precisely

$$\sigma d = \tau \;. \tag{2.76}$$

With eq. (2.75) we obtain for the potential difference or voltage

$$\delta\varphi = \frac{\sigma d}{\varepsilon_0} = \frac{\tau}{\varepsilon_0} \;, \tag{2.77}$$

which completes the proof.

Particularly simple is the case of two infinitely wide parallel planes with homogeneous surface charges (Fig. 2.24). For symmetry reasons, **D** has an x-component depending on x only. Fig. 2.25 illustrates (a) the field of a surface with the surface charge $+\sigma$, (b) the field of a surface with the surface charge $-\sigma$, and (c) the superposition of the two fields. For case (a) we get

$$D_{2x} - D_{1x} = \sigma \;.$$

For symmetry reasons

$$D_{2x} = -D_{1x} \;,$$

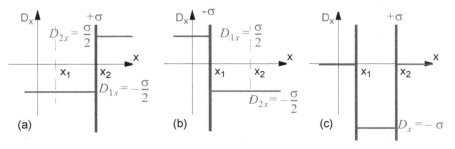

(a)   (b)   (c)

Fig. 2.25

or

$$D_{2x} = -D_{1x} = \frac{\sigma}{2} .$$ (2.78)

For case (b), in a similar way one obtains

$$D_{2x} = -D_{1x} = -\frac{\sigma}{2} .$$ (2.79)

The superposition gives a non-zero field only for the area between the two planes and it points from the positive plane to the negative one (Fig. 2.24)

$$D_x = -\sigma .$$ (2.80)

Therefore

$$E_x = -\frac{\sigma}{\varepsilon_0} .$$ (2.81)

and

$$\delta\varphi = -E_x d = \frac{\sigma d}{\varepsilon_0} = \frac{\tau}{\varepsilon_0} .$$ (2.82)

This equation is exact even for finite distances d, while for the general case, *i.e.*, when deriving eq. (2.75), an infinitesimal distance d was required.

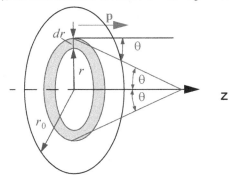

Fig. 2.26

Here is another **example** on how to apply eq. (2.72). To calculate the potential at the axis of a disk, uniformly coated with dipoles (Fig. 2.26). One needs to find the solid angle Ω. From eq. (2.70), follows for $z>0$:

$$\Omega = \int d\Omega = \int_0^{r_0} \frac{2\pi r}{r^2 + z^2} \cos\theta \, dr$$

$$= \int_0^{r_0} \frac{2\pi r}{r^2 + z^2} \frac{z \, dr}{\sqrt{r^2 + z^2}} = 2\pi z \int_0^{r_0} \frac{r \, dr}{\sqrt{r^2 + z^2}^3}$$

Now, substituting r^2 as a new variable. Then $dr^2 = 2r \, dr$ and thus

$$\Omega = \pi z \int_0^{r_0^2} \frac{dr^2}{\sqrt{r^2 + z^2}^3} = \pi z \left[-\frac{2}{\sqrt{r^2 + z^2}} \right]_0^{r_0^2}$$

$$= 2\pi \left(1 - \frac{z}{\sqrt{r_0^2 + z^2}} \right) = 2\pi(1 - \cos\theta_0)$$

For $z < 0$, on the other hand,

$$\Omega = -2\pi \left(1 - \frac{|z|}{\sqrt{r_0^2 + z^2}} \right) = 2\pi \left(-1 - \frac{z}{\sqrt{r_0^2 + z^2}} \right).$$

So

$$\varphi = \begin{cases} \dfrac{\tau}{2\varepsilon_0} \left(1 - \dfrac{z}{\sqrt{r_0^2 + z^2}} \right) & \text{for } z > 0 \\[4mm] \dfrac{\tau}{2\varepsilon_0} \left(-1 - \dfrac{z}{\sqrt{r_0^2 + z^2}} \right) & \text{for } z < 0 \end{cases}.$$

There is a discontinuity at $z = 0$, where φ jumps from $-\tau/(2\varepsilon_0)$ to $\tau/(2\varepsilon_0)$, which results in an overall discontinuity of τ/ε_0, as expected. The electric field on the axis is

$$E_z = -\frac{\partial \varphi}{\partial z}$$

and is calculated to be

$$E_z = \frac{\tau}{2\varepsilon_0} \frac{r_0^2}{\sqrt{r_0^2 + z^2}^3}.$$

E_z vanishes when r_0 approaches infinity, as necessary. Now, we come back to the case of Fig. 2.24, for which the field is non-zero only inside the layer.

2.5.4 Line Dipoles

It is possible to cover lines with dipoles, which is illustrated by a simple **example**. According to eq. (2.46), the potential of an infinitely long, straight line charge is

$$\varphi = -\frac{q}{2\pi\varepsilon_0}\ln\frac{r}{r_B} \; .$$

Two parallel line charges in close vicinity form a line dipole (Fig. 2.27), with the potential

$$\varphi = -\frac{q}{2\pi\varepsilon_0}\ln\frac{r_+}{r_B} + \frac{q}{2\pi\varepsilon_0}\ln\frac{r_-}{r_B}$$

$$= -\frac{q}{2\pi\varepsilon_0}\ln\frac{r_+}{r_-} = -\frac{q}{2\pi\varepsilon_0}\ln\frac{r_- - \delta}{r_-}$$

$$= -\frac{q}{2\pi\varepsilon_0}\ln\left(1 - \frac{\delta}{r_-}\right) \; ,$$

which holds as long as

$$r_{B1} = r_{B2} = r_B \; .$$

We now require that

$$d \ll r_+$$

and

$$d \ll r_- \; .$$

Then $\theta_+ \approx \theta_- \approx \theta$, $\delta = r_- - r_+ \approx d \cdot \cos\theta \ll r_-$. Furthermore $r_+ \approx r_- \approx r$, and because the power series of $\ln(1-x) = -(x + x^2/2 + \ldots)$ for $-1 \le x \le 1$, φ becomes

$$\boxed{\varphi \approx \frac{q}{2\pi\varepsilon_0 r}\frac{\delta}{r} \approx \frac{(qd)\cos\theta}{2\pi\varepsilon_0 r}} \; , \tag{2.83}$$

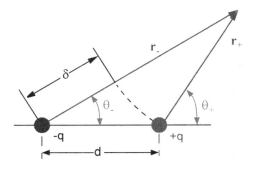

Fig. 2.27

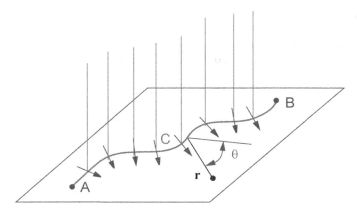

Fig. 2.28

where (qd) is the line dipole density (dipole moment per unit length) and φ is the potential of the infinitely long line dipole. The result should be compared to eq. (2.60), which represents the potential of a dipole. When comparing with eq. (2.60), replace p with (qd), use 2π instead of 4π, r instead of r^2, and let $\mathbf{r}_1 = 0$. We should keep in mind that r in eq. (2.60) represents the distance of the field point from the dipole, while in eq. (2.83) r represents the perpendicular distance to the line dipole.

From line dipoles that are parallel to each other, one can construct *cylindrical dipole layers* (Fig. 2.28 and Fig. 2.29).
The surface density of the dipole moment is

$$\tau(s) = \frac{d(qd)}{ds}$$

and thus the potential becomes

$$\varphi = \int_C \frac{\tau \cos\theta \, ds}{2\pi\varepsilon_0 r} \, ,$$

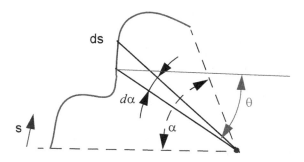

Fig. 2.29

where the integral from A to B is evaluated along the contour C. Now,

$$d\alpha = \frac{\cos\theta\,ds}{r}$$

is the angular element, under which the line element ds appears when looking from the field point. Therefore

$$\boxed{\varphi = \frac{1}{2\pi\varepsilon_0}\int_C \tau\,d\alpha}\,. \tag{2.84}$$

If τ is constant, this gives

$$\boxed{\varphi = \frac{\tau\alpha}{2\pi\varepsilon_0}}\,. \tag{2.85}$$

These two relations are equivalent to eq. (2.71) and eq. (2.72), respectively. There, we discussed general spatial problems, while here we are dealing with the *cylindrical case*, which is also called the *plane case* because it is independent of one of the spatial coordinates.

When the contour C is closed, the result is a closed cylinder. If, furthermore, τ is constant and the dipoles point outwardly, then

$$\alpha = \begin{cases} -2\pi & inside \\ 0 & outside \end{cases}$$

and thus

$$\varphi = \begin{cases} -\dfrac{\tau}{\varepsilon_0} & inside \\[2mm] 0 & outside. \end{cases}$$

As expected, there is the discontinuity of the potential τ/ε_0.

2.6   Behavior of a Conductor in an Electric Field

One finds in nature two quite distinct types of materials. There are materials containing charges which can move freely and there are materials where this is not the case. The former are called *conductors*, while latter are called *insulators* (or *dielectrics*). Consider the behavior of materials in the presence of an electric field for the case of a conductor. We refrain from discussing this broad classification further, but limit our discussion to the consequences for conductors in an electric field, and then tackle the problem of dielectrics in an electric field.

A conductor in an electric field experiences a force, which is actually exerted on the free charges within it. These start to move, and their motion will cease only if

$$\mathbf{E} = 0$$

everywhere inside the conductor and

$$\boxed{\varphi = const} \ . \tag{2.86}$$

The surface of the conductor has to have the same potential everywhere, *i.e.* it has to be an equipotential surface. Outside of the conductor, $\mathbf{E}$ will not vanish. Its tangential component has to be zero at the conductor surface

$$\boxed{E_t = 0} \ , \tag{2.87}$$

as otherwise the surface would not be an equipotential surface. The perpendicular component E_n of $\mathbf{E}$, however, will not vanish. There will be surface charges at the surface, such that the external field does not penetrate the conductor, *i.e.*, by eq. (2.74) we obtain

$$\boxed{D_n = \varepsilon_0 E_n = \sigma} \ . \tag{2.88}$$

To illustrate this, consider this simple **example**: We choose an infinitely wide, conducting plate within a uniform electric field which is perpendicular to the surface of the plate (Fig. 2.30). Depending on their sign, the free charges move in the direction of the field or opposite to it, until they reach the surface of the plate. This is so, regardless of whether there are only negative, positive, or both types of charges available. The result is a surface charge which is positive on one end, and

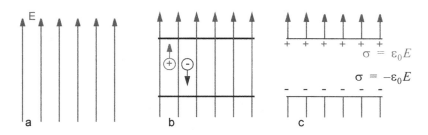

Fig. 2.30

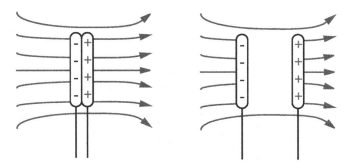

Fig. 2.31

negative on the other. The inside is free of any field if $\sigma = \pm\varepsilon_0 E$. The field of the surface charges exists only inside. It originates at the positive charges (sources) and ends at the negative charges (sinks). *i.e.*, it is exactly opposite to the external field but of the same magnitude. The external field is thus identically cancelled. The superposition of the fields of Fig. 2.30b and Fig. 2.24 results in Fig. 2.30c.

The thereby created surface charges are also called *influence charges*. They can be used to measure the electric field by magnitude and direction. A pair of conducting plates does the job. The plates are brought into the field while in contact with each other. In the field, they are separated, and by trying different orientations, one can find the direction of the field (Fig. 2.31).

To calculate the field created by a conductor, together with an external field in full generality is rather difficult. In the following, we will solve some problems that are, however, rather easy to solve.

2.6.1   Metallic Sphere in the Field of a Point Charge

By now, we have already calculated a number of fields and in principle, we know their equipotential surfaces. One may imagine each such equipotential surface as the surface of a conductor. In light of this, we have already solved many problems of this kind. In Section 2.4, we have found that the equipotential surface $\varphi = 0$ of two point charges with opposite sign is that of a sphere (Fig. 2.10). Take a sphere of radius r_s, centered at the origin of a Cartesian coordinate system, and a charge Q_1 to be located at $(0,0,z_1)$. Because of eq. (2.54) and eq. (2.55), a second charge Q_2 at location $(0,0,z_2)$ will make the sphere an equipotential surface provided,

$$z_2 = \frac{r_s^2}{z_1}, \tag{2.89}$$

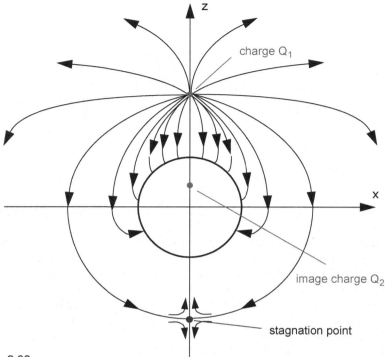

Fig. 2.32

$$Q_2 = -Q_1 \sqrt{\frac{z_2}{z_1}} \quad . \tag{2.90}$$

The charge Q_2 at location $(0,0,z_2)$ is a fictitious or image charge. Given Q_1 at location $(0,0,z_1)$, this image charge is necessary to create the very field outside of the sphere that we are looking for. There is no field inside the sphere. The field ends at the surface of the sphere at the correlating surface charges, which are determined by equation eq. (2.88). Integration of these charges over the surface of the sphere yields the charge Q_2. On the sphere's surface end all those field lines which would end at Q_2, the so-called *image charge*, if there were no sphere. The resulting configuration is illustrated in Fig. 2.32. The point $(0, 0, z_2 = r_s^2/z_1)$ is the image point of the point $(0, 0, z_1)$ with respect to the sphere. From this stems the term image charge and the method to solve this kind of problems is called the *method of images*.

One can modify this problem slightly, and require that the sphere holds a given charge **Q**. The solution results from recognizing that if one places an arbitrary charge at the center of the sphere its surface remains an equipotential

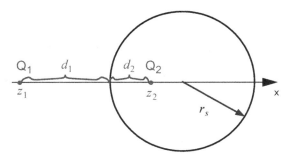

Fig. 2.33

surface. All we need to do is to superimpose the field of a point charge $(Q - Q_2)$ at the center to the initial field of Fig. 2.32

A charge in front of a plane, conducting wall represents the limit of the sphere with an infinite radius r_s. It results from eq. (2.89) that the mirror or image charge has to be located behind the wall, in the exact same distance as the real charge in front of it, *i.e.*, in its image point and that $Q_2 = -Q_1$. The charge location according to Fig. 2.33 is

$$z_1 = r_s + d_1 \,,$$

$$z_2 = r_s - d_2 \,,$$

and, therefore by equation eq. (2.89)

$$z_2 = r_s - d_2 = \frac{r_s^2}{r_s + d_1} \,,$$

$$= \frac{r_s}{1 + \dfrac{d_1}{r_s}} \,,$$

If $r_s \gg d_1$, then in the 1st approximation

$$z_2 = r_s - d_2 \approx r_s \left(1 - \frac{d_1}{r_s} \right) = r_s - d_1$$

that is

$$d_2 \approx d_1 \,.$$

It is plausible that thereby the boundary condition of constant potential or vanishing tangential field components is met at the wall (Fig. 2.34). It is also possible to apply this method to charges inside an angle as shown in Fig. 2.35. In this case there are the charges $+Q$ at *for example $(a,b,0)$* and *$(-a,-b,0)$* and the charges $-Q$ at *$(-a,b,0)$* and *$(a,-b,0)$*. One can think of the field in the 1st quadrant (there is no field in the other ones) being generated by those four charges and it is easy to verify that the planes xz and yz are equipotential surfaces, which is quite plausible.

Of course, it is also possible to place several charges, *for example*, in the vicinity of the sphere of Fig. 2.32. This requires multiple image charges, and all fields need to be added. In particular, it is possible to add another charge $-Q_1$ at $(0,0,-z_1)$ besides the charge Q_1 at $(0,0,z_1)$. This requires one to consider two image charges: Q_2 at $(0,0,z_2)$, and another one $-Q_2$ at $(0,0,-z_2)$. In the limit of Q_1 and z_1

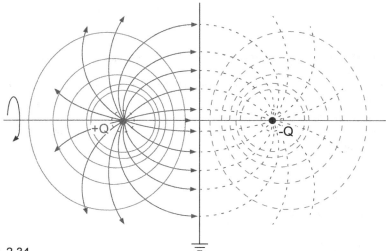

Fig. 2.34

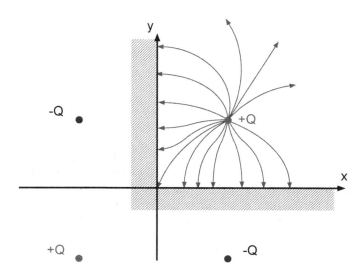

Fig. 2.35

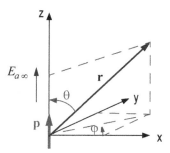

Fig. 2.36

approaching infinity then Q_2 needs to approach infinity in the same manner, while, z_2 goes to zero. That is to say, the two image charges in the said limit result in a dipole. The field of the two charges $\pm Q_1$ in the vicinity of the sphere can be regarded as being uniform. This suggests that the problem of a sphere in a uniform field can be solved by means of a fictitious (image) dipole at its center. This leads us to the next example.

2.6.2   Metallic Sphere in a Uniform Electric Field

Based on the just mentioned assumption and using the quantities from Fig. 2.36, one makes the following Ansatz:

$$\varphi = \frac{p\cos\theta}{4\pi\varepsilon_0 r^2} - E_{a,\infty} z$$

$$= \frac{p\cos\theta}{4\pi\varepsilon_0 r^2} - E_{a,\infty} r\cos\theta \ .$$

$E_{a,\infty}$ is the externally applied field, which at a sufficiently large distance, is not distorted by the metallic sphere. The potential is generated by the dipole according to eq. (2.60) and by a part that belongs to the uniform outside field. This assumption is confirmed if we can choose p such that φ is constant for all $r = r_s$:

$$\varphi = \varphi_0 = \frac{p\cos\theta}{4\pi\varepsilon_0 r_s^2} - E_{a,\infty} r_s\cos\theta \ .$$

φ will, in fact, be constant for $r = r_s$, provided one chooses

$$p = 4\pi\varepsilon_0 r_s^3 E_{a,\infty} \ .$$

Thus

$$\varphi = E_{a,\infty}\left(\frac{r_s^3}{r^2} - r\right)\cos\theta \ . \tag{2.91}$$

This allows one to calculate the components of $\mathbf{E}$:

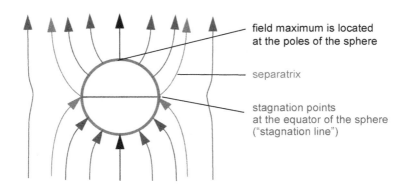

field maximum is located
at the poles of the sphere

separatrix

stagnation points
at the equator of the sphere
("stagnation line")

Fig. 2.37

$$E_r = E_{a,\infty}\left(2\frac{r_s^3}{r^3} + 1\right)\cos\theta \ , \tag{2.92}$$

$$E_\theta = E_{a,\infty}\left(\frac{r_s^3}{r^3} - 1\right)\sin\theta \ . \tag{2.93}$$

$E_\theta = 0$ at the surface of the sphere ($r = r_s$). E_r determines the surface charge:

$$\sigma = \varepsilon_0(E_r)_{r=r_s} = (D_r)_{r=r_s} = 3\varepsilon_0 E_{a,\infty}\cos\theta \ . \tag{2.94}$$

This configuration is illustrated in Fig. 2.37. The maximum field is $E = 3E_{a,\infty}$
and is located at the two poles of the sphere. The behavior at the equator is strange,
insofar as it consists entirely of many stagnation points forming a so-called
stagnation line. The field lines there form a tip, *i.e.* they have no unique direction,
which is, of course, only possible at stagnation points. Furthermore, one can show
that they form an angle of 45° against the equatorial plane (Fig. 2.38).

This problem can be generalized, which gives rise to the question how this
picture might change if the sphere carried the charge Q. Thus far, the effect of the
sphere was simulated by a fictitious dipole, *i.e.* the charge on the sphere vanishes,
which can also be obtained when integrating σ over the surface, eq. (2.94). So, one
only needs to place an additional charge in the center of the sphere. This solves the
problem because it also creates a constant potential on the sphere.
Instead of eq. (2.91), we now use

$$\varphi = E_{a,\infty}\left(\frac{r_s^3}{r^2} - r\right)\cos\theta + \frac{Q}{4\pi\varepsilon_0 r} \ .$$

Depending on Q, very different field configurations result, which are presented
here without proof. Consider:

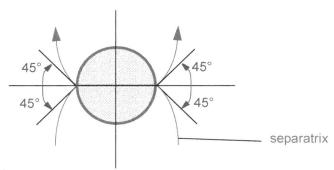

Fig. 2.38

Case 1: if $\left| \dfrac{Q}{4\pi\varepsilon_0 r_s^2 \cdot 3E_{a,\,\infty}} \right| < 1$,

then the stagnation lines are at circles of equal latitudes of the sphere, as shown in Fig. 2.39.; and

Case 2: if $\left| \dfrac{Q}{4\pi\varepsilon_0 r_s^2 \cdot 3E_{a,\,\infty}} \right| = 1$,

then the stagnation lines are degenerate and the stagnation points of Fig. 2.39 move to the poles of the sphere; and

Case 3: if $\left| \dfrac{Q}{4\pi\varepsilon_0 r_s^2 \cdot 3E_{a,\,\infty}} \right| > 1$

then the stagnation points detach from the sphere, and move out into the field along the axis through the poles (Fig. 2.40).

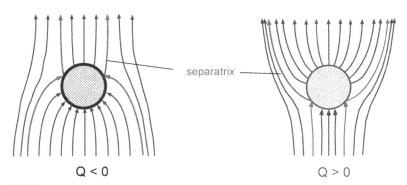

Fig. 2.39

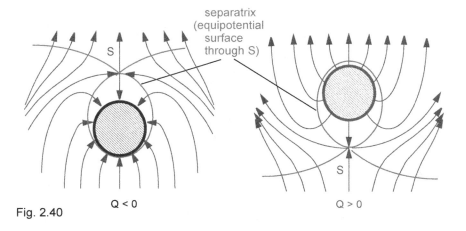

Fig. 2.40

2.6.3   Metallic Cylinder in the Field of a Line Charge

Consider a metallic cylinder to be located within the field of a uniform line charge with its axis oriented parallel to the line charge (see Fig. 2.41). One can think of the overall field outside the cylinder as being created by the given line charge q (outside the cylinder) and its image charge, also a line charge -q (inside). The product of the distances of the two line charges from the cylinder axis equals the square of the cylinder radius, *i.e.* the piercing points of the two line charges emerge by reflection at the circle $r = r_C$, (where r_C is the radius of the cylinder). Thus

$$x_1 \cdot x_2 = r_C^2$$

The proof is easy. Based on eq. (2.46), one first calculates the potential of the two line charges at a field point *(x,y,z)*

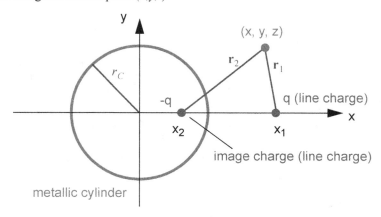

Fig. 2.41

$$\varphi = -\frac{q}{2\pi\varepsilon_0}\ln\frac{r_1}{r_B} + \frac{q}{2\pi\varepsilon_0}\ln\frac{r_2}{r_B} = \frac{q}{2\pi\varepsilon_0}\ln\frac{r_2}{r_1} \ .$$

where

$$r_1^2 = (x - x_1)^2 + y^2$$

and

$$r_2^2 = (x - x_2)^2 + y^2 = \left(x - \frac{r_C^2}{x_1}\right)^2 + y^2$$

On the cylinder wall we have

$$x^2 + y^2 = r_C^2$$

and thus

$$\frac{r_2^2}{r_1^2} = \frac{x^2 - \dfrac{2xr_C^2}{x_1} + \dfrac{r_C^4}{x_1^2} + y^2}{x^2 - 2xx_1 + x_1^2 + y^2}$$

$$= \frac{r_C^2 - \dfrac{2xr_C^2}{x_1} + \dfrac{r_C^4}{x_1^2}}{r_C^2 - 2xx_1 + x_1^2} = \frac{r_C^2}{x_1^2} = const$$

Therefore r_2/r_1, and thereby also φ are constant on the cylinder wall. The location of all the geometrical points for which the distance ratios r_2/r_1 from the two fixed points is constant, as shown in Fig. 2.41 These are known in geometry as the circles of Apollonius. The circular cross section of the cylinder constitutes one of those circles.

2.7   The Capacitor

Suppose there are two conductors (*for example* metals) with a charge of opposite sign (Q and $-Q$), then a field will form between them whose force lines originate on one surface and terminate on the other. Both surfaces are equipotential surfaces, *i.e.* a well defined voltage V between the two bodies is set up. This voltage is proportional to the charge Q. The ratio $|Q|/|V|$ is a geometric factor called the capacitance C. The whole configuration is termed the *capacitor*.

It is particularly simple to calculate the case of a plane, parallel plate capacitor when one makes the approximation that the plates extend to infinity, thereby neglecting fringing effects (Fig. 2.42). Then

$$E = \frac{\sigma}{\varepsilon_0}$$

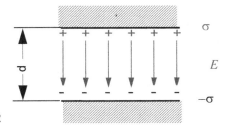

Fig. 2.42

and

$$|V| = \frac{\sigma d}{\varepsilon_0} .$$

The charge is

$$Q = \pm \sigma A ,$$

where A is the area of the plates. Therefore

$$\boxed{C = \frac{|Q|}{|V|} = \frac{\varepsilon_0 A}{d}} . \qquad\qquad (2.95)$$

One could also define the capacitance of a single conductor by using the value of its voltage against a point at infinity. Consider a sphere of radius r, with a voltage between its surface and infinity

$$V = \frac{Q}{4\pi\varepsilon_0 r}$$

so that

$$\boxed{C = \frac{|Q|}{|V|} = 4\pi\varepsilon_0 r} . \qquad\qquad (2.96)$$

Two concentric spheres form a spherical capacitor (Fig. 2.43). For this case, the voltage is

$$V = \frac{Q}{4\pi\varepsilon_0}\left(\frac{1}{r_i} - \frac{1}{r_o}\right)$$

and therefore

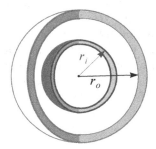

Fig. 2.43

$$\boxed{C = \frac{|Q|}{|V|} = 4\pi\varepsilon_0 \frac{r_i r_o}{r_o - r_i}} \,. \tag{2.97}$$

If one lets r_o and r_i become very large, but keep $r_o - r_i = d$ to be very small, then

$$C = 4\pi\varepsilon_0 \frac{r^2}{d} = \frac{\varepsilon_0 A}{d}$$

and the case of the parallel plane capacitor is recovered.

Two concentric cylinders form a *cylindrical capacitor*. Here

$$V = -\frac{Q}{2\pi\varepsilon_0 l}\ln\!\left(\frac{r_i}{r_B}\right) + \frac{Q}{2\pi\varepsilon_0 l}\ln\!\left(\frac{r_o}{r_B}\right) = \frac{Q}{2\pi\varepsilon_0 l}\ln\!\left(\frac{r_o}{r_i}\right)$$

and thus

$$\boxed{C = \frac{|Q|}{|V|} = 2\pi\varepsilon_0 l\left(\ln\frac{r_o}{r_i}\right)^{-1}} \,. \tag{2.98}$$

This is exact only for a cylinder of infinite length l, which would make C also infinite. Therefore, it is more practical to express the capacitance per unit length.

$$\frac{C}{l} = 2\pi\varepsilon_0\left(\ln\frac{r_o}{r_i}\right)^{-1} \,.$$

In spherical or cylindrical coordinates, the electric field has a spatial dependency according to eq. (2.2) and eq. (2.45):

	spherical	cylindrical
Electric field in general	$E = \dfrac{Q}{4\pi\varepsilon_0 r^2}$	$E = \dfrac{Q}{2\pi\varepsilon_0 l r}$
E has its maximum at the inner electrode	$E_{max} = \dfrac{Q}{4\pi\varepsilon_0 r_i^2}$	$E_{max} = \dfrac{Q}{2\pi\varepsilon_0 l r_i}$
Another way to write this is as follows:	$E_{max} = \dfrac{CV}{4\pi\varepsilon_0 r_i^2} = \dfrac{V r_o}{r_i(r_o - r_i)}$	$E_{max} = \dfrac{CV}{2\pi\varepsilon_0 l r_i} = \dfrac{V}{r_i \ln\dfrac{r_o}{r_i}}$

	spherical	cylindrical
For a given voltage and outer radius r_o, the maximum of the electric field E_{max} takes is lowest value if $\partial E_{max}/\partial r_i = 0$, *i.e.*, for	$r_i = \dfrac{r_o}{2}$	$r_i = \dfrac{r_o}{e} = \dfrac{r_o}{2.718...}$

This is of practical value when one wants to optimize capacitor structures. The unit of capacitance is the Farad. The definition of C determines

$$1F = \frac{1C}{1V} = 1\frac{As}{V} ,$$

as already mentioned in Section 1.13.

Two conductors form a capacitor not only when separated by vacuum, but also if the separating medium is an insulator. In this case, one finds that the presence of the insulator permeating the gap increases the capacitance by a characteristic factor. With the same charge, the result is a reduced voltage, or a reduced electric field. The voltage vanishes completely inside a conductor. Inside an insulator it is just reduced. Both situations have a similar cause. There are also charges inside an insulator. They, however, can not move about freely. The result is a limited shielding of the external field. This will be discussed in the next section.

The concept of capacitance can be generalized for systems that consist of several conductors. This will be covered in Chapter 3.

2.8   E and D inside Dielectrics

All matter consists of atoms, which themselves consist of a positively charged nucleus and negatively charged electrons. Inside a conductor, some of the electrons are free to move, and this leads to the effects described the last two sections. This is not the case for an insulator (dielectrics). Nevertheless, a certain displacement of positive charges versus the negative ones is still possible. If, in a medium, the centers of positive and negative charges of its atoms or molecules do not coincide, then they acquire a dipole moment. Two cases are of importance.

1.   Frequently, atoms and molecules have no initial dipole moment in the absence of an applied external field. However, an applied external field exerts a force on the charges, which deforms the atoms (or molecules), creating a dipole moment (Fig. 2.44). The so created dipole has its own field which tends to weaken the external field. This process is termed *polarization* of the dielectric (see also Section 2.5.2). The quantitative measure is the resulting dipole moment per unit volume. The general assumption is that the polarization is proportional to the electric field.

$$\mathbf{P} = \varepsilon_0\chi\mathbf{E}$$

(2.99)

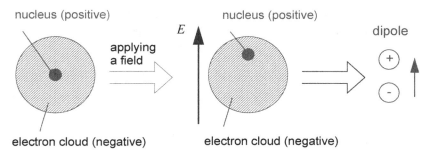

Fig. 2.44

This is not necessarily exact, but often times provides a suitable approximation, provided that the magnitude of electric field is not too large.

2. It is also possible that the atoms or molecules possess a "natural dipole moment", *i.e.* their respective centers of charge do not coincide, even in the absence of an externally applied field. Nevertheless, in the general case, the substance is still not polarized without an external field. The reason is that the dipoles are typically randomly oriented inside the material, and cancel each other, so that the system has no net dipole moment. When an external field is applied, a torque is exerted on the dipoles which tries to orient them along the external field. This may not be entirely successful. Thermal motion continually attempts to destroy this order created by the outside field. The orientation along the external field is therefore only partial, and less complete the higher the temperature. However, the polarization is still, approximately proportional to the external field. That is, equation eq. (2.99) for the polarization applies for this case as well.

There is also the case that the dipoles maintain their orientation without an external field. Such a substance is called *permanently polarized*. This is then also referred to as *electret*, in analogy to a permanent magnet.

The factor χ in eq. (2.99) is called the *electric susceptibility*. Depending on whether the material corresponds to case two (or one), χ will be (not be) a function of temperature.

We now turn to the case of a polarized, plane plate (Fig. 2.45). An outside field E_a, is applied perpendicular to the plate. As a reaction, an opposing field E_g is created inside. The resulting net field inside is weakened:

$$\mathbf{E}_i = \mathbf{E}_a + \mathbf{E}_g$$

Those fields are all uniform because of the plane geometry. Therefore, the polarization

$$\mathbf{P} = \varepsilon_0 \chi \mathbf{E}_i$$

is also uniform. Notice that $\mathbf{E}_i$ was used. The reason is that equation (2.99) uses the net field at the particular field point. The uniformly polarized dielectric carries the surface charge $\sigma = \pm P_n$ at its surface, as discussed in Section 2.5.2. Also, if we just take the magnitude

$$D_g = +\sigma = +P$$

or

$$E_g = \frac{D_g}{\varepsilon_0} = \frac{\sigma}{\varepsilon_0} = +\frac{P}{\varepsilon_0} = +\frac{\varepsilon_0 \chi E_i}{\varepsilon_0} = +\chi E_i$$

so that

$$E_i = E_a - \chi E_i$$

i.e.

$$\boxed{E_i = \frac{E_a}{1 + \chi}} \qquad (2.100)$$

and

$$E_g = \frac{\chi}{1 + \chi} E_a . \qquad (2.101)$$

Now, we may write

$$\varepsilon_0 E_a = \varepsilon_0 (1 + \chi) E_i = \varepsilon_0 \varepsilon_r E_i = \varepsilon E_i . \qquad (2.102)$$

Here, ε is the so-called *dielectric constant* or *permittivity* of the insulator and ε_r the so-called *relative dielectric constant,* defined by

$$\boxed{\varepsilon = \varepsilon_0 \varepsilon_r} . \qquad (2.103)$$

Note that ε_r is dimensionless and defined by

$$\boxed{\varepsilon_r = 1 + \chi} , \qquad \varepsilon_r > 1 \qquad (2.104)$$

Finally, one may complete the definition of $\mathbf{D}$, which up to now was only defined for vacuum. For linear media we define the electric displacement as

$$\boxed{\mathbf{D} = \varepsilon \mathbf{E}} \qquad (2.105)$$

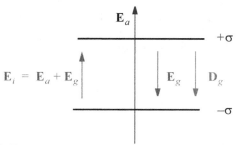

Fig. 2.45

Thus, because of eq. (2.102) $\mathbf{D}_a = \mathbf{D}_i$. In the general case, when $\mathbf{E}$ is not perpendicular to the insulator surfaces, we have to restrict this statement to apply to the normal components of $\mathbf{D}$:

$$\boxed{D_{na} = D_{ni}} \tag{2.106}$$

This is an important statement. It gives insight into the meaning of the definition of $\mathbf{D}$. The electric field intensity is discontinuous at the boundaries of the insulator, in such a way that only the normal component of $\mathbf{D}$ remains continuous. Thus, the influence of polarization is taken into account automatically.

There is no absolute necessity to make a distinction between , $\mathbf{E}$ and $\mathbf{D}$. Not to introduce $\mathbf{D}$ and solely work with the relations for vacuum is also possible. This requires an explicit consideration of all charges, including the surface charges due to the polarization. These cause a discontinuity in the normal component of $\mathbf{E}$. The above definition of $\mathbf{D}$, in contrast, considers these effects implicitly. If there are additional surface charges that are not caused by polarization, then these have to be taken into account explicitly in any case. To better distinguish those charges, the terms *free charges* and *bound charges* are used. Polarization causes bound charges. Accordingly, one introduces two charge densities

$$\rho = \rho_{\text{free}} + \rho_{\text{bound}} = \rho_f + \rho_b \; . \tag{2.107}$$

Then

$$\mathbf{D} = \varepsilon\mathbf{E} = \varepsilon_0\varepsilon_r\mathbf{E} = \varepsilon_0(1 + \chi)\mathbf{E}$$
$$= \varepsilon_0\mathbf{E} + \varepsilon_0\chi\mathbf{E} = \varepsilon_0\mathbf{E} + \mathbf{P} \; ,$$

i.e.

$$\boxed{\mathbf{D} = \varepsilon_0\mathbf{E} + \mathbf{P}} \; . \tag{2.108}$$

This relation shall generally apply, *i.e.* $\mathbf{D}$ is always defined by eq. (2.108), and is even done in the case of a permanent polarization (electret). Applying the general definition (2.108) to the special case of linear media results again in (2.105). Taking the divergence of (2.108) gives

$$\nabla\bullet(\varepsilon_0\mathbf{E}) = \rho = \rho_f + \rho_b = \nabla\bullet\mathbf{D} - \nabla\bullet\mathbf{P} \; ,$$

i.e., with (2.66)

$$\boxed{\nabla\bullet\mathbf{P} = -\rho_b} \tag{2.109}$$

and

$$\boxed{\nabla\bullet\mathbf{D} = \rho_f} \tag{2.110}$$

To avoid confusion here requires one to make a clear and conscious distinction between free and bound charges. This means to either calculate the electric field by considering all charges (free and bound), or to calculate the electric displacement from the free charges alone.

Note that one is still dealing with electrostatics (*i.e.* the time independent case) only. Nevertheless it shall be noted, that after applying an electric field, it takes some finite amount of time before the system reaches its final state. If one

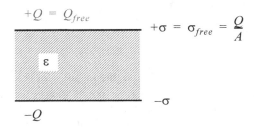

Fig. 2.46

applies alternating electric fields of sufficiently high frequency, then the equilibrium condition can not be reached anymore, and so χ and ε are actually frequency dependent. So far, we have been dealing only with χ and ε in the limit of zero frequency.

Moreover, many dielectrics are not isotropic, *i.e.* their polarization depends on the direction of the applied electric field and shows a preference for certain directions. In this case, however, ε is no longer a scalar, but a tensor. Eq. (2.105) is then replaced by the more complicated expression

$$D_x = \varepsilon_{xx}E_x + \varepsilon_{xy}E_y + \varepsilon_{xz}E_z$$
$$D_y = \varepsilon_{yx}E_x + \varepsilon_{yy}E_y + \varepsilon_{yz}E_z \qquad (2.111)$$
$$D_z = \varepsilon_{zx}E_x + \varepsilon_{zy}E_y + \varepsilon_{zz}E_z$$

or in tensor notation

$$\boxed{\mathbf{D} = \boldsymbol{\epsilon} \bullet \mathbf{E}} . \qquad (2.112)$$

$\boldsymbol{\epsilon}$ is a quantity with nine components whose individual components behave like products of vector components (*e.g.*, during a transformation). The scalar multiplication of a tensor of rank two (that is $\boldsymbol{\epsilon}$) with a vector results in a vector. Note that $\boldsymbol{\epsilon}$ is a symmetric tensor, thus

$$\varepsilon_{ik} = \varepsilon_{ki} \qquad (2.113)$$

If in (2.100) we set $\chi = \infty$, then $E_i = 0$. This suggests that in some respect, conductors behave like dielectrics with infinite susceptibility. The plausible reason is that a conductor has free charges which, when an electric field is applied, creates arbitrarily large dipole moments.

2.9   The Capacitor with a Dielectric

One is now able to understand why a dielectric increases the capacitance of a capacitor. As shown in Fig. 2.46, there are charges $\pm Q$ on the plates, and the space between them is filled with a dielectric of permittivity ε. Then

$$|\sigma| = \frac{|Q|}{A} = D = \varepsilon E = \varepsilon \frac{|V|}{d}$$

and therefore

$$C = \frac{|Q|}{|V|} = \varepsilon\frac{A}{d} = \varepsilon_0\varepsilon_r\frac{A}{d} \qquad (2.114)$$

Comparison with eq. (2.95) reveals that C has increased by the factor ε_r. This results quite plausibly from the fact, that for a given charge on the plate of the capacitor, the bound charges on the surfaces of the dielectric reduce the total charge and thereby also the electric field.

As another **example**, let us analyze a plane capacitor with a layered medium (Fig. 2.47). The voltage is

$$|V| = \sum_i E_i d_i \ .$$

On the other hand

$$\varepsilon_i E_i = D$$

is the same everywhere. Therefore

$$|V| = \sum_i \frac{D}{\varepsilon_i}d_i = D \sum_i \frac{d_i}{\varepsilon_i}$$

$$= \sigma \sum_i \frac{d_i}{\varepsilon_i} = \frac{|Q|}{A} \sum_i \frac{d_i}{\varepsilon_i} \ ,$$

or

$$C = \frac{|Q|}{|V|} = \frac{A}{\sum_i \frac{d_i}{\varepsilon_i}} \ . \qquad (2.115)$$

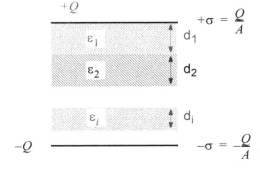

Fig. 2.47

2.10 Boundary Conditions for E and D and Refraction of Force Lines

Here, we analyze the boundary that separates two regions. Perhaps it is the boundary of two materials of different permittivities, or perhaps it carries a surface charge etc. The conditions that have to be met at such boundaries result from Maxwell's equations. We start with Faraday's law, eq. (1.68)

$$\nabla \times \mathbf{E} = -\frac{\partial}{\partial t}\mathbf{B} .$$

Integrating about the small area A shown in Fig. 2.48 gives

$$\int_A \nabla \times \mathbf{E} \bullet d\mathbf{A} = \oint \mathbf{E} \bullet d\mathbf{s} = ds(E_{2t} - E_{1t})$$

$$= -\frac{\partial}{\partial t}\int \mathbf{B} \bullet d\mathbf{A} = 0$$

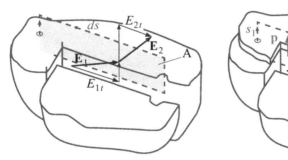

Fig. 2.48   Fig. 2.49

The reason is that the area becomes arbitrarily small when its perpendicular extent approaches zero. This implies that there is no voltage along the infinitesimal path element perpendicular to the boundary. This condition is not met for a dipole layer which is shown in Fig. 2.49, in which case the relation becomes:

$$\oint \mathbf{E} \bullet d\mathbf{s} = E_{2t}ds + \frac{\tau(s_2)}{\varepsilon_0} - E_{1t}ds - \frac{\tau(s_1)}{\varepsilon_0}$$

$$= 0 .$$

The reason is that the dipole layer causes a discontinuity of the potential in the direction of $\mathbf{p}$ by the factor τ/ε_0 (2.77). Thus

$$ds(E_{2t} - E_{1t}) = \frac{\tau(s_1) - \tau(s_2)}{\varepsilon_0} = \frac{\tau(s_1) - \tau(s_1 + ds)}{\varepsilon_0}$$

$$= -\frac{1}{\varepsilon_0}\frac{d\tau}{ds}ds$$

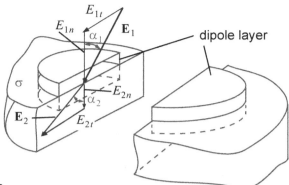

Fig. 2.50

i.e.,

$$E_{2t} - E_{1t} = -\frac{1}{\varepsilon_0}\frac{d\tau}{ds} \tag{2.116}$$

This is a special case. It describes the discontinuity of E_t along a given direction on the dipole layer as shown in Fig. 2.49. E_{1t}, E_{2t} are the tangential components of the electric field in this direction, where $d\tau/ds$ is the component of $\nabla\tau$ in this direction whereby $\nabla\tau$ is the two-dimensional gradient in the plane of the dipole layer. We can lift this restriction on the direction by writing

$$\boxed{E_{2t} - E_{1t} = -\frac{1}{\varepsilon_0}\nabla\tau} \,. \tag{2.117}$$

Equation (2.116) results from this by scalar multiplication with the unit vector in the chosen direction.

Without the dipole layer, one obtains as in the beginning of this section,

$$\boxed{E_{2t} = E_{t1}} \,. \tag{2.118}$$

The tangential component of **E** has to be continuous, as long as there is no dipole layer present.

A boundary condition for **D** follows from eq. (1.23) or its equivalent eq. (1.20) namely eq. (2.74), already derived in Section 2.5.3,

$$\boxed{D_{2n} - D_{1n} = \sigma} \,. \tag{2.119}$$

In agreement with our discussion of Section 2.8, this is generally true only, if σ merely represents the free, but not bound charges. The normal component of **D** is continuous only if $\sigma = 0$. Otherwise, it too is discontinuous.

The boundary conditions for **D** and **E** induce kinks (refraction) of the electric lines of force when entering another medium, or when passing through a dipole layer, or through a surface charge. With Fig. 2.50 it follows:

$$E_{2t} = E_{1t} - \frac{1}{\varepsilon_0}\frac{d\tau}{ds}$$

and

$$E_{2n} = \frac{D_{2n}}{\varepsilon_2} = \frac{D_{1n} + \sigma}{\varepsilon_2}$$

$$= \frac{\varepsilon_1 E_{1n} + \sigma}{\varepsilon_2} .$$

So

$$\tan\alpha_2 = \frac{E_{2t}}{E_{2n}} = \frac{E_{1t} - \frac{1}{\varepsilon_0}\frac{d\tau}{ds}}{\frac{\varepsilon_1}{\varepsilon_2}E_{1n} + \frac{\sigma}{\varepsilon_2}}$$

and

$$\tan\alpha_1 = \frac{E_{1t}}{E_{1n}}$$

i.e.

$$\boxed{\frac{\tan\alpha_2}{\tan\alpha_1} = \frac{1 - \frac{1}{\varepsilon_0 E_{1t}}\frac{d\tau}{ds}}{\frac{\varepsilon_1}{\varepsilon_2} + \frac{\sigma}{\varepsilon_2 E_{1n}}}} \qquad (2.120)$$

Specific cases of this relation are

1. Refraction at a boundary where $\varepsilon_1 \neq \varepsilon_2$, $\sigma = 0$

$$\frac{\tan\alpha_2}{\tan\alpha_1} = \frac{\varepsilon_2}{\varepsilon_1} \qquad (2.121)$$

2. Refraction at a surface with free surface charges where ($\varepsilon_1 = \varepsilon_2 = \varepsilon_0$)

$$\frac{\tan\alpha_2}{\tan\alpha_1} = \frac{1}{1 + \frac{\sigma}{\varepsilon_0 E_{1n}}} \qquad (2.122)$$

Generally, if there is a dipole layer, the electric field is not only refracted, but also rotated by the angle β with respect to the plane of incidence. Using eq. (2.117) and the illustration shown in Fig. 2.51:

$$\mathbf{E}_{2t} = \mathbf{E}_{1t} - \frac{1}{\varepsilon_0}\nabla\tau$$

and therefore

$$E_{2t\,\|} = E_{1t} - \frac{1}{\varepsilon_0}(\nabla\tau)_\|$$

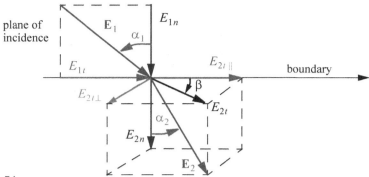

plane of incidence

boundary

Fig. 2.51

$$E_{2t\perp} = -\frac{1}{\varepsilon_0}(\nabla\tau)_\perp .$$

This results in

$$\tan\beta = \frac{E_{2t\perp}}{E_{2t\,\|}}$$

and

$$\tan\alpha_2 = \frac{E_{2t}}{E_{2n}} .$$

2.11 A Point Charge inside a Dielectric

2.11.1 Uniform Dielectric

Consider a point charge at the center of a hollow sphere made up of a dielectric material (Fig. 2.52).
For the entire space,

$$\mathbf{D} = \frac{Q}{4\pi r^3}\mathbf{r} .$$

For vacuum

$$\mathbf{E} = \frac{Q}{4\pi\varepsilon_0 r^3}\mathbf{r} ,$$

and inside the dielectric, hollow sphere

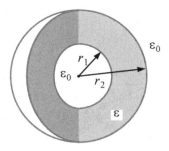

Fig. 2.52

$$\mathbf{E} = \frac{Q}{4\pi\varepsilon r^3}\mathbf{r} = \frac{Q}{4\pi\varepsilon_0\varepsilon_r r^3}\mathbf{r} \ .$$

The associated polarization is

$$\mathbf{P} = \varepsilon_0\chi\mathbf{E} = \frac{\varepsilon_r-1}{\varepsilon_r}\frac{Q}{4\pi r^3}\mathbf{r} \ .$$

and, as will be shown later (in 3.42)

$$\rho_b = -\nabla\bullet\mathbf{P} = -\frac{1}{r^2}\frac{\partial}{\partial r}\left(r^2\frac{\varepsilon_r-1}{\varepsilon_r}\frac{Q}{4\pi r^3}r\right) = 0 \ ,$$

which means that there are no bound volume charges inside the dielectric. Nevertheless, there are bound surface charges, specifically

$$\sigma_b = \begin{cases} -\dfrac{\varepsilon_r-1}{\varepsilon_r}\dfrac{Q}{4\pi r_1^2} & \text{for } r = r_1 \ , \\[2ex] +\dfrac{\varepsilon_r-1}{\varepsilon_r}\dfrac{Q}{4\pi r_2^2} & \text{for } r = r_2 \ . \end{cases}$$

Fig. 2.53 illustrates the behavior of $\mathbf{D}$ and $\mathbf{E}$ as functions of $\mathbf{r}$. Now we let r_1 approach zero, but r_2 shall go to infinity. As a result, one finds the net charge inside

$$Q' = Q - \frac{\varepsilon_r-1}{\varepsilon_r}\frac{Q}{4\pi r_1^2}4\pi r_1^2 = Q - \frac{\varepsilon_r-1}{\varepsilon_r}Q = \frac{Q}{\varepsilon_r} \ .$$

This means that the charge appears to be reduced as a result of the bound charges by the factor ε_r. The overall field, where now the dielectric fills the entire space, is

$$\mathbf{E} = \frac{Q}{4\pi\varepsilon_0\varepsilon_r r^3}\mathbf{r}$$

$$= \frac{Q'}{4\pi\varepsilon_0 r^3}\mathbf{r} \ ,$$

i.e., the field is reduced relative to vacuum by the same factor ε_r, as was the charge.

2.11.2 Plane Boundaries between two Dielectrics

Let a point charge reside in a space that is filled with two different dielectrics that are separated by a plane boundary (Fig. 2.54). Material 1 (ε_1) occupies the half-space $x > 0$; material 2 (ε_2) occupies the half-space $x < 0$. One can show that

1. The field in half-space 1 can be expressed as the superposition of the field of Q and the field of a fictitious charge (image charge) Q', located in half-space 2 at the same distance (a) from the boundary as Q.

2. The field in half-space 2 can be expressed as the field of the fictitious charge Q'', located at the same point as Q.

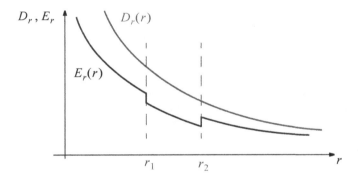

Fig. 2.53

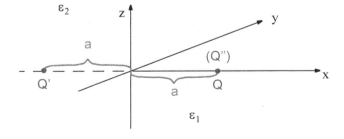

Fig. 2.54

This justifies the following Ansatz.

$$\varphi_1 = \frac{1}{4\pi\varepsilon_1}\left(\frac{Q}{\sqrt{(x-a)^2+y^2+z^2}} + \frac{Q'}{\sqrt{(x+a)^2+y^2+z^2}}\right)$$

$$\varphi_2 = \frac{1}{4\pi\varepsilon_2}\left(\frac{Q''}{\sqrt{(x-a)^2+y^2+z^2}}\right).$$

From this follow the electric fields

$$\mathbf{E}_1 = -\nabla\varphi_1 = \frac{1}{4\pi\varepsilon_1}\cdot\begin{bmatrix}\dfrac{Q(x-a)}{\sqrt{(x-a)^2+y^2+z^2}^3} + \dfrac{Q'(x+a)}{\sqrt{(x+a)^2+y^2+z^2}^3}\\[2ex] \dfrac{Qy}{\sqrt{(x-a)^2+y^2+z^2}^3} + \dfrac{Q'y}{\sqrt{(x+a)^2+y^2+z^2}^3}\\[2ex] \dfrac{Qz}{\sqrt{(x-a)^2+y^2+z^2}^3} + \dfrac{Q'z}{\sqrt{(x+a)^2+y^2+z^2}^3}\end{bmatrix},$$

$$\mathbf{E}_2 = -\nabla\varphi_2 = \frac{1}{4\pi\varepsilon_2}\cdot\begin{bmatrix}\dfrac{Q''(x-a)}{\sqrt{(x-a)^2+y^2+z^2}^3}\\[2ex] \dfrac{Q''y}{\sqrt{(x-a)^2+y^2+z^2}^3}\\[2ex] \dfrac{Q''z}{\sqrt{(x-a)^2+y^2+z^2}^3}\end{bmatrix}$$

The tangential components of $\mathbf{E}$, that is E_y and E_z, as well as the normal component of $\mathbf{D}$, that is εE_x, have to be continuous on the boundary $x=0$. E_y and E_z, are continuous if

$$\frac{Q+Q'}{\varepsilon_1} = \frac{Q''}{\varepsilon_2},$$

and εE_x is continuous if

$$Q'+Q'' = Q.$$

One may verify that the Ansatz is correct by showing that the appropriate choice of Q' and Q'' fulfills the boundary conditions everywhere on the boundary $x=0$, i.e., for all y and all z, which is not at all self-evident. Calculating Q' and Q'' from these two equations gives

$$Q' = Q\frac{\varepsilon_1-\varepsilon_2}{\varepsilon_1+\varepsilon_2},$$

$$Q'' = Q\frac{2\varepsilon_2}{\varepsilon_1+\varepsilon_2}.$$

Q'' always has the same sign as Q has, while Q' may have either one. In particular for $\varepsilon_1 = \varepsilon_2$, we find $Q' = 0$ and $Q'' = Q$, as expected. In the limit of ε_2

approaching infinity, one finds $Q' = -Q$, the same result as for the image charge at a conducting plane. As already mentioned, a conductor behaves in many ways like a dielectric with infinite permittivity. We will return to this topic in a later section.

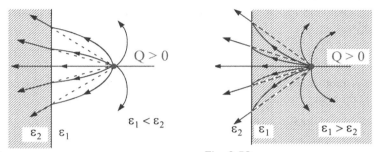

Fig. 2.55   Fig. 2.56

The field configuration for $\varepsilon_1 < \varepsilon_2$ is illustrated in Fig. 2.55 and for $\varepsilon_1 > \varepsilon_2$ in Fig. 2.56. The curvature of the force lines depends on the sign of Q'. For $\varepsilon_1 < \varepsilon_2$ and $Q > 0$ we have $Q' < 0$, *i.e.* Q and Q' attract each other, which results in field lines as shown in Fig. 2.55. For $\varepsilon_1 > \varepsilon_2$ and $Q > 0$, however, we have $Q' > 0$, *i.e.*, Q and Q' repel each other, which results in field lines as shown in Fig. 2.56.

2.12 A Dielectric Sphere in a Uniform Electric Field

2.12.1 The Field of a Uniformly Polarized Sphere

In order to solve the problem of a dielectric sphere, it is useful to consider the electric field of a uniformly polarized sphere. When r_s is the radius of the sphere and P its uniform polarization, then the overall dipole moment is

$$p = PV = \frac{4\pi r_s^3}{3} P .$$ (2.123)

One may think of a uniformly polarized sphere as being created by two spheres, charged with opposite charges that are slightly displaced against each other (Fig. 2.57). When $\pm\rho$ is its volume charge and d the displacement, then

$$P = \rho d .$$

Outside of the sphere, the field is that of a dipole at the origin, namely

$$\varphi = \frac{p\cos\theta}{4\pi\varepsilon_0 r^2} = \frac{Pr_s^3\cos\theta}{3\varepsilon_0 r^2} .$$ (2.124)

This is still correct for the surface of the sphere, where

$$\varphi = \varphi_s = \frac{Pr_s\cos\theta}{3\varepsilon_0} = \frac{Pz}{3\varepsilon_0} \ . \tag{2.125}$$

By means of theorems about the uniqueness of solutions of potentials, which we will deal with in Chapter 3, it is possible to show that the potential inside the sphere has to be

$$\varphi = \frac{Pz}{3\varepsilon_0} \ . \tag{2.126}$$

There is another way to prove this without using those theorems. The components of $\mathbf{E}$ at the sphere's outer surface are

$$\left.\begin{aligned}
E_r &= \frac{2p\cos\theta}{4\pi\varepsilon_0 r_s^3} = \frac{2P\cos\theta}{3\varepsilon_0} \\
E_\theta &= \frac{p\sin\theta}{4\pi\varepsilon_0 r_s^3} = \frac{P\sin\theta}{3\varepsilon_0} \\
E_\varphi &= 0
\end{aligned}\right\} , \tag{2.127}$$

which results from (2.63). The surface charge at the sphere's surface (bound charges due to polarization) is

$$\sigma_b = P\cos\theta \ . \tag{2.128}$$

This means that E_r decreases by $(P\cos\theta)/\varepsilon_0$ when passing through the surface from the outside towards the inside of the sphere, while the other components remain unchanged. At the inside surface of the sphere, one therefore obtains

$$\left.\begin{aligned}
E_r &= -\frac{P\cos\theta}{3\varepsilon_0} \\
E_\theta &= \frac{P\sin\theta}{3\varepsilon_0} \\
E_\varphi &= 0 \ .
\end{aligned}\right\} \tag{2.129}$$

Consequently the electric field $\mathbf{E}$ written in terms of its Cartesian components is

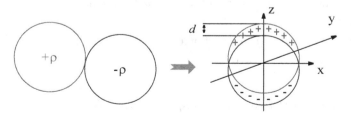

Fig. 2.57

$$
\left.
\begin{aligned}
E_x &= E_r \sin\theta\cos\varphi + E_\theta\cos\theta\cos\varphi = 0 \\
E_y &= E_r \sin\theta\sin\varphi + E_\theta\cos\theta\sin\varphi = 0 \\
E_z &= E_r\cos\theta - E_\theta\sin\theta = -\frac{P}{3\varepsilon_0}\quad.
\end{aligned}
\right\}
\tag{2.130}
$$

So, E has only a z-component which, furthermore, has the same value everywhere (on the inside surface). The inside of the sphere is free of (bound) volume charges and we therefore conclude that the whole inside space is filled with the same field. The related potential is $\varphi = Pz/3\varepsilon_0$, as was already suggested. Also notice that in connection with eq. (2.128), a conducting sphere within a uniform electric field carries a surface charge proportional to $\cos\theta$ (see .eq. (2.94)). Obviously, the field of the surface charge identically cancels the external field, $i.e.$, it creates a uniform field inside the sphere. Outside, it creates a dipole field, as we have also seen when analyzing the conducting sphere. We obtain the same results when we apply these results to the current case.

> *Summarizing, we may state that a uniformly polarized sphere (polarization **P**) creates an electric dipole field outside and a uniform electric field inside* $-(\mathbf{P}/3\varepsilon_0)$.

This provides the field of a permanently, uniform polarized sphere (*i.e.*, that of a uniform electret). The field in its interior is

$$
\mathbf{D} = \mathbf{P} + \varepsilon_0\mathbf{E} = \mathbf{P} - \frac{\mathbf{P}}{3} = \frac{2}{3}\mathbf{P} \quad.
$$

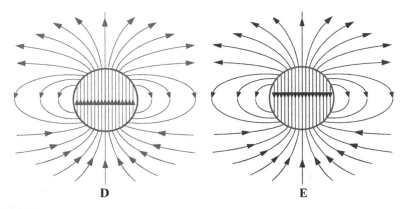

Fig. 2.58

Here (Fig. 2.58), **D** and **E** point in different directions: they are anti-parallel, not parallel as usual. Furthermore, a remarkable fact is that (as always in electrostatics) **E** is irrotational but not source free, while **D** is source free but not irrotational.

Besides spherical bodies, only ellipsoids have such simple characteristics. The general relation between **P** and **E** or **D** and **E**, respectively, is rather complicated. In particular, **D** and **E** may point in entirely different directions. Fig. 2.59 illustrates the fields of a uniformly polarized cuboid. On the inside of it, the fields of **D** and **E** have very different shapes. Of course, on the outside, there is the usual relation between **E** and **D**: $\mathbf{D} = \varepsilon_0 \mathbf{E}$.

So far in this section, we have not asked what causes polarization. It could be due to permanent polarization as illustrated in Fig. 2.58 and Fig. 2.59. It could also be due to an external, uniform field, which then would need to be considered as well.

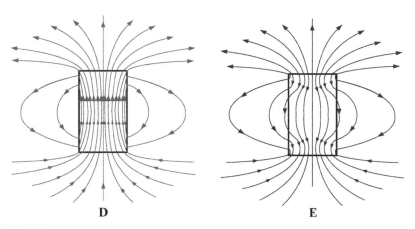

Fig. 2.59

2.12.2 An Externally Applied Uniform Field as the Cause of Polarization

When a sphere is exposed to a uniform field, an additional field is created as the sphere is polarized. This field needs to be superimposed onto the original field. As we have just seen for uniform polarization, the internal field is uniform as well. Uniform polarization would thus cause an overall uniform field inside, which on the other hand causes uniform polarization. We may therefore state that a uniform electric field uniformly polarizes a dielectric sphere. Using the relations from Fig. 2.60, one is now able to write

$$E_i = E_{a,\infty} - \frac{P}{3\varepsilon_0} = E_{a,\infty} - \frac{\varepsilon_0 \chi E_i}{3\varepsilon_0} \ .$$

from which one finds

$$\boxed{E_i = E_{a,\infty} \frac{3}{3+\chi} = E_{a,\infty} \frac{3}{2+\varepsilon_r}} \ . \tag{2.131}$$

Furthermore,

$$D_i = D_{a,\infty} \frac{3\varepsilon_r}{2+\varepsilon_r} \tag{2.132}$$

and

$$P = E_{a,\infty} \frac{3\varepsilon_0 \chi}{2+\varepsilon_r} = E_{a,\infty} 3\varepsilon_0 \frac{\varepsilon_r - 1}{\varepsilon_r + 2} \ . \tag{2.133}$$

2.12.3 Dielectric Sphere (ε_i) and Dielectric Space (ε_a)

Let us generalize the previous problem some more. The space outside the sphere shall now be a dielectric as well. Now, the field outside consists of a, so far, unknown dipole field and the uniform field $\mathbf{E}_{a,\infty}$, *i.e.*, we may write

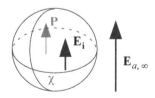

Fig. 2.60

$$E_{ra} = \frac{2C\cos\theta}{4\pi\varepsilon_0 r^3} + E_{a,\infty}\cos\theta \left.\right\}$$

$$E_{\theta a} = \frac{C\sin\theta}{4\pi\varepsilon_0 r^3} - E_{a,\infty}\sin\theta \left.\right\} \qquad (2.134)$$

$$E_{\varphi a} = 0 \quad .$$

C is a constant that is initially undetermined. It depends on the polarization of both, the outside space as well as the inside of the sphere. Inside, we have the yet undetermined uniform field $\mathbf{E_i}$, *i.e.*,

$$E_{ri} = E_i\cos\theta \left.\right\}$$

$$E_{\theta i} = -E_i\sin\theta \left.\right\} \qquad (2.135)$$

$$E_{\varphi i} = 0 \quad .$$

One can show that with the proper choice of C and E_i, all boundary conditions at the sphere's surface ($r = r_s$) can be fulfilled. Only this is the justification for the Ansatz in (2.134) and (2.135). E_θ has to be continuous for $r = r_s$, *i.e.*,

$$\frac{C\sin\theta}{4\pi\varepsilon_0 r_s^3} - E_{a,\infty}\sin\theta = -E_i\sin\theta \quad .$$

Furthermore, also for $r = r_s$, $D_r = \varepsilon E_r$ has to be continuous, *i.e.*,

$$\varepsilon_a\frac{2C\cos\theta}{4\pi\varepsilon_0 r_s^3} + \varepsilon_a E_{a,\infty}\cos\theta = \varepsilon_i E_i\cos\theta \quad .$$

Solving those equations gives

$$E_i = \frac{3\varepsilon_a}{\varepsilon_i + 2\varepsilon_a}E_{a,\infty} \qquad (2.136)$$

and

$$C = 4\pi\varepsilon_0\frac{\varepsilon_i - \varepsilon_a}{\varepsilon_i + 2\varepsilon_a}E_{a,\infty}\, r_s^3 \qquad (2.137)$$

(2.136) generalizes (2.131) and both coincide for $\varepsilon_a = \varepsilon_0$.

The polarization inside is

$$P_i = \varepsilon_0\chi_i E_i = (\varepsilon_i - \varepsilon_0)E_i = 3\varepsilon_a\frac{\varepsilon_i - \varepsilon_0}{\varepsilon_i + 2\varepsilon_a}E_{a,\infty} \qquad (2.138)$$

and uniform in z-direction. The polarization in the outside space is not uniform, but it is nevertheless divergenceless (source free), so that there are no bound charges in the outside space:

$$\rho_{b,a} = -\nabla\bullet\mathbf{P}_a = -(\varepsilon_a - \varepsilon_0)\nabla\bullet\boldsymbol{E}_a$$

$$= -(\varepsilon_a - \varepsilon_0)\left(\frac{1}{r^2}\frac{\partial}{\partial r}r^2 E_{ra} + \frac{1}{r\sin\theta}\left(\frac{\partial}{\partial\theta}\sin\theta E_{\theta a}\right)\right)$$

$$= 0 \ .$$

The expression used here for the divergence in spherical coordinates will be derived later. Bound charges exist solely at the surface of the sphere, namely

$$\sigma_b = \sigma_{b,\,i} + \sigma_{b,\,a}$$

$$= (\varepsilon_0\chi_i E_{ri} - \varepsilon_0\chi_a E_{ra})_{r\,=\,r_s}$$

$$= (\varepsilon_i - \varepsilon_0)E_i\cos\theta - (\varepsilon_a - \varepsilon_0)\left(\frac{2C}{4\pi\varepsilon_0 r_s^3} + E_{a,\,\infty}\right)\cos\theta$$

$$= 3\varepsilon_0\frac{\varepsilon_i - \varepsilon_a}{\varepsilon_i + 2\varepsilon_a}E_{a,\,\infty}\cos\theta \ .$$

Outside of the sphere, this charge brings about a dipole field which corresponds to the dipole moment

$$p = \left(\frac{4\pi}{3}r_s^3\right)3\varepsilon_0\frac{\varepsilon_i - \varepsilon_a}{\varepsilon_i + 2\varepsilon_a}E_{a,\,\infty}$$

$$= 4\pi\varepsilon_0\frac{\varepsilon_i - \varepsilon_a}{\varepsilon_i + 2\varepsilon_a}E_{a,\,\infty}\,r_s^3 \ .$$

Note eqs. (2.67) and (2.123). The constant C of the previous Ansatz is just the dipole moment, caused by the polarization of the inside and outside space, which justifies the Ansatz (2.134) and also illustrates the formal result of (2.137).

Together, the dipole field and, the uniform field bring about the potential in the outside space

$$\varphi_a = E_{a,\,\infty}\cos\theta\left(\frac{r_s^3}{r^2}\frac{\varepsilon_i - \varepsilon_a}{\varepsilon_i + 2\varepsilon_a} - r\right) \ . \tag{2.139}$$

To compare this potential with that of a conducting sphere in a uniform electric field as described by eq. (2.91) is an interesting exercise. It can be derived from the just obtained potential by the limit $\varepsilon_a/\varepsilon_i$ approaching zero. In some way, a conductor behaves like a dielectric in the limit of infinite permittivity. The law of refraction (2.121) illustrates this fact. Lines of force have to be perpendicular to the conductor surface. According to the law of refraction, this is also the case for a dielectric of infinite permittivity.

E_i in (2.136) also determines D_i,

$$\boxed{D_i = \frac{3\varepsilon_i}{\varepsilon_i + 2\varepsilon_a}D_{a,\,\infty}} \ . \tag{2.140}$$

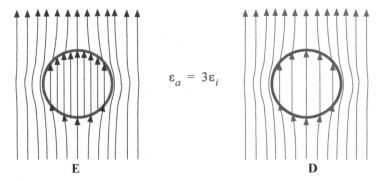

$\varepsilon_a = 3\varepsilon_i$

E   **D**

Fig. 2.61

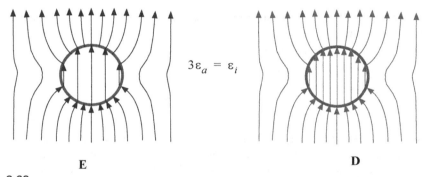

$3\varepsilon_a = \varepsilon_i$

E   **D**

Fig. 2.62

This result allows one to sketch the fields in Fig. 2.62 (for $\varepsilon_a > \varepsilon_i$) and Fig. 2.61 (for $\varepsilon_a < \varepsilon_i$). In all figures, **D** has no divergence. **E**, in contrast, because of the bound surface charges is not free of a divergence. For the case of Fig. 2.61, *i.e.* for ($\varepsilon_a > \varepsilon_i$), it is $E_i > E_{a,\infty}$ and $D_i < D_{a,\infty}$. In contrast, for the case of Fig. 2.62, *i.e.* for ($\varepsilon_a < \varepsilon_i$), it is $E_i < E_{a,\infty}$ and $D_i > D_{a,\infty}$.

2.12.4 Generalization: Ellipsoids

From our considerations of the plane disk discussions in Section 2.8, with uniform polarization perpendicular to its surface, one finds that

$$E_i = E_a - \frac{1}{\varepsilon_0}P \ . \tag{2.141}$$

The result for a sphere was

$$E_i = E_{a,\infty} - \frac{1}{3\varepsilon_0}P \; . \tag{2.142}$$

We may add the trivial case of a plane disk that is polarized parallel to its surface (Fig. 2.63), where the boundary condition causes

$$E_i = E_a - 0 \cdot P = E_a \; . \tag{2.143}$$

The factor in front of P in all those equations is called the *de-electrification* factor. In above three cases, this factor is $1/\varepsilon_0$, $1/3\varepsilon_0$, and 0, respectively.

We have already stressed the fact, that an arbitrarily shaped body of uniform polarization does by no means have a uniform field inside. This is only the case for ellipsoids and their limits (plane plates, cylinders, spheres). The proof shall not be provided here. The equation for an ellipsoid is

$$\frac{x^2}{a^2} + \frac{y^2}{b^2} + \frac{z^2}{c^2} = 1$$

An elliptical cylinder results from the limit of $c \to \infty$

$$\frac{x^2}{a^2} + \frac{y^2}{b^2} = 1 \; .$$

For this case, if also $a = b$, a circular cylinder results

$$x^2 + y^2 = a^2 \; .$$

For an ellipsoid with $a = b = c$ we get a sphere

$$x^2 + y^2 + z^2 = a^2 \; .$$

Two parallel plates emerge when two half-axes, *for example*, a and b approach infinity:

$$z^2 = c^2 \; ,$$

i.e.,

$$z = \pm c \; .$$

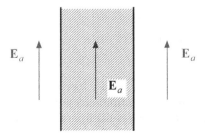

Fig. 2.63

We shall not calculate ellipsoids in any depth here, but merely provide the results, which may be obtained by means of the Ansatz, originated by Dirichlet, that a uniform polarization

$$\mathbf{P} = \langle P_x, P_y, P_z \rangle \tag{2.144}$$

creates a uniform internal field

$$\mathbf{E} = \langle -AP_x, -BP_y, -CP_z \rangle \ . \tag{2.145}$$

Therefore, the vectors $\mathbf{P}$ and $\mathbf{E}$ point generally in different directions. (2.145) could also be written in the following form

$$\mathbf{E} = -\begin{pmatrix} A & 0 & 0 \\ 0 & B & 0 \\ 0 & 0 & C \end{pmatrix} \mathbf{P} \ . \tag{2.146}$$

The three constants A, B, C are the *de-electrification factors for the ellipsoid*. A, B, C are different from each other for an ellipsoid with three distinct axes and determined by certain integrals, *for example*,

$$A = \frac{abc}{2\varepsilon_0} \int_0^\infty \frac{d\xi}{(a^2 + \xi^2)^{3/2}(b^2 + \xi^2)^{1/2}(c^2 + \xi^2)^{1/2}} \ .$$

Of course, the expressions for B and C are analogous. Remarkable is that in any case.

$$\boxed{A + B + C = \frac{1}{\varepsilon_0}} \tag{2.147}$$

For symmetry reasons, the relation for a sphere has to be $A = B = C = 1/3\varepsilon_0$, confirming our previous result. For a circular cylinder whose axis is oriented parallel to the z-axis, we have $C = 0$ and $A = B = 1/2\varepsilon_0$. This result can easily be derived by the method previously used for a sphere. It is an easy exercise to convince oneself that the field outside the cylinder is that of a line dipole at the axis of the cylinder. For a plane plate whose normal component is parallel to the z-axis, the constants are $A = B = 0$ and $C = 1/\varepsilon_0$, which again, is consistent with our previous result.

Later, in conjunction with problems of magnetism, we will meet similar factors, which are termed *de-magnetizing factors*.

2.13 Polarization Current

The chapter discussing electrostatic problems is not the most appropriate place to cover polarization currents. Nevertheless, we have introduced polarization and want to also introduce the polarization current, which results from time dependent polarization. We start from

$$\nabla \bullet \mathbf{P} = -\rho_b \ .$$

If **P** is time dependent, then ρ_b has to be time dependent as well. The charge conservation principle applies also to bound charges. If we call the related current density g_b, then according to the continuity equation (1.58), we obtain

$$\nabla \bullet \mathbf{g}_b + \frac{\partial \rho_b}{\partial t} = 0 .$$
(2.148)

or

$$\nabla \bullet \mathbf{g}_b - \nabla \bullet \left(\frac{\partial}{\partial t} \mathbf{P} \right) = 0 ,$$
(2.149)

i.e.,

$$\boxed{\mathbf{g}_b = \frac{\partial}{\partial t} \mathbf{P}} ,$$
(2.150)

if we also assume that any, by (2.149) still possible divergence-free, additional term vanishes. This current density of the bound charges is called the polarization current density.

The overall charge density is

$$\rho = \rho_b + \rho_f ,$$

and the total current density is

$$\mathbf{g} = \mathbf{g}_b + \mathbf{g}_f = \frac{\partial}{\partial t} \mathbf{P} + \mathbf{g}_f .$$

To prevent misconceptions, let us discuss how this applies, for instance, to Maxwell's first equation (1.61).

$$\nabla \times \mathbf{H} = \mathbf{g} + \frac{\partial \mathbf{D}}{\partial t} .$$

There are two possible approaches. Either we explicitly consider all charges and treat the given space as if it were vacuum, or we only consider the free charges and consider the space to be a dielectric. The first case gives

$$\mathbf{g} = \mathbf{g_f} + \mathbf{g_b} = \mathbf{g_f} + \frac{\partial}{\partial t} \mathbf{P}$$

and

$$\mathbf{D} = \varepsilon_0 \mathbf{E} ,$$

i.e.,

$$\nabla \times \mathbf{H} = \mathbf{g_f} + \frac{\partial}{\partial t} \mathbf{P} + \frac{\partial}{\partial t} \varepsilon_0 \mathbf{E} .$$

In the second case, the current density is

$$\mathbf{g} = \mathbf{g_f}$$

and

$$\mathbf{D} = \varepsilon_0 \mathbf{E} + \mathbf{P} ,$$

i.e.,

$$\nabla \times \mathbf{H} = \mathbf{g_f} + \frac{\partial}{\partial t}\mathbf{P} + \frac{\partial}{\partial t}\varepsilon_0 \mathbf{E} .$$

Both viewpoints lead to the same result, although care is needed in order not to confuse those two approaches.

2.14 Energy Principle

2.14.1 Energy Principle in its General Form

The energy principle of electrostatics is just a special case of the general energy principle of electrodynamics which we will cover here, although it is not strictly part of electrostatics. Starting point are the following two Maxwell's equations.

$$\nabla \times \mathbf{H} = \mathbf{g} + \frac{\partial \mathbf{D}}{\partial t} . \tag{2.151}$$

$$\nabla \times \mathbf{E} = -\frac{\partial}{\partial t}\mathbf{B} . \tag{2.152}$$

We define the so-called *Poynting vector*:

$$\boxed{\mathbf{S} = \mathbf{E} \times \mathbf{H}} , \tag{2.153}$$

whose significance we will recognize in the following. We take its divergence,

$$\nabla \bullet \mathbf{S} = \nabla \bullet (\mathbf{E} \times \mathbf{H}) = \mathbf{H} \bullet (\nabla \times \mathbf{E}) - \mathbf{E} \bullet (\nabla \times \mathbf{H}) , \tag{2.154}$$

then using Maxwell's equations:

$$\nabla \bullet \mathbf{S} = -\mathbf{H} \bullet \frac{\partial}{\partial t}\mathbf{B} - \mathbf{E} \bullet \frac{\partial}{\partial t}\mathbf{D} - \mathbf{E} \bullet \mathbf{g} . \tag{2.155}$$

The significance of this equation is illustrated by integrating over a volume V:

$$\int_V (\nabla \bullet \mathbf{S})d\tau = \oint_A \mathbf{S} \bullet d\mathbf{A} = -\int_V \left(\mathbf{H} \bullet \frac{\partial}{\partial t}\mathbf{B} + \mathbf{E} \bullet \frac{\partial}{\partial t}\mathbf{D}\right)d\tau - \int_V (\mathbf{E} \bullet \mathbf{g})d\tau \tag{2.156}$$

Although this equation will loose its generality, we will perform some more algebra, using the relations:

$$\left.\begin{aligned} \mathbf{D} &= \varepsilon \mathbf{E} \\ \mathbf{B} &= \mu \mathbf{H} \\ \mathbf{g} &= \kappa \mathbf{E} \end{aligned}\right\} \tag{2.157}$$

So far, we have used the equation $\mathbf{B} = \mu \mathbf{H}$ for vacuum only for which $\mu = \mu_0$. However, we will use its generalization, which will be discussed later. Then

$$\left. \begin{aligned} \mathbf{E} \bullet \frac{\partial \mathbf{D}}{\partial t} &= \frac{\partial}{\partial t}\left(\frac{1}{2}\varepsilon E^2\right) \\[4pt] \mathbf{H} \bullet \frac{\partial \mathbf{B}}{\partial t} &= \frac{\partial}{\partial t}\left(\frac{1}{2}\mu H^2\right) \\[4pt] \mathbf{E} \bullet \mathbf{g} &= \kappa E^2 = \frac{g^2}{\kappa} \end{aligned} \right\} ,$$

(2.158)

and finally, using (2.156) gives

$$\boxed{\frac{\partial}{\partial t}\int_V \left(\frac{1}{2}\varepsilon E^2 + \frac{1}{2}\mu H^2\right) d\tau + \int_V \frac{g^2}{\kappa}\,d\tau + \oint_A \mathbf{S} \bullet d\mathbf{A} = 0}$$

(2.159)

or using (2.155)

$$\boxed{\frac{\partial}{\partial t}\left(\frac{1}{2}\varepsilon E^2 + \frac{1}{2}\mu H^2\right) + \frac{g^2}{\kappa} + \nabla \bullet \mathbf{S} = 0}.$$

(2.160)

Those two, equivalent relations represent the energy principle in integral and differential from, respectively. To interpret it, we apply the following reasoning.

Imagine some system that contains the energy W in whatever form. This energy is distributed somehow within the space of that system, where it has the spatial density (energy density)

$$w = \frac{dW}{d\tau}$$

The energy may be distributed differently at different times, *i.e.*, it may flow from one point to another point. The energy per unit time and unit area that flows through an area element is called *energy flux density*. It is a vector and shall be named **v**. Energy W is not necessarily a conserved quantity, since one type of energy may be transformed into another type. Of course this same fraction of energy has to exist in that other form of energy. The transformed energy per unit time and unit volume shall be called u. The energy balance is then:

$$\oint_A \mathbf{v} \bullet d\mathbf{A} + \int_V u\,d\tau = -\frac{\partial}{\partial t}\int_V w\,d\tau$$

(2.161)

i.e., the energy that is lost from the overall volume consists of two parts. One flows away through the surface ($\oint_A \mathbf{v} \bullet d\mathbf{A}$) and one is transformed into another form of energy ($\int_V u\,d\tau$). This equation is comparable to eq. (2.159). When written in differential form it may be compared to (2.160). Applying Gauss' theorem we get

$$\int_V \left(\nabla \bullet \mathbf{v} + u + \frac{dw}{d\tau}\right) d\tau = 0$$

and thus

$$\nabla \bullet \mathbf{v} + u + \frac{dw}{d\tau} = 0.$$

(2.162)

Comparison allows identification of the different terms:

1.   **S** is the energy flux (density) of the electromagnetic field

2. $\frac{1}{2}\varepsilon E^2 + \frac{1}{2}\mu H^2$ is the electromagnetic energy density, which consists of an electric ($\varepsilon E^2/2$) and a magnetic ($\mu H^2/2$) part.

3. g^2/κ is the fraction of electromagnetic energy that is lost per unit time and unit volume, and is nothing else than the *heat loss* of electromagnetic energy converted to *heat energy due to the current* (resistance), as we shall show.

We will now analyze a cylindrical conductor with constant conductivity κ. It shall have the length l, cross section A, and shall carry the current of constant density **g**. Then the transformed power in its volume is

$$\int_V \frac{g^2}{\kappa}\,d\tau = \frac{g^2}{\kappa}Al = (gA)^2\frac{l}{\kappa A} = I^2 R$$

$$= (gA)\frac{g}{\kappa}l = (gA)(El) = IV \tag{2.163}$$

because the total current is

$$I = gA ,$$

and the voltage is

$$V = lE .$$

Furthermore, we have used

$$R = \frac{l}{\kappa A} . \tag{2.164}$$

R is the resistance of the conductor measured in Ohm (Ω) (see Section 1.13). Here, $I^2R = IV$ is the power transformed into heat due to the current. This confirms our hypothesis. Ohm's law is obtained in its usual form

$$V = RI . \tag{2.165}$$

This represents the integral form, of what has previously been introduced as Ohm's law

$$\mathbf{g} = \kappa\mathbf{E}$$

or

$$g = \kappa E = \frac{I}{A} = \kappa\frac{V}{l}$$

Multiplication by l/κ gives $IR = V$.

An important point to highlight is that eqs. (2.155) and (2.156) are much more general than eqs. (2.159) and (2.160), which we obtained using the more restrictive relations of eq.(2.157). The former also apply to nonlinear media as well as to moving charges (current densities) of any kind, which are possibly not caused by a conducting medium.

2.14.2 Electrostatic Energy

We focus now on electrostatic energy. Its spatial density has been determined in a rather formal way to be

$$w = \frac{\varepsilon E^2}{2} = \frac{\mathbf{D} \bullet \mathbf{E}}{2}.$$

It must be possible to also obtain this expression from purely electrostatic considerations.

By eq. (2.18), a point charge Q_1 at location $\mathbf{r}_1$ is the source of the potential

$$\varphi = \frac{Q_1}{4\pi\varepsilon_0|\mathbf{r} - \mathbf{r}_1|}$$

If we move a second charge Q_2 from infinity to point $\mathbf{r}_2$, the thereby stored energy is

$$W_{12} = \frac{Q_1 Q_2}{4\pi\varepsilon_0|\mathbf{r}_2 - \mathbf{r}_1|} = \frac{Q_1 Q_2}{4\pi\varepsilon_0 r_{12}}, \tag{2.166}$$

using

$$r_{12} = |\mathbf{r}_2 - \mathbf{r}_1|.$$

The potential created by both charges is

$$\varphi = \frac{Q_1}{4\pi\varepsilon_0|\mathbf{r} - \mathbf{r}_1|} + \frac{Q_2}{4\pi\varepsilon_0|\mathbf{r} - \mathbf{r}_2|}.$$

A third charge moved from infinity to point $\mathbf{r}_3$ adds to the stored energy

$$W_{13} + W_{23} = \frac{Q_1 Q_3}{4\pi\varepsilon_0 r_{13}} + \frac{Q_2 Q_3}{4\pi\varepsilon_0 r_{23}},$$

The total energy is now

$$W = \frac{Q_1 Q_2}{4\pi\varepsilon_0 r_{12}} + \frac{Q_1 Q_3}{4\pi\varepsilon_0 r_{13}} + \frac{Q_2 Q_3}{4\pi\varepsilon_0 r_{23}}.$$

If we add more charges the relation becomes

$$W = \frac{Q_1 Q_2}{4\pi\varepsilon_0 r_{12}} + \frac{Q_1 Q_3}{4\pi\varepsilon_0 r_{13}} + \dots + \frac{Q_1 Q_n}{4\pi\varepsilon_0 r_{1n}}$$

$$\frac{Q_2 Q_3}{4\pi\varepsilon_0 r_{23}} + \dots + \frac{Q_2 Q_n}{4\pi\varepsilon_0 r_{2n}}.$$

$$+ \dots\dots$$

$$+ \frac{Q_{n-1} Q_n}{4\pi\varepsilon_0 r_{n-1,n}}$$

Using the following abbreviation in eq. (2.166)

$$W_{ik} = \frac{Q_i Q_k}{4\pi\varepsilon_0 r_{ik}},$$

(2.167)

the energy can be written as

$$\boxed{W = \frac{1}{2} \sum_{i \neq k} W_{ik}}.$$

(2.168)

The sum extends over all indices i and k, where however, i and k have to be different. For $i = k$, the magnitude would be infinite because of $r_{ii} = 0$. Basically, every point charge stores an infinite amount of energy in its field (sometimes called self-energy), which we omit. This merely represents a certain normalization of the energy. The only contributions we need to consider are those that stem from interaction of the different point charges. The factor $1/2$ is necessary because of duplicate counting of contributions, as the sum in eq. (2.168) contains besides W_{12} also W_{21}, while only one, either W_{12} or W_{21} may be counted.

If instead the charge is continuously distributed over the space, the sum turns into an integral.

$$W = \frac{1}{2}\int_V\int_V \frac{dQ(\mathbf{r}')dQ(\mathbf{r}'')}{4\pi\varepsilon_0|\mathbf{r}'-\mathbf{r}''|},$$

(2.169)

or with

$$dQ(\mathbf{r}') = \rho(\mathbf{r}')d\tau' = \rho(\mathbf{r}')dx'dy'dz'$$

$$dQ(\mathbf{r}'') = \rho(\mathbf{r}'')d\tau'' = \rho(\mathbf{r}'')dx''dy''dz''$$

$$\boxed{W = \int_V\int_V \frac{\rho(\mathbf{r}')\cdot\rho(\mathbf{r}'')}{8\pi\varepsilon_0|\mathbf{r}'-\mathbf{r}''|}d\tau'd\tau''}.$$

(2.170)

The potential according to eq. (2.20) is

$$\varphi(\mathbf{r}'') = \frac{1}{4\pi\varepsilon_0}\int_V \frac{\rho(\mathbf{r}')d\tau'}{|\mathbf{r}''-\mathbf{r}'|}.$$

This allows to rewrite (2.170):

$$W = \frac{1}{2}\int_V \rho(\mathbf{r}'')\cdot\varphi(\mathbf{r}'')d\tau'' = \frac{1}{2}\int_V \varphi(\mathbf{r})(\nabla\bullet\mathbf{D})d\tau.$$

(2.171)

Because of the vector identity

$$\nabla\bullet(\mathbf{D}\varphi) = \mathbf{D}\bullet\nabla\varphi + \varphi\nabla\bullet\mathbf{D}$$

the energy can also be expressed as

$$W = -\frac{1}{2}\int_V \mathbf{D}\bullet\nabla\varphi\,d\tau + \frac{1}{2}\int_V \nabla\bullet(\mathbf{D}\varphi)d\tau$$

$$= \frac{1}{2}\int_V \mathbf{E}\bullet\mathbf{D}\,d\tau + \frac{1}{2}\oint\mathbf{D}\varphi\bullet d\mathbf{A}.$$

If we now consider the entire space whose surface has moved to infinity where $\varphi = 0$, then

$$W = \frac{1}{2}\int_V \mathbf{E} \cdot \mathbf{D}\, d\tau \ .$$ (2.172)

i.e., we obtain just the volume integral over the electrostatic energy density.

Of course, we may add surface charges. Then eq. (2.171) is replaced by

$$\boxed{W = \frac{1}{2}\int_V \rho(\mathbf{r}) \cdot \varphi(\mathbf{r})\, d\tau + \frac{1}{2}\oint_A \varphi(\mathbf{r})\sigma(\mathbf{r})\, dA}\ .$$ (2.173)

or, if there are only surface charges

$$W = \frac{1}{2}\oint_A \varphi(\mathbf{r})\sigma(\mathbf{r})\, dA\ .$$

For a plane capacitor (Fig. 2.64), *for instance*, the work becomes

$$W = \frac{1}{2}\oint_A \varphi_1 \sigma\, dA + \frac{1}{2}\oint_A \varphi_2(-\sigma)\, dA\ .$$

$$= \frac{\varphi_1 - \varphi_2}{2}\sigma A = \frac{QV}{2}\ .$$

Because of $Q = CV$ (2.95), one can express the energy that is stored in the field of the capacitor in several ways.

$$W = \frac{1}{2}QV = \frac{1}{2}CV^2 = \frac{1}{2}\frac{Q^2}{C}\ .$$ (2.174)

On the other hand, the energy is of course

$$W = \int_V \frac{\varepsilon E^2}{2}\, d\tau = \frac{\varepsilon}{2}\left(\frac{V}{d}\right)^2 Ad = \frac{1}{2}V^2\frac{\varepsilon A}{d} = \frac{1}{2}CV^2\ .$$

As necessary, the result is the same.

$+\sigma$ ——————— φ_1

$-\sigma$ ——————— φ_2

Fig. 2.64

2.15 Forces in the Electric Field

2.15.1 Force on the Plate of a Capacitor

Consider, a capacitor with the charge Q, insulated from its surroundings (*for example*, the charge Q remains unchanged). A charge Q within the field $\mathbf{E}$, experiences the force

$$\mathbf{F} = Q\mathbf{E} \ .$$

Here, $\mathbf{E}$ is the field that exists without the charge Q. The electric field inside the capacitor is

$$E = \frac{1}{\varepsilon}D = \frac{1}{\varepsilon}|\sigma| = \frac{1}{\varepsilon}\frac{|Q|}{A} \ .$$

It would be wrong to assume that the force that one plate exerts on the other could be calculated using this field. This field is created by the charges on both plates. We may conclude from our discussion of Fig. 2.25, that the field of one charged plate at the location of the other is exactly half that field, namely $|\sigma|/2\varepsilon$. The magnitude of the force is therefore

$$F = |Q|\frac{|\sigma|}{2\varepsilon} = |Q|\frac{|Q|}{2\varepsilon A} = \frac{Q^2}{2\varepsilon A} \ . \tag{2.175}$$

This force is attracting since the charges on the two plates have opposite signs.

To solve this problem in a different way is also possible. We take a capacitor with variable plate distance x. Its energy as a function of x is

$$W = \frac{Q^2}{2C} = \frac{Q^2}{2\varepsilon A}x \ .$$

It requires a force to increase the distance between the plates. *i.e.*, it requires mechanical energy to increase the plate's separation. Neglecting friction, this work must be found again in the electric field energy of the capacitor. For a *virtual displacement dx* we obtain with the notation of Fig. 2.65

$$-F_x dx = dW = \frac{Q^2}{2\varepsilon A}dx \ .$$

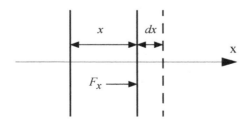

Fig. 2.65

i.e.,

$$F_x = -\frac{Q^2}{2\varepsilon A} = -\frac{dW}{dx} .$$ (2.176)

Besides the sign, which expresses the direction of the force, this confirms above expression. Both methods are thus equivalent. The second method is often more convenient. Written in a slightly different form:

$$F_x = -\frac{\sigma^2 A^2}{2\varepsilon A} = -\frac{1}{2}A\sigma\frac{\sigma}{\varepsilon} = -\frac{1}{2}ADE .$$

or per unit area the force is

$$\frac{F_x}{A} = -\frac{1}{2}ED .$$ (2.177)

2.15.2 Capacitor with two Dielectrics

Consider the capacitor shown in Fig. 2.66, which is filled with two different dielectrics. The question is whether or not those dielectrics exert some force on each other. To solve this problem is easy when using the method of virtual displacement. As before, the charge Q shall be kept constant, *i.e.*, the capacitor is insulated.
Then the energy of the capacitor is

$$W = \frac{Q^2}{2C} .$$

Using

$$Q = (\varepsilon_1 E)ax + (\varepsilon_2 E)a(l-x)$$

and

$$V = Ed$$

yields

$$C = \frac{Q}{V} = \frac{\varepsilon_1 ax + \varepsilon_2 a(l-x)}{d}$$

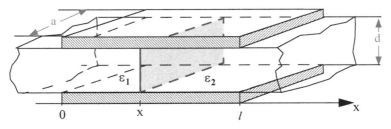

Fig. 2.66

and

$$W = \frac{Q^2 d}{2[\varepsilon_1 ax + \varepsilon_2 a(l-x)]} \; .$$

This results in

$$F_x = -\left(\frac{dW}{dx}\right)_{Q=\text{const}} = \frac{Q^2 d}{2[\varepsilon_1 ax + \varepsilon_2 a(l-x)]^2} a(\varepsilon_1 - \varepsilon_2)$$

$$= \frac{E^2[\varepsilon_1 ax + \varepsilon_2 a(l-x)]^2 d}{2[\varepsilon_1 ax + \varepsilon_2 a(l-x)]^2} a(\varepsilon_1 - \varepsilon_2)$$

$$= ad\frac{E^2}{2}(\varepsilon_1 - \varepsilon_2) \; ,$$

i.e., there is a force in positive x-direction when $\varepsilon_1 > \varepsilon_2$ and in negative x-direction when $\varepsilon_1 < \varepsilon_2$. Per unit area the force is

$$\boxed{\frac{F_x}{ad} = \frac{E^2}{2}(\varepsilon_1 - \varepsilon_2) = \frac{1}{2}E_1 D_1 - \frac{1}{2}E_2 D_2} \qquad (2.178)$$

where of course $E_1 = E_2$.

Both results, (2.177) and (2.178) indicate mechanical stress or pressure in the form $(1/2)ED$. Thus we can say, that electric fields cause mechanical stress $(1/2)ED$ in the direction parallel to their field and pressure $(1/2)ED$ perpendicular to them.

3 Formal Methods of Electrostatics

Having introduced the basic terminology in Chapter 2, we now discuss the formal methods by which electrostatic problems can be solved. Some problems were solved already in Chapter 2, but those problems were of such nature that they could be simplified by invoking symmetry or by plausibility arguments. This does not always work, and then we have to rely on formal methods having a general applicability. Even then, numerous problems can not always be solved analytically and one needs to use numerical methods (see Chapter 8). Here we will restrict ourselves to analytical methods and focus on the two of the more important ones:

    1.   the method of separation of variables

    2.   method of complex analysis for the case of plane fields

We will cover these here first in the context of electrostatics, even though they are of much more general nature and form the basis for the subsequent parts on current density fields, magnetostatics, and time dependent problems.

The first step in applying the separation method is to choose a convenient coordinate system, which allows a simple formulation of the boundary conditions. This calls for a coordinate transformation. With a few exceptions, we have thus far only used Cartesian coordinates. Also, the vector operators (grad (∇), div ($\nabla \bullet$), curl ($\nabla \times$), Laplacian (Δ or ∇^2)) have only been expressed in their Cartesian coordinates. Therefore, those will be discussed first in the subsequent sections, before returning to the electrostatic problems.

3.1 Coordinate Transformations

One defines a set of new coordinates based on Cartesian coordinates *(x,y,z)*:

$$\left. \begin{array}{l} u_1 = u_1(x, y, z) \\ u_2 = u_2(x, y, z) \\ u_3 = u_3(x, y, z) \end{array} \right\} \tag{3.1}$$

or if we solve for *(x,y,z)*:

$$\left. \begin{array}{l} x = x(u_1, u_2, u_3) \\ y = y(u_1, u_2, u_3) \\ z = z(u_1, u_2, u_3) \end{array} \right\} \tag{3.2}$$

The equation of a surface is obtained when holding one value fixed, *for example* u_1:

$$u_1(x, y, z) = c_1 . \tag{3.3}$$

G. Lehner, *Electromagnetic Field Theory for Engineers and Physicists*,
DOI 10.1007/978-3-540-76306-2_3, © Springer-Verlag Berlin Heidelberg 2010

Simultaneously fixing a second coordinate, *for example*, u_2 defines another surface. The intersection of both surfaces is defined by simultaneously meeting the equations

$$\left.\begin{aligned} u_1(x,y,z) &= c_1 \\ u_2(x,y,z) &= c_2 \end{aligned}\right\} . \tag{3.4}$$

Here, the only remaining variable is u_3. Its parameterized representation is

$$\left.\begin{aligned} x &= x(c_1, c_2, u_3) \\ y &= y(c_1, c_2, u_3) \\ z &= z(c_1, c_2, u_3) \end{aligned}\right\} \tag{3.5}$$

A point is obtained if we also fix u_3 ($u_3 = c_3$).

$$\left.\begin{aligned} u_1(x,y,z) &= c_1 \\ u_2(x,y,z) &= c_2 \\ u_3(x,y,z) &= c_3 \end{aligned}\right\} \tag{3.6}$$

One may view this point as the origin of a local, in general non Cartesian coordinate system (Fig. 3.1). Let us calculate the distance between this point $\mathbf{P}(u_1, u_2, u_3)$ and the point $\mathbf{P}'(u_1 + du_1, u_2 + du_2, u_3 + du_3)$. Using Cartesian coordinates we have

$$ds^2 = dx^2 + dy^2 + dz^2 . \tag{3.7}$$

Where of course

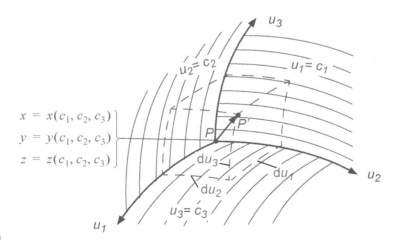

Fig. 3.1

$$dx = \frac{\partial x}{\partial u_1} du_1 + \frac{\partial x}{\partial u_2} du_2 + \frac{\partial x}{\partial u_3} du_3$$

$$dy = \frac{\partial y}{\partial u_1} du_1 + \frac{\partial y}{\partial u_2} du_2 + \frac{\partial y}{\partial u_3} du_3 \qquad (3.8)$$

$$dz = \frac{\partial z}{\partial u_1} du_1 + \frac{\partial z}{\partial u_2} du_2 + \frac{\partial z}{\partial u_3} du_3$$

Substituting (3.8) into (3.7) and after ordering we obtain

$$
\begin{aligned}
ds^2 = & \left[\left(\frac{\partial x}{\partial u_1}\right)^2 + \left(\frac{\partial y}{\partial u_1}\right)^2 + \left(\frac{\partial z}{\partial u_1}\right)^2 \right] du_1^2 \\
& + \left[\left(\frac{\partial x}{\partial u_2}\right)^2 + \left(\frac{\partial y}{\partial u_2}\right)^2 + \left(\frac{\partial z}{\partial u_2}\right)^2 \right] du_2^2 \\
& + \left[\left(\frac{\partial x}{\partial u_3}\right)^2 + \left(\frac{\partial y}{\partial u_3}\right)^2 + \left(\frac{\partial z}{\partial u_3}\right)^2 \right] du_3^2 \\
& + 2 \left[\begin{array}{l} \left[\frac{\partial x}{\partial u_1}\frac{\partial x}{\partial u_2} + \frac{\partial y}{\partial u_1}\frac{\partial y}{\partial u_2} + \frac{\partial z}{\partial u_1}\frac{\partial z}{\partial u_2}\right] du_1 du_2 \\[6pt] + \left[\frac{\partial x}{\partial u_1}\frac{\partial x}{\partial u_3} + \frac{\partial y}{\partial u_1}\frac{\partial y}{\partial u_3} + \frac{\partial z}{\partial u_1}\frac{\partial z}{\partial u_3}\right] du_1 du_3 \\[6pt] + \left[\frac{\partial x}{\partial u_2}\frac{\partial x}{\partial u_3} + \frac{\partial y}{\partial u_2}\frac{\partial y}{\partial u_3} + \frac{\partial z}{\partial u_2}\frac{\partial z}{\partial u_3}\right] du_2 du_3 \end{array} \right]
\end{aligned}
\qquad (3.9)
$$

We will not use this rather inconvenient expression in its full generality, but restrict ourselves to orthogonal coordinate systems. These are characterized by the fact that the three coordinate lines in Fig. 3.1 are mutually perpendicular at every point. We define the tangential vectors t_1, t_2, t_3. For instance, the vector tangent to the line u_3, which was given in eq. (3.5) is determined by:

$$t_3 = \langle \frac{\partial x}{\partial u_3}, \frac{\partial y}{\partial u_3}, \frac{\partial z}{\partial u_3} \rangle . \qquad (3.10)$$

Similarly for the remaining vectors:

$$t_1 = \langle \frac{\partial x}{\partial u_1}, \frac{\partial y}{\partial u_1}, \frac{\partial z}{\partial u_1} \rangle$$

$$t_2 = \langle \frac{\partial x}{\partial u_2}, \frac{\partial y}{\partial u_2}, \frac{\partial z}{\partial u_2} \rangle \qquad (3.11)$$

Fig. 3.2 shows the coordinate system of Fig. 3.1 with its tangent vectors.

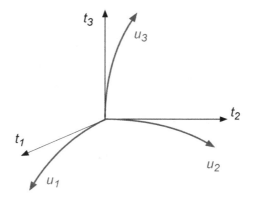

Fig. 3.2

A coordinate system is orthogonal if the following holds for every point:

$$\left.\begin{array}{l} \mathbf{t}_1 \bullet \mathbf{t}_2 = 0 \\ \mathbf{t}_2 \bullet \mathbf{t}_3 = 0 \\ \mathbf{t}_3 \bullet \mathbf{t}_1 = 0 \end{array}\right\} . \tag{3.12}$$

The vector from P to P' is

$$d\mathbf{r} = \mathbf{t}_1 du_1 + \mathbf{t}_2 du_2 + \mathbf{t}_3 du_3 . \tag{3.13}$$

Therefore the distance element (squared) is

$$\left.\begin{array}{rl} ds^2 =& d\mathbf{r} \bullet d\mathbf{r} \\ =& t_1^2 du_1^2 + t_2^2 du_2^2 + t_3^2 du_3^2 \\ & + 2\,t_1 \bullet t_2\, du_1 du_2 \\ & + 2\,t_1 \bullet t_3\, du_1 du_3 \\ & + 2\,t_2 \bullet t_3\, du_2 du_3 \end{array}\right\} . \tag{3.14}$$

This is a much shorter way to write eq. (3.9). For an orthogonal coordinate system, because of eq. (3.12), the (square of the) distance element simplifies even more

$$ds^2 = t_1^2 du_1^2 + t_2^2 du_2^2 + t_3^2 du_3^2 . \tag{3.15}$$

The only difference to the respective expression in Cartesian coordinates is the occurrence of the scale factors t_1, t_2, t_3 which are spatially dependent, *i.e.*, they are generally different at differing positions in space. The volume element in curvilinear coordinates is characterized by du_1, du_2, du_3. Its edges are $t_1 du_1, t_2 du_2, t_3 du_3$, and thus has the volume element

$$\boxed{d\tau = t_1 t_2 t_3 du_1 du_2 du_3} . \tag{3.16}$$

The infinitesimal line element or displacement $d\mathbf{s}$ has the components

$$\boxed{ds_i = t_i\, d u_i} \qquad (i = 1, 2, 3) \tag{3.17}$$

and an infinitesimal area element $d\mathbf{A}$ has the components

$$\boxed{dA_i = t_k t_l\, du_k\, du_l} \qquad (i, k, l\ \text{all different}) \tag{3.18}$$

The factors t_i^2 are the diagonal elements of the so-called metric tensor, which has off diagonal elements if the coordinate systems are not orthogonal.

3.2   Vector Analysis for Curvilinear, Orthogonal Coordinate Systems

3.2.1   Gradient

Starting form the definition

$$(\nabla\varphi)_{u_1} = \lim_{\Delta u_1 \to 0} \frac{\varphi(u_1 + \Delta u_1, u_2, u_3) - \varphi(u_1, u_2, u_3)}{\Delta s_1}$$

and because of

$$\Delta s_1 = t_1 \Delta u_1$$

one obtains

$$(\nabla\varphi)_{u_1} = \frac{1}{t_1} \lim_{t_1 \Delta u_1 \to 0} \frac{\varphi(u_1 + \Delta u_1, u_2, u_3) - \varphi(u_1, u_2, u_3)}{\Delta u_1}$$

$$= \frac{1}{t_1}\frac{\partial\varphi}{\partial u_1}$$

Similarly, for the other components

$$\boxed{\nabla\varphi = \langle \frac{1}{t_1}\frac{\partial\varphi}{\partial u_1}, \frac{1}{t_2}\frac{\partial\varphi}{\partial u_2}, \frac{1}{t_3}\frac{\partial\varphi}{\partial u_3} \rangle} . \tag{3.19}$$

3.2.2   Divergence

The starting point for the definition of the divergence is the limit of a surface integral of the type given by eq. (1.22). Using nomenclature established in Fig. 3.3, one finds the vector $\mathbf{a}$ with its components a_1, a_2, a_3

$$\nabla\bullet\mathbf{a} = \lim_{V \to 0} \frac{1}{V} \oint \mathbf{a}\, d\mathbf{A}$$

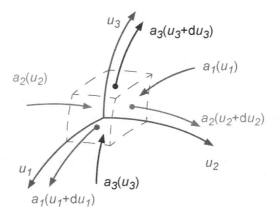

Fig. 3.3

$$\nabla \bullet \mathbf{a} \ = \ \lim_{du_1 du_2 du_3 \to 0} \frac{1}{t_1 t_2 t_3 du_1 du_2 du_3} \ \cdot$$

$$\begin{bmatrix} a_1(u_1 + du_1) t_2(u_1 + du_1) t_3(u_1 + du_1) du_2 du_3 \\ -a_1(u_1) t_2(u_1) t_3(u_1) du_2 du_3 \\ + a_2(u_2 + du_2) t_1(u_2 + du_2) t_3(u_2 + du_2) du_1 du_3 \\ -a_2(u_2) t_1(u_2) t_3(u_2) du_1 du_3 \\ + a_3(u_3 + du_3) t_1(u_3 + du_3) t_2(u_3 + du_3) du_1 du_2 \\ -a_3(u_3) t_1(u_3) t_2(u_3) du_1 du_2 \end{bmatrix}$$

$$= \ \lim_{du_1 du_2 du_3 \to 0} \frac{\left[\dfrac{\partial}{\partial u_1}(a_1 t_2 t_3) + \dfrac{\partial}{\partial u_2}(a_2 t_1 t_3) + \dfrac{\partial}{\partial u_3}(a_3 t_1 t_2) \right] du_1 du_2 du_3}{t_1 t_2 t_3 du_1 du_2 du_3},$$

i.e.,

$$\boxed{\nabla \bullet \mathbf{a} \ = \ \frac{1}{t_1 t_2 t_3} \left[\frac{\partial}{\partial u_1}(a_1 t_2 t_3) + \frac{\partial}{\partial u_2}(a_2 t_1 t_3) + \frac{\partial}{\partial u_3}(a_3 t_1 t_2) \right]}. \qquad (3.20)$$

3.2.3   Laplace Operator

Since

$$\Delta \varphi \ = \ \nabla \bullet \nabla \varphi \ = \ \nabla^2 \varphi$$

one may derive the Laplacian from both, eq. (3.19) and (3.20). One obtains

$$\nabla^2\varphi \;=\; \frac{1}{t_1 t_2 t_3}\left[\frac{\partial}{\partial u_1}\!\left(\frac{t_2 t_3}{t_1}\frac{\partial\varphi}{\partial u_1}\right) + \frac{\partial}{\partial u_2}\!\left(\frac{t_1 t_3}{t_2}\frac{\partial\varphi}{\partial u_2}\right) + \frac{\partial}{\partial u_3}\!\left(\frac{t_1 t_2}{t_3}\frac{\partial\varphi}{\partial u_3}\right)\right]. \qquad (3.21)$$

3.2.4 Circulation

To calculate the curl, we start from eq. (1.34) and use the right hand rule to establish the relation between the direction of the circulation (also called the rotation) and the direction of the line integral of Section 1.7. Using the notation from Fig. 3.4, one finds:

$$(\nabla\times\mathbf{a})_{u_1} \;=\; \lim_{du_2 du_3 \to 0}\; \frac{1}{t_2 t_3 du_2 du_3}\;\cdot$$

$$\begin{bmatrix} a_2(u_3)\cdot t_2(u_3)du_2 + a_3(u_2 + du_2)\cdot t_3(u_2 + du_2)du_3 \\ -a_2(u_3 + du_3)\cdot t_2(u_3 + du_3)du_2 \\ -a_3(u_2)\cdot t_3(u_2)du_3 \end{bmatrix}$$

$$=\; \lim_{du_2 du_3 \to 0}\; \frac{\left[\dfrac{\partial}{\partial u_2}(a_3 t_3) - \dfrac{\partial}{\partial u_3}(a_2 t_2)\right]du_2 du_3}{t_2 t_3 du_2 du_3}$$

$$=\; \frac{1}{t_2 t_3}\left[\frac{\partial}{\partial u_2}(a_3 t_3) - \frac{\partial}{\partial u_3}(a_2 t_2)\right].$$

The other two components of $\nabla\times\mathbf{a}$ are derived in the same way. The end result is

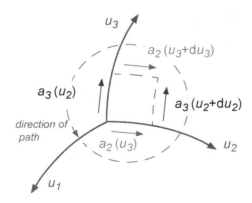

Fig. 3.4

$$\nabla \times \mathbf{a} = \begin{bmatrix} \dfrac{1}{t_2 t_3}\left[\dfrac{\partial}{\partial u_2}(a_3 t_3) - \dfrac{\partial}{\partial u_3}(a_2 t_2)\right] \\[2mm] \dfrac{1}{t_3 t_1}\left[\dfrac{\partial}{\partial u_3}(a_1 t_1) - \dfrac{\partial}{\partial u_1}(a_3 t_3)\right] \\[2mm] \dfrac{1}{t_1 t_2}\left[\dfrac{\partial}{\partial u_1}(a_2 t_2) - \dfrac{\partial}{\partial u_2}(a_1 t_1)\right] \end{bmatrix}. \tag{3.22}$$

Using the unit vectors $\mathbf{e}_{u_1}, \mathbf{e}_{u_2}, \mathbf{e}_{u_3}$ of a coordinate system,

$$\mathbf{e}_{u_1} = \frac{\mathbf{t}_1}{t_1},$$

$$\mathbf{e}_{u_2} = \frac{\mathbf{t}_2}{t_2},$$

$$\mathbf{e}_{u_3} = \frac{\mathbf{t}_3}{t_3},$$

we may also write curl $\mathbf{a}$ in its determinant form

$$\nabla \times \mathbf{a} = \begin{vmatrix} \dfrac{\mathbf{e}_{u_1}}{t_2 t_3} & \dfrac{\mathbf{e}_{u_2}}{t_3 t_1} & \dfrac{\mathbf{e}_{u_3}}{t_1 t_2} \\[2mm] \dfrac{\partial}{\partial u_1} & \dfrac{\partial}{\partial u_2} & \dfrac{\partial}{\partial u_3} \\[2mm] a_1 t_1 & a_2 t_2 & a_3 t_3 \end{vmatrix} \tag{3.23}$$

or

$$\nabla \times \mathbf{a} = \frac{1}{t_1 t_2 t_3} \begin{vmatrix} \mathbf{t}_1 & \mathbf{t}_2 & \mathbf{t}_3 \\[2mm] \dfrac{\partial}{\partial u_1} & \dfrac{\partial}{\partial u_2} & \dfrac{\partial}{\partial u_3} \\[2mm] a_1 t_1 & a_2 t_2 & a_3 t_3 \end{vmatrix}. \tag{3.24}$$

3.3   Some Important Coordinate Systems

Of the many interesting coordinate systems, this book will only employ three: Cartesian, cylindrical, and spherical coordinates. We will summarize the results of our previous discussion for these coordinate systems.

3.3.1   Cartesian Coordinates

For Cartesian coordinates, the scale factors are, of course, $t_1 = t_2 = t_3 = 1$ and from eq. (3.19) through eq. (3.24), we obtain the familiar expressions

$$\nabla \varphi = \langle \frac{\partial \varphi}{\partial x}, \frac{\partial \varphi}{\partial y}, \frac{\partial \varphi}{\partial z} \rangle \ .$$

$$\nabla \bullet \mathbf{a} = \frac{\partial a_x}{\partial x} + \frac{\partial a_y}{\partial y} + \frac{\partial a_z}{\partial z} \ .$$

$$\nabla^2 \varphi = \frac{\partial^2 \varphi}{\partial x^2} + \frac{\partial^2 \varphi}{\partial y^2} + \frac{\partial^2 \varphi}{\partial z^2} \ .$$

$$\nabla \times \mathbf{a} = \begin{vmatrix} \mathbf{e}_x & \mathbf{e}_y & \mathbf{e}_z \\ \dfrac{\partial}{\partial x} & \dfrac{\partial}{\partial y} & \dfrac{\partial}{\partial z} \\ a_x & a_y & a_z \end{vmatrix} = \begin{bmatrix} \dfrac{\partial a_z}{\partial y} - \dfrac{\partial a_y}{\partial z} \\ \dfrac{\partial a_x}{\partial z} - \dfrac{\partial a_z}{\partial x} \\ \dfrac{\partial a_y}{\partial x} - \dfrac{\partial a_x}{\partial y} \end{bmatrix} \ .$$

3.3.2   Cylindrical Coordinates

The case of cylindrical coordinates is illustrated in Fig. 3.5. The coordinates are

$$\left. \begin{aligned} u_1 &= r \\ u_2 &= \varphi \\ u_3 &= z \end{aligned} \right\} \ , \tag{3.25}$$

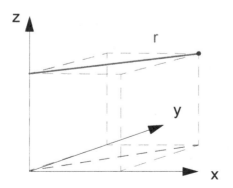

Fig. 3.5

or expressed in Cartesian coordinates

$$
\left.
\begin{aligned}
x &= r\cos\varphi \\
y &= r\sin\varphi \\
z &= z
\end{aligned}
\right\} .
\tag{3.26}
$$

With this, eq. (3.10), and (3.11) it follows for the tangent vectors in Cartesian coordinates:

$$
\left.
\begin{aligned}
\mathbf{t}_1 &= \langle \cos\varphi, \sin\varphi, 0 \rangle \\
\mathbf{t}_2 &= \langle -r\sin\varphi, r\cos\varphi, 0 \rangle \\
\mathbf{t}_3 &= \langle 0, 0, 1 \rangle
\end{aligned}
\right\} ,
\tag{3.27}
$$

i.e.,

$$
\left.
\begin{aligned}
t^2_1 &= 1 \\
t^2_2 &= r^2 \\
t^2_3 &= 1
\end{aligned}
\right\} ,
\tag{3.28}
$$

and consequently for the volume element and the (squared) distance element

$$
d\tau = r\,dr\,d\varphi\,dz ,
\tag{3.29}
$$

$$
ds^2 = dr^2 + r^2 d\varphi^2 + dz^2 ,
\tag{3.30}
$$

and furthermore

$$
\nabla\phi = \langle \frac{\partial}{\partial r}, \frac{1}{r}\frac{\partial}{\partial\varphi}, \frac{\partial}{\partial z} \rangle \phi .
\tag{3.31}
$$

$$
\nabla \bullet \mathbf{a} = \frac{1}{r}\frac{\partial}{\partial r}r a_r + \frac{1}{r}\frac{\partial}{\partial\varphi}a_\varphi + \frac{\partial}{\partial z}a_z .
\tag{3.32}
$$

$$
\nabla^2\phi = \left(\frac{1}{r}\frac{\partial}{\partial r}r\frac{\partial}{\partial r} + \frac{1}{r^2}\frac{\partial^2}{\partial\varphi^2} + \frac{\partial^2}{\partial z^2} \right)\phi .
\tag{3.33}
$$

$$
\nabla\times\mathbf{a} =
\begin{bmatrix}
(\nabla\times\mathbf{a})_r \\
(\nabla\times\mathbf{a})_\varphi \\
(\nabla\times\mathbf{a})_z
\end{bmatrix}
=
\begin{bmatrix}
\dfrac{1}{r}\dfrac{\partial}{\partial\varphi}a_z - \dfrac{\partial}{\partial z}a_\varphi \\[2mm]
\dfrac{\partial}{\partial z}a_r - \dfrac{\partial}{\partial r}a_z \\[2mm]
\dfrac{1}{r}\dfrac{\partial}{\partial r}(ra_\varphi) - \dfrac{1}{r}\dfrac{\partial}{\partial\varphi}a_r
\end{bmatrix}
\tag{3.34}
$$

Be careful not to confuse the angle φ with the potential ϕ, which was up to now also labeled with φ. We will identify the potential with some other symbol wherever confusion with the angle is anticipated.

3.3.3   Spherical Coordinates

Fig. 3.6 illustrates the spherical coordinate system.
The spherical coordinates are,

$$\left.\begin{aligned} u_1 &= r \\ u_2 &= \theta \\ u_3 &= \varphi \end{aligned}\right\} , \qquad (3.35)$$

or expressed in terms of the Cartesian coordinates

$$\left.\begin{aligned} x &= r\sin\theta\cos\varphi \\ y &= r\sin\theta\sin\varphi \\ z &= r\cos\theta \end{aligned}\right\} . \qquad (3.36)$$

For the tangent vectors

$$\left.\begin{aligned} \mathbf{t}_1 &= \langle \sin\theta\cos\varphi, \sin\theta\sin\varphi, \cos\theta\rangle \\ \mathbf{t}_2 &= \langle r\cos\theta\cos\varphi, r\cos\theta\sin\varphi, -r\sin\theta\rangle \\ \mathbf{t}_3 &= \langle -r\sin\theta\sin\varphi, -r\sin\theta\cos\varphi, 0\rangle \end{aligned}\right\} , \qquad (3.37)$$

and

$$\left.\begin{aligned} t^2_1 &= 1 \\ t^2_2 &= r^2 \\ t^2_3 &= r^2\sin^2\theta \end{aligned}\right\} , \qquad (3.38)$$

and consequently for the volume element and (squared) distance element

$$d\tau = r^2\sin\theta\, dr\, d\theta\, d\varphi , \qquad (3.39)$$

$$ds^2 = dr^2 + r^2 d\theta^2 + r^2\sin^2\theta\, d\varphi^2 , \qquad (3.40)$$

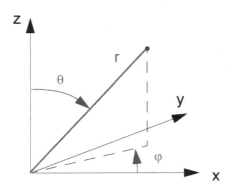

Fig. 3.6

and

$$\nabla\phi \ = \ \langle\frac{\partial}{\partial r}, \frac{1}{r}\frac{\partial}{\partial\theta}, \frac{1}{r\sin\theta}\frac{\partial}{\partial\varphi}\rangle\phi \ , \tag{3.41}$$

$$\nabla\bullet\mathbf{a} = \frac{1}{r^2}\frac{\partial}{\partial r}r^2 a_r + \frac{1}{r\sin\theta}\frac{\partial}{\partial\theta}\sin\theta a_\theta + \frac{1}{r\sin\theta}\frac{\partial}{\partial\varphi}a_\varphi \ , \tag{3.42}$$

$$\nabla^2\phi \ = \ \Big(\frac{1}{r^2}\frac{\partial}{\partial r}r^2\frac{\partial}{\partial r} + \frac{1}{r^2\sin\theta}\frac{\partial}{\partial\theta}\sin\theta\frac{\partial}{\partial\theta} + \frac{1}{r^2\sin^2\theta}\frac{\partial^2}{\partial\varphi^2}\Big)\phi \ , \tag{3.43}$$

$$\nabla\times\mathbf{a} \ = \ \begin{bmatrix} (\nabla\times\mathbf{a})_r \\ (\nabla\times\mathbf{a})_\theta \\ (\nabla\times\mathbf{a})_\varphi \end{bmatrix} = \begin{bmatrix} \dfrac{1}{r\sin\theta}\dfrac{\partial}{\partial\theta}\sin\theta a_\varphi - \dfrac{1}{r\sin\theta}\dfrac{\partial}{\partial\varphi}a_\theta \\[2mm] \dfrac{1}{r\sin\theta}\dfrac{\partial}{\partial\varphi}a_r - \dfrac{1}{r}\dfrac{\partial}{\partial r}r a_\varphi \\[2mm] \dfrac{1}{r}\dfrac{\partial}{\partial r}(r a_\theta) - \dfrac{1}{r}\dfrac{\partial}{\partial\theta}a_r \end{bmatrix} \tag{3.44}$$

3.4 Some Properties of Poisson's and Laplace's Equations (Potential Theory)

Poisson's and Laplace's equation (2.11) and (2.12), respectively are the basis for the formal treatment of electrostatics.

3.4.1 Problem Description

A large class of electrostatic problems are described in the following way: Consider an arbitrarily shaped region, with an arbitrary distribution of volume charges $\rho(\mathbf{r})$ inside. Given is the potential φ on its boundary or the perpendicular component of the electric field on the boundary, *i.e.*, $(\nabla\varphi)_n = \partial\varphi/\partial n$. This represents *Neumann's boundary value problem* when $\partial\varphi/\partial n$ is prescribed. On the other hand, when φ is prescribed, it is called *Dirichlet's boundary value problem*. Of course, it is possible to prescribe φ on part of the boundary and $\partial\varphi/\partial n$ on the other part, in which case one is dealing with a mixed boundary value problem.

The region in question may be bounded by any number of arbitrarily shaped surface areas.

One can uniquely solve these kind of boundary value problems using the methods of potential theory . Its proof involves Green's integral theorems.

3.4.2 Green's Theorems

The starting point is Gauss' integral theorem

$$\int_V \nabla \bullet \mathbf{a} \ d\tau = \oint_A \mathbf{a} \bullet d\mathbf{A} \ .$$

Define a vector $\mathbf{a}$ as

$$\mathbf{a} = \psi \nabla \phi \ ,$$

where ψ and ϕ are arbitrary scalar functions. When substituting, one obtains

$$\int_V \nabla \bullet \mathbf{a} \ d\tau \ = \ \int_V \nabla \bullet (\psi \nabla \phi) d\tau \ = \ \int_V [\psi(\nabla \bullet \nabla \phi) + (\nabla \psi) \bullet (\nabla \phi)] d\tau$$

$$= \ \oint_A \psi \nabla \phi \bullet d\mathbf{A} \qquad (3.45)$$

This is *Green's first identity*. It is permissible to exchange ψ and ϕ, as they are arbitrary anyway. This gives

$$\int_V \nabla \bullet (\phi \nabla \psi) d\tau \ = \ \int_V [\phi(\nabla \bullet \nabla \psi) + (\nabla \phi) \bullet (\nabla \psi)] d\tau$$

$$= \ \oint_A \phi \nabla \psi \bullet d\mathbf{A} \qquad (3.46)$$

Now we use:

$$\nabla \bullet (\nabla \phi) = \nabla^2 \phi \quad \text{and} \quad \nabla \bullet (\nabla \psi) = \nabla^2 \psi \ ,$$

as well as

$$\nabla \psi \bullet d\mathbf{A} = \nabla \psi \bullet \mathbf{n} dA = (\nabla \psi)_n dA$$

$$= \frac{\partial \psi}{\partial n} dA \ ,$$

and similarly

$$\nabla \phi \bullet d\mathbf{A} = \frac{\partial \phi}{\partial n} dA \ .$$

Using these relations and subtracting (3.46) from (3.45) yields *Green's second identity,* also known as *Green's theorem.*

$$\boxed{\int_V (\psi \nabla^2 \phi - \phi \nabla^2 \psi) d\tau \ = \ \oint \left(\psi \frac{\partial \phi}{\partial n} - \phi \frac{\partial \psi}{\partial n} \right) dA} \qquad (3.47)$$

On the other hand, if we let $\phi = \psi$ in one of the two equations (3.46) or (3.45), then we obtain one of Green's integral theorems:

$$\boxed{\int_V (\phi \nabla^2 \phi + (\nabla \phi)^2) d\tau \ = \ \oint \left(\phi \frac{\partial \phi}{\partial n} \right) dA} \ . \qquad (3.48)$$

If the functions ϕ and ψ depend on two variables only, then (3.47) and (3.48) reduce even more:

$$\int_A (\psi \nabla^2 \phi - \phi \nabla^2 \psi) dA \ = \ \oint \left(\psi \frac{\partial \phi}{\partial n} - \phi \frac{\partial \psi}{\partial n} \right) ds$$

and

$$\int_A (\phi \nabla^2 \phi + (\nabla \phi)^2) dA \ = \ \oint \left(\phi \frac{\partial \phi}{\partial n} \right) ds \ .$$

For the one dimensional case this gives

$$\int_{x_1}^{x_2}(\psi\phi'' - \phi\psi'')dx = [\psi\phi' - \phi\psi']_{x_1}^{x_2} \, ,$$

and

$$\int_{x_1}^{x_2}(\phi\phi'' + \phi'^2)dx = [\phi\phi']_{x_1}^{x_2} \, .$$

These are simple integrations by parts. Green's integral theorems are thus simply generalizations of integrations by parts to two or three dimensions, respectively.

3.4.3   Proof of Uniqueness

Suppose (for our Neumann, Dirichlet, or mixed problem) there are two solutions φ_1 and φ_2. This means that

$$\nabla^2\varphi_1 = -\frac{\rho(r)}{\varepsilon_0} \, ,$$

$$\nabla^2\varphi_2 = -\frac{\rho(r)}{\varepsilon_0}$$

and the boundary conditions shall also be satisfied.

We define a new function as the difference

$$\tilde{\varphi} = \varphi_1 - \varphi_2$$

then

$$\nabla^2\tilde{\varphi} = \nabla^2\varphi_1 - \nabla^2\varphi_2 = 0 \, ,$$

which means that it has to satisfy Laplace's equation. Furthermore, the potential along the boundary is

$$\tilde{\varphi} = 0$$

or

$$\frac{\partial}{\partial n}\tilde{\varphi} = 0$$

or – in case of mixed problems – one of the equations along part of the boundary and the other equation along the remainder of the boundary. We apply Green's theorem in its form (3.48) to $\tilde{\varphi}$ and obtain

$$\int_V (\nabla\tilde{\varphi})^2 d\tau = 0 \, .$$

Since $(\nabla\tilde{\varphi})^2$ is always positive, this can only be true if the integrand is identically zero

$$\nabla\tilde{\varphi} = 0$$

or

$\tilde{\varphi}$ = const .

The case of a Dirichlet, or even the mixed boundary value problem requires that $\tilde{\varphi}$ = 0 everywhere. For Neumann's problem, $\tilde{\varphi}$ is determined, except for a physically insignificant constant.

The problem may also be posed in a different way. Given is the charge of a conductor in an electric field. Then

$$\frac{\partial \varphi}{\partial n} = -E_n = \frac{\sigma}{\varepsilon_0}$$

(The negative sign in $E_n = -\sigma/\varepsilon_0$ is due to the fact that the normal component points outwardly, relative to the region that contains the field, which means it points into the conductor). Therefore

$$Q = \int \sigma dA = \varepsilon_0 \int \frac{\partial \varphi}{\partial n} dA .$$

Thus, it is not $\partial \varphi / dn$ that is prescribed along the boundary but the integral $\int (\partial \varphi / dn) dA$. Furthermore, φ is constant on the surface, although its value is yet unknown. If we now have two solutions φ_1 and φ_2, then, as before, Laplace's equation applies to both, and to their difference. Integrating both solutions over the boundary gives

$$Q = \varepsilon_0 \int \frac{\partial \varphi_1}{\partial n} dA = \varepsilon_0 \int \frac{\partial \varphi_2}{\partial n} dA .$$

Therefore

$$\int_V (\nabla \tilde{\varphi})^2 d\tau = \int_A \tilde{\varphi} \frac{\partial}{\partial n} \tilde{\varphi} dA$$

$$= \int (\varphi_1 - \varphi_2) \frac{\partial (\varphi_1 - \varphi_2)}{\partial n} dA = (\varphi_1 - \varphi_2) \int \frac{\partial (\varphi_1 - \varphi_2)}{\partial n} dA .$$

$$= (\varphi_1 - \varphi_2) \left(\frac{Q - Q}{\varepsilon_0} \right) = 0 ,$$

as before. Again we find

$$\nabla \tilde{\varphi} = 0$$

and

$\tilde{\varphi}$ = const .

These uniqueness proofs have a common theme, and are formally based on eq. (3.48), which concludes that if the boundary conditions require $\nabla^2 \varphi = 0$ on the surface, together with $\varphi = 0$, that then φ has to vanish everywhere. There is a plausible way to understand this. One can prove that in an area where $\nabla^2 \varphi = 0$, φ may neither have a maximum nor a minimum. If there were a maximum (or a minimum) of φ at any point inside the region, then in a neighborhood of this point, all lines of $\nabla \varphi$ had to point towards (or away from) it. Any surface in the vicinity of the maximum (or minimum) would be penetrated by a non-vanishing electrical flux. This is only possible if there are volume charges present, which would require $\nabla^2\varphi \neq 0$. Therefore, the assumption of a maximum or minimum inside of this area would lead to a contradiction. Then, if $\varphi = 0$ on the boundary, it can not be larger inside, but it can also not be smaller than zero inside and consequently $\varphi = 0$ everywhere in the region. In other words: *If in a region $\nabla^2\varphi = 0$, then the function φ may have its maximum or minimum values only on the boundary of that region.* Uniqueness of the solution of Dirichlet's boundary value problem is an immediate consequence of this statement.

3.4.4   Models

The equation $\nabla^2\varphi = 0$ has a frequent occurrence in physical science and it describes a vast number of problems. This enables one to frequently map physical problems onto a corresponding electrostatic problem. For example, the two-dimensional Laplace's equation

$$\left(\frac{\partial^2}{\partial x^2} + \frac{\partial^2}{\partial y^2}\right)\varphi = 0$$

also describes the displacement of a membrane suspended on a frame which is considered to be small. The boundary (frame) defines φ and inside $\nabla^2\varphi = 0$. Such a membrane can be considered a model for electrostatic problems.

3.4.5   Dirac's Delta Function (δ-Function)

The δ-function is particularly useful in the following, which is why we will introduce it here. It shall be noted that our exposition here does not substitute a rigorous mathematical introduction.

A rough, illustrative way to describe the character of the δ-function is to note that it vanishes everywhere except for one particular point of its argument (namely 0), where it takes an infinite value, exactly such that its integral equals 1.

$$\delta(x - x') = \begin{cases} 0 & \text{for} \quad x \neq x' \\ \infty & \text{for} \quad x = x' \end{cases} \tag{3.49}$$

$$\int_{-\infty}^{+\infty} \delta(x - x')\, dx = 1 \tag{3.50}$$

The δ-function is not a function in the usual sense. It belongs to a more general category of functions, which sometimes are called *improper functions, generalized functions,* or *distributions.* Another possibility is to imagine the δ-function as the limit of a series of functions. It can be constructed in various ways, *for example,*

1.   The limit of a series of rectangular functions as illustrated in Fig. 3.7. Thus

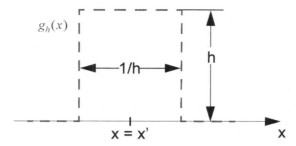

Fig. 3.7

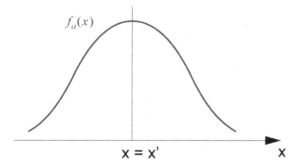

Fig. 3.8

$$g_h(x) = \begin{cases} h & \text{for } |x - x'| \le \dfrac{1}{2h} \\ 0 & \text{else} \end{cases}$$

and

$$\delta(x - x') = \lim_{h \to \infty} g_h(x)$$

2.   The limit of a series of Gaussian functions as illustrated in Fig. 3.8. Now

$$f_a(x) = \frac{1}{a\sqrt{\pi}} \exp\left[-\frac{(x - x')^2}{a^2}\right]$$

where

$$\int_{-\infty}^{+\infty} f_a(x)\, dx = 1$$

and

$$\delta(x - x') = \lim_{a \to 0} f_a(x).$$

The δ-function is an idealization and useful mathematical tool. Nature does not have δ-functions. It helps provide use with a formal analogue to the point charge, which is also an idealization. A point charge may formally be described by a charge density ρ, which vanishes everywhere but at one particular location, where it becomes infinitely large. We will generalize the one-dimensional δ-function for this purpose.

$$\boxed{\delta(\mathbf{r} - \mathbf{r}') = \delta(x - x')\delta(y - y')\delta(z - z')}. \tag{3.51}$$

This allows to describe a point charge Q at the location $\mathbf{r}'$ by

$$\rho(\mathbf{r}) = Q\delta(\mathbf{r} - \mathbf{r}').$$

Integrating over the entire space gives

$$\int_V \rho(\mathbf{r})d\tau = \int Q\delta(x - x')\delta(y - y')\delta(z - z')dx\,dy\,dz$$

$$= Q\int \delta(x - x')dx \int \delta(y - y')dy \int \delta(z - z')dz$$

$$= Q \cdot 1 \cdot 1 \cdot 1 = Q,$$

as we had expected.

An important property of the δ-function that follows from above discussion is:

$$\boxed{\int_{-\infty}^{+\infty} f(x)\delta(x - x')dx = f(x')} \tag{3.52}$$

Thus, the δ-function has the defining property of filtering a particular value from a smooth function when integrated.

One may write as well

$$\int_{-\infty}^{+\infty} f(x)\delta(x - x')dx = \int_{-\infty}^{+\infty} f(x')\delta(x - x')dx$$

$$= f(x')\int_{-\infty}^{+\infty} \delta(x - x')dx = f(x') \cdot 1.$$

Similarly for a function $f(\mathbf{r})$, defined in the entire space

$$\int_{entire \atop space} f(\mathbf{r})\delta(\mathbf{r} - \mathbf{r}')dx = f(\mathbf{r}'), \tag{3.53}$$

which results from multiple applications of (3.52).

The δ-function is symmetric

$$\delta(x - x') = \delta(-[x - x']) = \delta(x' - x), \tag{3.54}$$

It is possible to differentiate it as well as integrate it. Its indefinite integral is *Heaviside's step function*.

$$\boxed{H(x - x') = \begin{cases} 0 & \text{for} & x < x' \\ 1 & \text{for} & x > x' \end{cases}} \tag{3.55}$$

Indeed,

$$\int_{-\infty}^{x} \delta(x''-x')dx'' = \begin{cases} 0 & \text{for } x < x' \\ 1 & \text{for } x > x' \end{cases},$$

or

$$\int_{-\infty}^{x} \delta(x''-x')dx'' = H(x-x') .$$

conversely

$$\frac{dH(x-x')}{dx} = \delta(x-x') ,$$

i.e., differentiating the step function (which is not possible with ordinary functions) results in the δ-function.

To avoid confusion about the dimensions, we note that by eq. (3.50), the δ-function carries a dimension, namely the inverse of its argument, *i.e.*, x^{-1}. When the argument of the δ-function is a vector, then based on the definition of (3.51) its dimensions are *for example*, x^{-3} for the three-dimensional case.

3.4.6   Point Charge and δ-Function

Poisson's equation applies to the case of a point charge.

$$\nabla^2\varphi = -\frac{\rho}{\varepsilon_0} ,$$

where

$$\rho(\mathbf{r}) = Q\delta(\mathbf{r}-\mathbf{r}') ,$$

i.e.,

$$\nabla^2\varphi = -\frac{Q\delta(\mathbf{r}-\mathbf{r}')}{\varepsilon_0} ,$$

We know its solution already

$$\varphi = -\frac{Q}{4\pi\varepsilon_0} \cdot \frac{1}{|\mathbf{r}-\mathbf{r}'|} ,$$

i.e., we may now write for all locations, including the location of the point charge, (which was previously impossible)

$$\nabla^2 \frac{Q}{4\pi\varepsilon_0} \cdot \frac{1}{|\mathbf{r}-\mathbf{r}'|} = -\frac{Q}{\varepsilon_0} \cdot \delta(\mathbf{r}-\mathbf{r}') ,$$

or

$$\boxed{\nabla^2\frac{1}{|\mathbf{r}-\mathbf{r}'|} = -4\pi\delta(\mathbf{r}-\mathbf{r}')} . \tag{3.56}$$

In Section 2.3, we had to exclude the locations of the point charges, because we were unable to differentiate there. With the use of the δ-function, those difficulties or restrictions are now removed.

The potential for an arbitrary charge distribution $\rho(\mathbf{r})$ we previously found by superposition to be

$$\varphi(\mathbf{r}) = \frac{1}{4\pi\varepsilon_0} \cdot \int \frac{\rho(\mathbf{r}')}{|\mathbf{r}-\mathbf{r}'|} d\tau' \ .$$

It is worthwhile to also formally prove this result. Applying the Laplace operator to this equation gives

$$\nabla^2\varphi(\mathbf{r}) = \frac{1}{4\pi\varepsilon_0} \cdot \int \nabla_r^2 \frac{\rho(\mathbf{r}')}{|\mathbf{r}-\mathbf{r}'|} d\tau'$$

$$= \frac{1}{4\pi\varepsilon_0} \cdot \int \rho(\mathbf{r}')\nabla_r^2 \frac{1}{|\mathbf{r}-\mathbf{r}'|} d\tau'$$

$$= \frac{1}{4\pi\varepsilon_0} \cdot \int \rho(\mathbf{r}')(-4\pi)\delta(\mathbf{r}-\mathbf{r}') d\tau'$$

$$= -\frac{\rho(\mathbf{r})}{\varepsilon_0} \quad ,$$

this shows that the expression for φ fulfills Poisson's equation, proving its validity.

For completeness, we note that there exists a more general (nevertheless, not the most general) solution for

$$\nabla^2\varphi = -\frac{Q\delta(\mathbf{r}-\mathbf{r}')}{\varepsilon_0} \ ,$$

which is given by

$$\varphi = -\frac{Q}{4\pi\varepsilon_0} \cdot \frac{1}{|\mathbf{r}-\mathbf{r}'|} + const \ .$$

Only the boundary condition $\varphi = 0$ at infinity makes the solution unique and forces the constant to vanish.

3.4.7   Potential in a Bounded Region

When considering the entire space with all its charges, then φ is given by the usual integral

$$\varphi(\mathbf{r}) = \frac{1}{4\pi\varepsilon_0} \cdot \int \frac{\rho(\mathbf{r}')}{|\mathbf{r}-\mathbf{r}'|} d\tau' \ .$$

In contrast, if we just consider a finite space and only those charges inside this region, then Green's theorem (3.47) allows certain statements about this situation. For this purpose we use (3.47) and substitute

$$\phi = \varphi \quad \text{where} \quad \nabla^2\varphi = -\frac{\rho(\mathbf{r})}{\varepsilon_0}$$

and

$$\psi = \frac{1}{|\mathbf{r} - \mathbf{r'}|}$$

This gives

$$\int_V \left(\frac{1}{|\mathbf{r} - \mathbf{r'}|} \nabla^2 \varphi - \varphi \nabla^2 \frac{1}{|\mathbf{r} - \mathbf{r'}|} \right) d\tau = \int_V \left[\frac{-\rho(\mathbf{r})}{\varepsilon_0 |\mathbf{r} - \mathbf{r'}|} + \varphi(\mathbf{r}) 4\pi \delta(\mathbf{r} - \mathbf{r'}) \right] d\tau$$

$$= -\int_V \frac{\rho(\mathbf{r})}{\varepsilon_0 |\mathbf{r} - \mathbf{r'}|} d\tau + 4\pi \varphi(\mathbf{r'})$$

$$= \oint \left(\frac{1}{|\mathbf{r} - \mathbf{r'}|} \frac{\partial \varphi}{\partial n} - \varphi \frac{\partial}{\partial n} \frac{1}{|\mathbf{r} - \mathbf{r'}|} \right) dA$$

This result assumes that the point $\mathbf{r'}$ is inside the respective volume. If that point is at its surface, then the factor 4π needs to be replaced (in case of a "smooth surface" as in Section 8.2.1) by the factor 2π, and if that point is outside the volume, then the factor becomes 0. Exchanging $\mathbf{r}$ and $\mathbf{r'}$ gives

$$\boxed{\varphi(\mathbf{r}) = \frac{1}{4\pi\varepsilon_0} \int_V \frac{\rho(\mathbf{r'}) d\tau'}{|\mathbf{r} - \mathbf{r'}|} + \frac{1}{4\pi} \oint \frac{\frac{\partial}{\partial n'} \varphi(r')}{|\mathbf{r} - \mathbf{r'}|} dA' - \frac{1}{4\pi} \oint \varphi(r') \frac{\partial}{\partial n'} \frac{1}{|\mathbf{r} - \mathbf{r'}|} dA'} \qquad (3.57)$$

This result is peculiar and is sometimes called Kirchhoff theorem or Green's formula. It states that φ, besides its usual term

$$\frac{1}{4\pi\varepsilon_0} \int_V \frac{\rho(\mathbf{r'}) d\tau'}{|\mathbf{r} - \mathbf{r'}|} \ ,$$

which results from charges inside a volume, gets additional contributions from the boundary. Evidently, these represent charges that may possibly be located outside the volume. These formally derived contributions also have a plausible explanation.

1. The term

$$\frac{1}{4\pi} \oint \frac{\frac{\partial}{\partial n'} \varphi(r')}{|\mathbf{r} - \mathbf{r'}|} dA'$$

can be regarded as the potential of a distribution of surface charges (see (2.26)).

$$\frac{1}{4\pi\varepsilon_0} \int \frac{\sigma(\mathbf{r'})}{|\mathbf{r} - \mathbf{r'}|} dA'$$

with

$$\sigma = \varepsilon_0 \frac{\partial \varphi}{\partial n} .$$

2. The term

$$-\frac{1}{4\pi}\oint \varphi(r')\frac{\partial}{\partial n'}\frac{1}{|\mathbf{r}-\mathbf{r'}|}dA'$$

can be regarded as the potential of a dipole layer (see (2.69)),

$$-\frac{1}{4\pi\varepsilon_0}\oint \tau \frac{\partial}{\partial n'}\frac{1}{|\mathbf{r}-\mathbf{r'}|}dA'$$

with

$$\tau = \varepsilon_0 \varphi .$$

Thus, by choosing the proper combination of dipole layer and surface charges, one can simulate the effects that all possible charges outside the volume have on the inside. This result then does not describe the field outside the region of consideration. On the contrary, the dipole layer causes the boundary to have a potential $\varphi_a = 0$, while the surface charge causes the boundary to exhibit $(\partial\varphi/\partial n)_o = 0$. When we label the quantities at the inside with the index i and those at the outside with the index a, then

$$D_a - D_i = \sigma = -\varepsilon_0\left(\frac{\partial\varphi}{\partial n}\right)_a + \varepsilon_0\left(\frac{\partial\varphi}{\partial n}\right)_i ,$$

and because of

$$\sigma = \varepsilon_0\left(\frac{\partial\varphi}{\partial n}\right)_i$$

follows the claim that

$$\left(\frac{\partial\varphi}{\partial n}\right)_a = 0 .$$

Furthermore, it is

$$\varphi_a - \varphi_i = \frac{\tau}{\varepsilon_0} .$$

Because of

$$\tau = -\varepsilon_0 \varphi_i$$

follows our claim that

$$\varphi_a = 0 .$$

The surface charge and the dipole layer cause discontinuities of $\partial\varphi/\partial n$ and φ, respectively, which is of paramount importance in field theory. This point will be discussed in more detail (Section 8.2.1 and 8.2.2).

The quantities mentioned are closely related to the method of image charges which we have already discussed (Section 2.6.1). Our previous discussion used the reverse arguments, where the effects of surface charges were replaced by appropriate image charges.

In closing, we add a note of caution with regards to a potential misinterpretation of eq. (3.57). It does by no means allow, when $\rho(\mathbf{r})$ inside the region was already given, to arbitrarily choose φ and $\partial\varphi/\partial n$ on the surface, and from there calculate φ inside by (3.57). If we were to independently prescribe φ and $\partial\varphi/\partial n$ on the surface, then the problem would be overdetermined. To specify one of the two quantities, already makes the problem unique This means that eq. (3.57) merely expresses that φ is of this general form if the values for φ and $\partial\varphi/\partial n$ are compatible. Nevertheless, it is possible to use this relation as starting point for a solution, if by means of suitable Green functions, either one or the other of these two terms is eliminated. We will omit the details of this discussion, but will return to apply Green functions for concrete situations. Eq. (3.57) represents an important basis for analytical and numerical methods to solve boundary value problems, in particular the boundary element method (Sections 8.2 and 8.8).

Eq. (3.57) is also related to the Helmholtz theorem. We will come back to this in Appendix A.5, where the significance of these two surface integrals will become even more apparent.

3.5   Separation of Laplace's Equation in Cartesian Coordinates

3.5.1   Separation of Variables

In the following few sections, we will solve Laplace's equation by separation of variables for a number of different coordinate systems. We demonstrate the methodology by applying it to the simplest example, that is, to the Cartesian coordinate system. The equation to solve is

$$\nabla^2\varphi = \left(\frac{\partial^2}{\partial x^2} + \frac{\partial^2}{\partial y^2} + \frac{\partial^2}{\partial z^2}\right)\varphi = 0 \ . \tag{3.58}$$

Writing φ as

$$\varphi = X(x)Y(y)Z(z) \tag{3.59}$$

and inserting this in (3.58), gives

$$YZ\frac{\partial^2}{\partial x^2}X(x) + XZ\frac{\partial^2}{\partial y^2}Y(y) + XY\frac{\partial^2}{\partial z^2}Z(z) = 0 \ .$$

Divide this equation by $\varphi = XYZ$ to obtain

$$\frac{1}{X}\frac{\partial^2}{\partial x^2}X(x) + \frac{1}{Y}\frac{\partial^2}{\partial y^2}Y(y) + \frac{1}{Z}\frac{\partial^2}{\partial z^2}Z(z) = 0 \tag{3.60}$$

It is important to observe that the first term depends only on x, the second only on y, and the third only on z. The sum vanishes. This is possible only, if each of the three terms is constant. This enables one to write:

$$\frac{1}{X}\frac{\partial^2}{\partial x^2}X(x) = -k^2$$

$$\frac{1}{Y}\frac{\partial^2}{\partial y^2}Y(y) = -l^2$$

$$\frac{1}{Z}\frac{\partial^2}{\partial z^2}Z(z) = k^2 + l^2$$

$$(3.61)$$

This introduces two arbitrary constants, the so-called *constants of separation*. Slightly rewriting this yields

$$\frac{\partial^2}{\partial x^2}X(x) = -k^2 X(x)$$

$$\frac{\partial^2}{\partial y^2}Y(y) = -l^2 Y(y)$$

$$\frac{\partial^2}{\partial z^2}Z(z) = (k^2 + l^2)Z(z)$$

$$(3.62)$$

These are ordinary differential equations. Their general solutions are:

$$X = A \cdot \cos kx + B \cdot \sin kx$$
$$Y = C \cdot \cos ly + D \cdot \sin ly$$
$$Z = E \cdot \cosh\sqrt{k^2 + l^2}z + F \cdot \sinh\sqrt{k^2 + l^2}z$$

$$(3.63)$$

This gives

$$\varphi = (A \cdot \cos kx + B \cdot \sin kx)(C \cdot \cos ly + D \cdot \sin ly) \qquad (3.64)$$
$$(E \cdot \cosh\sqrt{k^2 + l^2}z + F \cdot \sinh\sqrt{k^2 + l^2}z)$$

The separation constants k and l may be chosen at will. The resulting solution for φ is thus only a very specific one. The general solution is, however, obtained by the superposition of all possible solutions (*i.e.*, using all possible values of k and l). We can choose other functions, *i.e.*, instead of *cos* and *sin*, we could have chosen *exp(ikx)* and *exp(-ikx)*, or *exp(kx)* and *exp(-kx)*, respectively, which resulted in writing X, Y, Z in a form different from above:

$$X = \bar{A} \cdot \exp(ikx) + \bar{B} \cdot \exp(-ikx)$$
$$Y = \bar{C} \cdot \exp(ily) + \bar{D} \cdot \exp(-ily)$$
$$Z = \bar{E} \cdot \exp(\sqrt{k^2 + l^2}z) + \bar{F} \cdot \exp((-\sqrt{k^2 + l^2})z)$$

$$(3.65)$$

This does not actually introduce anything new, because of

$$\left.\begin{array}{rcl} \exp(\pm ikx) &=& \cos kx \pm i\sin kx \\ \exp(\pm kx) &=& \cosh kz \pm \sinh kz \end{array}\right\} .$$

(3.66)

Whether one finds φ by superposing functions of type (3.63) or (3.65) is of lesser importance. For a specific case, however, there may be reasons to select one over the other. Depending on the type of problem, one needs to use functions with all values of k and l, or just specific values of k and l. This will be clarified when some specific examples are considered.

In very simple cases, one would find a quicker solution and not use the method of separation of variables. Let us consider the problem of finding the potential between two, infinite, parallel plates whose potentials are given constants (Fig. 3.9).

One may proceed as follows: First, observe that φ can only depend on x. Therefore

$$\nabla^2\varphi = \frac{\partial^2\varphi}{\partial x^2} = 0$$

with the general solution

$$\varphi = A + Bx$$

The integration constants A and B are determined by

$$\varphi_1 = A + B \cdot 0$$

$$\varphi_2 = A + B \cdot d$$

i.e.,

$$A = \varphi_1$$

$$B = \frac{\varphi_2 - \varphi_1}{d}$$

and

$$\varphi = \varphi_1 + \frac{\varphi_2 - \varphi_1}{d}x$$

or

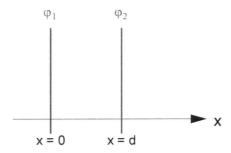

Fig. 3.9

$$E_x = -\frac{\partial \varphi}{\partial x} = -\frac{\varphi_2 - \varphi_1}{d}$$

which already solves this simple problem. Of course the formal route using separation is open as well, *i.e.*, using the Ansatz

$$\varphi = \tilde{A} \cdot \cos kx + \tilde{B} \cdot \sin kx \qquad k \to 0 \ .$$

The reason why k has to approach zero is, that because of the independence of y and z, both l, as well as $k^2 + l^2$, have to vanish. This results in

$$\varphi = \tilde{A} + \tilde{B}kx = A + Bx \ ,$$

when substituting $A = \tilde{A}$ and $B = \tilde{B}k$. Note that even for $k \to 0$, B may be finite, as $\tilde{B}$ may assume any, even infinitely large values.

We will apply this method of separation of variables to other coordinate systems as well. It shall be noted, however, that the method of separation of variables is not generally applicable, but rather is a specific property of certain orthogonal coordinate systems which allow for the separation of certain equations. Besides Cartesian, cylindrical, and spherical coordinates, there are 8 more, for a total of 11 coordinate systems which permit separation of the three-dimensional Laplace equation and the Helmholtz equation, yet to be introduced. Besides these, there are arbitrarily many coordinates systems that permit separation of the two-dimensional or plane Laplace equation. Finally, there is the possibility to expand the meaning of the term separability (now R separability, to distinguish it from simple separability), which then allows for the separation of Laplace's equation in a few more coordinate systems. A very useful summary of all these problems is provided by [2].

3.5.2   Examples

3.5.2.1 Dirichlet Boundary Value Problem without Charges Inside

The problem is to find the electric potential inside a cuboid whose sides are of lengths a, b, c (in x, y, z direction, as shown in Fig. 3.10). The boundary conditions are

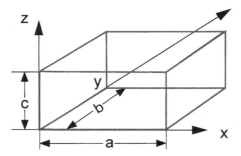

Fig. 3.10

$\varphi = 0$ on all faces except on one (*for example*, the top one where $z = c$),

$\varphi = \varphi_c(x, y)$ on this one face for $z = c$.

There shall be no charges inside the volume. For the solution, according to eq. (3.63), we make the Ansatz for the x-dependency

$X = A_k \cos kx + B_k \sin kx$ for $k \neq 0$

$X = A_0 + B_0 x$ for $k = 0$

Since $\varphi = 0$ for $x = 0$, it must be $A_k = 0$ and $A_0 = 0$, which leads to

$X = B_k \sin kx$ for $k \neq 0$

$X = B_0 x$ for $k = 0$

$\hspace{11cm}$ (3.67)

Since also $\varphi = 0$ for $x = a$

$B_k \sin kx = 0$

$B_0 x = 0$

Consequently

$B_0 = 0$ $\hspace{9.5cm}$ (3.68)

and

$ka = n\pi, \qquad k = \dfrac{n\pi}{a}$,

where n is a whole number. This means that the boundary conditions force k to take on very specific values:

$k = k_n = \dfrac{n\pi}{a}$. $\hspace{8cm}$ (3.69)

Those values are often called *eigenvalues* of the problem. Putting everything together, gives

$X = B_n \sin\dfrac{n\pi x}{a}, \qquad n = 1, 2, 3, ...$ $\hspace{4.5cm}$ (3.70)

In similar manner, we find

$Y = D_m \sin\dfrac{m\pi x}{b}, \qquad m = 1, 2, 3, ...$. $\hspace{4cm}$ (3.71)

To find the z-dependency, we start with the Ansatz

$Z = E \cdot \cosh\sqrt{k^2 + l^2}\, z + F \cdot \sinh\sqrt{k^2 + l^2}\, z$

$\varphi = 0$ for $z = 0$ so that $E = 0$ and therefore

$Z = F_{nm} \sinh\left[\sqrt{\left(\dfrac{n\pi}{a}\right)^2 + \left(\dfrac{m\pi}{b}\right)^2}\, z\right]$. $\hspace{4cm}$ (3.72)

Substituting

$$B_n D_m F_{nm} = C_{nm}$$

we get

$$\varphi = C_{nm} \sin\frac{n\pi x}{a} \sin\frac{m\pi y}{b} \sinh\left[\sqrt{\left(\frac{n\pi}{a}\right)^2 + \left(\frac{m\pi}{b}\right)^2}\, z\right] , \qquad (3.73)$$

with n and m being positive integers. That poses no restriction on the generality of this solution, because the sine function is anti-symmetric, the negative numbers are redundant. The general solution is therefore

$$\varphi = \sum_{n,\, m\, =\, 1}^{\infty} C_{nm} \sin\frac{n\pi x}{a} \sin\frac{m\pi y}{b} \sinh\left[\sqrt{\left(\frac{n\pi}{a}\right)^2 + \left(\frac{m\pi}{b}\right)^2}\, z\right] . \qquad (3.74)$$

It meets all boundary conditions, except for $z = c$. There, it has to be

$$\varphi_c(x, y) = \sum_{n,\, m\, =\, 1}^{\infty} C_{nm} \sin\frac{n\pi x}{a} \sin\frac{m\pi y}{b} \sinh\left[\sqrt{\left(\frac{n\pi}{a}\right)^2 + \left(\frac{m\pi}{b}\right)^2}\, c\right] . \qquad (3.75)$$

The task is now to find the coefficients C_{nm} to satisfy this boundary condition also. This is a known problem, namely to expand $\varphi_c(x, y)$ into a two-dimensional Fourier series. This may require some introductory explanation. A Fourier series is a means to expand a periodic function, over a given interval, as a linear combination of sines and cosines. This function may, in specific cases, be symmetric or anti-symmetric with respect to the center of the interval. There will only be cosines in the symmetric case and only sines in the anti-symmetric case. In our case, we may *for example.*, choose for x the interval $-a \leq x \leq a$, although our interest is only in the range $0 \leq x \leq a$. We may think of this function as periodic and anti-symmetric in the said interval, which justifies the expansion by the above method. The same is true for the y-dependency in the interval $-b \leq x \leq b$.

To find the coefficients C_{nm}, we make use of the orthogonality relations

$$\left.\begin{array}{ll} \displaystyle\int_0^a \sin\frac{n\pi x}{a} \sin\frac{n'\pi x}{a}\, dx = \frac{a}{2}\delta_{nn'} , & n, n' \geq 1 \\[3mm] \displaystyle\int_0^b \sin\frac{m\pi y}{b} \sin\frac{m'\pi y}{b}\, dx = \frac{b}{2}\delta_{mm'} , & m, m' \geq 1 \end{array}\right\} , \qquad (3.76)$$

where δ_{nm} is the so-called Kronecker symbol:

$$\delta_{nm} = \begin{cases} 1 & n = m \\ 0 & n \neq m \end{cases}$$

We multiply (3.75) by

$$\sin\frac{n'\pi x}{a} \sin\frac{m'\pi y}{b}$$

and then integrate x from 0 to a and y from 0 to b to get

$$\int_0^a \int_0^b \varphi_c(x,y) \sin\frac{n'\pi x}{a} \sin\frac{m'\pi y}{b} dx\,dy$$

$$= \sum_{n,m=1}^{\infty} C_{nm} \sinh\left[\sqrt{\left(\frac{n\pi}{a}\right)^2 + \left(\frac{m\pi}{b}\right)^2} \, c\right] \cdot$$

$$\cdot \int_0^a \sin\frac{n\pi x}{a} \sin\frac{n'\pi x}{a} n\,dx \cdot \int_0^b \sin\frac{m\pi y}{b} \sin\frac{m'\pi y}{b} dy$$

$$= \sum_{n,m=1}^{\infty} C_{nm} \sinh\left[\sqrt{\left(\frac{n\pi}{a}\right)^2 + \left(\frac{m\pi}{b}\right)^2} \, c\right] \cdot \delta_{nn'}\delta_{mm'}\frac{ab}{4}$$

$$= \left[\frac{ab}{4} \sinh\left[\sqrt{\left(\frac{n'\pi}{a}\right)^2 + \left(\frac{m'\pi}{b}\right)^2} \, c\right]\right] C_{n'm'} .$$

Now we have found C_{nm}:

$$C_{nm} = \frac{4\int_0^a \int_0^b \varphi_c(x,y) \sin\frac{n\pi x}{a} \sin\frac{m\pi y}{b} dx\,dy}{ab \sinh\left[\sqrt{\left(\frac{n\pi}{a}\right)^2 + \left(\frac{m\pi}{b}\right)^2} \, c\right]} . \tag{3.77}$$

And finally the solution to our problem is

$$\varphi(x,y,z) = \sum_{n,m=1}^{\infty} \frac{4}{ab} \int_0^a \int_0^b \varphi_c(x',y') \sin\frac{n\pi x'}{a} \sin\frac{m\pi y'}{b} dx'\,dy'$$

$$\cdot \sin\frac{n\pi x}{a} \sin\frac{m\pi x}{b} \frac{\sinh\left[\sqrt{\left(\frac{n\pi}{a}\right)^2 + \left(\frac{m\pi}{b}\right)^2} \, z\right]}{\sinh\left[\sqrt{\left(\frac{n\pi}{a}\right)^2 + \left(\frac{m\pi}{b}\right)^2} \, c\right]} . \tag{3.78}$$

This so derived function φ satisfies Laplace's equation and all the boundary conditions. Because of the uniqueness theorem, it is the only solution. For $z = c$ the result has to be of course $\varphi_c(x,y)$, which we may also write as

$$\varphi(x,y,c) = \varphi_c(x,y) = \frac{4}{ab} \int_0^a \int_0^b \varphi_c(x',y')$$

$$\cdot \left(\sum_{n=1}^{\infty} \sin\frac{n\pi x'}{a} \sin\frac{n\pi x}{a}\right)\left(\sum_{m=1}^{\infty} \sin\frac{m\pi y'}{b} \sin\frac{m\pi y}{b}\right) dx'\,dy' . \tag{3.79}$$

On the other hand, since x and y are inside the integration boundary

$$\varphi(x,y,c) = \frac{4}{ab} \int_0^a \int_0^b \varphi_c(x',y')\delta(x-x')\delta(y-y')(dx'\,dy') . \tag{3.80}$$

Comparison of the two equations for $\varphi_c(x, y)$ reveals that within the interval $0 \le x \le a$ and $0 \le x \le b$ the following holds

$$\frac{2}{a}\sum_{n=1}^{\infty} \sin\frac{n\pi x'}{a}\sin\frac{n\pi x}{a} = \delta(x-x')$$

(3.81)

$$\frac{2}{b}\sum_{m=1}^{\infty} \sin\frac{m\pi y'}{b}\sin\frac{m\pi y}{b} = \delta(y-y')$$

This important relation is called the *completeness* or *closure relation*. The reason for this name will be discussed. It is possible to derive this in a different way. Consider the task to expand the δ-function into a Fourier series relation. We attempt to expand

$$\delta(x-x') = \sum_{n=1}^{\infty} C_n(x')\sin\frac{n\pi x}{a}$$

and determine $C_n(x')$ by means of the orthogonality relation (3.76) to obtain

$$\int_0^a \delta(x-x')\sin\frac{n'\pi x}{a}dx = \sum_{n=1}^{\infty}\int_0^a C_n(x')\sin\frac{n\pi x}{a}\sin\frac{n'\pi x}{a}dx$$

$$= \frac{a}{2}\sum_{n=1}^{\infty} C_n(x')\delta_{nn'} = \frac{a}{2}C_{n'}(x')$$

or

$$C_n(x') = \frac{2}{a}\int_0^a \delta(x-x')\sin\frac{n\pi x}{a}dx = \frac{2}{a}\sin\frac{n\pi x'}{a}$$

and therefore as suggested

$$\delta(x-x') = \frac{2}{a}\sum_{n=1}^{\infty} \sin\frac{n\pi x}{a}\sin\frac{n\pi x'}{a} \ .$$

If we consider this as a function of the entire x space, then the sum represents an anti-symmetric function in the interval $-a \le x \le a$, which repeats itself periodically, *i.e.*, it represents positive δ-functions at all values where $x = x' + 2pa$ and negative δ-functions at all values where $x = -x' + 2pa$ and p is a whole number but otherwise arbitrary. This means that as long as we do not restrict ourselves to $0 \le x \le a$, the following is true:

$$\frac{2}{a}\sum_{n=1}^{\infty} \sin\frac{n\pi x}{a}\sin\frac{n\pi x'}{a} = \sum_{p=-\infty}^{+\infty} \delta(x-x'-2pa)-\delta(x+x'-2pa) \ .$$

The more general Dirichlet problem, where the potential is prescribed on the entire surface, can be reduced to the problem discussed above. Observe that it is possible to choose another of the faces where to prescribe the potential (to be non-zero), while it is zero on the other five faces. Overall, there are thus six such solutions and the general solution is obtained by linear superposition of those solutions.

For the most part, the methodology with which the current problem was solved can also be applied to Neumann or mixed type problems.

3.5.2.2 Dirichlet Boundary Value Problem with Charges Inside the Volume

The task is to find the electric potential inside a cuboid with the edges a, b, c where on the total surface $\varphi = 0$, while there are surface charges $\sigma(x, y)$ on the plane $z = z_0$ located inside the cuboid ($0 < z_0 < c$) (Fig. 3.11).

In this case, we need to write a separate Ansatz for each volume, $0 \le z \le z_0$ and $z_0 \le z \le c$. We use the results from the previous example, $i.e.$, use (3.74) to write φ in the following way: In volume 1 where $0 \le z \le z_0$ we found

$$\varphi_1 = \sum_{n,m=1}^{\infty} C_{nm} \sin\frac{n\pi x}{a} \sin\frac{m\pi y}{b} \frac{\sinh\left[\sqrt{\left(\frac{n\pi}{a}\right)^2 + \left(\frac{m\pi}{b}\right)^2}\, z\right]}{\sinh\left[\sqrt{\left(\frac{n\pi}{a}\right)^2 + \left(\frac{m\pi}{b}\right)^2}\, z_0\right]}, \qquad (3.82)$$

and in volume 2 where $z_0 \le z \le c$ the potential is

$$\varphi_2 = \sum_{n,m=1}^{\infty} C_{nm} \sin\frac{n\pi x}{a} \sin\frac{m\pi y}{b} \frac{\sinh\left[\sqrt{\left(\frac{n\pi}{a}\right)^2 + \left(\frac{m\pi}{b}\right)^2}\, (c-z)\right]}{\sinh\left[\sqrt{\left(\frac{n\pi}{a}\right)^2 + \left(\frac{m\pi}{b}\right)^2}\, (c-z_0)\right]}. \qquad (3.83)$$

Those trial functions fulfill the required boundary condition $\varphi = 0$ on the entire surface. Furthermore, it has the property to make φ continuous at the area $z = z_0$. Therefore

$$\left(\frac{\partial\varphi_1}{\partial x}\right)_{z=z_0} = \left(\frac{\partial\varphi_2}{\partial x}\right)_{z=z_0}$$

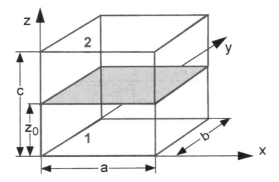

Fig. 3.11

and

$$\left(\frac{\partial \varphi_1}{\partial y}\right)_{z=z_0} = \left(\frac{\partial \varphi_2}{\partial y}\right)_{z=z_0}$$

i.e., the tangential components of the electric field are continuous, as they have to be. A further requirement is

$$\left(\frac{\partial \varphi_1}{\partial z} - \frac{\partial \varphi_2}{\partial z}\right)_{z=z_0} = \frac{\sigma(x,y)}{\varepsilon_0} , \qquad (3.84)$$

from which one can find the coefficients C_{nm}. For this purpose we also need to expand $\sigma(x,y)$:

$$\sigma(x,y) = \sum_{n,m=1}^{\infty} \sigma_{nm} \sin\frac{n\pi x}{a} \sin\frac{m\pi y}{b} , \qquad (3.85)$$

We use the orthogonality relation (3.76) to calculate

$$\int_0^a \int_0^b \sigma(x,y)\sin\frac{n'\pi x}{a}\sin\frac{m'\pi y}{b}dx\,dy$$

$$= \sum_{n,m=1}^{\infty} \sigma_{nm}\int_0^a \sin\frac{n\pi x}{a}\sin\frac{n'\pi x}{a}dx\int_0^b \sin\frac{m\pi y}{b}\sin\frac{m'\pi y}{b}dy$$

$$= \sum_{n,m=1}^{\infty} \sigma_{nm}\frac{a}{2}\delta_{nn'}\frac{b}{2}\delta_{mm'} = \sigma_{n'm'}\frac{ab}{4}$$

that is

$$\sigma_{nm} = \frac{4}{ab}\int_0^a \int_0^b \sigma(x',y')\sin\frac{n\pi x'}{a}\sin\frac{m\pi y'}{b}dx'\,dy' . \qquad (3.86)$$

The condition (3.84) can now be formulated as

$$\sum_{n,m=1}^{\infty} C_{nm}\sin\frac{n\pi x}{a}\sin\frac{m\pi y}{b}\left[\frac{\sqrt{\left(\frac{n\pi}{a}\right)^2+\left(\frac{m\pi}{b}\right)^2}\cosh\left[\sqrt{\left(\frac{n\pi}{a}\right)^2+\left(\frac{m\pi}{b}\right)^2}z_0\right]}{\sinh\left[\sqrt{\left(\frac{n\pi}{a}\right)^2+\left(\frac{m\pi}{b}\right)^2}z_0\right]}\right.$$

$$\left. + \frac{\sqrt{\left(\frac{n\pi}{a}\right)^2+\left(\frac{m\pi}{b}\right)^2}\cosh\left[\sqrt{\left(\frac{n\pi}{a}\right)^2+\left(\frac{m\pi}{b}\right)^2}(c-z_0)\right]}{\sinh\left[\sqrt{\left(\frac{n\pi}{a}\right)^2+\left(\frac{m\pi}{b}\right)^2}(c-z_0)\right]}\right]$$

$$= \sum_{n,m=1}^{\infty} \frac{\sigma_{nm}}{\varepsilon_0}\sin\frac{n\pi x}{a}\sin\frac{m\pi y}{b} .$$

Comparing coefficients yields

$$C_{nm} = \frac{\sigma_{nm}}{\varepsilon_0} \frac{\sinh\left[\sqrt{\left(\frac{n\pi}{a}\right)^2 + \left(\frac{m\pi}{b}\right)^2}\, z_0\right] \sinh\left[\sqrt{\left(\frac{n\pi}{a}\right)^2 + \left(\frac{m\pi}{b}\right)^2}\,(c - z_0)\right]}{\sqrt{\left(\frac{n\pi}{a}\right)^2 + \left(\frac{m\pi}{b}\right)^2}\, \sinh\left[\sqrt{\left(\frac{n\pi}{a}\right)^2 + \left(\frac{m\pi}{b}\right)^2}\, c\right]} ,\quad (3.87)$$

where the trigonometric identity $\sinh(x + y) = \sinh x \cosh y + \cosh x \sinh y$ was used to rewrite the hyperbolic functions.

Of particular interest is a special case of this result. Let us assume that there is a point charge Q at location x_0, y_0, z_0. This is equivalent to a surface charge

$$\sigma(x, y) = Q\delta(x - x_0)\delta(y - y_0) . \tag{3.88}$$

Based on (3.86), this gives

$$\sigma_{nm} = \frac{4Q}{ab} \sin\frac{n\pi x_0}{a} \sin\frac{m\pi y_0}{b} . \tag{3.89}$$

Combining this with eqs. (3.82), (3.83), and (3.87) yields the somewhat lengthy result:

$$\begin{aligned}
\varphi_{1,2} = \ &\frac{4Q}{ab\varepsilon_0} \sum_{n,m=1}^{\infty} \sin\frac{n\pi x}{a}\sin\frac{n\pi x_0}{a}\sin\frac{m\pi y}{b}\sin\frac{m\pi y_0}{b} \cdot \\
&\cdot \frac{1}{\sqrt{\left(\frac{n\pi}{a}\right)^2 + \left(\frac{m\pi}{b}\right)^2}\,\sinh\left[\sqrt{\left(\frac{n\pi}{a}\right)^2 + \left(\frac{m\pi}{b}\right)^2}\,c\right]} \cdot \\
&\cdot \begin{cases}
\sinh\left[\sqrt{\left(\frac{n\pi}{a}\right)^2 + \left(\frac{m\pi}{b}\right)^2}\,z\right]\sinh\left[\sqrt{\left(\frac{n\pi}{a}\right)^2 + \left(\frac{m\pi}{b}\right)^2}\,(c-z_0)\right] & \text{vol. 1} \\[2ex]
\sinh\left[\sqrt{\left(\frac{n\pi}{a}\right)^2 + \left(\frac{m\pi}{b}\right)^2}\,(c-z)\right]\sinh\left[\sqrt{\left(\frac{n\pi}{a}\right)^2 + \left(\frac{m\pi}{b}\right)^2}\,z_0\right] & \text{vol. 2}
\end{cases}
\end{aligned}$$

$$(3.90)$$

Although lengthy, $\varphi = \varphi_{1,2}$ is basically a simple function and the solution to Poisson's equation

$$\nabla^2\frac{\varphi}{Q} = -\frac{\delta(x - x_0)\delta(y - y_0)\delta(z - z_0)}{\varepsilon_0} = -\frac{\delta(\mathbf{r} - \mathbf{r}_0)}{\varepsilon_0} , \tag{3.91}$$

where the boundary condition $\varphi = 0$ is prescribed on the whole surface of the cuboid. This is referred to as its Green's function:

$$\frac{\varphi}{Q} = G(\mathbf{r}, \mathbf{r}_0) = G(x, y, z, x_0, y_0, z_0) . \tag{3.92}$$

Its significance lies in the fact, that it allows one to reduce every charge distribution $\rho(\mathbf{r})$ in the cuboid to this equation, for which one has

$$\varphi(\mathbf{r}) = \int_V G(\mathbf{r}, \mathbf{r}_0)\rho(\mathbf{r}_0)d\tau_0 \ . \tag{3.93}$$

This is apparent from the superposition principle but may also be proven formally by inserting it into Poisson's equation.

$$\nabla^2 \varphi = -\frac{\rho}{\varepsilon_0}$$

$$\nabla^2 \varphi = \nabla^2 \int_V G(\mathbf{r}, \mathbf{r}_0)\rho(\mathbf{r}_0)d\tau_0$$

$$= \int_V \nabla_r^2 G(\mathbf{r}, \mathbf{r}_0)\rho(\mathbf{r}_0)d\tau_0$$

$$= -\int_V \frac{\delta(\mathbf{r}-\mathbf{r}_0)}{\varepsilon_0}\rho(\mathbf{r}_0)d\tau_0$$

$$= -\frac{\rho(\mathbf{r}_0)}{\varepsilon_0}$$

Green's function $G(\mathbf{r}, \mathbf{r}_0)$ has another, very interesting property. If in Green's theorem (3.47), we choose

$$\phi = G \quad \text{where} \quad \nabla^2 G = -\frac{\delta(\mathbf{r}-\mathbf{r}_0)}{\varepsilon_0}, \qquad G = 0 \quad \text{on the boundary}$$

and

$$\psi = \varphi \quad \text{where} \quad \nabla^2 \varphi = 0 , \qquad \varphi \quad \text{arbitrary on the boundary} ,$$

then it follows from (3.47) that

$$\int_V (\varphi \nabla^2 G - G\nabla^2 \varphi)d\tau = \oint \left(\varphi \frac{\partial G}{\partial n} - G\frac{\partial \varphi}{\partial n} \right) dA$$

and since G vanishes on the surface

$$-\int_V \varphi \frac{\delta(\mathbf{r}-\mathbf{r}_0)}{\varepsilon_0}d\tau = \oint \varphi \frac{\partial G}{\partial n} dA \ .$$

Therefore

$$\varphi(\mathbf{r}_0) = -\varepsilon_0 \oint \varphi(\mathbf{r})\frac{\partial}{\partial n}G(\mathbf{r}, \mathbf{r}_0)dA \ . \tag{3.94}$$

Notice that φ in eq. (3.94) is a solution of Laplace's equation. This reveals that Green's function solves at the same time Dirichlet's boundary value problem for Laplace's equation. If one prescribes arbitrary values for φ on the surface then φ in the whole volume is determined by eq. (3.94). It is revealing to compare it with the previously derived eq. (3.57). There, the first term vanishes because of $\rho(\mathbf{r}) = 0$. However, the difference is that (3.57), besides the surface integral with φ, also contains a term with $\partial\varphi/\partial n$, which is not the case in (3.94). This is the reason, why (3.57) is not suitable to solve this kind of boundary value problem. Eq.

(3.94) also allows one to solve the problem of Section 3.5.2.1 by means of Green's function, derived in the current section (3.5.2.2). In fact, one can prove that Green's function $G = \varphi_{1,2}/Q$ as of (3.90) is a solution to the example of Section 3.5.2.1, yielding the solution given in (3.78).

The results from these examples may be generalized to solve any problem of this kind. The derivation of (3.94) is not specific to that particular example but is valid for arbitrary surfaces, provided that G is suitable for that surface.

> *Green's function (or more precisely, Green's function for Dirichlet's problem) plays a peculiar double role. It yields the solution of Poisson's equation when the potential vanishes everywhere on the surface, and also the solution of Laplace's equation when the potential is prescribed everywhere on the surface. Eqs. (3.93) and (3.94) are the respective defining relations.*

Correspondingly, there is a Green function for Neumann's problem that plays a similar role. We will prove this here by comparing the two cases side by side. Consider a region with an arbitrary surface whose surface area we will call A. There is a unit charge inside the region at point $\mathbf{r}_0$. The solution of the related Poisson equation with the boundary condition that the potential vanishes everywhere on the surface, or the normal derivative of the potential $\partial\varphi/\partial n$ is kept constant on the surface, respectively. These are referred to as Green's function of the first kind or second kind, respectively or also Green's function of Dirichlet's problem or Neumann's problem.

Dirichlet problems	Neumann problems	
$\nabla_r^2 G_D(\mathbf{r};\mathbf{r}_0) = -\dfrac{\delta(\mathbf{r}-\mathbf{r}_0)}{\varepsilon_0}$	$\nabla_r^2 G_N(\mathbf{r};\mathbf{r}_0) = -\dfrac{\delta(\mathbf{r}-\mathbf{r}_0)}{\varepsilon_0}$	(3.95)
$G_D(\mathbf{r};\mathbf{r}_0) = 0$	$\dfrac{\partial}{\partial n}G_N(\mathbf{r};\mathbf{r}_0) = -\dfrac{1}{\varepsilon_0 A}$	(3.96)
for $\mathbf{r}$ on the surface	for $\mathbf{r}$ on the surface	

For the derivative of G_N in the normal direction to be constant (for the inside Neumann problem), it has to exactly assume the given value. The reason is that $(1/Q)\int \mathbf{D} \bullet d\mathbf{A} = -\varepsilon_0\int(\partial G_N/\partial n)dA = 1$, which is achieved by the above choice of the constant.

The solution of the general Poisson equation

$$\nabla^2\varphi(\mathbf{r}) = -\frac{\rho(\mathbf{r}_0)}{\varepsilon_0} \tag{3.97}$$

with the respective boundary conditions is thus

$$\varphi(\mathbf{r}) = \int_V \rho(\mathbf{r}_0)G_D(\mathbf{r};\mathbf{r}_0)d\tau_0 \quad \bigg| \quad \varphi(\mathbf{r}) = \int_V \rho(\mathbf{r}_0)G_N(\mathbf{r};\mathbf{r}_0)d\tau_0 \qquad (3.98)$$

The solution of the Laplace equation

$$\nabla^2\varphi(\mathbf{r}) = 0 \qquad (3.99)$$

with the prescribed values at the surface

$$\varphi(\mathbf{r}) \quad \bigg| \quad \frac{\partial\varphi(\mathbf{r})}{\partial n}, \quad \oint\frac{\partial\varphi(\mathbf{r}_0)}{\partial n}dA_0 = 0 \qquad (3.100)$$

can also be obtained from Green's functions, namely

$$\varphi(\mathbf{r}) = -\varepsilon_0\oint\varphi(\mathbf{r}_0)\frac{\partial G_D(\mathbf{r};\mathbf{r}_0)}{\partial n}dA_0 \quad \bigg| \quad \varphi(\mathbf{r}) = \varepsilon_0\oint\frac{\partial\varphi(\mathbf{r}_0)}{\partial n}G_N(\mathbf{r};\mathbf{r}_0)dA_0 + C$$

$$(3.101)$$

The values of $\partial\varphi/\partial n$ for the inner Neumann problem can not be to prescribed arbitrarily. Since there are no volume charges, the electric flux through the surface has to vanish, which provides for an additional constraint. While (3.101) was already proven for Dirichlet's problem, this is still to do for Neumann's problem. We achieve this by Green's theorem eq. (3.47), using

$$\phi = G_N$$

and

$$\psi = \varphi$$

where eq. (3.99) and eq. (3.100) define φ. Therefore

$$\int_V(\varphi\nabla_r^2 G_N - G_N\nabla^2\varphi)d\tau = \oint\left(\varphi\frac{\partial G_N}{\partial n} - G_N\frac{\partial\varphi}{\partial n}\right)dA \ ,$$

and with eq. (3.95) and (3.96)

$$-\frac{1}{\varepsilon_0}\int_V\varphi(\mathbf{r})\delta(\mathbf{r}-\mathbf{r}_0)d\tau = -\oint\frac{\varphi(\mathbf{r})}{\varepsilon_0 A}dA - \oint\frac{\partial\varphi(\mathbf{r})}{\partial n}G_N(\mathbf{r},\mathbf{r}_0)dA$$

$$-\frac{\varphi(\mathbf{r}_0)}{\varepsilon_0} = -\frac{1}{\varepsilon_0 A}\oint\varphi(\mathbf{r})dA - \oint\frac{\partial\varphi(\mathbf{r})}{\partial n}G_N(\mathbf{r};\mathbf{r}_0)dA \ ,$$

which, after exchanging $\mathbf{r}_0$ and $\mathbf{r}$, yields eq. (3.101). The constant is determined to be:

$$C = \frac{\oint\varphi(\mathbf{r})dA}{A} \ .$$

3.5.2.3 Point Charge in Infinite Space

Consider the infinite plane $z = z_0$ with a surface charge $\sigma(x, y)$. We make the simplifying assumption that $\sigma(x, y)$ is symmetric with respect to the point (x_0, y_0), that is

$$\left.\begin{array}{l} \sigma(x - x_0, y) = \sigma(x_0 - x, y_0) \\ \sigma(x, y - y_0) = \sigma(x, y_0 - y) \end{array}\right\} . \tag{3.102}$$

We choose the following Ansatz for the potential

$$\varphi_{1,2} = \int_0^\infty \int_0^\infty f(k, l) \cos[k(x - x_0)] \cos[l(y - y_0)] \tag{3.103}$$

$$\cdot \exp[-\sqrt{k^2 + l^2}|z - z_0|] dk dl .$$

The two cosines are a consequence of the symmetry assumed for $\sigma(x, y)$. Because of its infinite extend, there are no special values for k or l. Since we have to accept any value for k and l, the sum is replaced by the integral, or rather, because of the two dimensions, a double integral (*Fourier integral*). The trial function for the z-dependency is chosen such that φ vanishes for $z \to \pm\infty$, as required by physics. Exponential functions thereby provide us with more convenient expansion functions, as compared with hyperbolic functions. As in Section 3.5.2.2, we deal with two regions, region 1 where $z \leq z_0$ and region 2 where $z \geq z_0$. Certain boundary conditions have to be met for $z = z_0$, where we require that

$$|z - z_0| = \begin{cases} z - z_0 & \text{for} \quad z \geq z_0 \quad (\text{region 2}) \\ z_0 - z & \text{for} \quad z \leq z_0 \quad (\text{region 1}) . \end{cases} \tag{3.104}$$

The Ansatz (3.103) guarantees that the tangential components of the electric field are continuous. Furthermore, because of the surface charges, the normal component of **D** has to be discontinuous by

$$\left(\frac{\partial\varphi_1}{\partial z}\right)_{z = z_0} - \left(\frac{\partial\varphi_2}{\partial z}\right)_{z = z_0} = \frac{\sigma(x, y)}{\varepsilon_0} . \tag{3.105}$$

To solve this, we expand $\sigma(x, y)$

$$\sigma(x, y) = \int_0^\infty \int_0^\infty \tilde{\sigma}(k, l) \cos[k(x - x_0)] \cos[l(y - y_0)] dk dl . \tag{3.106}$$

This asks for applying the orthogonality relation

$$\boxed{\int_{-\infty}^{+\infty} \cos[k(x - x_0)] \cos[k'(x - x_0)] dx = \pi\delta(k - k') + \pi\delta(k + k')} . \tag{3.107}$$

From (3.106) and (3.107) we obtain

$$\int_{-\infty}^{+\infty} \int_{-\infty}^{+\infty} \sigma(x, y) \cos[k'(x - x_0)] \cos[l'(y - y_0)] dx dy$$

$$= \int_{-\infty}^{+\infty}\int_{-\infty}^{+\infty}\left\{\int_0^{\infty}\int_0^{\infty}\tilde{\sigma}(k,l)\cos[k(x-x_0)]\cos[l(y-y_0)]dk\,dl\right\}$$

$$\cdot \cos[k'(x-x_0)]\cos[l'(y-y_0)]dx\,dy$$

$$= \int_0^{\infty}\int_0^{\infty}\tilde{\sigma}(k,l)\left\{\int_{-\infty}^{+\infty}\cos[k(x-x_0)]\cos[k'(x-x_0)]dx\right.$$

$$\left.\cdot\int_{-\infty}^{+\infty}\cos[l(y-y_0)]\cos[l'(y-y_0)]dy\right\}dk\,dl$$

$$= \int_0^{\infty}\int_0^{\infty}\tilde{\sigma}(k,l)\pi^2[\delta(k-k')+\delta(k+k')][\delta(l-l')+\delta(l+l')]dk\,dl$$

$$= \pi^2\tilde{\sigma}(k',l')\ .$$

renaming $k' \Rightarrow k$ and $l' \Rightarrow l$ yields

$$\tilde{\sigma}(k,l) = \frac{1}{\pi^2}\int_{-\infty}^{+\infty}\int_{-\infty}^{+\infty}\sigma(x,y)\cos[k(x-x_0)]\cos[l(y-y_0)]dx\,dy\ . \tag{3.108}$$

For now, we are only interested in a point charge Q at location x_0, y_0, z_0, for which

$$\tilde{\sigma}(k,l) = \frac{Q}{\pi^2}\int_{-\infty}^{+\infty}\int_{-\infty}^{+\infty}\delta(x-x_0)\delta(y-y_0)\cos[k(x-x_0)]\cos[l(y-y_0)]dx\,dy$$

$$= \frac{Q}{\pi^2}\ . \tag{3.109}$$

The boundary condition (3.105) can now be written in the following way:

$$\int_0^{\infty}\int_0^{\infty}f(k,l)\cos[k(x-x_0)]\cos[l(y-y_0)]\sqrt{k^2+l^2}2\,dk\,dl$$

$$= \frac{Q}{\pi^2\varepsilon_0}\int_0^{\infty}\int_0^{\infty}\cos[k(x-x_0)]\cos[l(y-y_0)]dk\,dl$$

i.e.,

$$f(k,l) = \frac{Q}{2\pi^2\varepsilon_0\sqrt{k^2+l^2}}\ . \tag{3.110}$$

This solves the problem:

$$\boxed{\varphi_{1,2} = \frac{Q}{2\pi^2\varepsilon_0}\int_0^{\infty}\int_0^{\infty}\frac{\cos[k(x-x_0)]\cos[l(y-y_0)]\exp[-\sqrt{k^2+l^2}|z-z_0|]}{\sqrt{k^2+l^2}}dk\,dl}$$

$$\tag{3.111}$$

Since also

$$\varphi = \frac{Q}{4\pi\varepsilon_0\sqrt{(x-x_0)^2+(y-y_0)^2+(z-z_0)^2}},$$

a comparison reveals that

$$\frac{1}{\sqrt{(x-x_0)^2+(y-y_0)^2+(z-z_0)^2}}$$

$$= \frac{2}{\pi}\int_0^\infty\int_0^\infty\frac{\cos[k(x-x_0)]\cos[l(y-y_0)]\exp[-\sqrt{k^2+l^2}|z-z_0|]dkdl}{\sqrt{k^2+l^2}}$$

(3.112)

This is the Fourier integral for the inverse distance.

We have solved a quite simple problem with rather complicated means. We had already known the potential of a point charge in the infinite space when we calculated it earlier by much simpler means. The purpose of this example was to illustrate the methodology. As a by-product did we obtain the important relation (3.112) for which there may not exist a simpler derivation.

3.5.2.4 Appendix to Section 3.5: Fourier Series and Fourier Integrals

The topic of Fourier series and Fourier integrals with respect to one or two dimensions automatically arose when separating Laplace's equation in Cartesian coordinates. This is a specific case of the more general expansion of functions by certain orthogonal and complete systems of functions, whose occurrence is typical for problems of this kind. We will begin the next section (3.6) with general observations on such systems of functions, which will retroactively highlight the role of Fourier series and Fourier integrals from a more general point of view. Sections 3.7 and 3.8 will provide some examples for the expansion by orthogonal and complete systems of functions. Here, we summarize the most important formulas for Fourier series and Fourier integrals.

a)   Fourier Series

A periodic function $f(x)$ may be represented as a Fourier series. If c is the period, then the series is represented by

$$f(x) = \sum_{n=0}^\infty a_n\cos\left(\frac{2\pi n}{c}x\right) + \sum_{n=1}^\infty b_n\sin\left(\frac{2\pi n}{c}x\right)$$

(3.113)

It describes the value of $f(x)$ everywhere, except where f is discontinuous. Where f is discontinuous the series takes the value

$$\sum_{n=0}^\infty a_n\cos\left(\frac{2\pi n}{c}x\right) + \sum_{n=1}^\infty b_n\sin\left(\frac{2\pi n}{c}x\right) = \lim_{\varepsilon\to 0}\frac{f(x-\varepsilon)+f(x+\varepsilon)}{2}$$

(3.114)

That is, it assumes the average of the left and right sided limit at that point.

The coefficients a_n and b_n of the expansion (3.113) can be found by means of an orthogonality relation for trigonometric functions.

$$
\int_0^c \cos\left(\frac{2\pi n}{c}x\right)\cos\left(\frac{2\pi m}{c}x\right) dx = \begin{cases} \frac{c}{2}\delta_{nm} & \text{for } n \text{ or } m \geq 1 \\ c & \text{for } n = m = 0 \end{cases}
\tag{3.115}
$$

$$
\int_0^c \sin\left(\frac{2\pi n}{c}x\right)\sin\left(\frac{2\pi m}{c}x\right) dx = \begin{cases} \frac{c}{2}\delta_{nm} & \text{for } n \text{ or } m \geq 1 \\ c & \text{for } n = m = 0 \end{cases}
\tag{3.116}
$$

The integration is carried out over a full period, but does not require to span from 0 to c, it could be any x_0 to $x_0 + c$. Multiply the expansion Ansatz (3.113) by $\cos(2\pi nx/c)$ and $\sin(2\pi nx/c)$, respectively, then integrate the resulting equation over a full period while making use of (3.115) and (3.116). This gives

$$
a_n = \begin{cases} \frac{2}{c}\int_0^c f(x)\cos\left(\frac{2\pi n}{c}x\right) dx & \text{for } n \geq 1 \\ \frac{1}{c}\int_0^c f(x)\, dx & \text{for } n = 0 \end{cases}
\tag{3.117}
$$

$$
b_n = \frac{2}{c}\int_0^c f(x)\sin\left(\frac{2\pi n}{c}x\right) dx \qquad \text{for } n \geq 1
\tag{3.118}
$$

In particular, when $f(x)$ is an "even" or "symmetric" function [$f(x) = f(-x)$] then by (3.118) all $b_n = 0$, that is, the result is a pure "cosine series". Conversely when $f(x)$ is an "odd" or "anti-symmetric" function [$f(x) = -f(-x)$] then by (3.118) all $a_n = 0$, that is, the result is a pure "sine series".

$$
f(x) = \sum_{n=0}^{\infty} a_n \cos\left(\frac{2\pi n}{c}x\right) \qquad [f(x) \text{ even}]
\tag{3.119}
$$

$$
f(x) = \sum_{n=1}^{\infty} b_n \sin\left(\frac{2\pi n}{c}x\right) \qquad [f(x) \text{ odd}]
\tag{3.120}
$$

Another option is to expand $f(x)$ as a complex Fourier series

$$
f(x) = \sum_{n=-\infty}^{+\infty} d_n \exp\left(i\frac{2\pi n}{c}x\right).
\tag{3.121}
$$

Because of

$$\exp(i\alpha) = \cos\alpha + i\sin\alpha$$

we may also write

$$f(x) = \sum_{n=-\infty}^{+\infty}\left[d_n\cos\left(\frac{2\pi n}{c}x\right) + id_n\sin\frac{2\pi n}{c}x\right]$$

$$= d_0 + \sum_{n=1}^{\infty}\left[(d_n + d_{-n})\cos\left(\frac{2\pi n}{c}x\right) + i(d_n - d_{-n})\sin\frac{2\pi n}{c}x\right]$$

This establishes the relationship between the expansions (3.113) and (3.121).

$$\left.\begin{aligned} d_0 &= a_0 \\[4pt] d_n + d_{-n} &= a_n \\ i(d_n - d_{-n}) &= b_n \end{aligned}\right\} n \geq 1 \tag{3.122}$$

or the inverse relation

$$\left.\begin{aligned} d_0 &= a_0 \\[4pt] d_{-n} &= \frac{a_n + ib_n}{2} \\[4pt] d_{+n} &= \frac{a_n - ib_n}{2} = d_{-n}^* \end{aligned}\right\} n \geq 1 \tag{3.123}$$

To immediately calculate the coefficients d_n use Ansatz (3.121) and apply the orthogonality relation

$$\boxed{\int_0^c \exp\left(i\frac{2\pi n}{c}x\right)\exp\left(-i\frac{2\pi m}{c}x\right)dx = c\delta_{nm}} \tag{3.124}$$

which yields

$$\boxed{d_n = \frac{1}{c}\int_0^c f(x)\exp\left(-i\frac{2\pi n}{c}x\right)dx} \tag{3.125}$$

b)   Fourier Integrals

Under certain, very general conditions which we will not discuss here, a function may be represented as a Fourier integral:

$$\boxed{f(x) = C\int_{-\infty}^{+\infty}\tilde{f}(k)\exp(ikx)dk} \,. \tag{3.126}$$

We have introduced an arbitrary factor C because the Fourier integral is defined differently by different authors. The inverse is the Fourier transform of $f(x)$

$$\boxed{\tilde{f}(k) = \frac{1}{2\pi C}\int_{-\infty}^{+\infty}f(x)\exp(-ikx)dx} \,. \tag{3.127}$$

This is a result of (3.126) and the orthogonality relation

$$\int_{-\infty}^{+\infty} \exp(ikx)\exp(-ik'x)dx = 2\pi\delta(k-k') \;.$$
(3.128)

Multiplying (3.126) by $\exp(-ik'x)$ and then integrate over x from $-\infty$ through $+\infty$ just yields (3.127), when also considering (3.128).
When $f(x) = f(-x)$ i.e., $f(x)$ is an even function, then

$$\tilde{f}(k) = \frac{1}{2\pi C}\int_{-\infty}^{+\infty} f(x)\cos(kx)dx$$

that is

$$\tilde{f}(k) = \frac{1}{\pi C}\int_{0}^{+\infty} f(x)\cos(kx)dx \;.$$
(3.129)

and

$$f(x) = 2C\int_{0}^{+\infty} \tilde{f}(k)\cos(kx)dk \;.$$
(3.130)

On the other hand, when $f(x) = -f(-x)$ i.e., $f(x)$ is an odd function, then

$$\tilde{f}(k) = -\frac{i}{2\pi C}\int_{-\infty}^{+\infty} f(x)\sin(kx)dx$$

that is

$$\tilde{f}(k) = -\frac{i}{\pi C}\int_{0}^{+\infty} f(x)\sin(kx)dx \;.$$
(3.131)

and

$$f(x) = 2iC\int_{0}^{+\infty} \tilde{f}(k)\sin(kx)dk \;.$$
(3.132)

Thus there are basically three Fourier integrals
1.   the exponential Fourier integral (3.126) and its inverse (3.127)
2.   the cosine Fourier integral (3.130) and its inverse (3.129)
3.   the sine Fourier integral (3.132) and its inverse (3.131)
To directly derive eqs. (3.129) and (3.130), or (3.131) and (3.132) is possible when using the orthogonality relations

$$\int_{0}^{+\infty} \cos(kx)\cos(k'x)dx = \frac{\pi}{2}\delta(k-k') + \frac{\pi}{2}\delta(k+k')$$
(3.133)

$$\int_{0}^{+\infty} \sin(kx)\sin(k'x)dx = \frac{\pi}{2}\delta(k-k') - \frac{\pi}{2}\delta(k+k')$$
(3.134)

The choice of the factor C is a mere matter of convenience. Frequently, it is set to $C = 1$ in (3.126), which causes a factor of $1/2\pi$ in (3.127). Another frequent choice is $C = \sqrt{1/2\pi}$ which makes (3.126) and (3.127) "symmetric", that is, the same factor $\sqrt{1/2\pi}$ occurs in (3.127). Similarly, the sine and cosine transforms are by no means uniquely defined. Frequently the definition of $C = 1/2$ yields the factor in (3.130) to become one, or when defining $C = -i/2$, the factor in (3.132) becomes one.

As a practical matter, because of the different definitions, it is advisable not to define this factor too early in the calculations. Errors can be excluded when selecting the best fitting approach (while leaving the factor open), and then calculate the coefficients by means of the orthogonality relation.

3.6   Complete Orthogonal Systems of Functions

Before starting Separation of Variables for Laplace's equation in cylindrical and spherical coordinates, we will generalize the terminology, derived previously from rather specific examples. Furthermore, their rather hidden analogy to vector calculus shall be illustrated. Solving boundary value problems, under suitable conditions, leads to systems of functions which are orthogonal to each other and are complete in the sense that all possible functions, defined in the given region can be constructed from them by superposition. Our analysis be in one dimension, as multiple ones will not change the principle nature of the relations which we are interested in.

A function $f(x)$, defined in an interval $c \leq x \leq d$, can be thought of as a vector. So to speak, x acts as continuous index which identifies the components of the vector $f(x)$. An integral

$$\int_c^d f^*(x)g(x)dx = \langle f \mid g \rangle \tag{3.135}$$

can be regarded as the scalar product of the two vectors $f(x)$ and $g(x)$. f^* is the conjugate complex function to f. Oftentimes we deal with real functions, in which case it is of course

$$\int_c^d f^*(x)g(x)dx = \langle f \mid g \rangle = \int_c^d f(x)g(x)dx .$$

In Vector Calculus one introduces a number of basis vectors, depending on the spatial dimensions, to construct every vector

$$\mathbf{a} = \sum_{i=1}^n a_i \mathbf{e}_i . \tag{3.136}$$

The system of basis vectors is orthogonal and normalized if

$$\mathbf{e}_i \bullet \mathbf{e}_k = \delta_{ik} . \tag{3.137}$$

The expansion of the vector $\mathbf{a}$ by its basis system $\mathbf{e}_i$ is achieved by scalar multiplication of eq. (3.136) by $\mathbf{e}_k$. This yields

$$\mathbf{a} \bullet \mathbf{e}_k = \sum_{i=1}^n a_i \mathbf{e}_i \bullet \mathbf{e}_k = \sum_{i=1}^n a_i \delta_{ik} = a_k .$$

i.e.,

$$a_k = \mathbf{a} \bullet \mathbf{e}_k , \tag{3.138}$$

and thus

$$\mathbf{a} = \sum_{i=1}^{n} (\mathbf{a} \bullet \mathbf{e}_i)\mathbf{e}_i = \sum_{i=1}^{n} (\mathbf{e}_i\mathbf{e}_i) \bullet \mathbf{a} \quad , \tag{3.139}$$

It is worthwhile to analyze the parts of this sum.

$$(\mathbf{e}_i\mathbf{e}_i) \bullet \mathbf{a}$$

is the projection of $\mathbf{a}$ onto the direction given by $\mathbf{e}_i$. One may think of this *undetermined product* $\mathbf{e}_i\mathbf{e}_i$ (frequently called *dyadic product*) to be an operator which transforms $\mathbf{a}$ into another vector, namely the vector that results form projecting it on $\mathbf{e}_i$. Thus $\mathbf{e}_i\mathbf{e}_i$ is also called projection operator. The resulting vector is

$$\mathbf{a}_i = a_i\mathbf{e}_i = \mathbf{a} \bullet \mathbf{e}_i\mathbf{e}_i = \mathbf{e}_i\mathbf{e}_i \bullet \mathbf{a} \ . \tag{3.140}$$

Of course, the sum of all projections has to yield the vector itself, if the system of basis vectors is complete. In this case, the sum of all basis vectors has to result in the identity operator (identity tensor):

$$\sum_{i=1}^{n} \mathbf{e}_i\mathbf{e}_i = \underline{\mathbf{1}} \ . \tag{3.141}$$

This represents the *completeness relation*. Eqs. (3.139) and (3.141) are equivalent and both express completeness of the basis system. The dyadic product is an operator that has the form of a matrix when writing it in component form. The dyadic product $\mathbf{ab}$ has the components (matrix elements)

$$(\mathbf{ab})_{ik} = (a_ib_k)$$

The unit vectors of, *for example*, a three dimensional Cartesian coordinate system are

$$\mathbf{e}_1 = \langle 1, 0, 0 \rangle$$
$$\mathbf{e}_2 = \langle 0, 1, 0 \rangle \ .$$
$$\mathbf{e}_3 = \langle 0, 0, 1 \rangle$$

The three projection operators result thereof

$$\mathbf{e}_1\mathbf{e}_1 = \begin{bmatrix} 1 & 0 & 0 \\ 0 & 0 & 0 \\ 0 & 0 & 0 \end{bmatrix} \qquad \mathbf{e}_2\mathbf{e}_2 = \begin{bmatrix} 0 & 0 & 0 \\ 0 & 1 & 0 \\ 0 & 0 & 0 \end{bmatrix} \qquad \mathbf{e}_3\mathbf{e}_3 = \begin{bmatrix} 0 & 0 & 0 \\ 0 & 0 & 0 \\ 0 & 0 & 1 \end{bmatrix}$$

Their sum is indeed the identity operator

$$\sum_{i=1}^{3} \mathbf{e}_i\mathbf{e}_i = \begin{bmatrix} 1 & 0 & 0 \\ 0 & 1 & 0 \\ 0 & 0 & 1 \end{bmatrix} = (\delta_{ik}) = \underline{\mathbf{1}} \ .$$

Applied to a vector

$$\mathbf{a} = \langle a_1, a_2, a_3 \rangle$$

gives

$$e_1 e_1 \bullet a = \begin{bmatrix} 1 & 0 & 0 \\ 0 & 0 & 0 \\ 0 & 0 & 0 \end{bmatrix} \begin{bmatrix} a_1 \\ a_2 \\ a_3 \end{bmatrix} = \langle a_1, 0, 0 \rangle$$

which is the respective projection.

All this can be applied to functions treated as vectors of an infinite dimensional space. We first build $f(x)$ from a complete system of orthogonal functions. There are two cases, depending on whether the eigenvalues are discrete or continuous (or expressed in different words: whether they form a discrete or a continuous spectrum). Examples for both were given in Section 3.5.
We have therefore the following expansions for $f(x)$:

Discrete spectrum	Continuous spectrum

$$f(x) = \sum_{n=1}^{\infty} a_n \varphi_n(x) \qquad\qquad f(x) = \int a(k)\varphi(k;x)\,dk \qquad\qquad (3.142)$$

For the continuous case, the integration has to be over all possible values of k. The expansions are based on the orthogonality relations for the basis functions which are assumed to be normalized.

$$\int_c^d \varphi_n(x)\varphi_m^*(x)\,dx = \delta_{nm} \qquad \int_c^d \varphi(k;x)\varphi^*(k;x)\,dx = \delta(k-k') \quad (3.143)$$

where

$$\int_c^d f(x)\varphi_m^*(x)\,dx = \qquad\qquad \int_c^d f(x)\varphi^*(k;x)\,dx =$$

$$= \sum_{n=1}^{\infty} a_n \int_c^d \varphi_n(x)\varphi_m^*(x)\,dx \qquad = \iint_c^d a(k)\varphi(k;x)\varphi^*(k';x)\,dx\,dk$$

$$= \sum_{n=1}^{\infty} a_n \delta_{nm} = a_m \qquad\qquad = \int a(k)\delta(k-k')\,dk = a(k')$$

$$a_n = \int_c^d f(x)\varphi_n^*(x)\,dx \qquad\qquad a(k) = \int_c^d f(x)\varphi^*(k;x)\,dx$$
$$\qquad\qquad\qquad\qquad\qquad\qquad\qquad\qquad\qquad (3.144)$$
$$= \langle \varphi_n(x) | f(x) \rangle \qquad\qquad\qquad = \langle \varphi(k;x) | f(x) \rangle$$

This determines the coefficients of the expansion, the components of the vector so to speak, where relation (3.144) is entirely equivalent to (3.138). The projection onto the "direction" of $\varphi_n(x)$ and $\varphi(k;x)$, respectively is then

$$a_n\varphi_n(x) =$$

$$= \int_c^d \varphi_n(x)\varphi_n^*(x')f(x')dx'$$

$$= \varphi_n(x)\langle\varphi_n(x')|f(x')\rangle$$

$$a(k)\varphi(k;x) =$$

$$= \int_c^d \varphi(k;x)\varphi^*(k;x')f(x')dx'$$

$$= \varphi(k;x)\langle\varphi(k;x')|f(x')\rangle$$

For completeness, the sum of all those projections has to result in the function itself, *i.e.*,

$$f(x) = \sum_{n=1}^{\infty} a_n\varphi_n(x) =$$

$$= \int_c^d \sum_{n=1}^{\infty} \varphi_n(x)\varphi_n^*(x')f(x')dx'$$

$$f(x) = \int a(k)\varphi(k;x)dk =$$

$$= \int\int_c^d \varphi(k;x)\varphi^*(k;x')dk \cdot f(x')dx' .$$

At the same time, it also is

$$f(x) = \int_c^d \delta(x-x')f(x')dx'$$

By comparison, we obtain the completeness relations

$$\sum_{n=1}^{\infty} \varphi_n(x)\varphi_n^*(x') = \delta(x-x') \quad\bigg|\quad \int\varphi(k;x)\varphi^*(k;x')dk = \delta(x-x') \qquad (3.145)$$

On the other hand, to calculate the expansion coefficients by means of the completeness relation is also possible. First, one may apply the identity operator, that is in our case basically the δ-function, to any function. So we start with

$$f(x) = \int_c^d \delta(x-x')f(x')dx'$$

Substituting gives

$$f(x) = \int_c^d \sum_{n=1}^{\infty} \varphi_n(x)\varphi_n^*(x')f(x')dx' \quad\bigg|\quad f(x) = \int_c^d \int\varphi(k;x)\varphi^*(k;x')dkf(x')dx'$$

what immediately leads to the coefficients of the expansion

$$a_n = \int_c^d \varphi_n^*(x')f(x')dx' \quad\bigg|\quad a(k) = \int_c^d \varphi^*(k;x')f(x')dx'$$

Let us now consider two functions $f(x)$ and $g(x)$ whose scalar product we aim to find. $f(x)$ shall have the components

$$a_n \quad\bigg|\quad a(k),$$

while the components of $g(x)$ shall be

$$b_n \quad \bigg| \quad b(k)\,.$$

Then

$$f(x) = \sum_{n=1}^{\infty} a_n \varphi_n(x) \qquad\qquad f(x) = \int a(k)\varphi(k;x)\,dk$$

$$g(x) = \sum_{n=1}^{\infty} b_n \varphi_n(x) \qquad\qquad g(x) = \int b(k)\varphi(k;x)\,dk$$

Therefore

$$\langle g|f \rangle = \int_c^d f(x)g^*(x)\,dx \;= \qquad\qquad \langle g|f \rangle = \int_c^d f(x)g^*(x)\,dx \;=$$

$$= \sum_{n,\,m=1}^{\infty} \int_c^d a_n b_m^* \varphi_n(x)\varphi_m^*(x)\,dx \qquad = \iiint_c a(k)b^*(k')\varphi(k;x)\varphi^*(k';x)\,dx\,dk\,dk'$$

$$= \sum_{n=1}^{\infty} a_n b_m^* \delta_{nm} \qquad\qquad = \iint a(k)b^*(k')\delta(k-k')\,dk\,dk'$$

$$= \sum_{n=1}^{\infty} a_n b_n^* \qquad\qquad\qquad = \int a(k)b^*(k)\,dk \qquad\qquad (3.146)$$

which is in complete analogy to Vector Calculus, especially for the case of the discrete spectrum.

In closing this section we shall provide a few explanatory words on integral operators. We have already met some of those integral operators when studying the various Green's functions in the previous Section 3.5. They cause a transformation of a function $f(x)$ into another function $\tilde{f}(x)$ in the form of some integral transform:

$$\tilde{f}(x) = \int_c^d O(x, x')f(x')\,dx'\,. \qquad\qquad (3.147)$$

The δ-function is just one such integral operator. It projects a function onto itself. The image function $\tilde{f}(x)$ is then equal to $f(x)$

$$\tilde{f}(x) = \int_c^d \delta(x-x')f(x')\,dx' = f(x)$$

The δ-function thus acts as the identity operator. The function $O(x, x')$ is called the kernel of the integral transform. It is also possible to expand these kernels, *for example*, for the discrete case we have

$$O(x, x') = \sum_{i,k=1}^{\infty} O_{ik}\varphi_i(x)\varphi_k^*(x') \quad .$$ (3.148)

We obtain

$$\int_c^d\int_c^d O(x, x')\varphi_{i'}^*(x)\varphi_{k'}(x')dx\,dx' =$$

$$= \sum_{i,k=1}^{\infty} O_{ik}\int_c^d\varphi_i(x)\varphi_{i'}^*(x)dx\int_c^d\varphi_k^*(x')\varphi_{k'}(x')dx'$$

$$= \sum_{i,k=1}^{\infty} O_{ik}\,\delta_{ii'}\delta_{kk'} = O_{i'k'}$$

i.e.,

$$O_{ik} = \int_c^d\int_c^d O(x, x')\varphi_i^*(x)\varphi_k(x')dx\,dx'$$ (3.149)

Therefore, it is on one hand

$$\tilde{f}(x) = \int_c^d O(x, x')f(x')dx' =$$

$$= \int_c^d\left[\sum_{i,k=1}^{\infty} O_{ik}\,\varphi_i(x)\varphi_k^*(x')\right]\left[\sum_{l=1}^{\infty} a_l\varphi_l(x')\right]dx'$$

$$= \sum_{i,k,l}^{\infty} O_{ik}\,\varphi_i(x)a_l\delta_{kl} = \sum_{i,k}^{\infty} O_{ik}\,a_k\varphi_i(x)$$

and on the other, we can expand $\tilde{f}(x)$

$$\tilde{f}(x) = \sum_i \tilde{a}_i\varphi_i(x) \quad .$$

The components of $\tilde{f}(x)$ result from f in the following way:

$$\tilde{a}_i = \sum_k O_{ik}\,a_i \,,$$ (3.150)

that is by matrix multiplication, again in total harmony with standard vector calculus. The situation for the continuous spectrum is quite similar.

$$O(x, x') = \iint O(k, k')\varphi(k;x)\varphi^*(k';x')dk\,dk'$$ (3.151)

where

$$O(k, k') = \int_c^d \int_c^d O(x, x') \varphi^*(k;x) \varphi(k';x') dx\, dx' \tag{3.152}$$

and the projection of $f(x)$ onto $\tilde{f}(x)$ is performed by

$$\tilde{a}(k) = \int O(k, k') a(k') dk' , \tag{3.153}$$

so that we also obtain an integral transformation for the continuous case. Integral transformations are the continuous analogues of matrix multiplications.

3.7 Separation of Variables of Laplace's Equation in Cylindrical Coordinates

3.7.1 Separation of Variables

Laplace's equation for the potential F is given by eq. (3.33)

$$\left(\frac{1}{r} \frac{\partial}{\partial r} r \frac{\partial}{\partial r} + \frac{1}{r^2} \frac{\partial^2}{\partial \varphi^2} + \frac{\partial^2}{\partial z^2} \right) F = 0 . \tag{3.154}$$

To find its solution, the following separation Ansatz shall be used

$$F(r, \varphi, z) = R(r) \phi(\varphi) Z(z) \tag{3.155}$$

to obtain

$$\frac{1}{R(r)} \frac{1}{r} \frac{\partial}{\partial r} r \frac{\partial}{\partial r} R(r) + \frac{1}{r^2 \phi(\varphi)} \frac{\partial^2}{\partial \varphi^2} \phi(\varphi) + \frac{1}{Z(z)} \frac{\partial^2}{\partial z^2} Z(z) = 0 . \tag{3.156}$$

The first two terms solely depend on r and φ, the third depends only on z. This allows us to introduce the separation constant k^2

$$\frac{1}{Z(z)} \frac{\partial^2}{\partial z^2} Z(z) = k^2 . \tag{3.157}$$

There are several representations to express the result, *for example,*

$$Z = A_1 \cosh kz + A_2 \sinh kz \tag{3.158}$$

or

$$Z = \tilde{A}_1 \exp(kz) + \tilde{A}_2 \exp(-kz) . \tag{3.159}$$

Now we tackle the r-φ dependent part of eq. (3.156) which takes the form

$$\frac{1}{R(r)} \frac{1}{r} \frac{\partial}{\partial r} r \frac{\partial}{\partial r} R(r) + \frac{1}{r^2 \phi(\varphi)} \frac{\partial^2}{\partial \varphi^2} \phi(\varphi) + k^2 = 0$$

Multiplying by r^2 gives

$$\frac{r}{R(r)} \frac{\partial}{\partial r} r \frac{\partial}{\partial r} R(r) + k^2 r^2 + \frac{1}{\phi(\varphi)} \frac{\partial^2}{\partial \varphi^2} \phi(\varphi) = 0 .$$

This allows to continue the separation process. Substitute

$$\frac{1}{\phi}\frac{\partial^2}{\partial\varphi^2}\phi \; = \; -m^2 \; , \tag{3.160}$$

and obtain

$$\phi \; = \; B_1\cos m\varphi + B_2\sin m\varphi \; , \tag{3.161}$$

or alternatively

$$\phi \; = \; \tilde{B}_1\exp(im\varphi) + \tilde{B}_2\exp(-im\varphi) \; . \tag{3.162}$$

Where m has to be a whole number because of the necessary periodicity of φ. What remains for R is

$$r\frac{\partial}{\partial r}r\frac{\partial}{\partial r}R(r) + (k^2r^2 - m^2)R \; = \; 0$$

After dividing by k^2r^2, and introducing the dimensionless coordinate

$$\xi \; = \; kr \; , \text{ this becomes} \tag{3.163}$$

$$\boxed{\frac{1}{\xi}\frac{\partial}{\partial\xi}\xi\frac{\partial}{\partial\xi}R(\xi) + \left(1 - \frac{m^2}{\xi^2}\right)R(\xi) \; = \; 0} \; . \tag{3.164}$$

This is one of the most famous equations in mathematical physics, the so-called *Bessel differential equation*. Its solution is a linear combination of two different, linearly independent functions, the so-called *cylindrical functions*, which may be chosen in various ways. Such a pair of functions consists of *for example*, the so-called *Bessel function* $J_m(\xi) \; = \; J_m(kr)$ and the so-called *Neumann function* $N_m(\xi) \; = \; N_m(kr)$, that is,

$$\boxed{R(r) \; = \; C_1J_m(kr) + C_2N_m(kr)} \; . \tag{3.165}$$

Another possibility is to separate the equations in a different manner by substituting:

$$\frac{1}{Z(z)}\frac{\partial^2}{\partial z^2}Z(z) \; = \; -k^2 \; , \tag{3.166}$$

so that

$$Z \; = \; A_1\cos kz + A_2\sin kz \tag{3.167}$$

or

$$Z \; = \; \tilde{A}_1\exp(ikz) + \tilde{A}_2\exp(-ikz) \; . \tag{3.168}$$

This, together with the dimensionless variable

$$\eta \; = \; ikr \; , \tag{3.169}$$

yields for R again the Bessel differential equation

$$\frac{1}{\eta}\frac{\partial}{\partial\eta}\eta\frac{\partial}{\partial\eta}R(\eta) + \left(1 - \frac{m^2}{\eta^2}\right)R(\eta) \; = \; 0$$

and its solution is

$$\boxed{R(r) \; = \; C_1J_m(ikr) + C_2N_m(ikr)} \; . \tag{3.170}$$

It depends on the type of problem to determine which kind of separation is more appropriate. We will discuss this in the context of several examples. Both approaches are equivalent and it is possible to transpose from one to the other by substituting k by ik and vice versa. If the problem is independent of z, then $k = 0$. This special case leads to fundamental functions $R(r)$, which we will discuss in Section 3.7.3.5 (see eqs. (3.263) and (3.264)).

The functions $J_m(kr)$ and $N_m(kr)$ have significantly different properties than the functions $J_m(ikr)$ and $N_m(ikr)$. The argument (ikr) occurs so frequently that the so-called modified Bessel functions were introduced. The *modified Bessel functions of the first kind* are defined by

$$I_m(x) = i^{-m} J_m(ix)$$ (3.171)

and the *modified Bessel functions of the second kind* are defined by

$$K_m(x) = \frac{\pi}{2} i^{m+1} [J_m(ix) + i N_m(ix)]$$ (3.172)

This makes it therefore possible to also write the general solution (3.170) in the following way:

$$\boxed{R(r) = \check{C}_1 I_m(kr) + \check{C}_2 K_m(kr)}$$ (3.173)

It shall also be noted that there are specific problems where the separation constants might have to be chosen differently, *e.g.*, if one wants to specify the potential at a certain surfaces to be $\varphi = \text{const}$. We will not discuss this any further, but just note that in this case, m will not be a whole number.

3.7.2   Some Properties of Cylindrical Functions

Here we can only sketch some of the most important characteristics of the cylindrical functions. Properties of those functions are found in suitable reference books [3 - 7].

For small arguments $J_m(x)$ behaves like x^m,

$$J_m(x) \approx \left(\frac{x}{2} \right)^m \frac{1}{m!} \qquad \text{for } |x| \ll 1 \,,$$ (3.174)

while $N_m(x)$ diverges for small arguments, namely

$$N_m(x) \approx -\frac{(m-1)!}{\pi} \left(\frac{2}{x} \right)^m \qquad \text{for } |x| \ll 1 \quad \text{and} \quad m = 1, 2, \dots \,,$$ (3.175)

$$N_0(x) \approx \frac{2}{\pi} \ln \frac{\gamma x}{2} \approx \frac{2}{\pi} \ln x \qquad \text{for } |x| \ll 1, \quad (\gamma \approx 1.781) \,.$$ (3.176)

The graph of some of those functions is illustrated in Figures 3.12 through 3.15.

For very large arguments (asymptotic behavior), J_m and N_m behave basically like damped trigonometric functions, that is

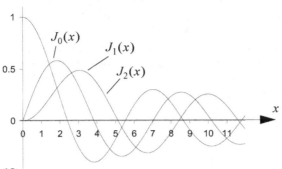

Fig. 3.12

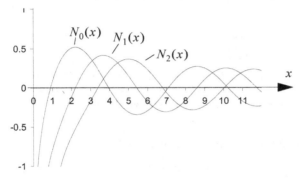

Fig. 3.13

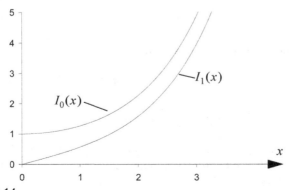

Fig. 3.14

$$J_m(x) \approx \sqrt{\frac{2}{\pi x}} \cos\left(x - \frac{\pi}{4} - \frac{m\pi}{2}\right)$$
$$N_m(x) \approx \sqrt{\frac{2}{\pi x}} \sin\left(x - \frac{\pi}{4} - \frac{m\pi}{2}\right)$$
$$\left.\vphantom{\begin{array}{c}1\\1\\1\end{array}}\right\} \qquad \text{for } x \to \infty \ . \qquad (3.177)$$

For small arguments the modified Bessel functions behave somewhat like J_m and N_m. Here we have

$$I_m(x) \approx \left(\frac{x}{2}\right)^m \frac{1}{m!} \qquad \text{for } 0 < x \ll 1 \ , \qquad (3.178)$$

$$K_m(x) \approx \frac{(m-1)!}{2}\left(\frac{2}{x}\right)^m \qquad \text{for } 0 < x \le 1 \text{ and } m = 1, 2, \dots, \ , \qquad (3.179)$$

$$K_0(x) \approx -\ln\frac{\gamma x}{2} \approx -\ln x \qquad \text{for } 0 < x \le 1, \ (\gamma \approx 1.781) \ . \qquad (3.180)$$

For large arguments, however, they differ significantly from J_m and N_m where they behave like exponential functions, namely

$$I_m(x) \approx \frac{\exp(x)}{\sqrt{2\pi x}}$$
$$K_m(x) \approx \frac{\sqrt{\pi}\exp(-x)}{\sqrt{2x}}$$
$$\left.\vphantom{\begin{array}{c}1\\1\\1\end{array}}\right\} \qquad \text{for } x \to \infty \ . \qquad (3.181)$$

The different asymptotic behavior for large arguments of J_m, N_m (like trigonometric functions) and I_m, K_m (like exponential functions) is important and will be significant in the following examples. It will also be important that J_m and N_m are finite at the origin, while I_m and K_m diverge there. In closing, a few important relations for cylindrical functions are summarized:

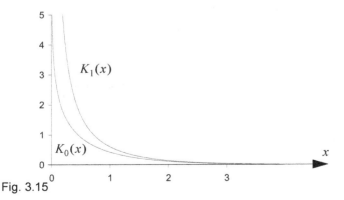

Fig. 3.15

Bessel function:	Neumann function:
$J_{n-1}(x) + J_{n+1}(x) = \dfrac{2n}{x}J_n(x)$	$N_{n-1}(x) + N_{n+1}(x) = \dfrac{2n}{x}N_n(x)$
$J_{n-1}(x) - J_{n+1}(x) = 2J_n{}'(x)$	$N_{n-1}(x) - N_{n+1}(x) = 2N_n{}'(x)$
$J_{-n}(x) = (-1)^n J_n(x)$	$N_{-n}(x) = (-1)^n N_n(x)$
$\dfrac{\mathrm{d}}{\mathrm{d}x}J_n(x) = J_{n-1}(x) - \dfrac{n}{x}J_n(x)$	$\dfrac{\mathrm{d}}{\mathrm{d}x}N_n(x) = N_{n-1}(x) - \dfrac{n}{x}N_n(x)$
$\int x^{n+1}J_n(x)dx = x^{n+1}J_{n+1}(x)$	$\int x^{n+1}N_n(x)dx = x^{n+1}N_{n+1}(x)$
$\int x^{-n+1}J_n(x)dx = -x^{-n+1}J_{n-1}(x)$	$\int x^{-n+1}N_n(x)dx = -x^{-n+1}N_{n-1}(x)$
Specifically: $J_0{}'(x) = -J_1(x)$	Specifically: $N_0{}'(x) = -N_1(x)$

and $J_n(x)N_{n+1}(x) - J_{n+1}(x)N_n(x) = -\dfrac{2}{\pi x}$

Modified Bessel function of the first kind:	Modified Bessel function of the second kind
$I_{n-1}(x) + I_{n+1}(x) = 2I_n{}'(x)$	$K_{n-1}(x) + K_{n+1}(x) = -2K_n{}'(x)$
$I_{n-1}(x) - I_{n+1}(x) = \dfrac{2n}{x}I_n(x)$	$-K_{n-1}(x) + K_{n+1}(x) = \dfrac{2n}{x}K_n(x)$
$I_{-n}(x) = I_n(x)$	$K_{-n}(x) = K_n(x)$
$\dfrac{\mathrm{d}}{\mathrm{d}x}I_n(x) = I_{n-1}(x) - \dfrac{n}{x}I_n(x)$	$\dfrac{\mathrm{d}}{\mathrm{d}x}K_n(x) = -K_{n-1}(x) - \dfrac{n}{x}K_n(x)$
$\int x^{n+1}I_n(x)dx = x^{n+1}I_{n+1}(x)$	$\int x^{n+1}K_n(x)dx = -x^{n+1}K_{n+1}(x)$
$\int x^{-n+1}I_n(x)dx = x^{-n+1}I_{n-1}(x)$	$\int x^{-n+1}K_n(x)dx = -x^{-n+1}K_{n-1}(x)$
Specifically: $I_0{}'(x) = I_1(x)$	Specifically: $K_0{}'(x) = -K_1(x)$

and $K_n(x)I_{n+1}(x) + K_{n+1}(x)I_n(x) = \dfrac{1}{x}$

3.7.3 Examples

3.7.3.1 A Cylinder with Surface Charges

An infinitely tall circular cylinder of radius r_0 (Fig. 3.16) carries a rotationally symmetric surface charge, mirrored at the x-y axis ($z = 0$), that is

$$\sigma(z) = \sigma(-z) . \tag{3.182}$$

We want to find its potential. This requires us to solve Laplace's equation for region 1 (inside the cylinder surface) and for region 2 (outside the cylinder surface). Both solutions will be joined together via the boundary conditions. There is no azimuthal dependency because of the rotational symmetry. Consequently m, one of the separation constants vanishes

$$m = 0 . \tag{3.183}$$

This leaves a dependency on z and r, for which we choose the Ansatz according to (3.167) and (3.173):

$$\left.\begin{array}{l} Z = A_1 \cos kz + A_2 \sin kz \\ R = C_1 I_0(kr) + C_2 K_0(kr) \end{array}\right\} \tag{3.184}$$

The cosine sum suffices for this problem because of the symmetry ($A_2 = 0$). There are no restrictions on k. In region 1, C_2 has to vanish because of the divergence of K_0 at the origin, which would cause the potential to diverge. Conversely, in region 2, C_1 has to vanish because I_0 diverges at infinity. This allows for an Ansatz like this:

$$\left.\begin{array}{l} \varphi_1 = \int_0^\infty f_1(k) \cos(kz) \cdot I_0(kr) dk \\[2mm] \varphi_2 = \int_0^\infty f_2(k) \cos(kz) \cdot K_0(kr) dk \end{array}\right\} \tag{3.185}$$

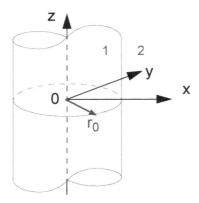

Fig. 3.16

On the surface $r = r_0$ it has to be

$$(E_{z1})_{r = r_0} = (E_{z2})_{r = r_0}$$

that is

$$\left(\frac{\partial \varphi_1}{\partial z}\right)_{r = r_0} - \left(\frac{\partial \varphi_2}{\partial z}\right)_{r = r_0} = 0 \tag{3.186}$$

and

$$(D_{r2})_{r = r_0} - (D_{r1})_{r = r_0} = \sigma$$

that is

$$\left(\frac{\partial \varphi_1}{\partial r}\right)_{r = r_0} - \left(\frac{\partial \varphi_2}{\partial r}\right)_{r = r_0} = \frac{\sigma}{\varepsilon_0} \ . \tag{3.187}$$

The condition (3.186) gives

$$\int_0^\infty k \sin(kz)[f_2(k)K_0(kr_0) - f_1(k)I_0(kr_0)]dk = 0 \ , \tag{3.188}$$

and the condition (3.187), when using

$$I_0{}'(\xi) = \frac{\mathrm{d}}{\mathrm{d}\xi}I_0(\xi) = +I_1(\xi) \tag{3.189}$$

and

$$K_0{}'(\xi) = \frac{\mathrm{d}}{\mathrm{d}\xi}K_0(\xi) = -K_1(\xi) \tag{3.190}$$

gives

$$\int_0^\infty k \cos(kz)[f_1(k)I_1(kr_0) + f_2(k)K_1(kr_0)]dk = \frac{\sigma}{\varepsilon_0} \ . \tag{3.191}$$

To continue, we need the Fourier transform of $\sigma(z)$,

$$\sigma(z) = \int_0^\infty \tilde{\sigma}(z)\cos(kz)dk \ . \tag{3.192}$$

It follows from (3.188) that

$$f_1(k)I_0(kr_0) - f_2(k)K_0(kr_0) = 0 \ ,$$

and by (3.191), using (3.192) gives

$$f_1(k)I_1(kr_0) + f_2(k)K_1(kr_0) - \frac{\tilde{\sigma}(k)}{\varepsilon_0 k} = 0 \ ,$$

which yields f_1 and f_2:

$$f_1(k) = \frac{\tilde{\sigma}(k)K_0(kr_0)}{\varepsilon_0 k[K_0(kr_0)I_1(kr_0) + K_1(kr_0)I_0(kr_0)]}$$

$$f_2(k) = \frac{\tilde{\sigma}(k)I_0(kr_0)}{\varepsilon_0 k[K_0(kr_0)I_1(kr_0) + K_1(kr_0)I_0(kr_0)]} \ .$$

By means of the relation (see Section 3.7.2)

$$K_0(kr_0)I_1(kr_0) + K_1(kr_0)I_0(kr_0) = \frac{1}{kr_0} .$$

(3.193)

this simplifies to

$$f_1(k) = \frac{\tilde{\sigma}(k)r_0 K_0(kr_0)}{\varepsilon_0}$$

$$f_2(k) = \frac{\tilde{\sigma}(k)r_0 I_0(kr_0)}{\varepsilon_0}$$

(3.194)

This solves our problem:

$$\varphi_1 = \frac{r_0}{\varepsilon_0}\int_0^\infty \tilde{\sigma}(k)\cos(kz)K_0(kr_0)I_0(kr)dk$$

$$\varphi_2 = \frac{r_0}{\varepsilon_0}\int_0^\infty \tilde{\sigma}(k)\cos(kz)I_0(kr_0)K_0(kr)dk$$

(3.195)

To continue further is generally only possible with numerical methods. The very general result contains some interesting special cases. *For example*, consider the case of a uniformly charged circular loop for which

$$\sigma(z) = q\delta(z) .$$

(3.196)

Where q is its line charge with units *[C/m]*. We use the orthogonality relation (3.107) to calculate σ

$$\int_{-\infty}^\infty \sigma(z)\cos(k'z)dz = \int_{-\infty}^\infty \int_0^\infty \tilde{\sigma}(k)\cos(kz)\cos(k'z)dkdz$$

$$= \pi\int_0^\infty \tilde{\sigma}(k)\delta(k-k')dk = \pi\tilde{\sigma}(k') ,$$

that is

$$\tilde{\sigma}(k) = \frac{1}{\pi}\int_{-\infty}^\infty \sigma(z)\cos(kz)dz$$

$$= \frac{1}{\pi}\int_{-\infty}^\infty q\delta(z)\cos(kz)dz = \frac{q}{\pi}\cos 0 = \frac{q}{\pi}$$

(3.197)

and

$$q\delta(z) = \frac{q}{\pi}\int_0^\infty \cos(kz)dk ,$$

or

$$\delta(z) = \frac{1}{\pi}\int_0^\infty \cos(kz)dk .$$

(3.198)

This reveals another important fact about the δ-function: Its Fourier transform is a constant (*i.e.*, its spectrum contains all frequencies with the same amplitude). This is expressed in eq. (3.198), which represents at the same time, both, an interesting and an important representation of the δ-function. Eq. (3.197) in (3.195) allows to represent the potential of the uniformly charged circular loop as:

$$
\left.
\begin{aligned}
\varphi_1 &= \frac{qr_0}{\pi\varepsilon_0} \int_0^\infty \cos(kz) K_0(kr_0) I_0(kr)\,dk \\[2mm]
\varphi_2 &= \frac{qr_0}{\pi\varepsilon_0} \int_0^\infty \cos(kz) I_0(kr_0) K_0(kr)\,dk
\end{aligned}
\right\}. \tag{3.199}
$$

Of course, to calculate the potential of the circular loop by much simpler means is also possible. Integrating according to (2.29) gives.

$$
\varphi = \frac{qr_0}{2\pi\varepsilon_0} \frac{l}{\sqrt{rr_0}} K\!\left(\frac{\pi}{2}, l\right), \tag{3.200}
$$

where $K\!\left(\frac{\pi}{2}, l\right)$ represents the total elliptic integral of the first kind:

$$
K\!\left(\frac{\pi}{2}, l\right) = \int_0^{\pi/2} \frac{d\psi}{\sqrt{1 - l^2\sin^2\psi}}, \tag{3.201}
$$

and

$$
l^2 = \frac{4rr_0}{r^2 + z^2 + r_0^2 + 2rr_0}. \tag{3.202}
$$

In fact, one can prove that the results in (3.199) and (3.200) are identical. One needs to show that the Fourier transform of $K_0 I_0$ basically yields the elliptic integral of the first kind. For more details refer to [5], volume 1, p 49, eq. (46).

One might further specialize this problem. For $r_0 \to 0$ and $q \to \infty$, in such a way that $2\pi r_0 q = Q$. Then the result has to be again that of a point charge at the origin. (This has to be the potential of φ_2 while φ_1 becomes irrelevant):

$$
\varphi = \frac{Q}{2\pi^2\varepsilon_0} \int_0^\infty \cos(kz) K_0(kr)\,dk. \tag{3.203}
$$

On the other hand, this must be

$$
\varphi = \frac{Q}{4\pi\varepsilon_0\sqrt{r^2 + z^2}},
$$

so that

$$
\boxed{\frac{1}{\sqrt{r^2 + z^2}} = \frac{2}{\pi} \int_0^\infty \cos(kz) K_0(kr)\,dk}. \tag{3.204}
$$

The two relations (3.203) and (3.204) represent the series expansion for the inverse distance in cylindrical coordinates and correspond to the similar equations (3.111) and (3.112). In both cases, the calculation of a simple, known problem by means of a complicated approach is not a superficial luxury. Rather, sometimes to solve particular problems, requires one to have certain expansions at ones disposal. The next example shall demonstrate this.

3.7.3.2 Point Charge on the Axis of a Dielectric Cylinder

We want to find the potential caused by a point charge Q at the axis of a dielectric, circular cylinder (ε_1). For the remaining space, we assume a different dielectric (ε_2) (see Fig. 3.17).
This is by no means a trivial problem, but it can be solved with the results obtained from the previous example. We need to solve Laplace's equation in region 2, what is certainly possible with an approach similar to (3.185):

$$\varphi_2 = \frac{Q}{2\pi^2\varepsilon_1}\int_0^\infty \cos(kz)K_0(kr)g_2(k)dk \ . \tag{3.205}$$

The factor in front of the integral is just for convenience and could as well be regarded as belonging to $g_2(k)$. In region 1, we have to solve Poisson's equation for a point charge. This may be done by superposition of the point charge potential according to (3.203) and the general solution of Laplace's equation according to (3.185), *i.e.*, by

$$\varphi_1 = \frac{Q}{2\pi^2\varepsilon_1}\int_0^\infty \cos(kz)[K_0(kr) + I_0(kr)g_1(k)]dk \tag{3.206}$$

Remember that the general solution of the inhomogeneous (Poisson) equation is obtained from the superposition of the specific solution of the inhomogeneous equation and the general solution of the homogeneous (Laplace) equation. Significant is now, to know the potential of the point charge in the specific form that fits this problem. Another way to view the trial functions of (3.205) and (3.206) is by superposition of the point charge potential and the potential of bound surface charges at the cylinder surface (wall). The following boundary conditions apply for $r = r_0$:

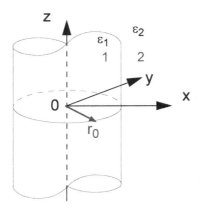

Fig. 3.17

$$\left(\frac{\partial \varphi_1}{\partial z}\right)_{r=r_0} - \left(\frac{\partial \varphi_2}{\partial z}\right)_{r=r_0} = 0$$

$$\varepsilon_1\left(\frac{\partial \varphi_1}{\partial r}\right)_{r=r_0} - \varepsilon_2\left(\frac{\partial \varphi_2}{\partial r}\right)_{r=r_0} = 0 .$$

This gives

$$\frac{Q}{2\pi^2\varepsilon_1}\int_0^\infty k\sin(kz)[K_0(kr_0) + I_0(kr_0)g_1(k) - K_0(kr_0)g_2(k)]dk = 0$$

and

$$\frac{Q}{2\pi^2\varepsilon_1}\int_0^\infty k\cos(kz)[-\varepsilon_1 K_1(kr_0) + \varepsilon_1 I_1(kr_0)g_1(k) + \varepsilon_2 K_1(kr_0)g_2(k)]dk = 0$$

or

$$K_0(kr_0) + I_0(kr_0)g_1(k) - K_0(kr_0)g_2(k) = 0$$

$$-\varepsilon_1 K_1(kr_0) + \varepsilon_1 I_1(kr_0)g_1(k) + \varepsilon_2 K_1(kr_0)g_2(k) = 0 .$$

Solved for g_1 and g_2 yields

$$\left.\begin{array}{l}
g_1(k) = \dfrac{\left(1 - \dfrac{\varepsilon_2}{\varepsilon_1}\right)kr_0 K_0(kr_0)K_1(kr_0)}{1 + \left(\dfrac{\varepsilon_2}{\varepsilon_1} - 1\right)kr_0 I_0(kr_0)K_1(kr_0)} \\[3em]
g_2(k) = \dfrac{1}{1 + \left(\dfrac{\varepsilon_2}{\varepsilon_1} - 1\right)kr_0 I_0(kr_0)K_1(kr_0)}
\end{array}\right\} \qquad (3.207)$$

Hereby we used equations (3.189), (3.190), and (3.193). For $\varepsilon_1 = \varepsilon_2$, we find that $g_1 = 0$ and $g_2 = 1$, that is, $\varphi_1 = \varphi_2 = \varphi$, agreeing with (3.203), as expected.

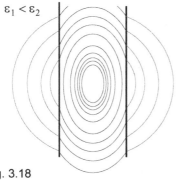

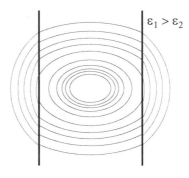

Fig. 3.18   Fig. 3.19

The solution obtained by substituting (3.207) in (3.205) and (3.206) needs to be solved numerically. Qualitative illustrations are presented in Fig. 3.18 ($\varepsilon_1 < \varepsilon_2$) and Fig. 3.19 ($\varepsilon_1 > \varepsilon_2$). Those figures compare to Fig. 2.55 and Fig. 2.56, whereby it is to note that field lines are shown there, while the equipotential surfaces are drawn here.

3.7.3.3 Dirichlet's Boundary Value Problem and the Fourier-Bessel Series

Consider the cylinder shown in Fig. 3.20 with radius r_0 and height h. We want to find the potential inside the charge-free cylinder with the following boundary conditions.

$$\varphi = \varphi_h(r) \qquad \text{for } z = h$$

$$\varphi = 0 \qquad \text{on the remaining surface .}$$

This is the cylindrical analogue to the example in Section 3.5.2.1. Again, $m = 0$ because of the rotational symmetry. For the radial part R, we may choose either J_0 or I_0, but not N_0 or K_0 since those let the potential at the axis diverge. The potential has to also vanish for $r = r_0$. J_0 has zeros for real arguments, I_0 does not. The convenient choice is thus J_0, For the z-dependency, we may choose an approach, *for example*, based on (3.158) where only the *sinh* is an option because of the restriction $\varphi = 0$ for $z = 0$. We suggest that solving the problem is possible by superposing expressions of the form:

$$J_0(kr)\sinh kz .$$

Hereby, the condition

$$J_0(kr_0) = 0 \tag{3.208}$$

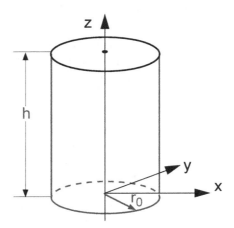

Fig. 3.20

has to be fulfilled. J_0 has an infinite number of zeros, like a trigonometric function, which it approaches for large arguments. Suppose the zeros are at λ_{0n} then

$$J_0(\lambda_{0n}) = 0 \qquad n = 1, 2, \ldots \tag{3.209}$$

where

$$\lambda_{01} < \lambda_{02} < \lambda_{03} < \ldots , \tag{3.210}$$

and

$$kr_0 = \lambda_{0n}$$

or

$$k = k_n = \frac{\lambda_{0n}}{r_0} . \tag{3.211}$$

This determines the eigenvalues of our problem. Note the similarity to the example of Section 3.5.2.1, in particular eq. (3.69). This determines the Ansatz for this problem:

$$\varphi = \sum_{n=1}^{\infty} C_n J_0(k_n r) \sinh(k_n z) . \tag{3.212}$$

This introduces us to a new kind of series expansions, $i.e.$, expansions by functions $J_0(k_n r)$, which are typical for cylindrical problems, just as Fourier series are for Cartesian problems. This is known as the Fourier-Bessel series – a fitting name, which simultaneously expresses its similarity to Fourier series and its relation to Cylindrical problems.

Before continuing to solve this problem, we would like to discuss Fourier-Bessel series some more. They are generally understood as being series expansions of the function $f(x)$, defined within the interval $0 \le x \le 1$ and are of the form

$$\boxed{f(x) = \sum_{n=1}^{\infty} C_n J_m(\lambda_{mn} x)} , \tag{3.213}$$

where λ_{mn} is the nth zero of J_m. Such expansions are useful only if the functions $J_m(\lambda_{mn} x)$ form a complete system within this interval. This is the case. They are also orthogonal to each other in the following sense

$$\boxed{\int_0^1 x J_m(\lambda_{mn} x) J_m(\lambda_{mn'} x) dx = \frac{1}{2} [J_m'(\lambda_{mn})]^2 \delta_{nn'}} . \tag{3.214}$$

Strictly speaking, the functions $\sqrt{x} J_m(\lambda_{mn} x)$ are orthogonal to each other. This can also be expressed in the following form: In the interval $0 \le x \le 1$, the functions $J_m(\lambda_{mn} x)$ are orthogonal with the weight function x. Using eq. (3.214) enables us to write the coefficients C_n of the expansion given in eq. (3.213):

$$\int_0^1 f(x)xJ_m(\lambda_{mn'}x)dx \;=\; \sum_{n=1}^{\infty} C_n \int_0^1 xJ_m(\lambda_{mn}x)J_m(\lambda_{mn'}x)dx$$

$$= \sum_{n=1}^{\infty} C_n \frac{1}{2}[J_m{'}(\lambda_{mn})]^2 \delta_{nn'}$$

$$= C_{n'}\frac{1}{2}[J_m{'}(\lambda_{mn'})]^2$$

so that

$$C_n = \frac{2\int_0^1 xf(x)J_m(\lambda_{mn}x)dx}{[J_m{'}(\lambda_{mn})]^2} .$$ (3.215)

Substituting C_n in the expansion (3.213) and comparison with $f(x) = \int \delta(x-x')f(x')dx'$ yields the completeness relation. In this case we have

$$\sum_{n=1}^{\infty} \frac{2x'J_m(\lambda_{mn}x)J_m(\lambda_{mn}x')}{[J_m{'}(\lambda_{mn})]^2} = \delta(x-x') .$$ (3.216)

With this result, we return to our example, *i.e.* the statement in eq. (3.212). It fulfills all boundary conditions, except the one at $z = h$. There, it shall be $\varphi = \varphi_h(r)$, thus

$$\varphi_h(r) = \sum_{n=1}^{\infty} C_n J_0(k_n r)\sinh(k_n h)$$

$$= \sum_{n=1}^{\infty} C_n J_0\!\left(\lambda_{0n}\frac{r}{r_0}\right)\sinh\!\left(\lambda_{0n}\frac{h}{r_0}\right)$$

and substituting

$$\xi = \frac{r}{r_0}$$

gives

$$\varphi_h(r_0\xi) = \sum_{n=1}^{\infty} C_n \sinh\!\left(\lambda_{0n}\frac{h}{r_0}\right)J_0(\lambda_{0n}\xi)$$

where $0 \le r \le r_0$, *i.e.*, $0 \le \xi \le 1$. By eq. (3.215) it is now

$$C_n \sinh\!\left(\lambda_{0n}\frac{h}{r_0}\right) = \frac{2\int_0^1 \xi'J_0(\lambda_{0n}\xi')\varphi_h(r_0\xi')d\xi'}{[J_0{'}(\lambda_{0n})]^2}$$

Since

$$J_0{'} = -J_1 ,$$ (3.217)

we finally obtain

$$\varphi(\xi, z) = \sum_{n=1}^{\infty} \frac{2\left[\int_0^1 \xi' J_0(\lambda_{0n}\xi')\varphi_h(r_0\xi')d\xi'\right]J_0(\lambda_{0n}\xi)\sinh\left(\frac{\lambda_{0n}z}{r_0}\right)}{[J_1(\lambda_{0n})]^2 \sinh\left(\lambda_{0n}\frac{h}{r_0}\right)} . \quad (3.218)$$

As an **example**, consider the simple special case where

$$\varphi_h(r) = \varphi_0 .$$

Here, one has to evaluate the integral

$$\varphi_0 \int_0^1 \xi' J_0(\lambda_{0n}\xi')d\xi' = \frac{\varphi_0}{\lambda_{0n}^2}\int_0^{\lambda_{0n}} z J_0(z)dz = \frac{\varphi_0}{\lambda_{0n}^2}[z J_1(z)]_0^{\lambda_{0n}}$$

$$= \frac{\varphi_0}{\lambda_{0n}^2}\lambda_{0n}J_1(\lambda_{0n}) = \frac{\varphi_0 J_1(\lambda_{0n})}{\lambda_{0n}} .$$

Where the relation $\int z J_0(z)dz = z J_1(z)$ was used. The final result is

$$\varphi(r, z) = \varphi_0 \sum_{n=1}^{\infty} \frac{2 J_0\left(\lambda_{0n}\frac{r}{r_0}\right)\sinh\left(\lambda_{0n}\frac{z}{r_0}\right)}{\lambda_{0n}J_1(\lambda_{0n})\sinh\left(\lambda_{0n}\frac{h}{r_0}\right)} . \quad (3.219)$$

Table 3.1 lists the first four zeros of J_0 and the corresponding values of J_1 :

Table 3.1

n	λ_{0n}	$J_1(\lambda_{0n})$
1	2.4048	+ 0.5192
2	5.5201	- 0.3403
3	8.6537	+ 0.2715
4	11.7915	- 0.2325

It follows, by the way, from eq. (3.177) that for large arguments

$$\lambda_{0n} \approx \left(n - \frac{1}{4}\right)\pi \quad (3.220)$$

where

$$J_1(\lambda_{0n}) \approx (-1)^{n-1}\sqrt{\frac{2}{\pi\lambda_{0n}}} \quad (3.221)$$

which is already rather accurate for $n = 4$.

3.7.3.4 Rotationally Symmetric Surface Charges in the Plane z = 0 and the Hankel Transformation

Consider the surface charges $\sigma(r)$ that are distributed in the plane $z = 0$ in a rotationally symmetric way. The task is to find the thereby created potential. Obviously, the potential may not diverge, neither at $r = 0$ nor in the limit $z \to \pm\infty$. The Ansatz

$$
\begin{aligned}
\varphi_1 &= \int_0^\infty f_1(k)J_0(kr)\exp(+kz)\,dk && \text{for } (z < 0) \\[2mm]
\varphi_2 &= \int_0^\infty f_2(k)J_0(kr)\exp(-kz)\,dk && \text{for } (z > 0)
\end{aligned}
\right\}
\tag{3.222}
$$

meets these restrictions. For $z = 0$ we have to fulfill

$$
\left(\frac{\partial\varphi_1}{\partial r}\right)_{z=0} - \left(\frac{\partial\varphi_2}{\partial r}\right)_{z=0} = 0
$$

$$
\left(\frac{\partial\varphi_1}{\partial z}\right)_{z=0} - \left(\frac{\partial\varphi_2}{\partial z}\right)_{z=0} = \frac{D_2 - D_1}{\varepsilon_0} = \frac{\sigma(r)}{\varepsilon_0} .
$$

The first restriction immediately yields
$$
f_1(k) = f_2(k) = f(k) .
$$

With this result, the second restriction then gives

$$
\int_0^\infty f(k)J_0(kr)2k\,dk = \frac{\sigma(r)}{\varepsilon_0} .
\tag{3.223}
$$

The task is now to solve this equation for $f(k)$, that is, to express $f(k)$ as a function of $\sigma(r)$. We have dealt with a similar example previously, which could be solved by Fourier transformation in Cartesian coordinates (Sect. 3.5.2.3, but see also Sect. 3.7.3.1). However, the integral equation, which eq. (3.223) represents and needs evaluation, can not be solved by Fourier transforms. On the other hand, we were able to solve finite problems of similar kind by the Fourier series for the Cartesian case, and by the analogous Fourier-Bessel series in case of cylindrical problems. What we therefore need is, in analogy to the Fourier transform, an integral transform for cylindrical problems, where *for example*, Bessel functions replace the exponential functions. Such transforms do indeed exist. Those are the so-called *Hankel transforms*. This relation is better expressed in the seldom used name Fourier-Bessel transform. The Hankel transforms assigns to the function $f(x)$ a new function $\tilde{f}(k)$ (its Hankel transform) in the following form:

$$
\boxed{\tilde{f}(k) = \int_0^\infty x\,f(x)J_m(kx)\,dx}
\tag{3.224}
$$

and its inverse

$$
\boxed{f(x) = \int_0^\infty k\,\tilde{f}(k)J_m(kx)\,dk} .
\tag{3.225}
$$

This relation rests upon the orthogonality relation

$$\int_0^\infty kx J_m(kx) J_m(k'x)\, dx = \delta(k-k') \, .$$

(3.226)

Multiplying (3.225) by $x J_m(k'x)$, then integrating it from 0 to ∞ yields

$$\int_0^\infty f(x) x J_m(k'x)\, dx = \int_0^\infty \int_0^\infty k \tilde{f}(k) J_m(kx) x J_m(k'x)\, dk\, dx$$

$$= \int_0^\infty \tilde{f}(k) \delta(k-k')\, dk = \tilde{f}(k') \, .$$

This is just (3.224). The result of specifically expanding the δ-function $\delta(x-x')$ is

$$\tilde{f}(k) = x' J_m(kx')$$

and thus

$$\delta(x-x') = \int_0^\infty kx' J_m(kx') J_m(kx)\, dx \, .$$

(3.227)

That is just the completeness relation. Just as before for the Fourier transform, this is an example of expanding by a basis system with a continuous eigenvalue spectrum. The general formalism was explained in Sect. 3.6, It shall be noted that it is possible (and actually done) to define the Hankel transform differently – that is, with different factors. We have chosen a definition that makes the eqs. (3.224) and (3.225) entirely symmetric.

Now, we are enabled to solve eq. (3.223). We multiply it by $r J_0(k'r)$ and then integrate over r from 0 to ∞

$$\int_0^\infty \int_0^\infty f(k) J_0(kr) 2kr J_0(k'r)\, dk\, dr = \frac{1}{\varepsilon_0} \int_0^\infty \sigma(r) r J_0(k'r)\, dr$$

i.e.,

$$\int_0^\infty 2f(k) \delta(k-k')\, dk = 2f(k') = \frac{1}{\varepsilon_0} \tilde{\sigma}(k')$$

$$f(k) = \frac{1}{2\varepsilon_0} \tilde{\sigma}(k)$$

and

$$\varphi_{1,2} = \frac{1}{2\varepsilon_0} \int_0^\infty \tilde{\sigma}(k) J_0(kr) \exp(-k|z|)\, dk$$

$$\tilde{\sigma}(k) = \int_0^\infty r\sigma(r) J_0(kr)\, dr$$

(3.228)

which solves our problem.

A simple **example** of how to apply these results is that of the uniformly charged circular loop with

$$\sigma(r) = q\delta(r-r_0) \, ,$$

where q is its line charge. Then we write

$$\tilde{\sigma}(k) = q\int_0^\infty r\delta(r - r_0)J_0(kr)dr = qr_0J_0(kr_0)$$

and

$$\varphi_{1,2} = \frac{qr_0}{2\varepsilon_0}\int_0^\infty J_0(kr_0)J_0(kr)\exp(-k|z|)dk \quad . \tag{3.229}$$

This represents a potential which we have already solved in a different way. As previously (3.199), so eq. (3.229) is also identical to (3.200) (refer to [5], Vol. II, p 14, eq. (17)).
We may proceed with a point charge ($Q = 2\pi r_0 q$, $r_0 \to 0$, $q \to \infty$) and obtain

$$\varphi_{1,2} = \frac{Q}{4\pi\varepsilon_0}\int_0^\infty J_0(kr)\exp(-k|z|)dk \quad , \tag{3.230}$$

having used

$$J_0(0) = 1 \quad .$$

From this follows the requirement that

$$\frac{1}{\sqrt{r^2 + z^2}} = \int_0^\infty J_0(kr)\exp(-k|z|)dk \tag{3.231}$$

(refer to [5], Vol. II, p 9, eq. (18)).

3.7.3.5 Charge Distributions that are not Rotationally Symmetric

So far, we discussed problems with rotational symmetry. For this reason, the parameter m was always zero ($m = 0$). Now, we want to solve problems by separation in cylindrical coordinates that lack rotational symmetry. Consider a cylinder of radius r_0 with a surface charge density of

$$\sigma(\varphi, z) = \frac{Q}{r_0}\delta(z)\delta(\varphi - \varphi_0) \tag{3.232}$$

that is a point charge at $z = 0$, $\varphi = \varphi_0$. The potential F can be written in the form

$$F_{i,a} = \sum_{m=0}^\infty \int_0^\infty \begin{Bmatrix} I_m(kr)K_m(kr_0) \\ I_m(kr_0)K_m(kr) \end{Bmatrix} [f_m(k)\cos m\varphi + g_m(k)(\sin m\varphi)]\cos(kz)dk \tag{3.233}$$

The expression at the top is for $r \leq r_0$ while the bottom one is for $r \geq r_0$. This Ansatz was chosen in such a way that F, together with the tangential component of the electric field, is continuous for $r = r_0$. Furthermore, it has to be

$$(E_{ar} - E_{ir})_{r=r_0} = \left(\frac{\partial F_i}{\partial r} - \frac{\partial F_a}{\partial r}\right)_{r=r_0} = \frac{Q}{\varepsilon_0 r_0}\delta(z)\delta(\varphi - \varphi_0) \tag{3.234}$$

i.e.,

$$\left.\begin{array}{l}\sum\limits_{m=0}^{\infty}\int\limits_{0}^{\infty}k\cos(kz)[I_m{}'(kr_0)K_m(kr_0)-I_m(kr_0)K_m{}'(kr_0)] \\[1em] \cdot\,[f_m(k)\cos m\varphi+g_m(k)(\sin m\varphi)]dk = \dfrac{Q}{r_0}\delta(z)\delta(\varphi-\varphi_0)\end{array}\right\} \qquad (3.235)$$

By means of the relations of Section 3.7.2, we obtain

$$I_m{}'(kr_0)K_m(kr_0)-I_m(kr_0)K_m{}'(kr_0) = \frac{1}{kr_0} . \qquad (3.236)$$

Multiply (3.235) with $\cos k'z\cos m'\varphi$ and $\cos k'z\sin m'\varphi$, respectively. Then integrate over z and φ, using the orthogonality relation to obtain

$$f_0(k) = \frac{Q}{2\pi^2\varepsilon_0} \qquad (3.237)$$

$$f_m(k) = \frac{Q}{\pi^2\varepsilon_0}\cos m\varphi_0, \qquad m \geq 1 \qquad (3.238)$$

$$g_m(k) = \frac{Q}{\pi^2\varepsilon_0}\sin m\varphi_0 . \qquad (3.239)$$

Now we substitute $z - z_0$ for z, that is, the charge is located at $\mathbf{r}_0(r_0, \varphi_0, z_0)$. With these coefficients the result is

$$\boxed{\begin{array}{l} F_{i,a} = \dfrac{Q}{2\pi^2\varepsilon_0}\sum\limits_{m=0}^{\infty}\int\limits_{0}^{\infty}\begin{Bmatrix} I_m(kr)K_m(kr_0) \\ I_m(kr_0)K_m(kr) \end{Bmatrix} \\[1.5em] \qquad \cdot\cos[k(z-z_0)]\cos[m(\varphi-\varphi_0)](2-\delta_{0m})dk . \end{array}} \qquad (3.240)$$

We obtain for the inverse distance between the points $\mathbf{r}(r, \varphi, z)$ and $\mathbf{r}_0(r_0, \varphi_0, z_0)$

$$\boxed{\begin{array}{l} \dfrac{1}{|\mathbf{r}-\mathbf{r}_0|} = \dfrac{2}{\pi}\sum\limits_{m=0}^{\infty}\int\limits_{0}^{\infty}\begin{Bmatrix} I_m(kr)K_m(kr_0) \\ I_m(kr_0)K_m(kr) \end{Bmatrix} \\[1.5em] \qquad \cdot\cos[k(z-z_0)]\cos[m(\varphi-\varphi_0)](2-\delta_{0m})dk . \end{array}} \qquad (3.241)$$

This formula is more general than the previous result of eq. (3.204). which may be obtained in the limit $r_0 = 0$ and $z_0 = 0$. The rational behind this is that $I_0(0) = 1$ and $I_m(0) = 0$ for $m \geq 1$, that is, $I_m(0) = \delta_{0m}$. Now, only the bottom expression is of interest since $r \geq r_0 = 0$.

A different approach is also possible. Consider the plane $z = 0$ carrying the point charge Q at $\mathbf{r}_0(r_0, \varphi_0, 0)$, which is represented by the surface charge density

$$\sigma(r, \varphi) = \frac{Q}{r_0}\delta(r-r_0)\delta(\varphi-\varphi_0) . \qquad (3.242)$$

The Ansatz for the potential is now

$$F = \sum_{m=0}^{\infty} \int_0^{\infty} J_m(kr)[f_m(k)\cos m\varphi + g_m(k)(\sin m\varphi)]\exp(-k|z|)dk \quad . \tag{3.243}$$

This Ansatz was particularly chosen so that at the boundary $z = 0$, F and its tangential derivatives are continuous. Furthermore we have

$$[E_z(z>0) - E_z(z<0)]_{z=0} = \sum_{m=0}^{\infty} \int_0^{\infty} 2kJ_m(kr)$$

$$\cdot [f_m(k)\cos m\varphi + g_m(k)(\sin m\varphi)]dk$$

$$= \frac{Q}{\varepsilon_0 r_0}\delta(r-r_0)\delta(\varphi-\varphi_0) \quad . \tag{3.244}$$

One uses the same procedure again: multiply with $r \cdot J_m(k'r)\cos(m'\varphi)$ and $r \cdot J_m(k'r)\sin(m'\varphi)$, respectively, then integrate over r and φ, apply the orthogonality relation, and finally obtain

$$f_0(k) = \frac{Q}{4\pi\varepsilon_0}J_0(kr_0) \tag{3.245}$$

$$f_m(k) = \frac{Q}{2\pi\varepsilon_0}J_m(kr_0)\cos m\varphi_0, \qquad m \geq 1 \tag{3.246}$$

$$g_m(k) = \frac{Q}{2\pi\varepsilon_0}J_m(kr_0)\sin m\varphi_0 \quad . \tag{3.247}$$

All that remains now for a charge located at point $\mathbf{r}_0(r_0, \varphi_0, z_0)$, is to replace z by $z - z_0$. Applying the just obtained coefficients gives

$$F(\mathbf{r}) = \frac{Q}{4\pi\varepsilon_0} \sum_{m=0}^{\infty} \int_0^{\infty} (2-\delta_{0m})J_m(kr)J_m(kr_0)$$

$$\cdot \exp[-k|z-z_0|]\cos[m(\varphi-\varphi_0)]dk \tag{3.248}$$

and the inverse distance is

$$\frac{1}{|\mathbf{r}-\mathbf{r}_0|} = \sum_{m=0}^{\infty} \int_0^{\infty} (2-\delta_{0m})J_m(kr)J_m(kr_0)\exp[-k|z-z_0|]\cos[m(\varphi-\varphi_0)]dk. \tag{3.249}$$

These equations represent a generalization of the previously found eqs. (3.230) and (3.231). One obtains these equations by setting $r_0 = 0, z_0 = 0$, and making use of the fact that $J_m(kr_0) = J_m(0) = \delta_{0m}$.

We were able to be rather brief here, as the procedure is similar to that of Sections (3.7.3.1) and (3.7.3.4). Dividing the potentials (3.240) and (3.248) by Q yields the respective Green's functions $G(\mathbf{r}, \mathbf{r}_0)$. If there are arbitrary surface charges

$$\sigma = \sigma(z, \varphi) \tag{3.250}$$

or

$$\sigma = \sigma(r, \varphi) \tag{3.251}$$

prescribed on the cylinder at $r = r_0$ or on the plane surface $z = z_0$, then the related potentials are

$$F(\mathbf{r}) = \int G(\mathbf{r}, \mathbf{r}_0)\sigma(\mathbf{r}_0)\,dA_0 \ , \tag{3.252}$$

where those integrals have to be evaluated over the cylinder surface or the plane where

$$dA_0 = r_0\,d\varphi_0\,dz_0 \tag{3.253}$$

or

$$dA_0 = r_0\,d\varphi_0\,dr_0 \ . \tag{3.254}$$

Notice that the rotationally symmetric charge distributions of Sections 3.7.3.1 and 3.7.3.4 are contained in this solution as special cases. Another special case shall be discussed here as an **example**. Suppose that on the surface of a cylinder at $r = r_0$, there exists a surface charge

$$\sigma = \sigma_0 \cos n\varphi \ . \tag{3.255}$$

This allows one to calculate $F_{i,\,a}(\mathbf{r})$ based on (3.240) to obtain

$$F_{i,\,a} = \int_0^{2\pi} d\varphi_0 \int_{-\infty}^{+\infty} dz_0 \int_0^{+\infty} dk \sum_{m=0}^{\infty} \frac{\sigma_0}{2\pi^2\varepsilon_0} \begin{Bmatrix} I_m(kr)K_m(kr_0) \\ I_m(kr_0)K_m(kr) \end{Bmatrix}$$
$$\cdot \cos[k(z-z_0)]\cos[m(\varphi-\varphi_0)](2-\delta_{0m})r_0\cos n\varphi_0 \tag{3.256}$$

Since the charge distribution (3.255) is independent of z_0, integration over z_0 yields

$$\int_{-\infty}^{+\infty}\cos[k(z-z_0)]dz_0 = 2\pi\delta(k) \ , \tag{3.257}$$

and one needs to investigate the Bessel functions for vanishing arguments. Using eqs. (3.178) through (3.180) gives

$$\left. \begin{matrix} \lim\limits_{k\to 0} I_n(kr)K_n(kr_0) = \dfrac{1}{2n}\left(\dfrac{r}{r_0}\right)^n \\[3mm] \lim\limits_{k\to 0} I_n(kr_0)K_n(kr) = \dfrac{1}{2n}\left(\dfrac{r_0}{r}\right)^n \end{matrix} \right\} \quad \text{for } n \ne 0 \tag{3.258}$$

and

$$\left. \begin{matrix} \lim\limits_{k\to 0} I_0(kr)K_0(kr_0) = C - \ln r_0 \\[3mm] \lim\limits_{k\to 0} I_0(kr_0)K_0(kr) = C - \ln r \end{matrix} \right\} \quad \text{for } n = 0 \ , \tag{3.259}$$

where C, although it becomes infinitely large, is still insignificant and may be arbitrarily chosen. Finally, one obtains

$$F_{i,a} = \frac{\sigma_0 r_0}{2n\varepsilon_0}\left\{\begin{array}{c}\left(\dfrac{r}{r_0}\right)^n\\[2ex]\left(\dfrac{r_0}{r}\right)^n\end{array}\right\}\cos n\varphi \qquad \text{for } n \neq 0 \qquad (3.260)$$

and

$$F_{i,a} = -\frac{\sigma_0 r_0}{\varepsilon_0}\left\{\begin{array}{c}0\\[1ex]\ln\dfrac{r}{r_0}\end{array}\right\} \qquad \text{for } n = 0 , \qquad (3.261)$$

where $C = \ln r_0$ has been chosen. For the current case, it is easier to derive this result by not using the general formulas. The independence of z, makes it possible to start from the differential equation

$$r\frac{\partial}{\partial r}r\frac{\partial}{\partial r}R(r) = m^2 R(r) , \qquad (3.262)$$

which results from eqs. (3.156) and (3.160) with $k = 0$. We can immediately write its solutions

$$R = Ar^m + B\frac{1}{r^m} \qquad \text{for } m \neq 0 , \qquad (3.263)$$

$$R = A + B\ln r \qquad \text{for } m = 0 . \qquad (3.264)$$

Thus, for a cylinder of radius r_0 (carrying arbitrary surface charges) we obtain

$$F_i = \sum_{m=1}^{\infty}\left(\frac{r}{r_0}\right)^m(A_m\cos m\varphi + B_m\sin m\varphi) + 0, \qquad (3.265)$$

$$F_a = \sum_{m=1}^{\infty}\left(\frac{r_0}{r}\right)^m(A_m\cos m\varphi + B_m\sin m\varphi) + A_0\ln\frac{r}{r_0} . \qquad (3.266)$$

Again, the coefficients were chosen such that the potential and thereby the tangential components of the electric field are continuous at the surface of the cylinder $r = r_0$. If the surface charge (3.255) is prescribed, then the following must hold

$$(E_{ra} - E_{ri})_{r=r_0} = \left(\frac{\partial F_i}{\partial r} - \frac{\partial F_a}{\partial r}\right)_{r=r_0}$$

$$= \sum_{m=1}^{\infty}\frac{2m}{r_0}(A_m\cos m\varphi + B_m\sin m\varphi) - \frac{A_0}{r_0} \qquad (3.267)$$

$$= \frac{\sigma_0}{\varepsilon_0}\cos n\varphi$$

Obviously, except for A_n, all terms A_m and B_m vanish:

$$A_n = \frac{\sigma_0 r_0}{2n\varepsilon_0} \qquad \text{for } n \neq 0 \; , \tag{3.268}$$

$$A_o = -\frac{\sigma_0 r_0}{\varepsilon_0} \qquad \text{for } n = 0 \; , \tag{3.269}$$

which just leads to the potentials provided above (3.260) and (3.261). Thus it becomes now obvious that the direct calculation of the potentials is indeed simpler than by using the general formulas. The reason for this is that here, the modified Bessel functions converge onto the elementary solutions (3.263) for $m \neq 0$ and (3.264) for $m = 0$, which is obvious from eqs. (3.178) through (3.180). For $n = 0$, we obtain the solutions we already know, i.e., the logarithmic potential outside and a constant potential inside, where the reference point was chosen such that the potential vanishes at the cylinder surface $r = r_0$. Since,

$$2\pi r_0 \sigma_0 = q \; , \tag{3.270}$$

thus

$$F_a = -\frac{\sigma_0 r_0}{\varepsilon_0} \ln\frac{r}{r_0} = -\frac{q}{2\pi\varepsilon_0} \ln\frac{r}{r_0} \; . \tag{3.271}$$

3.8   Separation of Variables for Laplace's Equation in Spherical Coordinates

3.8.1   Separation of Variables

By eq. (3.43), the potential equation for the potential F in spherical polar coordinates is

$$\left(\frac{1}{r^2}\frac{\partial}{\partial r}r^2\frac{\partial}{\partial r} + \frac{1}{r^2\sin\theta}\frac{\partial}{\partial\theta}\sin\theta\frac{\partial}{\partial\theta} + \frac{1}{r^2\sin^2\theta}\frac{\partial^2}{\partial\varphi^2} \right) F = 0 \; . \tag{3.272}$$

We attempt to separate this equation via the product solution

$$F(r, \theta, \varphi) = R(r)D(\theta)\phi(\varphi) \; . \tag{3.273}$$

Multiplying eq. (3.272) by

$$\frac{r^2\sin^2\theta}{R(r)D(\theta)\phi(\varphi)}$$

yields

$$\frac{\sin^2\theta}{R(r)}\frac{\partial}{\partial r}r^2\frac{\partial}{\partial r}R(r) + \frac{\sin\theta}{D(\theta)}\frac{\partial}{\partial\theta}\sin\theta\frac{\partial}{\partial\theta}D(\theta) + \frac{1}{\phi(\varphi)}\frac{\partial^2}{\partial\varphi^2}\phi(\varphi) = 0 \; . \tag{3.274}$$

The first two terms solely depend on r and θ, the last solely on φ. This allows us to separate the equations, *for example*, by substituting

$$\frac{1}{\phi(\phi)}\frac{\partial^2}{\partial\phi^2}\phi(\phi) = -m^2 , \tag{3.275}$$

with its general solution

$$\phi = A_1 \exp(im\phi) + A_2 \exp(-im\phi) \tag{3.276}$$

or equivalently

$$\phi = \tilde{A}_1 \cos(m\phi) + \tilde{A}_2 \sin(m\phi) . \tag{3.277}$$

Here, m is an integer if the dependency on ϕ is periodic with the period 2π. Then, after dividing eq. (3.274) by $\sin^2\theta$ we obtain

$$\frac{1}{R(r)}\frac{\partial}{\partial r}r^2\frac{\partial}{\partial r}R(r) + \frac{1}{\sin\theta D(\theta)}\frac{\partial}{\partial\theta}\sin\theta\frac{\partial}{\partial\theta}D(\theta) - \frac{m^2}{\sin^2\theta} = 0 . \tag{3.278}$$

Now the first term solely depends on r, while the others solely depend on θ. This permits further substitution

$$\frac{1}{R(r)}\frac{\partial}{\partial r}r^2\frac{\partial}{\partial r}R(r) = n(n+1) , \tag{3.279}$$

from which we obtain the general solution

$$R(r) = B_1 r^n + \frac{B_2}{r^{n+1}} , \tag{3.280}$$

which may instantly be verified by substitution. Substituting (3.279) into (3.278) and multiplying by $D(\theta)$ gives

$$\frac{1}{\sin\theta}\frac{\partial}{\partial\theta}\sin\theta\frac{\partial}{\partial\theta}D(\theta) + \left[n(n+1) - \frac{m^2}{\sin^2\theta}\right]D(\theta) = 0 . \tag{3.281}$$

This important differential equation is identified as the generalized Legendre equation, and its solutions are the *Associated Legendre functions*. For the special case of $m = 0$, they are called Legendre functions. Since this is a differential equation of second order, there must be two linearly independent solutions, *i.e.*, the associated Legendre functions of the first and second kind. Only functions of the first kind are finite everywhere on the sphere, *i.e.*, for all values of θ, while the ones of the second kind have singularities at the poles (at $\theta = 0$ and $\theta = \pi$). Therefore, when solving a problem that includes the poles, one has to exclude the associated Legendre functions of the second kind, since they would cause the potential to diverge. For this reason, we will not study them further, realizing that we exclude certain boundary value problems, namely those which do not contain the poles $\theta = 0$ or $\theta = \pi$. (Notice also that our choice of integer values for m represents another restriction on the generality, as we always take the whole space of angles $0 \leq \phi \leq 2\pi$ and not just a fraction thereof). So, we only study the associated Legendre functions of the first kind. These are finite only for n being whole numbers. They are labelled by P_n^m, and we have

$$D(\theta) = P_n^m(\cos\theta) . \tag{3.282}$$

Specifically for $m = 0$

$$D(\theta) = P_n^0(\cos\theta) = P_n(\cos\theta) \ .$$ (3.283)

Frequently, the variable

$$\xi = \cos\theta, \quad d\xi = -\sin\theta\, d\theta \ .$$ (3.284)

is introduced. This changes the differential equation (3.281) into

$$\frac{\partial}{\partial\xi}(1-\xi^2)\frac{\partial}{\partial\xi}D(\xi) + \left[n(n+1) - \frac{m^2}{1-\xi^2}\right]D(\xi) = 0 \ .$$ (3.285)

and

$$D(\xi) = P_n^m(\xi) \ .$$ (3.286)

The functions

$$P_n^m(\cos\theta)\cos(m\varphi)$$ (3.287)

or

$$P_n^m(\cos\theta)\sin(m\varphi)$$ (3.288)

are called spherical harmonics, which is the same name as for the functions

$$Y_n^m = P_n^m(\cos\theta)\exp(im\varphi) \ .$$ (3.289)

The name reflects on the fact that when holding r fixed, the angles θ, φ address all points on the sphere with $r = \mathrm{const}$. The various sorts of spherical harmonics are all solutions of the differential equation

$$\left[\frac{1}{\sin\theta}\frac{\partial}{\partial\theta}\sin\theta\frac{\partial}{\partial\theta} + n(n+1) + \frac{1}{\sin^2\theta}\frac{\partial^2}{\partial\varphi^2}\right]F(\theta,\varphi) = 0 \ ,$$ (3.290)

which emerges from (3.274), when one separates only the radial part.

The P_n^m can be expressed in the following way:

$$P_n^m(\xi) = \frac{(1-\xi^2)^{m/2}}{2^n n!}\frac{d^{n+m}}{d\xi^{n+m}}(\xi^2-1)^n \ .$$ (3.291)

This makes the simplest spherical harmonics (Legendre polynomials)

$$\left.\begin{array}{l} P_0^0 = P_0 = 1 \\[2mm] P_1^0 = P_1 = \xi = \cos\theta \\[2mm] P_2^0 = P_2 = \frac{1}{2}(3\xi^2-1) = \frac{1}{2}(3\cos^2\theta-1) = \frac{1}{4}(3\cos2\theta+1) \end{array}\right\} \ .$$ (3.292)

etc. The simplest associated Legendre functions are

$$P_1^1 = \sqrt{1-\xi^2} = \sin\theta$$

$$P_2^1 = 3\xi\sqrt{1-\xi^2} = 3\sin\theta\cos\theta = (3/2)\sin 2\theta$$

$$P_2^2 = 3(1-\xi^2) = 3\sin^2\theta = \frac{3}{2}(1-\cos 2\theta)$$

(3.293)

etc. It is only necessary to evaluate m-values which are less or equal to n, as

$$P_n^m = 0 \qquad \text{for } m > n .$$

(3.294)

To expand an arbitrary function within the interval $-1 \le \xi \le 1$ by P_n^m is possible. For this purpose, use the orthogonality relation

$$\int_{-1}^{+1} P_n^m(\xi)P_{n'}^m(\xi)d\xi = \frac{2(n+m)!}{(2n+1)(n-m)!}\delta_{nn'} .$$

(3.295)

As we have already seen before, the angular functions are orthogonal. For the given interval $0 \le \varphi \le 2\pi$, the orthogonality relations are

$$\int_0^{2\pi} \cos(m\varphi)\cos(m'\varphi)d\varphi = \begin{cases} 2\pi\delta_{mm'} & \text{for } m = 0 \\ \pi\delta_{mm'} & \text{for } m = 1, 2, \dots \end{cases}$$

$$\int_0^{2\pi} \sin(m\varphi)\sin(m'\varphi)d\varphi = \pi\delta_{mm'} \qquad \text{for } m = 1, 2, \dots$$

$$\int_0^{2\pi} \cos(m\varphi)\sin(m'\varphi)d\varphi = 0$$

(3.296)

The exponential functions $\exp(im\varphi)$ are also orthogonal

$$\int_0^{2\pi} \exp(im\varphi)\exp(-im'\varphi)d\varphi = 2\pi\delta_{mm'} .$$

(3.297)

Note that according to the definition (3.135), the scalar product of complex functions requires the use of the complex conjugate, that is, $\exp(-im'\varphi)$. The relations (3.296) and (3.297) are equivalent because of

$$\exp(\pm im\varphi) = \cos(m\varphi) \pm i\sin(m\varphi)$$

Of course, there is a continuum analogue for (3.297), which shall be given for comparison:

$$\int_{-\infty}^{+\infty} \exp(ikx)\exp(-ik'x)dx = \int_{-\infty}^{+\infty} \exp[i(k-k')x]dx = 2\pi\delta(k-k') .$$

(3.298)

This formula is the basis for the exponential Fourier transformation. It also reveals an important expression for the δ-function.

$$\delta(k-k') = \frac{1}{2\pi}\int_{-\infty}^{+\infty} \exp[i(k-k')x]dx .$$

(3.299)

Because of the symmetry of the cosine function and the anti-symmetry of the sine function, this allows one to write

$$\delta(k - k') = \frac{1}{2\pi}\int_{-\infty}^{+\infty}\{\cos[(k - k')x] + i\sin[(k - k')x]\}\,dx$$

$$= \frac{1}{2\pi}\int_{-\infty}^{+\infty}\cos[(k - k')x]\,dx = \frac{1}{\pi}\int_{0}^{+\infty}\cos[(k - k')x]\,dx$$

This represents a slightly different form from the previously used eq. (3.198).

The orthogonality of the respective product functions, *i.e.* of the spherical harmonics, is a consequence of eqs. (3.295) through (3.297). For instance, it is

$$\int_{0}^{2\pi}\cos(m\varphi)\cos(m'\varphi)\,d\varphi\int_{-1}^{+1}P_{n}^{m}(\xi)P_{n'}^{m'}(\xi)\,d\xi$$

$$= \int_{0}^{2\pi}\int_{-1}^{+1}\cos(m\varphi)P_{n}^{m}(\cos\theta)\cos(m'\varphi)P_{n'}^{m'}(\cos\theta)\,d(\cos\theta)\,d\varphi$$

$$= \int_{0}^{2\pi}\int_{-1}^{+1}\cos(m\varphi)P_{n}^{m}(\cos\theta)\cos(m'\varphi)P_{n'}^{m'}(\cos\theta)\sin\theta\,d\theta\,d\varphi$$

$$= \int\cos(m\varphi)P_{n}^{m}(\cos\theta)\cos(m'\varphi)P_{n'}^{m'}(\cos\theta)\,d\Omega$$

$$= \begin{cases} \dfrac{2\pi}{2n+1}\dfrac{(n+m)!}{(n-m)!}\delta_{nn'}\delta_{mm'} & \text{for } m \geq 1 \\[2ex] \dfrac{4\pi}{2n+1}\dfrac{(n+m)!}{(n-m)!}\delta_{nn'}\delta_{mm'} & \text{for } m = 0 \end{cases} = \frac{2\pi(1 + \delta_{0m})(n+m)!}{2n+1}\frac{(n+m)!}{(n-m)!}\delta_{nn'}\delta_{mm'}$$

$$(3.300)$$

where we used the solid angle on the sphere's surface

$$\sin\theta\,d\theta\,d\varphi = d\Omega . \tag{3.301}$$

Furthermore, integrating over the entire solid angle gives

$$\int Y_{n}^{m}Y_{n'}^{m'*}\,d\Omega = \frac{4\pi}{2n+1}\cdot\frac{(n+m)!}{(n-m)!}\cdot\delta_{nn'}\delta_{mm'} . \tag{3.302}$$

For the general solution of the potential we may try the form

$$F = \sum_{n=0}^{\infty}\sum_{m=0}^{+n}\left(A_{n}r^{n} + B_{n}\frac{1}{r^{n+1}}\right)P_{n}^{m}(\cos\theta)[C_{nm}\cos(m\varphi) + D_{nm}\sin(m\varphi)] \tag{3.303}$$

as well as this form

$$F = \sum_{n=0}^{\infty}\sum_{m=-n}^{\infty}\left(A_{nm}r^{n} + B_{nm}\frac{1}{r^{n+1}}\right)Y_{n}^{m}(\theta, \varphi) . \tag{3.304}$$

The properties of the spherical harmonics are compiled in the previously mentioned books [3-7] (see Section 3.7.2).

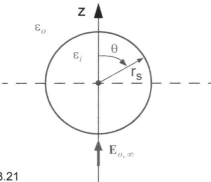

Fig. 3.21

3.8.2   Examples

3.8.2.1 Dielectric Sphere in a Uniform Electric Field

Let us start with a simple problem, whose solution is already known (Sect. 2.12), namely the problem of a sphere in a uniform electric field (Fig. 3.21). The potential of the outside applied electric field $E_{a,\infty}$ is

$$F_{a,\infty} = -E_{a,\infty} z = -E_{a,\infty} r \cos\theta \quad . \tag{3.305}$$

The potential due to the bound charges at the sphere's surface has to be added. For the inside potential we write

$$F_i = \sum_{n=0}^{\infty} A_n r^n P_n(\cos\theta) \quad . \tag{3.306}$$

This is a result of the Ansatz (3.303) and the rotational symmetry $m = 0$. B_n has to be zero to avoid the singularity at the center of the sphere. Conversely, for the outside space we write

$$F_a = \sum_{n=0}^{\infty} B_n \frac{1}{r^{n+1}} P_n(\cos\theta) + F_{a,\infty} \quad .$$

Comparison with (3.292) reveals that $\cos\theta = P_1$, *i.e.*, one might as well write:

$$F_a = \sum_{n=0}^{\infty} \left(\frac{B_n}{r^{n+1}} - \delta_{n1} E_{a,\infty} r \right) P_n(\cos\theta) \quad . \tag{3.307}$$

F has to be continuous at $r = r_s$ so that the tangential components of **E** will be continuous there. This requires that

$$A_n r_s^n = \frac{B_n}{r_s^{n+1}} - \delta_{n1} E_{a, \infty} r_s .$$

(3.308)

Furthermore, the normal component of **D** has to be continuous, which requires that

$$\varepsilon_i \frac{\partial F_i}{\partial r} = \varepsilon_a \frac{\partial F_a}{\partial r} .$$

that is

$$\varepsilon_i n A_n r_s^{n-1} = -\varepsilon_a (n+1) \frac{B_n}{r_s^{n+2}} - \varepsilon_a \delta_{n1} E_{a, \infty} .$$

(3.309)

The pair of equations for $n \neq 1$ is

$$A_n r_s^n = \frac{B_n}{r_s^{n+1}}$$

$$\varepsilon_i A_n r_s^{n-1} = -\varepsilon_a (n+1) \frac{B_n}{r_s^{n+2}} .$$

It has only trivial solutions

$$A_n = 0, B_n = 0$$

For $n = 1$, however, the solutions are

$$A_1 r_s = \frac{B_1}{r_s^2} - E_{a, \infty} r_s$$

$$\varepsilon_i A_1 = -2\varepsilon_a \frac{B_1}{r_s^3} - \varepsilon_a E_{a, \infty}$$

which gives

$$\left. \begin{array}{l} A_1 = -E_{a, \infty} \dfrac{3\varepsilon_a}{\varepsilon_i + 2\varepsilon_a} \\[4mm] B_1 = E_{a, \infty} r_s^3 \dfrac{\varepsilon_i - \varepsilon_a}{\varepsilon_i + 2\varepsilon_a} \end{array} \right\} .$$

(3.310)

This gives the potential inside

$$F_i = -E_{a, \infty} \frac{3\varepsilon_a}{\varepsilon_i + 2\varepsilon_a} r \cos\theta$$

$$= -E_{a, \infty} \frac{3\varepsilon_a}{\varepsilon_i + 2\varepsilon_a} z$$

and the electric field inside

$$E_i = E_{a,\infty} \frac{3\varepsilon_a}{\varepsilon_i + 2\varepsilon_a} \tag{3.311}$$

in accordance with our previous result (2.136). Thus for F_a we obtain

$$F_a = E_{a,\infty} \cos\theta \left(\frac{r_s^3}{r^2} \frac{\varepsilon_i - \varepsilon_a}{\varepsilon_i + 2\varepsilon_a} - r \right) \tag{3.312}$$

which is in complete agreement with our previous result (2.139)

3.8.2.2 A Sphere Carrying an arbitrary Surface Charge

Consider a sphere of radius r_0 and the surface charge

$$\sigma = \sigma(\theta, \varphi) . \tag{3.313}$$

By eq. (3.303) the potentials inside and outside are

$$F_{i,a} = \sum_{n=0}^{\infty} \sum_{m=0}^{n} \left\{ \begin{matrix} \left(\dfrac{r}{r_0}\right)^n \\ \left(\dfrac{r_0}{r}\right)^{n+1} \end{matrix} \right\} P_n^m(\cos\theta)[A_{nm}\cos m\varphi + B_{nm}\sin m\varphi]. \tag{3.314}$$

The coefficients were chosen such that the potential is continuous at the sphere's surface at $r = r_0$

$$F_i(r_0, \theta, \varphi) = F_a(r_0, \theta, \varphi) . \tag{3.315}$$

This forces the tangential component of the electric field to be continuous. The boundary condition

$$(E_{ra} - E_{ri})_{r=r_0} = \left(\frac{\partial F_i}{\partial r} - \frac{\partial F_a}{\partial r} \right)_{r=r_0} = \frac{\sigma(\theta, \varphi)}{\varepsilon_0} \tag{3.316}$$

has to be met as well. Therefore

$$\sum_{n=0}^{\infty} \sum_{m=0}^{n} P_n^m(\cos\theta) \frac{2n+1}{r_0}[A_{nm}\cos m\varphi + B_{nm}\sin m\varphi] = \frac{\sigma(\theta, \varphi)}{\varepsilon_0} . \tag{3.317}$$

Multiplying this equation by $P_n^{m'}(\cos\theta)\cos m'\varphi$ and $P_{n'}^{m'}(\cos\theta)\sin m'\varphi$, respectively, then integrating over the solid angle $\int \sin\theta \, d\theta \, d\varphi$ and using the orthogonality relation (3.300) gives the coefficients A_{nm} and B_{nm}, and thereby solves the problem. We illustrate with a further **example**

$$\sigma(\theta, \varphi) = \frac{Q}{r_0^2 \sin\theta_0}[\delta(\theta - \theta_0)\delta(\varphi - \varphi_0)] \tag{3.318}$$

i.e., a point charge Q on the surface of the sphere at location $(r_0, \theta_0, \varphi_0)$. We thus obtain from Green's function for this problem which can be used to solve the general problem of eq. (3.313). We obtain

$$\sum_{n=0}^{\infty} \sum_{m=0}^{n} \frac{2n+1}{r_0} \oint P_n^m(\cos\theta) P_{n'}^{m'}(\cos\theta)$$

$$\cdot [A_{nm}\cos m\varphi + B_{nm}\sin m\varphi] \begin{Bmatrix} \cos m'\varphi \\ \sin m'\varphi \end{Bmatrix} \sin\theta\, d\theta\, d\varphi$$

$$= \frac{Q}{\varepsilon_0 r_0^2} \oint \delta(\theta - \theta_0)\delta(\varphi - \varphi_0) P_{n'}^{m'}(\cos\theta) \begin{Bmatrix} \cos m'\varphi \\ \sin m'\varphi \end{Bmatrix} d\theta\, d\varphi$$

$$= \frac{Q}{\varepsilon_0 r_0^2} P_{n'}^{m'}(\cos\theta_0) \begin{Bmatrix} \cos m'\varphi_0 \\ \sin m'\varphi_0 \end{Bmatrix}$$

This gives for $m = 0$

$$A_{n0} = \frac{Q}{4\pi\varepsilon_0 r_0} P_n^0(\cos\theta_0) \tag{3.319}$$

while B_{n0} is insignificant. For $m \neq 0$ we obtain

$$A_{nm} = \frac{Q}{2\pi\varepsilon_0 r_0} \frac{(n-m)!}{(n+m)!} P_n^m(\cos\theta_0)\cos m\varphi_0 \tag{3.320}$$

$$B_{nm} = \frac{Q}{2\pi\varepsilon_0 r_0} \frac{(n-m)!}{(n+m)!} P_n^m(\cos\theta_0)\sin m\varphi_0 \ . \tag{3.321}$$

Finally we obtain

$$F_{i,a} = \frac{Q}{4\pi\varepsilon_0 r_0} \sum_{n=0}^{\infty} \sum_{m=0}^{n} (2-\delta_{0m}) \begin{Bmatrix} \left(\dfrac{r}{r_0}\right)^n \\ \left(\dfrac{r_0}{r}\right)^{n+1} \end{Bmatrix}$$

$$\cdot \frac{(n-m)!}{(n+m)!} P_n^m(\cos\theta) P_n^m(\cos\theta_0)\cos[m(\varphi - \varphi_0)] \tag{3.322}$$

having used the trigonometric identity

$$\cos m\varphi \cos m\varphi_0 + \sin m\varphi \sin m\varphi_0 = \cos m(\varphi - \varphi_0) \ .$$

The factor $(2-\delta_{0m})$ ensures that the special case of $m = 0$ is covered correctly. Notice the term in braces in eqs. (3.314) and (3.322): The upper term applies to the case $r \leq r_0$, while the lower one for covers $r \geq r_0$. The potential is of course

$$F_{i,a} = \frac{Q}{4\pi\varepsilon_0 |\mathbf{r} - \mathbf{r}_0|}$$

$$= \frac{1}{4\pi\varepsilon_0} \frac{Q}{\sqrt{r^2 + r_0^2 - 2rr_0[\sin\theta\cos\theta_0\cos(\varphi - \varphi_0) + \cos\theta\cos\theta_0]}} \ . \tag{3.323}$$

As a result of this, we obtain the oftentimes useful series of the inverse distance as an expansion of spherical harmonics:

$$
\frac{1}{|\mathbf{r}-\mathbf{r}_0|} = \frac{1}{r_0}\sum_{n=0}^{\infty}\sum_{m=0}^{n}(2-\delta_{0m})\left\{\begin{array}{l}\left(\dfrac{r}{r_0}\right)^{n}\\[2mm]\left(\dfrac{r_0}{r}\right)^{n+1}\end{array}\right\}
$$

$$
\cdot\frac{(n-m)!}{(n+m)!}P_n^m(\cos\theta)P_n^m(\cos\theta_0)\cos\left[m(\varphi-\varphi_0)\right].
$$

(3.324)

For a field point r on the positive z-axis, we have $\theta = 0$ and $\cos\theta = 1$. Thus

$$
P_n^m(\cos\theta) = P_n^m(1) = 0 \qquad \text{for } m \geq 1, P_n^0(1) = 1 ,
$$

(3.325)

which simplifies eq. (3.324)

$$
\frac{1}{|\mathbf{r}-\mathbf{r}_0|} = \frac{1}{r_0}\sum_{n=0}^{\infty}\left\{\begin{array}{l}\left(\dfrac{r}{r_0}\right)^{n}\\[2mm]\left(\dfrac{r_0}{r}\right)^{n+1}\end{array}\right\}P_n^0(\cos\theta_0) .
$$

(3.326)

We obtain the Green's function $G(\mathbf{r}, \mathbf{r}_0)$ when dividing $F_{i,a}$ (eq. (3.322) by Q. Then, the potential for an arbitrary distribution of surface charges (3.313) is

$$
F(\mathbf{r}) = \oint G(\mathbf{r}, \mathbf{r}_0)\sigma(\mathbf{r}_0)r_0^2\sin\theta_0 d\theta_0 d\varphi_0 .
$$

(3.327)

A specifically simple example is that of a constant surface charge

$$
\sigma = \sigma_0 .
$$

(3.328)

The only term left in this case is the one where $P_0^0 = 1$ which yields

$$
F_{i,a} = \frac{1}{4\pi\varepsilon_0 r_0}\sigma_0 r_0^2\left\{\begin{array}{l}1\\[2mm]\left(\dfrac{r_0}{r}\right)\end{array}\right\}\oint\sin\theta_0 d\theta_0 d\varphi_0 = \frac{\sigma_0 r_0}{\varepsilon_0}\left\{\begin{array}{l}1\\[2mm]\dfrac{r_0}{r}\end{array}\right\}
$$

$$
F_{i,a} = \left\{\begin{array}{ll}\dfrac{\sigma_0 r_0}{\varepsilon_0} = \dfrac{Q}{4\pi\varepsilon_0 r_0} & \text{for } r \leq r_0\\[4mm]\dfrac{\sigma_0 r_0^2}{\varepsilon_0 r} = \dfrac{Q}{4\pi\varepsilon_0 r} & \text{for } r \geq r_0\end{array}\right\},
$$

(3.329)

as required.

The power series of the inverse distance as an expansion by spherical harmonics is of significant interest for electromagnetic field theory. Consider an arbitrary distribution of volume charges. We aim to compute the potential at an arbitrary point outside the region of this volume charge distribution. Without

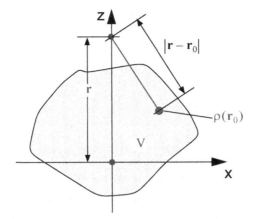

Fig. 3.22

limiting the generality, we simplify by selecting that point to be on the z-axis, where $\theta = 0$ (see Fig. 3.22).

Then by eq. (3.326)

$$F(\mathbf{r}) = \frac{1}{4\pi\varepsilon_0 r} \sum_{n=0}^{\infty} \int_V \left(\frac{r_0}{r}\right)^n \rho(\mathbf{r}_0) P_n^0(\cos\theta_0) d\tau_0$$

$$F(\mathbf{r}) = \frac{1}{4\pi\varepsilon_0 r} \int_V \rho(\mathbf{r}_0) d\tau_0 + \frac{1}{4\pi\varepsilon_0 r^2} \int_V r_0 \rho(\mathbf{r}_0) P_1^0(\cos\theta_0) d\tau_0$$

$$+ \frac{1}{4\pi\varepsilon_0 r^3} \int_V r_0^2 \rho(\mathbf{r}_0) P_2^0(\cos\theta_0) d\tau_0 + \dots$$

(3.330)

This is the so-called *multipole expansion* of the potential for a charge distribution. It sorts these into contributions which decrease by powers of $r^{-(n+1)}$. These are the moments of charge distribution and the individual terms (potentials) are called *monopole, dipole, quadrupole, octopole moment*, etc. If the total charge is zero, that is, if

$$\int \rho(\mathbf{r}_0) d\tau_0 = 0 \ ,$$

(3.331)

then the dipole becomes the leading term of the series in this asymptotic charge distribution, given that it does not also vanish. Now, consider for instance, a dipole with charge $\pm Q$ at $r_0 = d/2$, $\theta_0 = 0$ or $\theta_0 = \pi$, then the dipole potential becomes

$$F = \frac{1}{4\pi\varepsilon_0 r^2}\left[\frac{d}{2}Q + \left(-\frac{d}{2}\right)\cdot(-Q)\right] = \frac{Qd}{4\pi\varepsilon_0 r^2} = \frac{p}{4\pi\varepsilon_0 r^2}$$

(3.332)

as required for a point at the axis for which $\cos\theta = 1$. If the dipole potential also vanishes, then the next significant term is the quadrupole potential, etc. Notice that even a point charge has multipole moments if it is not located at the origin. Using eq. (3.330) for a point charge yields for the potential

$$F = \frac{Q}{4\pi\varepsilon_0 r} \sum_{n=0}^{\infty} \left(\frac{r_0}{r}\right)^n P_n^0(\cos\theta_0) , \tag{3.333}$$

which corresponds to the inverse distance eq. (3.326) for $r > r_0$.

  If the point for which the potential should be determined is not on the z-axis, then the more general eq. (3.324) for the inverse distance has to be used, which leads to inconvenient expressions for the individual multipole moments.

3.8.2.3 Dirichlet's Problem for a Sphere

Consider the point charge $Q(r_0 < r_s)$ at location r_0 inside a sphere with radius r_s. At the sphere's surface, the potential is specified to be zero. Without any loss of generality, we may assume that the charge is located on the z-axis ($\theta_0 = 0$). One obtains the potential by superposition of the specific solution of the inhomogeneous Poisson equation, and the general solution of the homogeneous Laplace equation, i.e., we try

$$F = \frac{Q}{4\pi\varepsilon_0 r_0} \sum_{n=0}^{\infty} \begin{Bmatrix} \left(\dfrac{r}{r_0}\right)^n \\[2mm] \left(\dfrac{r_0}{r}\right)^{n+1} \end{Bmatrix} P_n^0(\cos\theta) + \sum_{n=0}^{\infty} A_n \left(\frac{r}{r_s}\right)^n P_n^0(\cos\theta) . \tag{3.334}$$

The first term represents the potential of the point charge in the infinite space for $r \le r_0$ (top) and $r \ge r_0$ (bottom), which is eq. (3.322) and (3.325) applied to the current case ($\theta_0 = 0$). On the sphere's surface is $F = 0$ if

$$A_n = -\frac{Q}{4\pi\varepsilon_0 r_0} \left(\frac{r_0}{r_s}\right)^{n+1} . \tag{3.335}$$

Therefore

$$F = \frac{Q}{4\pi\varepsilon_0} \sum_{n=0}^{\infty} \left[\frac{1}{r_0} \begin{Bmatrix} \left(\dfrac{r}{r_0}\right)^n \\[2mm] \left(\dfrac{r_0}{r}\right)^{n+1} \end{Bmatrix} - \left(\frac{r_0^n r^n}{r_s^{2n+1}}\right) \right] P_n^0(\cos\theta) . \tag{3.336}$$

The second term of (3.334) is the solution of Laplace's equation for inside the sphere ($r \le r_s$). It already considers the rotational symmetry of the field, which is established by the charge on the z-axis. This second term represents the effect of the charges on the sphere's surface. This term on its own, using (3.335), gives

$$F = -\frac{Q \cdot \dfrac{r_s}{r_0}}{4\pi\varepsilon_0} \cdot \frac{1}{\dfrac{r_s^2}{r_0}} \sum_{n=0}^{\infty} \frac{r^n}{\left(\dfrac{r_s^2}{r_0}\right)^n} P_n^0(\cos\theta) = -\frac{Q}{4\pi\varepsilon_0} \sum_{n=0}^{\infty} \left[\frac{r_0^n r^n}{r_s^{2n+1}}\right] P_n^0(\cos\theta) .$$

(3.337)

This is nothing else than the potential of the charge

$$Q' = -Q\frac{r_s}{r_0} ,$$

(3.338)

located on the z-axis at the point

$$r_0' = \frac{r_s^2}{r_0} .$$

(3.339)

As it has to be, one sees again that one can solve this problem by means of this image charge. The radial electric field at the sphere's surface is

$$(E_r)_{r=r_s} = -\left(\frac{\partial F}{\partial r}\right)_{r=r_s} = -\frac{Q}{4\pi\varepsilon_0} \sum_{n=0}^{\infty} \frac{r_0^n}{r_s^{n+2}}[2n+1]P_n^0(\cos\theta) .$$

(3.340)

This represents a surface charge

$$\sigma = -\frac{Q}{4\pi} \sum_{n=0}^{\infty} \frac{r_0^n}{r_s^{n+2}}[2n+1]P_n^0(\cos\theta) .$$

(3.341)

By (3.327), this surface charge on its own produces the potential inside the sphere

$$F_\sigma = -\frac{Q}{4\pi\varepsilon_0} \sum_{n=0}^{\infty} \frac{r_0^n r^n}{r_s^{2n+1}}P_n^0(\cos\theta) .$$

(3.342)

This is exactly the potential given by (3.337), which represents the additional potential due to the image charge.

3.9   Multi-Conductor Systems

The previous sections illustrated the use of the method of separation of variables to solve electrostatic problems by means of some examples. Although there are a number of additional coordinate systems for which this separation is possible, it has become clear that there are many problems where a solution using this approach is not possible.

Analytical methods may not always be used, such as for systems consisting of many charged conductors with a complicated geometry. Nevertheless, general statements on arbitrary multi-conductor systems are possible. Fig. 3.23 shows such a system with five conductors. We will now analyze a system with n conductors. They shall carry the charge Q_i, $i = 1, 2, ..., n$, and their surface shall have the potential φ_i, $i = 1, 2, ..., n$. Then, the potential becomes

$$\varphi_k = \frac{1}{4\pi\varepsilon_0} \sum_{i=1}^{\infty} \oint_{A_i} \frac{\sigma_i}{r_{ik}} dA_i \qquad (3.343)$$

where r_{ik} is the distance from a variable point on conductor i to a fixed point on conductor k. Furthermore, we use the relation

$$\oint_{A_k} \varphi_k \sigma_k dA_k = \varphi_k \oint_{A_k} \sigma_k dA_k = \varphi_k Q_k = \frac{1}{4\pi\varepsilon_0} \sum_{i=1}^{n} \oint_{A_i A_k} \frac{\sigma_i \sigma_k}{r_{ik}} dA_i dA_k$$

or

$$\varphi_k = \sum_{i=1}^{n} \frac{1}{4\pi\varepsilon_0 Q_i Q_k} \oint_{A_i A_k} \frac{\sigma_i \sigma_k}{r_{ik}} dA_i dA_k Q_i \ .$$

If we define

$$p_{ki} = \frac{1}{4\pi\varepsilon_0 Q_i Q_k} \oint_{A_i A_k} \frac{\sigma_i \sigma_k}{r_{ik}} dA_i dA_k \qquad (3.344)$$

we may as well write

$$\boxed{\varphi_k = \sum_{i=1}^{\infty} p_{ki} Q_i} \ . \qquad (3.345)$$

The coefficients p_{ki} of this linear relation do not depend on the charges, but solely on the geometry of the conductors. The definition reveals, that they are symmetric and nonnegative:

$$\left.\begin{array}{l} p_{ki} = p_{ik} \\ p_{ki} \geq 0 \end{array}\right\} \ . \qquad (3.346)$$

The fact that $p_{ki} \geq 0$ is not immediately obvious from (3.344). However, if we consider a system of several conductors where only one caries a charge (*for example*, the k-th one), then it becomes clear that

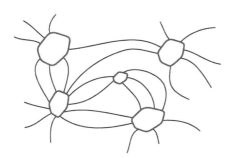

Fig. 3.23

$$p_{ki} \geq 0 .$$

The p_{ki} are called *potential coefficients*. We can imagine solving eq. (3.345) for the charges:

$$\boxed{Q_i = \sum_{i=1}^{\infty} c_{ik}\varphi_k} , \qquad (3.347)$$

where

$$c_{ik} = \frac{P_{ik}}{\Delta} . \qquad (3.348)$$

Δ is the determinant of the coefficients p_{ik} and P_{ik} is the minor determinant of p_{ki}, i.e., the $(-1)^{i+k}$ multiple of the determinant that results from $\Delta = |p_{ki}|$ by deleting the k-th row and the i-th column. This symmetry transfers from the p_{ik} to the c_{ik}, the so-called *influence coefficients*.

$$c_{ki} = c_{ik} . \qquad (3.349)$$

Beyond this, they differ from the p_{ik} by the following properties:

$$\left. \begin{aligned} & c_{ii} \geq 0 \\ & c_{ik} \leq 0 \qquad \text{for } i \neq k \\ & \textstyle\sum_{i=1}^{n} c_{ik} \geq 0 \end{aligned} \right\} . \qquad (3.350)$$

This can be seen by the following reasoning. All conductors except for one (*for example*, i-th conductor) shall have the potential $\varphi_k = 0, (k \neq i)$, while the i-th conductor shall have the potential $\varphi_i = 1V$. Since the potential in the charge-free space may not have any maxima or minima (Sect. 3.4), the configuration has to be qualitatively as illustrated in Fig. 3.24. All lines of force that originate from conductor i have to terminate at another conductor or extend toward infinity. However, there may not be any line originating at a conductor with potential 0 and ending at another conductor with zero potential.

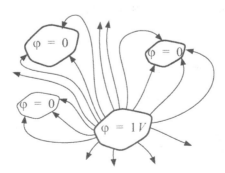

Fig. 3.24

The conductor i carries the positive charge Q_i. All other charges are negative. Now, with eq. (3.347) we obtain

$$Q_k = c_{ki}\varphi_i \leq 0, \quad (i \neq k)$$

i.e.,

$$c_{ki} \leq 0, \quad (i \neq k)$$

as claimed. On the other hand

$$Q_i = c_{ii}\varphi_i \geq 0$$

and therefore as claimed

$$0 \leq c_{ii} \, .$$

The overall charge is also positive, that is

$$0 \leq \sum_{k=1}^{n} Q_k = \sum_{k=1}^{n} c_{ki}\varphi_i = \varphi_i \sum_{k=1}^{n} c_{ki}$$

therefore

$$\sum_{k=1}^{n} c_{ki} \geq 0 \, ,$$

which proves the last of our claims (3.350), where the coefficients only depend on the geometry and not on the charge. Written more explicitly:

$$Q_1 = c_{11}\varphi_1 + c_{12}\varphi_2 + c_{13}\varphi_3 + \dots$$
$$Q_2 = c_{21}\varphi_1 + c_{22}\varphi_2 + c_{23}\varphi_3 + \dots$$
$$Q_3 = c_{31}\varphi_1 + c_{32}\varphi_2 + c_{33}\varphi_3 + \dots$$

$$\vdots \qquad \dots\dots\dots\dots\dots\dots\dots\dots$$

Expressed differently:

$$Q_1 = (c_{11} + c_{12} + c_{13} + \dots)\varphi_1 + c_{12}(\varphi_2 - \varphi_1) + c_{13}(\varphi_3 - \varphi_1) + \dots$$
$$Q_2 = c_{21}(\varphi_1 - \varphi_2) + (c_{22} + c_{21} + c_{23} + \dots)\varphi_2 + c_{23}(\varphi_3 - \varphi_2) + \dots$$
$$Q_3 = c_{31}(\varphi_1 - \varphi_3) + c_{32}(\varphi_2 - \varphi_3) + (c_{33} + c_{31} + c_{32} + \dots)\varphi_3 + \dots$$

$$\vdots \qquad \dots\dots\dots\dots\dots\dots\dots\dots$$

or

$$
\left.
\begin{aligned}
Q_1 &= C_{11}\varphi_1 + C_{12}(\varphi_1 - \varphi_2) + C_{13}(\varphi_1 - \varphi_3) + \dots \\
Q_2 &= C_{21}(\varphi_2 - \varphi_1) + C_{22}\varphi_2 + C_{23}(\varphi_2 - \varphi_3) + \dots \\
Q_3 &= C_{31}(\varphi_3 - \varphi_1) + C_{32}(\varphi_3 - \varphi_2) + C_{33}\varphi_3 + \dots \\
\vdots \quad &\quad \dots\dots\dots\dots\dots\dots\dots\dots\dots
\end{aligned}
\right\}
\tag{3.351}
$$

where

$$
\left.
\begin{aligned}
C_{ii} &= \sum_{k=1}^{n} c_{ik} \geq 0 \\
C_{ik} &= -c_{ik} \geq 0 \qquad (i \neq k)
\end{aligned}
\right\}.
\tag{3.352}
$$

Equation (3.351) establishes a relation between charges and potential differences (Voltages). The C_{ii} are therefore called *capacitance coefficients*. They are a generalization of the capacitance of a capacitor as defined in Sect. 2.7.

The electrostatic energy of the conductor system can be calculated as follows. Using eq. (2.173) shows

$$
W = \frac{1}{2}\sum_{k=1}^{n} \oint_A \varphi_i \sigma_i dA_i = \frac{1}{2}\sum_{k=1}^{n} \varphi_i Q_i ,
$$

which, when using (3.345) and (3.347) yields

$$
\boxed{W = \frac{1}{2}\sum_{i,k=1}^{n} p_{ik}Q_iQ_k = \frac{1}{2}\sum_{i,k=1}^{n} c_{ik}\varphi_i\varphi_k }.
\tag{3.353}
$$

Another useful theorem, sometimes called the *reciprocity theorem*, is the result of another interesting application of the influence coefficients. Let us consider two different states of a system consisting of n conductors. The potentials φ_i shall belong to the charges Q_i and the potentials $\tilde{\varphi}_i$ shall belong to the charges $\tilde{Q}_i$. Then we write

$$
\sum_{i=1}^{n} \tilde{\varphi}_i Q_i = \sum_{i=1}^{n} \tilde{\varphi}_i \sum_{k=1}^{n} c_{ik}\varphi_k = \sum_{i,k=1}^{n} c_{ki}\varphi_k\tilde{\varphi}_i = \sum_{i,k=1}^{n} c_{ik}\varphi_i\tilde{\varphi}_k
$$

$$
= \sum_{i=1}^{n} \varphi_i \sum_{k=1}^{n} c_{ik}\tilde{\varphi}_k = \sum_{i=1}^{n} \varphi_i \tilde{Q}_i ,
$$

then

$$
\boxed{\sum_{i=1}^{n} \tilde{\varphi}_i Q_i = \sum_{i=1}^{n} \varphi_i \tilde{Q}_i }.
\tag{3.354}
$$

The following **example** shall serve as a simple application. The charge Q_1 shall be distributed on a sphere with arbitrarily small radius. This small sphere shall be located a distance r away from the center of a larger, grounded sphere of radius

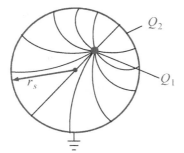

Fig. 3.25

$r_s(r > r_s)$. What are the influence charges on the large (grounded) sphere? We will call these influence charges Q_2. Let us consider two different states of the two-conductor system. The first state is just an arbitrary state to be used for comparison but should be easy to calculate. The second state is the one we would like to solve.

1) $\tilde{Q}_1 = 0, \quad \tilde{Q}_2 = \tilde{Q}_2$

$$\tilde{\varphi}_1 = \frac{\tilde{Q}_2}{4\pi\varepsilon_0 r}, \quad \tilde{\varphi}_2 = \frac{\tilde{Q}_2}{4\pi\varepsilon_0 r_s}$$

2) $Q_1 = Q_1, \quad Q_2 = Q_2$

$$\varphi_1 = \varphi_1, \quad \varphi_2 = 0$$

From (3.354), we find

$$\left.\begin{array}{l} \tilde{Q}_1\varphi_1 + \tilde{Q}_2\varphi_2 = 0 = Q_1\tilde{\varphi}_1 + Q_2\tilde{\varphi}_2 \\[2mm] Q_2 = -Q_1\dfrac{\tilde{\varphi}_1}{\tilde{\varphi}_2} = -Q_1\dfrac{r_s}{r} \end{array}\right\} . \qquad (3.355)$$

This solves the problem. Of course, Q_2 is just the image charge, which is known from Sect. 2.6, when substituting $z_1 = r$. For the case where $r < r_s$, the problem becomes trivial. The influence charge then has to be $Q_2 = -Q_1$, because all force lines have to terminate on the grounded sphere (see Fig. 3.25). This can be obtained formally because of

$$\tilde{\varphi}_1 = \tilde{\varphi}_2 = \frac{\tilde{Q}_2}{4\pi\varepsilon_0 r_s} .$$

3.10 Plane Electrostatic Problems and the Flux Function

Oftentimes, there are problems that only depend on two Cartesian coordinates, *i.e.*, on x and y, but not z. Such planar problems have a number of special properties which we will discuss in this section.

If the field $\mathbf{E}$ is independent of z, then all derivatives of z vanish, *i.e.*, we can write the strictly formal statement

$$\frac{\partial}{\partial z} = 0 \qquad\qquad (3.356)$$

applied to $\mathbf{E}$ gives

$$\nabla \times \mathbf{E} = \left(\frac{\partial E_z}{\partial y} - \frac{\partial E_y}{\partial z}, \frac{\partial E_x}{\partial z} - \frac{\partial E_z}{\partial x}, \frac{\partial E_y}{\partial x} - \frac{\partial E_x}{\partial y} \right)$$

$$= \left(\frac{\partial E_z}{\partial y}, -\frac{\partial E_z}{\partial x}, \frac{\partial E_y}{\partial x} - \frac{\partial E_x}{\partial y} \right) = 0 \; .$$

In particular

$$\frac{\partial E_z}{\partial y} = \frac{\partial E_z}{\partial x} = 0$$

or

$$E_z = \text{const} \; . \qquad\qquad (3.357)$$

We conclude that the component E_z is of no particular significance. It may assume any value, but has to be constant in space. An interesting equation comes from the third component of $\nabla \times \mathbf{E}$, namely

$$\frac{\partial E_y}{\partial x} - \frac{\partial E_x}{\partial y} = 0 \; . \qquad\qquad (3.358)$$

This equation can be satisfied by means of an arbitrary function φ if we define

$$\left. \begin{aligned} E_x &= -\frac{\partial \varphi}{\partial x} \\[2mm] E_y &= -\frac{\partial \varphi}{\partial y} \end{aligned} \right\} \; . \qquad\qquad (3.359)$$

Of course, this should not come as a surprise. It only shows that the electrostatic field, just as in the general, three-dimensional case, can be obtained from a potential.

For the charge-free space we also have

$$\nabla \cdot \mathbf{E} = \frac{\partial E_x}{\partial x} + \frac{\partial E_y}{\partial y} = 0 \; . \qquad\qquad (3.360)$$

If we now, by means of an arbitrary scalar function, define

$$E_x = -\frac{\partial \psi}{\partial y}$$
$$E_y = \frac{\partial \psi}{\partial x}$$
(3.361)

then, eq. (3.360) is obviously satisfied:

$$\frac{\partial}{\partial x}\left(-\frac{\partial \psi}{\partial y}\right) + \frac{\partial}{\partial y}\left(\frac{\partial \psi}{\partial x}\right) = 0 \ .$$

This function ψ is called *flux function*. Thus, one may calculate the field from the potential φ as well as from the flux function ψ. According to eqs. (3.359) and (3.361), φ and ψ belong to the same field if

$$\frac{\partial \varphi}{\partial x} = \frac{\partial \psi}{\partial y}$$
$$\frac{\partial \varphi}{\partial y} = -\frac{\partial \psi}{\partial x}$$
(3.362)

These are the *Cauchy-Riemann differential equations*. We will discuss their fundamental significance for function theory (complex analysis) in the next section. The considerable consequences of φ and ψ meeting these conditions will then become obvious. Prior, we want to mention a few properties of flux functions.

1. Eqs. (3.359) and (3.360) yield for the charge-free space

$$\frac{\partial}{\partial x}\left(-\frac{\partial \varphi}{\partial x}\right) + \frac{\partial}{\partial y}\left(-\frac{\partial \varphi}{\partial y}\right) = -\nabla^2 \varphi = 0$$
(3.363)

which we already learned in Section 2.1. Furthermore – and this is new – the consequence of eqs. (3.358) and (3.361) is

$$\frac{\partial}{\partial x}\left(\frac{\partial \psi}{\partial x}\right) - \frac{\partial}{\partial y}\left(-\frac{\partial \psi}{\partial y}\right) = \nabla^2 \psi = 0 \ .$$
(3.364)

Both, φ and ψ satisfy Laplace's equation and thus, are *harmonic functions*.

2. ψ *is constant along a force line*. We also calculate

$$\mathbf{E} \bullet \nabla \psi = E_x \frac{\partial \psi}{\partial x} + E_y \frac{\partial \psi}{\partial y}$$
$$= E_x E_y + E_y(-E_x) = 0 \ .$$
(3.365)

The vector $\mathbf{E}$ is therefore perpendicular to the vector $\nabla \psi$. Since $\nabla \psi$ is perpendicular to the surface $\psi = \text{const}$, $\mathbf{E}$ has to lie in this plane, which concludes the proof of our claim. On the other hand, $\mathbf{E}$ is perpendicular to the planes $\varphi = \text{const}$, that is, the planes $\varphi = \text{const}$ and $\psi = \text{const}$ intersect everywhere with a right angle (Fig. 3.26). The surfaces $\varphi = \text{const}$ and

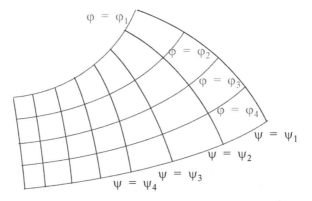

Fig. 3.26

ψ = const form, *for example*, in the plane $z = 0$ an *orthogonal grid* of two perpendicular arrays of curves (*orthogonal trajectories*).

3. The flux function gets is name, among other things, from the fact that it is closely related to the electric flux (current) that "flows" between two points in the plane (Fig. 3.27). Let the points A and B be connected by some contour. The flux passing through this contour per unit length is then

$$\frac{\Omega}{L} = \varepsilon_0 \int_A^B E_\perp \, ds = \varepsilon_0 \int_A^B (\mathbf{E} \times d\mathbf{s})_z = \varepsilon_0 \int_A^B (E_x dy - E_y dx)$$

$$= \varepsilon_0 \int_A^B \left(-\frac{\partial \psi}{\partial y} \, dy - \frac{\partial \psi}{\partial x} \, dx \right) = -\varepsilon_0 \int_A^B d\psi$$

$$\frac{\Omega}{L} = -\varepsilon_0 [\psi(B) - \psi(A)] \ . \tag{3.366}$$

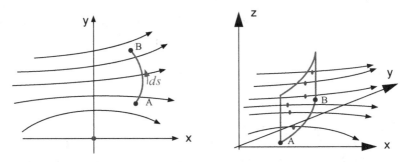

Fig. 3.27

Notice that, except for the factor ε_0, the difference between the flux function at two different points represents the flux that passes between them (per unit length in z direction).

Formally speaking, the flux function was introduced by statement (3.361) to solve eq. (3.360). This approach is not restricted to plane problems, but applicable more generally to any problem which is two dimensional by any kind of symmetry. If one, for instance, takes a cylindrical problem that is rotationally symmetric (*i.e.*, independent of the azimuthal angle), then

$$\varphi = \varphi(r, z) \,, \tag{3.367}$$

with

$$\left.\begin{aligned} E_r &= -\frac{\partial \varphi}{\partial r} \\[2ex] E_z &= -\frac{\partial \varphi}{\partial z} \end{aligned}\right\} . \tag{3.368}$$

Furthermore, by (3.32) for cylindrical coordinates we have

$$\nabla \bullet \mathbf{E} = \frac{1}{r}\frac{\partial}{\partial r} r E_r + \frac{\partial}{\partial z} E_z \,. \tag{3.369}$$

The Ansatz

$$\left.\begin{aligned} E_r &= -\frac{1}{r}\frac{\partial \psi}{\partial z} \\[2ex] E_z &= \frac{1}{r}\frac{\partial \psi}{\partial r} \end{aligned}\right\} , \tag{3.370}$$

satisfies eq. (3.369) for every ψ. Comparison of (3.368) and (3.370) reveals

$$\left.\begin{aligned} E_r &= -\frac{\partial \varphi}{\partial r} = -\frac{1}{r}\frac{\partial \psi}{\partial z} \\[2ex] E_z &= -\frac{\partial \varphi}{\partial z} = +\frac{1}{r}\frac{\partial \psi}{\partial r} \end{aligned}\right\} . \tag{3.371}$$

These equations replace (3.362). Despite a great similarity, there is also a significant difference: Cauchy-Riemann equations emerge only from the plane case. The consequence is that the function-theoretical methods which we will discuss in the next section may only be applied to plane problems – a regrettable restriction.

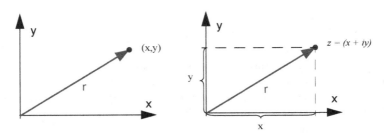

Fig. 3.28

3.11 Analytic Functions and Conformal Mappings

A point in an (x, y) plane can be addressed "uniquely" (*i.e.*, unique in both directions) by a complex number

$$z = x + iy .$$

Notice that z here has no relation to the third Cartesian coordinate, which is anyway irrelevant for plane problems. The complex number is a two-dimensional quantity, whose properties are quite similar to those of a two-dimensional vector. This, therefore, allows to identify z with the vectors $\langle x, y \rangle$ (see Fig. 3.29).

$$z = x + iy \Leftrightarrow \langle x, y \rangle = \mathbf{r} . \tag{3.372}$$

In any case, the complex number z and the vector $\mathbf{r}$ address the same point. It is also possible to use Vector Calculus on complex numbers, where those operations simplify in some sense. The scalar product, for instance,

$$\mathbf{r}_1 \bullet \mathbf{r}_2 = \langle x_1, y_1 \rangle \bullet \langle x_2, y_2 \rangle = x_1 x_2 + y_1 y_2 , \tag{3.373}$$

and the vector product, or more precisely, its only component that does not vanish in the plane case is

$$\mathbf{r}_1 \times \mathbf{r}_2 = \langle x_1, y_1 \rangle \times \langle x_2, y_2 \rangle = x_1 y_2 - y_1 x_2 . \tag{3.374}$$

Now, multiplying two complex numbers,

$$z_1 = x_1 + iy_1$$

and

$$z_2 = x_2 + iy_2 ,$$

gives

$$\begin{aligned} z_1^* z_2 &= (x_1 - iy_1)(x_2 + iy_2) \\ &= x_1 x_2 + y_1 y_2 + i(x_1 y_2 - y_1 x_2) \\ &= \mathbf{r}_1 \bullet \mathbf{r}_2 + i\mathbf{r}_1 \times \mathbf{r}_2 . \end{aligned} \tag{3.375}$$

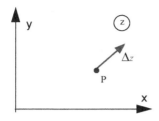

Fig. 3.29

z^* is the complex conjugate, or simply conjugate of the complex number z

$$z^* = x - iy \ . \tag{3.376}$$

Eq. (3.375) states that the product $z_1^* z_2$ can be interpreted in the following way: The real part is the scalar product and the imaginary part is the vector product of the two "vectors" z_1 and z_2.

We may now study any function of complex numbers, $i.e.$, complex functions, *for example*,

$$f(z) = z^2 = x^2 - y^2 + i2xy \ ,$$

or

$$f(z) = z^* = x - iy \ ,$$

etc. Each such function can be split into a real part and an imaginary part, that is, we can always express it in the following form

$$f(z) = u(x, y) + iv(x, y) \ . \tag{3.377}$$

Differentiating such functions does not always provide a unique result. We start by defining the differential quotient, as we know it from real functions (Fig. 3.29)

$$f'(z) = \frac{df(z)}{dz} = \lim_{\Delta z \to 0} \frac{f(z + \Delta z) - f(z)}{\Delta z} \ . \tag{3.378}$$

The general result will depend on the direction of the vector Δz. In any case, the definition (3.378) yields

$$f'(z) = \lim_{\Delta x, \Delta y \to 0} \frac{u(x + \Delta x, y + \Delta y) + iv(x + \Delta x, y + \Delta y) - u(x, y) - iv(x, y)}{\Delta x + i\Delta y}$$

$$= \lim_{\Delta x, \Delta y \to 0} \frac{u(x,y) + \frac{\partial u}{\partial x}\Delta x + \frac{\partial u}{\partial y}\Delta y + iv(x,y) + i\frac{\partial v}{\partial y}\Delta x + i\frac{\partial v}{\partial y}\Delta y - u(x,y) - iv(x,y)}{\Delta x + i\Delta y}$$

$$= \lim_{\Delta x, \Delta y \to 0} \frac{\left(\frac{\partial u}{\partial x} + i\frac{\partial v}{\partial x}\right)\Delta x + \left(\frac{\partial u}{\partial y} + i\frac{\partial v}{\partial y}\right)\Delta y}{\Delta x + i\Delta y} \ .$$

If

$$\Delta y = c \Delta x \ ,$$

then the parameter c defines the direction which we take from point P when differentiating. This gives

$$f'(z) = \lim_{\Delta x \to 0} \frac{\left(\frac{\partial u}{\partial x} + i\frac{\partial v}{\partial x}\right)\Delta x + \left(\frac{\partial u}{\partial y} + i\frac{\partial v}{\partial y}\right)c\Delta x}{\Delta x + ic\Delta x}$$

$$f'(z) = \lim_{\Delta x \to 0} \frac{\left(\frac{\partial u}{\partial x} + i\frac{\partial v}{\partial x}\right) + \left(-i\frac{\partial u}{\partial y} + \frac{\partial v}{\partial y}\right)ic}{1 + ic} \ .$$

If the two expressions in parenthesis are equal, then

$$f'(z) = \frac{\partial u}{\partial x} + i\frac{\partial v}{\partial x} = -i\frac{\partial u}{\partial y} + \frac{\partial v}{\partial y} \ , \tag{3.379}$$

which also means that $f'(z)$ is independent of c, *i.e.* the direction. It is important to realize that eq. (3.379) is necessary for the direction-independence of the derivative. Eq. (3.379) therefore provides a set of necessary and sufficient conditions for a function to be uniquely differentiable – the previously mentioned Cauchy-Riemann differential equations:

$$\boxed{\begin{aligned} \frac{\partial u}{\partial x} &= \frac{\partial v}{\partial y} \\[2mm] \frac{\partial u}{\partial y} &= -\frac{\partial v}{\partial x} \ . \end{aligned}} \tag{3.380}$$

If these conditions are satisfied, then the function $u + iv$ is uniquely differentiable. The function $f'(z) = u + iv$ is then termed to be an *analytic function*. It is furthermore necessary that $f'(z) \neq 0$ and finite. If, for so-called singular points, $f'(z) = 0$ or $f'(z) \Rightarrow \infty$, then this function is not analytic there.

For instance, the function z^2 is analytic because

$$\frac{\partial u}{\partial x} = \frac{\partial}{\partial x}(x^2 - y^2) = 2x = \frac{\partial v}{\partial y}$$

$$\frac{\partial u}{\partial y} = \frac{\partial}{\partial y}(x^2 - y^2) = -2y = -\frac{\partial v}{\partial x} \ ,$$

whereby we have to exclude the origin, a singular point. The function z^*, on the other hand, is not analytic since

$$\frac{\partial u}{\partial x} = 1 \neq \frac{\partial v}{\partial y} = -1 \ .$$

The square of the absolute value zz^* is also not analytic

$$zz^* = x^2 + y^2 \ ,$$

i.e.

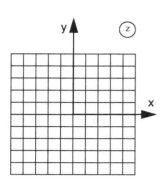

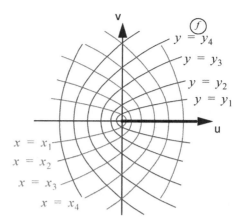

Fig. 3.30

$$u(x, y) = x^2 + y^2$$
$$v(x, y) = 0$$

and

$$\frac{\partial u}{\partial x} = 2x \neq \frac{\partial v}{\partial y} = 0$$

$$\frac{\partial u}{\partial y} = 2y \neq -\frac{\partial v}{\partial x} = 0 \ .$$

Any complex function $f(z)$, whether analytic or not, can be understood as a mapping of the complex plane z onto the plane f or vice versa. As illustration, we choose the example

$$f(z) = z^2 = x^2 - y^2 + 2ixy$$

where

$$u(x, y) = x^2 - y^2$$
$$v(x, y) = 2xy \ .$$

The straight line $x = x_i$ in the z-plane corresponds to a curve in the f-plane (Fig. 3.30), whose parameter representation is

$$u = x_i^2 - y^2$$
$$v = 2x_i y \ .$$

Eliminating y, results in the equation for the curve

$$u = x_i^2 - v^2/(4x_i^2) \ .$$

This is the equation of a parabola that opens to the left and whose focal point is at the origin. Conversely, the straight line $y = y_i$ corresponds to the curve

$$u = x^2 - y_i^2$$
$$v = 2xy_i \quad .$$

or

$$u = v^2/(4y_i^2) - y_i^2 \ ,$$

that is now a parabola, which opens to the right and whose focal point is also at the origin. Collectively, one obtains 2 sets of parabolas. All have their focal points at the origin, *i.e.*, they are *confocal*. As a side note, u and v, together with z from a curvilinear orthogonal coordinate system in which the three-dimensional Laplace equation can be separated. The "coordinates of this parabolic cylinder" generate one of the 11 "separable" coordinate systems. The two sets of parabolas intersect each other at a right angle, which as we shall see, is no accident. This mapping has a number of peculiar characteristics. Each half of the x-axis transforms into the positive half of the u-axis; each half of the y-axis transforms into the negative half of the u-axis. Mapping of only half of the x-y-plane results in a complete coverage of the u-v-plane. One might imagine the u-v-plane as a result of a distortion of the x-y-plane in such a way that the negative x-axis is bent over towards the positive one.

Conversely, the lines $u = u_i$ transform into hyperbolas

$$x^2 - y^2 = u_i \ ,$$

as well as the lines $v = v_i$ (Fig. 3.31), where the whole u-v-plane (f-plane) is mapped onto one half of the x-y-plane.
A little further on, we will discuss what really happens here (see Sect. 3.12, Example 5).
The v-axis ($u = 0$) transforms into the pair of lines

$$x^2 - y^2 = (x+y)(x-y) = 0 \ ,$$

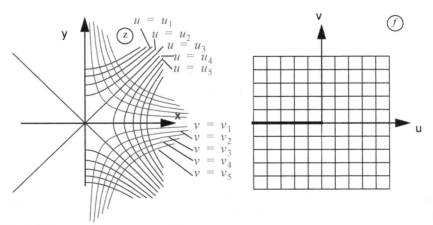

Fig. 3.31

i.e.,

$$x = \mp y$$

and the *u*-axis ($v = 0$) transforms into the pair of lines

$$x \cdot y = 0 \; .$$

i.e.,

$$x = 0$$

or

$$y = 0 \; .$$

If the mapping function is analytic, as in the just discussed example, then the mapping is called *conformal*. This term shall express an important property of the thereby created mappings, namely their angle preserving property (magnitude and sense). In our prove, we will use the fact that every complex number can be written in the form

$$z = x + iy = r(\cos\varphi + i\sin\varphi) = r\exp(i\varphi) \; , \tag{3.381}$$

where

$$\left.\begin{aligned}|z| &= r = \sqrt{x^2 + y^2}\\[4pt] \tan\varphi &= \frac{y}{x}\end{aligned}\right\} \; . \tag{3.382}$$

Given two complex numbers

$$z_1 = r_1\exp(i\varphi_1)$$

and

$$z_2 = r_2\exp(i\varphi_2) \; ,$$

then their product is

$$z_1 z_2 = r_1 r_2\exp[i(\varphi_1 + \varphi_2)] \; . \tag{3.383}$$

Calling *r* the magnitude and φ the argument of a complex number, then the multiplication of complex numbers is carried out by multiplying their magnitudes and adding their arguments

$$\left.\begin{aligned}|z_1 z_2| &= r_1 r_2\\[4pt] \arg(z_1 z_2) &= \varphi_1 + \varphi_2\end{aligned}\right\} \; . \tag{3.384}$$

Consider a point and its neighborhood before and after the conformal mapping as illustrated in Fig. 3.32. Because the mapping is conformal, *i.e., f'* has the same value for all directions, it is

$$df_1 = f'dz_1$$

$$df_2 = f'dz_2 \; .$$

and furthermore

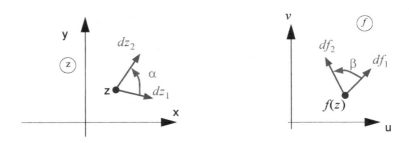

Fig. 3.32

$$\alpha = \arg(z_2) - \arg(z_1)$$

as well as

$$\beta = \arg(df_2) - \arg(df_1)$$
$$= \arg(f'dz_2) - \arg(f'dz_1)$$
$$= \arg(f') + \arg(dz_2) - \arg(f') - \arg(dz_1)$$
$$= \arg(dz_2) - \arg(dz_1)$$

that is

$$\boxed{\alpha = \beta}. \tag{3.385}$$

This concludes the proof of the angle preserving property. Furthermore, this clarifies why the two sets of parabolas of Fig. 3.30 and the two sets of hyperbolas of Fig. 3.31 intersect with a right angle.

From the angle preserving property it follows that conformal mappings in a "small scale" are similar, i.e., mapping transforms infinitesimally small figures into similar ones, whereby the linear dimensions are scaled by a factors $|f'(z)|$ and its surfaces by a factor $|f'(z)|^2$. The scale factor can magnify or shrink the result. Overall, we can say that small squares are mapped into small squares, right angles remain right angles, and therefore, an orthogonal grid transforms into an orthogonal grid, as shown in Fig. 3.30 and Fig. 3.31. Despite the likeness in the small scale, there is no likeness in the large scale, as is also obvious from Fig. 3.30 and Fig. 3.31. Finite figures are not mapped as like figures, but are distorted.

Real part and imaginary part of an analytic function are harmonic functions. This is a consequence of the Cauchy-Riemann equations (3.380). Because of

$$\nabla^2 u = \frac{\partial}{\partial x}\left(\frac{\partial u}{\partial x}\right) + \frac{\partial}{\partial y}\left(\frac{\partial u}{\partial y}\right)$$
$$= \frac{\partial}{\partial x}\frac{\partial v}{\partial y} + \frac{\partial}{\partial y}\left(-\frac{\partial v}{\partial x}\right) = 0 . \tag{3.386}$$

and

$$\nabla^2 v = \frac{\partial}{\partial x}\left(\frac{\partial v}{\partial x}\right) + \frac{\partial}{\partial y}\left(\frac{\partial v}{\partial y}\right)$$

$$= \frac{\partial}{\partial x}\left(-\frac{\partial u}{\partial y}\right) + \frac{\partial}{\partial y}\left(\frac{\partial u}{\partial x}\right) = 0 \ .$$

(3.387)

By the coordinate transformation

$$\left. \begin{array}{l} u = u(x, y) \\ v = v(x, y) \end{array} \right\}$$

(3.388)

in a plane, one can introduce a new coordinate system. These new coordinates are orthogonal as long as u and v satisfy the Cauchy-Riemann equations. We shall note without proof, that a consequence of the Cauchy-Riemann equations is, that Laplace's equation

$$\frac{\partial^2 \varphi}{\partial x^2} + \frac{\partial^2 \varphi}{\partial y^2} = 0$$

(3.389)

transforms into

$$\frac{\partial^2 \varphi}{\partial u^2} + \frac{\partial^2 \varphi}{\partial v^2} = 0 \ ,$$

(3.390)

that is, it maintains its form. Obviously, Laplace's equation is also separable in u and v as in Cartesian coordinates x and y. This justifies the previous claim that the plane two-dimensional Laplace equation is separable in any number of coordinate systems (Sect. 3.5). Every analytic function provides such a system of coordinates. This is remarkable because the situation in the three-dimensional space is quite different, where only 11 coordinate systems are "separable".

3.12 The Complex Potential

When comparing the statements of the last two sections, one realizes that real part and imaginary part of an analytic function behave exactly like the potential and flux function of an electrostatic field. The reader is encouraged to compare eqs. (3.362) and (3.380), as well as (3.363), (3.364) and (3.386), (3.387). One may conclude that every analytic function can be regarded in an electrostatic way. Its real part u can be identified with the potential φ and its imaginary part v with the flux function ψ of the related field. In light of this point of view, the analytic function $w(z)$ is called *complex potential*.

$$\boxed{w(z) = u(x, y) + iv(x, y)}$$

$$\qquad\quad \text{potential} \quad \text{flux function}$$

(3.391)

The accompanying field is

$$E_x = -\frac{\partial u}{\partial x} = -\frac{\partial v}{\partial y} \left.\vphantom{\frac{\partial u}{\partial x}}\right\}$$
$$E_y = -\frac{\partial u}{\partial y} = +\frac{\partial v}{\partial x} \left.\vphantom{\frac{\partial u}{\partial y}}\right\}$$

(3.392)

Naturally, with $w(z)$ so is $iw(z)$ an analytic function. The potential here is

$$\tilde{w}(z) = iw(z) = -v(x, y) + iu(x, y)$$

potential   flux function

(3.393)

and the accompanying field is now

$$\tilde{E}_x = \frac{\partial v}{\partial x} = -\frac{\partial u}{\partial y} \left.\vphantom{\frac{\partial v}{\partial x}}\right\}$$
$$\tilde{E}_y = \frac{\partial v}{\partial y} = \frac{\partial u}{\partial x} \left.\vphantom{\frac{\partial v}{\partial y}}\right\}$$

(3.394)

Thus, multiplication by i results basically in the exchange of potential and flux function (disregarding the sign). This is also expressed by

$$\mathbf{E} \bullet \tilde{\mathbf{E}} = 0 ,$$

(3.395)

that is, the fact that $\mathbf{E}$ is perpendicular to $\tilde{\mathbf{E}}$. This allows to interpret every analytic function in two ways:

1.	$u \leftrightarrow$ potential	$v \leftrightarrow$ flux function
2.	$-v \leftrightarrow$ potential	$u \leftrightarrow$ flux function

We might as well define a complex electric field

$$E = E_x + iE_y .$$

(3.396)

Because of $\partial z / \partial x = 1$, we find

$$\frac{dw(z)}{dz} = \frac{\partial w(z)}{\partial x} = \frac{\partial u}{\partial x} + i\frac{\partial v}{\partial x}$$
$$= -E_x + iE_y = -E^*$$

or

$$\frac{dw^*(z)}{dz^*} = \frac{\partial u}{\partial x} - i\frac{\partial v}{\partial x} = -E_x - iE_y = -E$$

and

$$\frac{dw(z)}{dz}\frac{dw^*(z)}{dz^*} = EE^* = |E|^2 .$$

(3.397)

Therefore, singular points of the potential are also singular points of the electric field. In particular, the electric field is infinitely large at points where the derivative $w' = dw/dz$ is infinite.

Every analytic function solves a set of electrostatic problems. The following shall serve to study a number of such complex potentials. It is easy to establish a catalog of complex potentials, along with their accompanying fields (which one may picture as being a result of the conformal mapping of the uniform field), and then determine from the result which boundary value problems they solve. The reverse approach, *i.e.* to start from a boundary value problem and then calculate its potential, is much more difficult. We will limit ourselves to a small catalog of interesting mappings.

Example 1:

$$w = -\frac{q}{2\pi\varepsilon_0} \ln\frac{z}{z_B} .$$

Using

$$z = r\exp(i\varphi), \quad z_B = r_B\exp(i\varphi_B) .$$

gives

$$w = -\frac{q}{2\pi\varepsilon_0} \ln[r\exp(i\varphi)] + \frac{q}{2\pi\varepsilon_0} \ln z_B .$$

or

$$w = -\frac{q}{2\pi\varepsilon_0} \ln r - \frac{q}{2\pi\varepsilon_0} \ln\exp(i\varphi) + \frac{q}{2\pi\varepsilon_0} \ln r_B + \frac{q}{2\pi\varepsilon_0} \ln\exp(i\varphi_B) .$$

$$w = -\frac{q}{2\pi\varepsilon_0} \ln\frac{r}{r_B} - i\frac{q}{2\pi\varepsilon_0}(\varphi - \varphi_B) .$$

Therefore

$$u = -\frac{q}{2\pi\varepsilon_0} \ln\frac{r}{r_B} .$$

$$v = -\frac{q}{2\pi\varepsilon_0}(\varphi - \varphi_B) .$$

The result is that of the straight, uniform line-charge q, which we know already. Its equipotentials are circles around the line-charge and the field lines spread radially from it:

$$\mathbf{E} = -\nabla u .$$

i.e.,

$$E_r = -\frac{\partial u}{\partial r} = \frac{q}{2\pi\varepsilon_0 r} ,$$

$$E_\varphi = -\frac{1}{r}\frac{\partial u}{\partial\varphi} = 0 .$$

Conversely, one may also start with

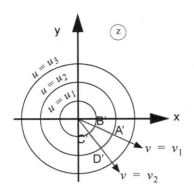 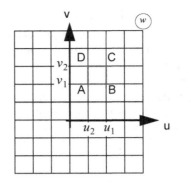

Fig. 3.33

$$\frac{dw}{dz} = -\frac{q}{2\pi\varepsilon_0 z} = -E_x + iE_y$$

$$= -\frac{qz^*}{2\pi\varepsilon_0 zz^*} = -\frac{q}{2\pi\varepsilon_0}\frac{x - iy}{x^2 + y^2} \quad .$$

i.e.,

$$E_x = \frac{q}{2\pi\varepsilon_0}\frac{x}{x^2 + y^2} = \frac{q}{2\pi\varepsilon_0}\frac{x}{r^2}$$

$$E_y = \frac{q}{2\pi\varepsilon_0}\frac{y}{x^2 + y^2} = \frac{q}{2\pi\varepsilon_0}\frac{y}{r^2} \quad ,$$

which also results in

$$E_r = E_x\cos\varphi + E_y\sin\varphi$$

$$= E_x\frac{x}{r} + E_y\frac{y}{r} = \frac{q}{2\pi\varepsilon_0}\left(\frac{x^2}{r^3} + \frac{y^2}{r^3}\right) = \frac{q}{2\pi\varepsilon_0}\frac{1}{r} \quad .$$

Fig. 3.33 illustrates some properties of the conformal mapping $w(z)$. To simplify, φ_B was chosen to be zero. Then one gets

$$u = -\frac{q}{2\pi\varepsilon_0}\ln\frac{r}{r_B}, \qquad \frac{r}{r_B} = \exp\left(-\frac{2\pi\varepsilon_0 u}{q}\right) \quad ,$$

$$v = -\frac{q}{2\pi\varepsilon_0}\varphi \quad .$$

The tetragon $ABCD$ in the w-plane is mapped in the z-plane onto the shape $A'B'C'D'$. This shape closes to a full annulus when $v_2 - v_1 = q/\varepsilon_0$. If the difference $v_2 - v_1$ increases further, then the annulus may be covered multiple times. The situation is simpler with respect to u. r tends to infinity ($r \to \infty$) when

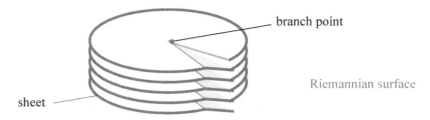

Fig. 3.34

$u_1 \to -\infty$, and $r \to 0$ when $u_1 \to +\infty$. Overall, we find that the u-v-plane covers the x-y-plane arbitrarily may times.

Every stripe

$$\left.\begin{array}{l} -\infty < u_1 < \infty \\ v_1 \le v \le v_1 + q/\varepsilon_0 \end{array}\right\} . \tag{3.398}$$

produces the entire x-y-plane. The resulting plane is a connected surface with infinitely many sheets, and is called a *Riemannian surface*. These sheets turn, so to speak, infinitely many times around the origin which is called its *branch point* (Fig. 3.34). The mapping is not conformal at this point.

Example 2

$$w(z) = iC\ln(z/r_B) .$$

Now, we have

$$u = -C\varphi$$
$$v = C\ln(r/r_B)$$

from which we find (Fig. 3.35)

$$E_r = 0$$
$$E_\varphi = C/r .$$

Comparison with Example 1 shows that this result is very similar, but just has potential and flux function exchanged. Fields with such properties are actually possible. The only thing to do is to implement the equipotentials (these are surfaces where φ is constant) as actual conductor surfaces.

Example 3

$$w(z) = \frac{q}{2\pi\varepsilon_0} \ln \frac{z + \dfrac{d}{2}}{z - \dfrac{d}{2}} .$$

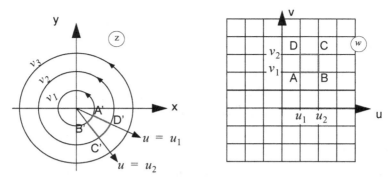

Fig. 3.35

This is the field of two line charges, one is positive $(+q)$ at location $(x = d/2, y = 0)$ and one is negative $(-q)$ at location $(x = -d/2, y = 0)$, for which we write:

$$w(z) = -\frac{q}{2\pi\varepsilon_0}\ln\frac{z-\dfrac{d}{2}}{z_B} + \frac{q}{2\pi\varepsilon_0}\ln\frac{z+\dfrac{d}{2}}{z_B} = \frac{q}{2\pi\varepsilon_0}\ln\frac{z+\dfrac{d}{2}}{z-\dfrac{d}{2}}.$$

We use some algebra in order to separate real and imaginary part

$$w(z) = \frac{q}{2\pi\varepsilon_0}\ln\frac{x+\dfrac{d}{2}+iy}{x-\dfrac{d}{2}+iy} = \frac{q}{2\pi\varepsilon_0}\left\{ \begin{array}{l} \ln\left[\sqrt{\left(x+\dfrac{d}{2}\right)^2+y^2}\cdot\exp(i\varphi_-)\right] \\ -\ln\left[\sqrt{\left(x-\dfrac{d}{2}\right)^2+y^2}\cdot\exp(i\varphi_+)\right] \end{array}\right\}$$

$$= \frac{q}{2\pi\varepsilon_0}\ln\sqrt{\frac{\left(x+\dfrac{d}{2}\right)^2+y^2}{\left(x-\dfrac{d}{2}\right)^2+y^2}} + i\frac{q}{2\pi\varepsilon_0}(\varphi_- - \varphi_+),$$

where φ_+ and φ_- are given by Fig. 3.36. Therefore

$$u = \frac{q}{4\pi\varepsilon_0}\ln\frac{\left(x+\dfrac{d}{2}\right)^2+y^2}{\left(x-\dfrac{d}{2}\right)^2+y^2}$$

and

$$v = \frac{q}{2\pi\varepsilon_0}(\varphi_- - \varphi_+).$$

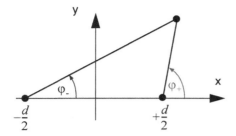

Fig. 3.36

The equipotential surfaces $u = u_i$ are circles (Circles of Apollonius). From

$$u_i = \frac{q}{4\pi\varepsilon_0} \ln \frac{\left(x + \frac{d}{2}\right)^2 + y^2}{\left(x - \frac{d}{2}\right)^2 + y^2}$$

follows

$$\frac{\left(x + \frac{d}{2}\right)^2 + y^2}{\left(x - \frac{d}{2}\right)^2 + y^2} = \exp\left(\frac{4\pi\varepsilon_0 u_i}{q}\right) = C_i$$

from which, when using

$$\left[x + \frac{d(1 + C_i)}{2(1 - C_i)}\right]^2 + y^2 = \frac{d^2 C_i}{(1 - C_i)^2}$$

one derives the equation for a circle whose center is at the x-axis. From $v = v_i$ follows

$$\frac{1}{d_i} = \tan\frac{2\pi\varepsilon_0 v_i}{q} = \tan(\varphi_- - \varphi_+)$$

$$= \frac{\tan\varphi_- - \tan\varphi_+}{1 + \tan\varphi_- \cdot \tan\varphi_+} = \frac{\dfrac{y}{x + d/2} - \dfrac{y}{x - d/2}}{1 + \dfrac{y^2}{x^2 - \dfrac{d^2}{2}}} = \frac{-yd}{x^2 + y^2 - \dfrac{d^2}{2}}$$

or

$$x^2 + \left(y + \frac{d_i d}{2}\right)^2 = \frac{d^2}{4}(1 + d_i^2) \ ,$$

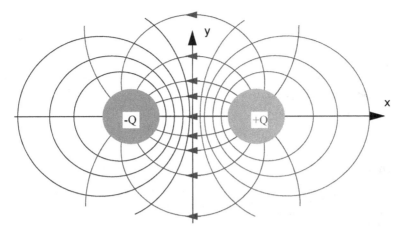

Fig. 3.37

that is, the equation for a circle whose center is at the y-axis. Both sets of curves
consist of circles (Fig. 3.37). The thereby described field has many applications.
By appropriate choice of the parameters, one can obtain *for example*, the field of an
eccentric cylindrical capacitor (Fig. 3.38a) or that of a conductor made from two
cylinders (Fig. 3.38b). This resembles the previously discussed problem of
imaging a line charge on a cylinder (Sect. 2.6.3).

Example 4

$$w(z) = \frac{qd}{2\pi\varepsilon_0 z}$$

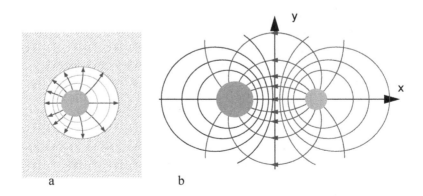

Fig. 3.38

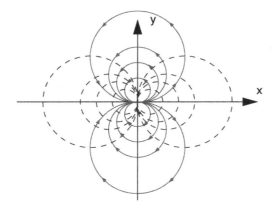

Fig. 3.39

If we start from Example 3 and take the limit to a line dipole ($d \to 0$, $q \to \infty$, qd finite) then one finds

$$w(z) = \frac{q}{2\pi\varepsilon_0} \ln \frac{1 + \dfrac{d}{2z}}{1 - \dfrac{d}{2z}} \approx \frac{q}{2\pi\varepsilon_0} \ln\left(1 + \frac{d}{2z}\right)\left(1 + \frac{d}{2z}\right)$$

$$\approx \frac{q}{2\pi\varepsilon_0} \ln\left(1 + \frac{d}{z}\right) \approx \frac{qd}{2\pi\varepsilon_0 z} \ .$$

This is the complex potential of a line dipole.

$$w(z) = \frac{qd}{2\pi\varepsilon_0 z} = \frac{qdz^*}{2\pi\varepsilon_0 zz^*} = \frac{qd(x - iy)}{2\pi\varepsilon_0(x^2 + y^2)},$$

$$u(x, y) = \frac{qd}{2\pi\varepsilon_0} \frac{x}{x^2 + y^2},$$

$$v(x, y) = -\frac{qd}{2\pi\varepsilon_0} \frac{y}{x^2 + y^2} \ .$$

From this one finds the equipotentials to be circles through the origin (centered on the x-axis) and the flux lines are also circles through the origin (centered on y-axis) as illustrated in Fig. 3.39, (which may also be regarded as being a result of taking the limit of Fig. 3.37).

Example 5

$$w = Cz^p$$

$$w = Cz^p = C(x + iy)^p = C[r\exp(i\varphi)]^p = Cr^p \exp(ip\varphi)$$

i.e.,

$$u = Cr^p \cos(p\varphi)$$

$$v = Cr^p \sin(p\varphi)$$

The lines $\cos p\varphi = 0$, *i.e.*,

$$p\varphi = \frac{2n-1}{2}\pi$$

characterize special equipotential surfaces ($u = 0$), *for example*, for $n = 0$ and $n = 1$, this gives

$$\varphi = \pm\frac{\pi}{2p} .$$

The lines $\sin p\varphi = 0$, *i.e.*,

$$p\varphi = n\pi ,$$

are special flux lines ($v = 0$), for instance

$$\varphi = 0 .$$

When using surfaces of conductors, which are a possible implementation of equipotential surfaces, one obtains fields like those illustrated in Fig. 3.40 (p > 1) and Fig. 3.41 (p < 1).

For the case of Fig. 3.41 (p < 1), the derivative

$$\frac{\mathrm{d}w}{\mathrm{d}z} = Cpz^{p-1} = \frac{Cp}{z^{1-p}} ,$$

diverges at the origin. Therefore, the electric field becomes infinitely large. This is a typical characteristic of tips, and is also of great practical significance. Figs. 3.30 and 3.31 are special cases where $p = 1/2$ and $p = 2$, respectively.

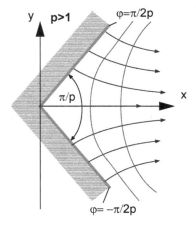

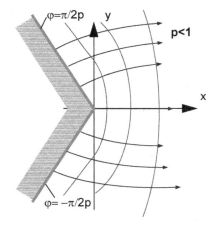

Fig. 3.40   Fig. 3.41

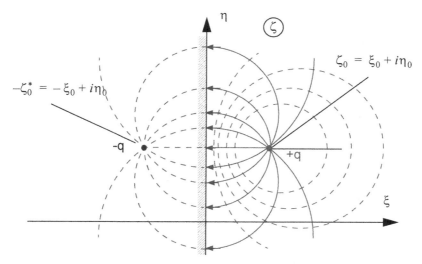

Fig. 3.42

Example 6

$$w(z) = \frac{q}{2\pi\varepsilon_0} \ln\left(\frac{z^p + z_0^{*p}}{z^p - z_0^p}\right)$$

We may stage conformal mappings. For instance, we want to calculate the field of a line charge inside a wedge-shaped area, like that of Fig. 3.40. As long as 2π is an even number multiple of the opening angle, we are able to solve this problem by multiple imaging steps, as indicated for the for point charges in Figs. 2.34 and 2.35, where the opening angle is π and $\pi/2$.

The imaging method can not be used for arbitrary angles. However, one may approach this problem as follows. First, by writing

$$w = \frac{q}{2\pi\varepsilon_0} \ln\left(\frac{\zeta + \zeta_0^{*}}{\zeta - \zeta_0}\right),$$

one obtains the field of two line charges in the ζ-plane, which differ from Example 3 only by a shift parallel to the η-axis (Fig. 3.42), where the η-axis has to be seen as the equipotential surface (conductor surface) and $-q$ is the image charge. Now applying the mapping z^p of Example 5 already solves the problem (Fig. 3.43):

$$\zeta = z^p,$$

and

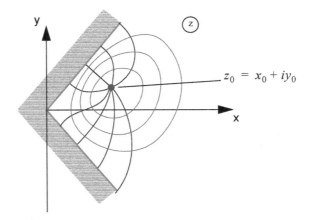

Fig. 3.43

$$w(z) = \frac{q}{2\pi\varepsilon_0} \ln\left(\frac{z^p + z_0^{*p}}{z^p - z_0^p}\right)$$

where

$$\zeta_0 = z_0^p,$$

is the location of the line charge in Fig. 3.42 and z_0 its location in Fig. 3.43.

Example 7

$$z = a\cosh\frac{w}{w_0},$$

$$\cosh\frac{w}{w_0} = \cosh(u + iv) = \frac{1}{2}[\exp(u + iv) + \exp(-u - iv)]$$

$$= \frac{1}{2}\{\exp(u)[\cos v + i\sin v] + \exp(-u)[\cos v - i\sin v]\}$$

$$= \cos v \cosh u + i\sin v \sinh u \ ,$$

i.e.,

$$z = x + iy = a\cos v \cosh u + ia\sin v \sinh u,$$

or

$$x = a\cos v \cosh u$$
$$y = a\sin v \sinh u \ ,$$

so that

$$\cos^2 v + \sin^2 v = 1 = \frac{x^2}{a^2\cosh^2 u} + \frac{y^2}{a^2\sinh^2 u},$$

and

$$\cosh^2 u - \sinh^2 u \ = \ 1 \ = \ \frac{x^2}{a^2 \cos^2 v} - \frac{y^2}{a^2 \sin^2 v} \ .$$

This shows us that the equipotentials (u = const) are elliptical cylinders and the flux lines (v = const) are hyperbolic cylinders. They are all confocal because for an ellipse, we have the relation

$$e^2 \ = \ a^2 \cosh^2 u - a^2 \sinh^2 u \ = \ a^2 \ .$$

and the relation for a hyperbola is

$$e^2 \ = \ a^2 \cos^2 v + a^2 \sin^2 v \ = \ a^2 \ .$$

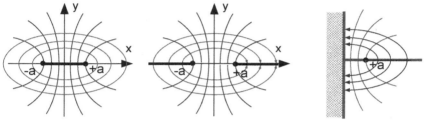

Fig. 3.44   Fig. 3.45   Fig. 3.46

Their graph is given in Fig. 3.44. Special cases are, *for example*, the field between two opposite edges (Fig. 3.45), or the field of an edge opposite a plane (Fig. 3.46).
    Furthermore, the mapping

$$u \ = \ u(x, y)$$
$$v \ = \ v(x, y)$$

is remarkable because u and v, together with z constitute an orthogonal system in the three-dimensional space, which allows for the separation of variables for Laplace's equation (coordinates of the *elliptical cylinder*).

Example 8

$$w(z) \ = \ \sqrt{z^2 + B^2}$$

This rather simple looking mapping exhibits peculiar properties and shall serve here as an example for many of similar kind.
Squaring gives

$$w^2 \ = \ u^2 - v^2 + i2uv \ = \ z^2 + B^2 \ = \ x^2 - y^2 + B^2 + i2xy \ ,$$

i.e.,

$$u^2 - v^2 \ = \ x^2 - y^2 + B^2$$

and

$$uv \ = \ xy \ .$$

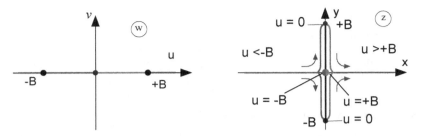

Fig. 3.47

If $v = 0$ (u-axis) then

$$\left.\begin{array}{l} 0 \le u^2 = x^2 - y^2 + B^2 \\ xy = 0 \end{array}\right\}$$

Thus, it has to be either $x = 0$ or $y = 0$. The case $x = 0$ yields

$$0 \le -y^2 + B^2$$
$$y^2 \le B^2 \qquad (x = 0)$$

and the case $y = 0$ yields

$$0 \le x^2 + B^2 \qquad (y = 0) \,,$$

which is always the case. This now, results in a peculiar behavior of the u-axis in the z-plane. The range $-\infty < u \le -B$, (i.e., that part of the negative u-axis), maps to the negative x-axis. Similarly, for the range $B \le u < +\infty$, (i.e., that part of the positive u-axis), maps to the positive x-axis ($0 \le x < +\infty$). Both points $u = \pm B$ map onto the origin of the z-plane. The section of the u-axis between -B and +B maps onto the points between -B and +B on the y-axis. For $-B \le u \le 0$, the image points on the y-axis move from the origin towards the points +B and -B. For $0 \le u \le +B$, the image points on the y-axis move back towards the origin

Conversely, we find for $u = 0$ (v-axis)

$$\left.\begin{array}{l} 0 \ge -v^2 = x^2 - y^2 + B^2 \\ xy = 0 \end{array}\right\} \,.$$

Here, $y = 0$ is impossible because this would require

$$-v^2 = x^2 + B^2 \,,$$

which is not possible since v has to be real. Hence, $x = 0$ and

$$-v^2 = -y^2 + B^2 \,.$$

i.e.,

$$y^2 = B^2 + v^2 \ge B^2 \,.$$

This means either

$$y > B$$

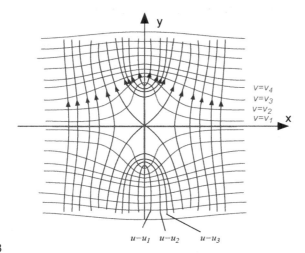

$v=v_4$
$v=v_3$
$v=v_2$
$v=v_1$

$u=u_1$ $u=u_2$ $u=u_3$

Fig. 3.48

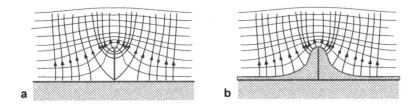

a   b

Fig. 3.49

or

$$y < -B .$$

Now, the v-axis maps onto the part of the y-axis that lacks the range from $-B$ to $+B$. The current conformal mapping is an example of an interesting category of mappings, the Schwarz-Christoffel mappings.

Fig. 3.48 illustrates this configuration, which is suitable to determine the fields for a more or less sharp edge (Fig. 3.49).

4 The Stationary Current Density Field

The following shall discuss the field of the stationary electric currents, or more precisely, the field generated by the current density **g**. By certain simplifying assumptions, we may reduce the current density problems to electrostatic problems of Chapter 2 and 3.

4.1 The Basic Equations

For electric currents inside metallic conductors to exist, Ohm's law requires that there be an electric field present, such that

$$\mathbf{g} = \kappa \mathbf{E} \, , \tag{4.1}$$

where κ is the *specific* electric conductivity. It shall be emphasized that this simple form of Ohm's law is not always valid. Oftentimes, it has to be replaced by much more complicated relations. Even if eq. (4.1) is applicable in the given form, κ may still depend on the location, *i.e.*, for some heterogeneous material, or even in a homogeneous material, if a non-uniform magnetic field is applied, which is a property that is frequently exploited to measure magnetic fields. This shall not be our concern here, however. We assume that κ is at least piece wise constant.

We have already derived the continuity equation, or the principle of charge conservation (1.58):

$$\nabla \bullet \mathbf{g} + \frac{\partial \rho}{\partial t} = 0 \, , \tag{4.2}$$

which reduces in the stationary case to

$$\nabla \bullet \mathbf{g} = 0 \, . \tag{4.3}$$

From (4.1) and (4.2) follows

$$\nabla \bullet (\kappa \mathbf{E}) = 0 \, .$$

Letting κ be constant and using

$$\mathbf{E} = -\nabla \varphi$$

gives

$$\kappa \nabla \bullet \mathbf{E} = -\kappa \nabla \bullet (\nabla \varphi) = 0 \, ,$$

or

$$\nabla^2 \varphi = 0 \, . \tag{4.4}$$

For an arbitrary volume, this results in

$$\int_V \nabla \bullet \mathbf{g} \, d\tau = \oint_A \mathbf{g} \bullet d\mathbf{A} = 0 \, . \tag{4.5}$$

G. Lehner, *Electromagnetic Field Theory for Engineers and Physicists*,
DOI 10.1007/978-3-540-76306-2_4, © Springer-Verlag Berlin Heidelberg 2010

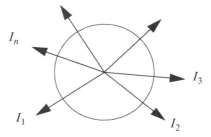

Fig. 4.1

This means that for the stationary case, the sum of all currents emerging from the volume have to vanish. A special case of this is Kirchhoff's first theorem (Fig. 4.1).

$$\sum_{i=1}^{n} I_i = 0 \ . \tag{4.6}$$

This is a fundamental relation in electric circuits theory.

The question whether stationary currents actually exist arises. Based on our previous knowledge, one might get the impression that they did not. To find the answer, consider a charged capacitor within which one introduces a conductor (Fig. 4.2). Initially, there is a field inside the conductor that causes currents. This represents moving charges, which will cease moving only, when the field inside the conductor has vanished, that is, when influence charges appear. The current in the conductor decays. The time for this to occur is the relaxation time and depends on the conductivity (see Sect. 4.2). Similar considerations lead to decay of currents in every "electrostatic" situation. Since

$$\nabla \times \mathbf{E} = 0$$

or

$$\oint \mathbf{E} \bullet d\mathbf{s} = 0 \tag{4.7}$$

and therefore

$$\oint \mathbf{g} \bullet d\mathbf{s} = 0 \ . \tag{4.8}$$

In a stationary, closed circuit, this can only be satisfied if

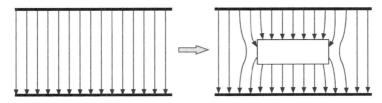

Fig. 4.2

$$\mathbf{g} = 0 \ . \tag{4.9}$$

This is a consequence of the Helmholtz theorem (which will be discussed in Appendix A.5)

However, to create stationary currents is possible by using an impressed electric field such as batteries produce. This results in

$$V = \oint \mathbf{E}_e \bullet d\mathbf{s} \neq 0 \ , \tag{4.10}$$

i.e., there must be a voltage source which provides an *electromotive force* (EMF). Putting things together gives

$$g = \kappa(\mathbf{E} + \mathbf{E}_e) \ , \tag{4.11}$$

and therefore

$$\oint \mathbf{g} \bullet d\mathbf{s} = \kappa \oint \mathbf{E} \bullet d\mathbf{s} + \kappa \oint \mathbf{E}_e \bullet d\mathbf{s}$$

i.e.,

$$V = \frac{1}{\kappa} \oint \mathbf{g} \bullet d\mathbf{s} \ . \tag{4.12}$$

It is then

$$\frac{1}{\kappa} \oint \mathbf{g} \bullet d\mathbf{s} = IR \ , \tag{4.13}$$

which may be regarded as the definition for resistance R. R depends on the geometry and the material, but not on the current. If we separate the resistance into an external part R_x and an internal one for the voltage source R_i, then

$$V = IR = I(R_x + R_i) \ . \tag{4.14}$$

The definition

$$I = \oint_A \mathbf{g} \bullet d\mathbf{A} \ , \tag{4.15}$$

and eq. (4.13) yields

$$R = \frac{1}{\kappa} \frac{\oint \mathbf{g} \bullet d\mathbf{s}}{\int_A \mathbf{g} \bullet d\mathbf{A}} \ . \tag{4.16}$$

If we do not integrate over the closed circuit, but just the section outside the voltage source, then we find for R_x

$$R_x = \frac{1}{\kappa} \frac{\int_x \mathbf{g} \bullet d\mathbf{s}}{\int_A \mathbf{g} \bullet d\mathbf{A}} \ . \tag{4.17}$$

Fig. 4.3, makes it obvious that the actual integration path is irrelevant, since in any case

$$\frac{1}{\kappa} \int_x \mathbf{g} \bullet d\mathbf{s} = \int_1^2 E \bullet d\mathbf{s} = \varphi_2 - \varphi_1 \ .$$

Also irrelevant is the choice of the cross section, since the current

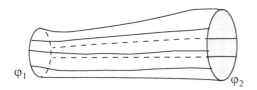

Fig. 4.3

$$I = \int_A \mathbf{g} \bullet d\mathbf{A}$$

in the stationary case is the same for every cross section. This simplifies for a conductor with uniform cross section and uniform current density:

$$R_x = \frac{1}{\kappa} \frac{\oint_x dl}{\int_A dA} = \frac{l}{\kappa A} \,, \tag{4.18}$$

which is the familiar formula for the resistance.

This shows that one can create a stationary current density by means of an EMF. The fact that this is not strictly true for an actual EMF (*i.e.* after its depletion) is immaterial. What matters is that it should be possible to maintain the current density for a time period that is longer than the characteristic times of the system and the relaxation time in particular, which we will discuss in the next section (the relaxation time is very short for good conductors).

The unit of resistance in the MKSA system is 1 Ohm (1Ω):

$$1\Omega = \frac{1V}{1A} \,. \tag{4.19}$$

One finds the unit for conductivity from eq. (4.18)

$$\frac{1m}{1\Omega \cdot 1m^2} = \frac{1}{1\Omega \cdot 1m} \,. \tag{4.20}$$

Good conductors are metals such as Silver and Copper with values

$$\kappa_{Ag} = 6.17 \cdot 10^7 (\Omega m)^{-1}$$

$$\kappa_{Cu} = 5.80 \cdot 10^7 (\Omega m)^{-1} \,,$$

4.2   Relaxation Time

Charges and their related fields decay in a conductive medium. In the pure electrostatic case, the final state is one where the inside of the conductor is free of any field, and all charges are located at its surface. Consequently, ρ is a function of $\mathbf{r}$ and t

$$\rho = \rho(\mathbf{r}, t) \,.$$

From

$$\nabla \bullet \mathbf{g} + \frac{\partial \rho}{\partial t} = 0 \ ,$$

$$\mathbf{g} = \kappa \mathbf{E} = \frac{\kappa \mathbf{D}}{\varepsilon} \ ,$$

and

$$\nabla \bullet \mathbf{D} = \rho \ ,$$

follows that

$$\frac{\kappa}{\varepsilon} \rho + \frac{\partial \rho}{\partial t} = 0 \ . \tag{4.21}$$

This allows for the introduction of the dimensionless time

$$\tau = t \frac{\kappa}{\varepsilon} = \frac{t}{\varepsilon / \kappa} = \frac{t}{t_r} \ , \tag{4.22}$$

which gives time in units of the so-called relaxation time

$$t_r = \frac{\varepsilon}{\kappa} \ . \tag{4.23}$$

Thus

$$\rho + \frac{\partial \rho}{\partial \tau} = 0 \ . \tag{4.24}$$

Trying

$$\rho = h(\mathbf{r})f(\tau)$$

yields

$$f(\tau) + \frac{\partial}{\partial \tau} f(\tau) = 0$$

i.e.,

$$f(\tau) = C \exp(-\tau)$$

and

$$\rho = C \, h(\mathbf{r}) \exp(-\tau)$$

For $\tau = 0$ this simplifies to

$$\rho(\mathbf{r}, 0) = C \, h(\mathbf{r})$$

and then

$$\rho(\mathbf{r}, t) = \rho(\mathbf{r}, 0) \exp(-\tau) = \rho(\mathbf{r}, 0) \exp(-t/t_r) \ . \tag{4.25}$$

Consequently, the charges decay with the relaxation time t_r. Depending on the material, t_r may take on a wide range of values. The following examples shall illustrate this

good conductor: $(\varepsilon = \varepsilon_0)\begin{cases} \text{Silver} & t_r \approx 1.4 \cdot 10^{-19}\text{s} \\ \text{Copper} & t_r \approx 1.5 \cdot 10^{-19}\text{s} \end{cases}$

poor conductor: distilled water: $t_r \approx 10^{-6}\text{s}$

Insulator: molten quartz : $t_r \approx 10^6\text{s} \approx 10\,\text{days}$

4.3 Boundary Conditions

The continuity equation for charges (eq. (4.2)) applied to the boundary between two media yields a relation between the normal components of **g** and the surface charge (Fig. 4.4). Integrating this equation over a small slice of the volume (*i.e.*, an area) gives

$$g_{2n}dA - g_{1n}dA + \frac{\partial\sigma}{dt}dA = 0 \ ,$$

or when cancelling the dA

$$\boxed{g_{2n} - g_{1n} + \frac{\partial\sigma}{dt} = 0} \ . \tag{4.26}$$

It was assumed here that there is no surface current, which means that all conductivities are finite.

The relation for the stationary case is

$$\frac{\partial\sigma}{dt} = 0 \ ,$$

and therefore the boundary condition for g

$$\boxed{g_{2n} = g_{1n}} \ . \tag{4.27}$$

Notice that this does not require for σ to be zero and, as we shall see soon, sometimes this has to be different from zero. A consequence of (4.27) is that

$$\kappa_2 E_{2n} = \kappa_1 E_{1n} \tag{4.28}$$

Furthermore, it is necessary that

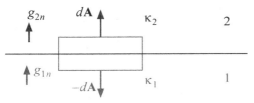

Fig. 4.4

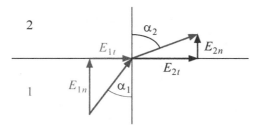

Fig. 4.5

$$E_{2t} = E_{1t} \, . \tag{4.29}$$

Putting the two together yields the law of refraction for **g** and **E** lines (Fig. 4.5):

$$\frac{\tan\alpha_1}{\tan\alpha_2} = \frac{\dfrac{E_{1t}}{E_{1n}}}{\dfrac{E_{2t}}{E_{2n}}} = \frac{E_{2n}}{E_{1n}}$$

or

$$\boxed{\frac{\tan\alpha_1}{\tan\alpha_2} = \frac{\kappa_1}{\kappa_2}} \, . \tag{4.30}$$

On the other hand, eq. (2.120) with a surface charge on the boundary between the two media 1 and 2, and letting $\dfrac{d\tau}{ds} = 0$ gives

$$\frac{\tan\alpha_1}{\tan\alpha_2} = \frac{\varepsilon_1}{\varepsilon_2} + \frac{\sigma}{\varepsilon_2 E_{1n}} \, . \tag{4.31}$$

The two eqs eq. (4.30) and eq. (4.31) are compatible only, if

$$\boxed{\frac{\kappa_1}{\kappa_2} = \frac{\varepsilon_1}{\varepsilon_2} + \frac{\sigma}{\varepsilon_2 E_{1n}}}$$

that is, if

$$\sigma = \varepsilon_2 E_{1n}\left(\frac{\kappa_1}{\kappa_2} - \frac{\varepsilon_1}{\varepsilon_2}\right) = \frac{\varepsilon_2}{\kappa_1} g_{1n}\left(\frac{\kappa_1}{\kappa_2} - \frac{\varepsilon_1}{\varepsilon_2}\right)$$

$$\boxed{\sigma = g_{1n}\left(\frac{\varepsilon_2}{\kappa_2} - \frac{\varepsilon_1}{\kappa_1}\right) = g_{1n}(t_{r2} - t_{r1})} \tag{4.32}$$

This is a rather peculiar result. It says that a stationary condition without surface charges on the boundary, can only occur if both relaxation times are identical. Conversely, if $t_{r1} \neq t_{r2}$, then the stationary condition is reached only after σ reaches the value mandated by eq. (4.32) (if it was initially different from that). The time it takes to reach equilibrium depends on the geometry of the set-up and the quantities ε_1, ε_2, κ_1, and κ_2. The law of refraction (4.30) does not apply before

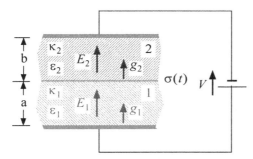

Fig. 4.6

the equilibrium is reached. During the transition, eq. (4.31) and the actual values of σ have to be used. As an illustrative **example**, consider that at $t = 0$, a voltage is abruptly applied to a layered resistor, extending to infinity (Fig. 4.6). For $t \geq 0$ this voltage is constant. The initial surface charge shall be $\sigma(0) = 0$ The following three formulas apply:

$$aE_1(t) + bE_2(t) = V \tag{4.33}$$

$$-\varepsilon_1 E_1(t) + \varepsilon_2 E_2(t) = \sigma(t) \tag{4.34}$$

$$-\kappa_1 E_1(t) + \kappa_2 E_2(t) + \frac{\partial \sigma(t)}{\partial t} = 0 \ . \tag{4.35}$$

The first two equations yield $E_1(t)$ and $E_2(t)$

$$\left. \begin{aligned} E_1(t) &= \frac{V\varepsilon_2 - b\sigma(t)}{a\varepsilon_2 + b\varepsilon_1} \\ E_2(t) &= \frac{V\varepsilon_1 + a\sigma(t)}{a\varepsilon_2 + b\varepsilon_1} \end{aligned} \right\} . \tag{4.36}$$

Substituting this in eq. (4.35) results in the differential equation

$$\frac{\partial \sigma(t)}{\partial t} + \frac{\sigma(t)}{t_{12}} = V\frac{\varepsilon_2 \kappa_1 - \varepsilon_1 \kappa_2}{a\varepsilon_2 + b\varepsilon_1} \ , \tag{4.37}$$

where

$$t_{12} = \frac{a\varepsilon_2 + b\varepsilon_1}{a\kappa_2 + b\kappa_1} \ , \tag{4.38}$$

with the solution

$$\sigma(t) = V\frac{\varepsilon_2 \kappa_1 - \varepsilon_1 \kappa_2}{a\kappa_2 + b\kappa_1}\left[1 - \exp\left(-\frac{t}{t_{12}}\right)\right] \ . \tag{4.39}$$

After a very long time, this becomes

$$\sigma_\infty = V\frac{\varepsilon_2\kappa_1 - \varepsilon_1\kappa_2}{a\kappa_2 + b\kappa_1} \ .$$

By (4.36), the related electric fields are

$$\left.\begin{aligned} E_{1\infty} &= V\frac{\kappa_2}{a\kappa_2 + b\kappa_1} \\[2mm] E_{2\infty} &= V\frac{\kappa_1}{a\kappa_2 + b\kappa_1} \end{aligned}\right\} , \tag{4.40}$$

and the related current density is

$$g_\infty = g_{1\infty} = g_{2\infty} = V\frac{\kappa_1\kappa_2}{a\kappa_2 + b\kappa_1} \tag{4.41}$$

and

$$\begin{aligned} \sigma(t) &= g_\infty\left(\frac{\varepsilon_2}{\kappa_2} - \frac{\varepsilon_1}{\kappa_1}\right)\left[1 - \exp\left(-\frac{t}{t_{12}}\right)\right] \\[2mm] &= g_\infty(t_{r2} - t_{r1})\left[1 - \exp\left(-\frac{t}{t_{12}}\right)\right] . \end{aligned} \tag{4.42}$$

which is illustrated in Fig. 4.7.

The stationary state described by eqs. (4.27), (4.30), and (4.32), is reached only after the typical time t_{12} when the relaxation process has subsided. A circuit of capacitors and resistors as shown in Fig. 4.8 allows one to simulate this transition. If A is the cross sectional area (assumed to be very large) between the layered resistors (or layered capacitor) of Fig. 4.6, then one can write

$$\left.\begin{aligned} C_1 &= \frac{\varepsilon_1 A}{a} ; & C_2 &= \frac{\varepsilon_2 A}{b} \\[2mm] R_1 &= \frac{a}{\kappa_1 A} ; & R_2 &= \frac{b}{\kappa_2 A} \end{aligned}\right\} . \tag{4.43}$$

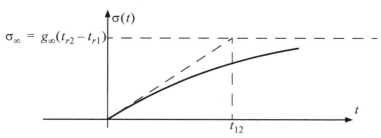

Fig. 4.7

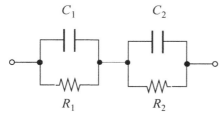

Fig. 4.8

Calculating the circuit of Fig. 4.8 leads to the same results as the field-theoretical derivation. Without going into the details, let it be noted that the relaxation time results in the expression

$$t_{12} = \frac{C_1 + C_2}{\dfrac{1}{R_1} + \dfrac{1}{R_2}} = \frac{\dfrac{\varepsilon_1 A}{a} + \dfrac{\varepsilon_2 A}{b}}{\dfrac{\kappa_1 A}{a} + \dfrac{\kappa_2 A}{b}} = \frac{\varepsilon_1 b + \varepsilon_2 a}{\kappa_1 b + \kappa_2 a} \ . \tag{4.44}$$

which is again eq. (4.38).

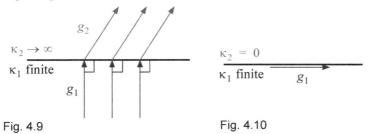

Fig. 4.9   Fig. 4.10

An important special case of the law of refraction (4.30) results from

$$\frac{\kappa_1}{\kappa_2} \to 0 \ ,$$

which occurs when $\kappa_2 \to \infty$ while κ_1 remains finite. This forces $\tan \alpha_1 = 0$ and the **g**-lines become perpendicular to the surface (Fig. 4.9). Conversely, if $\kappa_2 = 0$ then the **g**-lines become parallel to the boundary (Fig. 4.10). Thus, the boundary to a conductor with infinite conductivity is an equipotential surface, *i.e.*,

$$\varphi = \text{const.} \tag{4.45}$$

Conversely, the boundary towards an insulator is characterized by

$$\frac{\partial \varphi}{\partial n} = 0 \ . \tag{4.46}$$

4.4   The formal analogy between D and g

We have learned that

$$\mathbf{E} = -\nabla\varphi .$$

Furthermore

$\mathbf{D} = \varepsilon\mathbf{E}$	$\mathbf{g} = \kappa\mathbf{E}$
$\mathbf{D} = -\varepsilon\nabla\varphi$	$\mathbf{g} = -\kappa\nabla\varphi .$
In the charge-free space	In the stationary case
$\nabla\bullet\mathbf{D} = 0$	$\nabla\bullet\mathbf{g} = 0$

therefore in a uniform space we have

$$\nabla^2\varphi = 0 .$$

The normal components at the boundary are then

$$D_{2n} = D_{1n} \qquad g_{2n} = g_{1n},$$

while because of $\mathbf{E}_{2t} = \mathbf{E}_{1t}$, the tangential components are

$$\frac{D_{2t}}{\varepsilon_2} = \frac{D_{1t}}{\varepsilon_1} \qquad \frac{g_{2t}}{\kappa_2} = \frac{g_{1t}}{\kappa_1} .$$

This results in the law of refraction

$$\frac{\tan\alpha_1}{\tan\alpha_2} = \frac{\varepsilon_1}{\varepsilon_2} \qquad \frac{\tan\alpha_1}{\tan\alpha_2} = \frac{\kappa_1}{\kappa_2} .$$

One concludes that there is complete formal analogy between electrostatic fields $\mathbf{D}$, on one hand, and the stationary current density fields $\mathbf{g}$, on the other hand. This is an important observation because it allows to apply most of the results from Chapters 2 and 3 to current density fields. This is particularly true for the formal methods developed in Chapter 3, which may now be directly reused to solve boundary value problems of current density fields by separation or conformal mapping. Based on eqs. (4.45) and (4.46), there is a twofold application of these results. First, we obtain a Dirichlet problem when conductors are confined by ideal conductors. Second, for conductors that are terminated by ideal non-conductors (insulators), we obtain a Neumann problem. Of course, a combination of the two gives a mixed boundary value problem.

4.5 Some Current Density Fields

4.5.1 Point-like Current Source in Space

A current density emerging from a point-like source inside an infinitely large, uniform medium is illustrated in Fig. 4.11.
For symmetry reasons it must be

$$\int_A \mathbf{g} \bullet d\mathbf{A} = 4\pi r^2 g_r = I$$

that is

$$g_r = \frac{I}{4\pi r^2} \tag{4.47}$$

and

$$E_r = \frac{g_r}{\kappa} = \frac{I}{4\pi\kappa r^2} . \tag{4.48}$$

All other components of $\mathbf{g}$ and $\mathbf{E}$ vanish. Of course, the assumption of spherical symmetry is an idealization because of the current supply. For the potential, we get

$$\varphi(r) = \frac{I}{4\pi\kappa r} . \tag{4.49}$$

The image method allows to analyze the case of a point-like source in a volume that consists of two half spaces, where each of which has a different conductivity (Fig. 4.12).
We might try to solve this for region 1 by considering the (real) source I and the image source I' in region 2, while the solution for region 2 uses the image source I'' located in region 1. One writes:

$$\varphi_1 = \frac{1}{4\pi\kappa_1}\left(\frac{I}{\sqrt{(x-a)^2 + y^2 + z^2}} + \frac{I'}{\sqrt{(x+a)^2 + y^2 + z^2}}\right) ,$$

$$\varphi_2 = \frac{1}{4\pi\kappa_2}\left(\frac{I''}{\sqrt{(x-a)^2 + y^2 + z^2}}\right) .$$

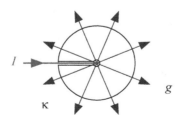

Fig. 4.11

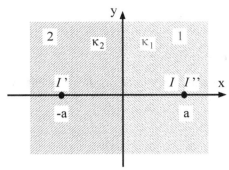

Fig. 4.12

Notice that this formula is a complete analogue to Sect. 2.11.2. We therefore find by just replacing ε by κ

$$I' = I\frac{\kappa_1 - \kappa_2}{\kappa_1 + \kappa_2} ,$$

(4.50)

$$I'' = I\frac{2\kappa_2}{\kappa_1 + \kappa_2} ,$$

(4.51)

The current density fields correspond to the figures of the electric field of Sect. 2.11.2. Two limits are most important. For an insulator $\kappa = 0$ in region 2 we have:

$$\left. \begin{array}{l} I' = I \\ I'' = 0 \end{array} \right\} .$$

This case is illustrated in Fig. 4.13. Conversely, Fig. 4.14 shows the case when region 2 is filled with a perfect conductor. $\kappa_2 \to \infty$ results in

$$\left. \begin{array}{l} I' = -I \\ I'' = 0 \end{array} \right\} .$$

Fig. 4.13 represents a Neumann boundary condition $\partial\varphi / \partial n = 0$ and Fig. 4.14 the Dirichlet boundary condition $\varphi = \text{const.}$

Using these results allows to solve an abundance of problems by superposition of appropriate sources. For instance, Fig. 4.15 suggests how to solve the problem of a point source in a quadrant where the boundary on one side consists of a perfect conductor, while the other side borders a perfect insulator. Conversely, if both media are perfect conductors, then Fig. 4.16 suggests a solution path.

By taking a limit one may transition to a dipole source and, for instance, solve the problem of a sphere embedded in some uniform medium of different conductivity, by superposition of dipole current density and uniform current density (this is analog to Section 2.12).

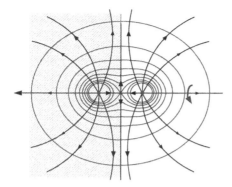

Fig. 4.13

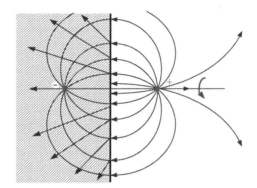

Fig. 4.14

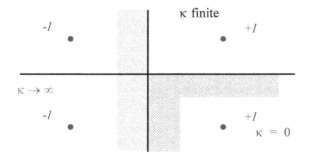

Fig. 4.15

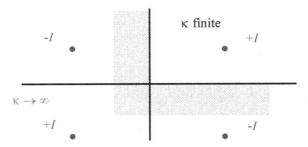

Fig. 4.16

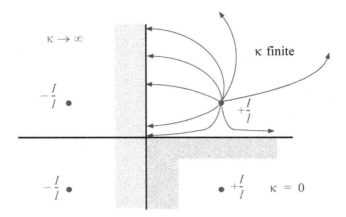

Fig. 4.17

4.5.2   Line Sources

Basically, all that was said for line charges applies to line sources as well. If r is the perpendicular distance to a uniform line source I/l in the infinite space, then

$$g_r = \frac{I}{2\pi l r} \,, \tag{4.52}$$

$$E_r = \frac{I}{2\pi \kappa l r} \,, \tag{4.53}$$

and

$$\varphi(r) = -\frac{I}{2\pi \kappa l} \ln \frac{r}{r_B} \,, \tag{4.54}$$

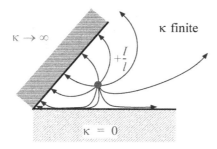

Fig. 4.18

Furthermore, we may apply all examples of Section 4.5.1 to line sources by substituting I by I/l in Figs. 4.12 through 4.16. This also holds for the two eqs. (4.50) and (4.51).

The solutions resulting thereof can be used as a starting point to solve additional plane problems by conformal mapping. For instance, the field shown in Fig. 4.18 results from Fig. 4.17 by the conformal mapping $\xi = z^p$ (similar to Example 6 of Sect. 3.12).

4.5.3   Mixed Boundary Value Problem

To demonstrate the separation of variable method, we choose for simplicity reasons a plane and, mixed boundary value problem for Laplace's equation $\nabla^2\varphi = 0$. Consider the rectangular piece of a uniform conductor with the sides a and b where a voltage is applied as shown in Fig. 4.19. The two sides $y = 0$ and $y = b$ are coated with a very good conductor (like silver) and are grounded. The side $x = a$

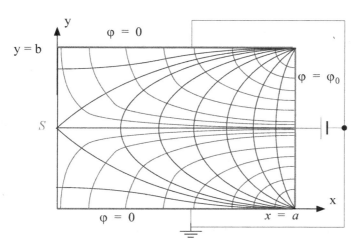

Fig. 4.19

shall also be coated with silver but the potential there shall be $\varphi = \varphi_0$. There is no conducting material at the side $x = 0$ and therefore, the current density lines have to be parallel to that boundary, which means that $\partial\varphi/dn = 0$. We summarize the boundary conditions:

$$
\left.
\begin{aligned}
\varphi = 0 \qquad &\text{for} \quad \begin{cases} y = 0 \\ y = b \end{cases} \\[2mm]
\varphi = \varphi_0 \qquad &\text{for} \quad x = a \\[2mm]
\frac{\partial\varphi}{\partial n} = \frac{\partial\varphi}{\partial x} = 0 \quad &\text{for} \quad x = 0
\end{aligned}
\right\} .
\tag{4.55}
$$

In order to meet the boundary conditions for y, the y-dependent part is written as
$$A\cos(ky) + B\sin(ky) .$$
This determines the form of the x-dependency, since the problem is independent of z.
$$C\cosh(kx) + D\sinh(kx) .$$
One needs to let $A = 0$ in order to satisfy the condition $\varphi = 0$ for $y = 0$. For $y = b$, this results in the requirement
$$\sin(kb) = 0$$
i.e.,
$$kb = n\pi$$
or
$$k = k_n = \frac{n\pi}{b} .$$
Furthermore, letting $D = 0$, satisfies $\partial\varphi/dn = 0$ for $x = 0$. Then

$$
\varphi(x, y) = \sum_{n=1}^{\infty} C_n \cosh\left(\frac{n\pi x}{b}\right) \sin\left(\frac{n\pi y}{b}\right) .
\tag{4.56}
$$

There is potentially an additional term $(A + By)(C + Dx)$ stemming from the case $k = 0$. However, it vanishes because of the boundary conditions. Obviously eq. (4.56) satisfies all boundary conditions, except that at $x = a$. This needs to be taken care of. One has to find the coefficients C_n in such a way that

$$
\varphi_0 = \sum_{n=1}^{\infty} C_n \cosh\left(\frac{n\pi a}{b}\right) \sin\left(\frac{n\pi y}{b}\right) .
\tag{4.57}
$$

Using the usual trick, one multiplies this equation by $\sin(n'\pi y/b)$ and integrates y from 0 to b:

$$\int_0^b \varphi_0 \sin\left(\frac{n'\pi y}{b}\right) dy = \sum_{n=1}^{\infty} C_n \cosh\left(\frac{n\pi a}{b}\right) \int_0^b \sin\left(\frac{n\pi y}{b}\right) \sin\left(\frac{n'\pi y}{b}\right) dy .$$

With the orthogonality relation (3.76) one obtains

$$\frac{b}{2} C_{n'} \cosh\left(\frac{n'\pi a}{b}\right) = \varphi_0 \int_0^b \sin\left(\frac{n'\pi y}{b}\right) dy .$$

Integrating the right side gives

$$\int_0^b \sin\left(\frac{n'\pi y}{b}\right) dy = \frac{b}{n'\pi}[\cos 0 - \cos(n'\pi)] = \frac{b}{n'\pi}[1 - (-1)^{n'}] .$$

This makes

$$C_n = \frac{2\varphi_0}{n\pi \cosh\left(\frac{n\pi a}{b}\right)}[1 - (-1)^n] \tag{4.58}$$

i.e.,

$$C_n = \begin{cases} \dfrac{4\varphi_0}{n\pi \cosh\left(\dfrac{n\pi a}{b}\right)} & \text{for} \quad n = 1, 3, 5 \ldots \\[2em] 0 & \text{for} \quad n = 2, 4, 6 \ldots \end{cases} \tag{4.59}$$

and with (4.56), the solution is finally

$$\varphi_0 = \frac{4\varphi_0}{\pi} \sum_{n=0}^{\infty} \frac{1}{2n+1} \frac{\cosh\left(\dfrac{(2n+1)\pi x}{b}\right)}{\cosh\left(\dfrac{(2n+1)\pi a}{b}\right)} \sin\left(\frac{(2n+1)\pi y}{b}\right) . \tag{4.60}$$

The problem may also be interpreted as a boundary value problem for ψ, because

$$\nabla^2 \psi = 0 \tag{4.61}$$

applies to ψ as well. However the boundary conditions are different from those for φ in (4.55). Now one has the situation as illustrated in Fig. 4.20.

$$\frac{\partial \psi}{\partial y} = -E_x = 0 \qquad\qquad \text{for} \quad \begin{cases} y = 0 \\ y = b \\ x = 0 \end{cases} . \tag{4.62}$$

$$\frac{\partial \psi}{\partial x} = E_y = 2\varphi_0[\delta(y-b) - \delta(y)] \quad \text{for} \quad x = a$$

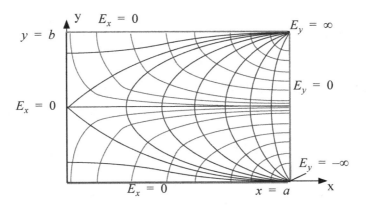

Fig. 4.20

The factor $2\varphi_0$ in front of the δ-function is due to the behavior of the potential along $x = a$. If we view this potential as a function of y, then

$$\varphi(a, y) = \varphi_0 \qquad \text{for} \quad 0 < y < b$$

$$\varphi(a, y) = -\varphi_0 \qquad \text{for} \quad b < y < 2b, \quad -b < y < 0 .$$

This implies that there is a discontinuity by $2\varphi_0$ at both locations, $y = 0$ and $y = b$

The Ansatz

$$\psi(x, y) = \sum_{n=1}^{\infty} d_n \sinh\left(\frac{n\pi x}{b}\right) \cos\left(\frac{n\pi y}{b}\right) \tag{4.63}$$

satisfies the first three conditions of eq. (4.62), but not the last one. This requires

$$\sum_{n=1}^{\infty} d_n \frac{n\pi}{b} \cosh\left(\frac{n\pi a}{b}\right) \cos\left(\frac{n\pi y}{b}\right) = 2\varphi_0[\delta(y - b) - \delta(y)] ,$$

from which one gets

$$d_n = \frac{2\varphi_0}{n\pi \cosh\left(\frac{n\pi a}{b}\right)} [(-1)^n - 1] . \tag{4.64}$$

And finally

$$\psi(x, y) = -\frac{4\varphi_0}{\pi} \sum_{n=0}^{\infty} \frac{1}{2n+1} \frac{\sinh\left(\frac{(2n+1)\pi x}{b}\right)}{\cosh\left(\frac{(2n+1)\pi a}{b}\right)} \cos\left(\frac{(2n+1)\pi y}{b}\right) . \tag{4.65}$$

The two results (4.60) and (4.65) are equivalent, and enable one to find the field. Both φ and ψ have to satisfy the Cauchy-Riemann condition:

$$\frac{\partial \varphi}{\partial x} = \frac{\partial \psi}{\partial y}$$

$$\frac{\partial \varphi}{\partial y} = -\frac{\partial \psi}{\partial x} \; ,$$

which is the case. Of course, one could have used this relation to calculate ψ, which would have been easier than solving the boundary value problem again.

Knowing φ and ψ allows to write the complex potential of the field:

$$w(z) = \varphi(x, y) + i\psi(x, y) \; .$$

Because of

$$\sinh(z) = \sinh(x + iy) = \sinh(x)\cosh(iy) + \cosh(x)\sinh(iy)$$
$$= \sinh(x)\cos(y) + i\cosh(x)\sin(y)$$

we re-write $w(z)$

$$w(z) = \frac{4\varphi_0}{\pi} \cdot$$

$$\cdot \sum_{n=0}^{\infty} \frac{\cosh\frac{(2n+1)\pi x}{b}\sin\frac{(2n+1)\pi y}{b} - i\sinh\frac{(2n+1)\pi x}{b}\cos\frac{(2n+1)\pi y}{b}}{(2n+1)\cosh\left(\frac{(2n+1)\pi a}{b}\right)} \; .$$

which gives

$$w(z) = -\frac{4\varphi_0 i}{\pi} \sum_{n=0}^{\infty} \frac{\sinh\frac{(2n+1)\pi z}{b}}{(2n+1)\cosh\left(\frac{(2n+1)\pi a}{b}\right)} \; . \tag{4.66}$$

This represents the conformal mapping which solves this particular problem.

The current density function allows to calculate the resistance of the setup or rather its inverse the conductance:

$$G = \frac{1}{R} = \frac{I}{V} = \frac{I}{\varphi_0 - 0} \; .$$

If d represents the thickness of a conductive layer, it is then

$$I = \left| d \int_{(a,0)}^{(a,b)} g_n dy \right| = \left| \kappa \, d \int_{(a,0)}^{(a,b)} E_x dy \right|$$

$$= \left| -\kappa \, d \int_{(a,0)}^{(a,b)} \frac{\partial \psi}{\partial y} dy \right| = \left| -\kappa \, d \int_{(a,0)}^{(a,b)} d\psi \right|$$

$$= \left| -\kappa \, d \, [\psi(a, b) - \psi(a, 0)] \right|$$

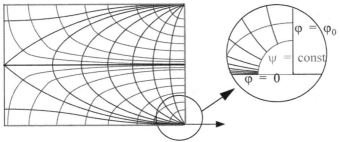

Fig. 4.21

$$I = \frac{4\kappa\varphi_0 d}{\pi} \sum_{n=0}^{\infty} 2\frac{1}{2n+1} \tanh\frac{(2n+1)\pi a}{b} \quad .$$

And then for the conductance

$$G = \frac{8\kappa d}{\pi} \sum_{n=0}^{\infty} \frac{\tanh\dfrac{(2n+1)\pi a}{b}}{(2n+1)} \quad \Rightarrow \infty \; . \tag{4.67}$$

The Conductance becomes infinitely large, which is a consequence of the idealized assumptions. The potential is discontinuous at the corners $x = a$, $y = 0$ and $y = b$, respectively. This is the reason why there, the electric field is infinitely large, as we have seen from the boundary conditions (4.62). Furthermore, the current density, and even the integrated total current becomes infinite, $i.e.$, $\psi(a, b)$ and $\psi(a, 0)$ are not finite. The singularity can be eliminated if we remove small pieces from the corners, while following a current density line (Fig. 4.21).

Notice that in solving this problem, one may consider a number of other problems as well. Fig. 4.22 shows a larger picture of the field. Fig. 4.22 allows a number of different interpretations. For instance, one may pick the range from $y = -b/2$ through $y = +b/2$ as shown in Fig. 4.23 and regard it as the solution of a different boundary value problem. Fig. 4.24 shows the solution when taking only the range form 0 through $b/2$. Furthermore, one can find a new meaning by exchanging the roles of φ and ψ. For instance, this transforms Fig. 4.23 into a point like current source, injected at $x = a$ and $y = 0$, while the current is drained off on the three sides $y = \pm b/2$ and $x = 0$, while the side with $x = a$ borders an insulator (disregarding the point like source), illustrated in Fig. 4.25. Calculating the resistance in this case, one finds that it diverges. The reason is the singularity of the current injection at $x = a$ and $y = 0$. We have discussed this situation already in conjunction with the diverging conductance of eq. (4.67). In the present interpretation, the current remains finite, while the quantity which we interpret as being the potential $\psi(a, 0)$ diverges. Finite current when voltage is infinite means that resistance has to be infinite. As before, this divergence is of

formal nature and can only be eliminated by removing a small piece along a line
ψ = const. (which here is represented by an equipotential line), as shown in
Fig. 4.21.
When using the term "equipotential line", it has to be understood that this means
the intersection of the equipotential surfaces with the observed plane. This is
permissible in case of plane problems because the equipotential lines uniquely
characterize the equipotential surfaces.

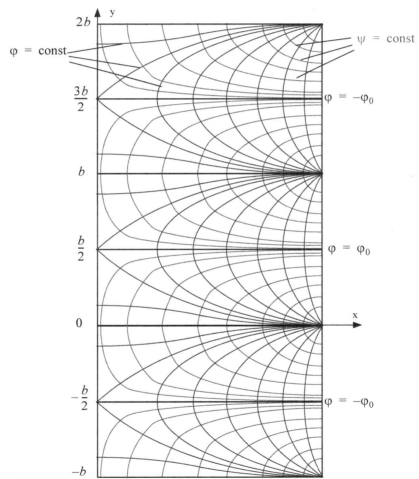

Fig. 4.22

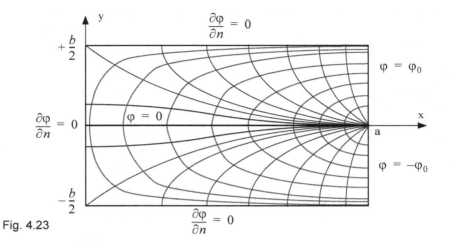

Fig. 4.23

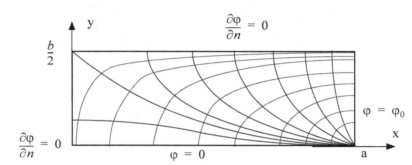

Fig. 4.24

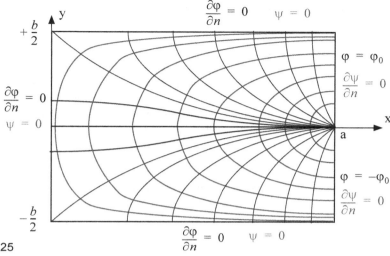

Fig. 4.25

5 Basics of Magnetostatics

5.1 Basic Equations

Maxwell's equations were introduced in Chapter 1 eq. (1.72). In the limit of just time-independent problems, the system of Maxwell's equations nicely splits into two electrostatic and two magnetostatic equations. The latter consists of Ampere's law, and the fact that the magnetic field is always source free (*i.e.* solenoidal).

$$\boxed{\nabla \times \mathbf{H} = \mathbf{g}} \tag{5.1}$$

$$\boxed{\nabla \bullet \mathbf{B} = 0} \, . \tag{5.2}$$

Beyond that, we have to establish a relation between $\mathbf{B}$ and $\mathbf{H}$

$$\mathbf{B} = \mathbf{B}(\mathbf{H}) \, . \tag{5.3}$$

The relation for vacuum is

$$\mathbf{B} = \mu_0 \mathbf{H} \, . \tag{5.4}$$

The $\mathbf{B}$-field is perceivable because it exhibits a velocity dependent force on charged particles (*Lorentz force*). If an electric field exists simultaneously, the force becomes

$$\mathbf{F} = Q(\mathbf{E} + \mathbf{v} \times \mathbf{B}) \, . \tag{5.5}$$

Integrating eq. (5.1) over an arbitrary area gives its integral representation

$$\int_A (\nabla \times \mathbf{H}) d\mathbf{A} = \int_A \mathbf{g} \, d\mathbf{A} \, .$$

Applying Stokes's theorem gives

$$\boxed{\oint \mathbf{H} \bullet d\mathbf{s} = I} \, , \tag{5.6}$$

where I is the current through the respective area.

In the following, we calculate the magnetic fields for a variety of arrangements. Simple cases with a high degree of symmetry allow one to almost directly write down the magnetic fields from eq. (5.6). However, frequently more laborious, formal methods need to be applied. This requires the introduction of the *vector potential* $\mathbf{A}$. Since the divergence of the curl vanishes for every vector

$$\nabla \bullet (\nabla \times \mathbf{a}) = 0 \, ,$$

it is possible to write $\mathbf{B}$ in the form

$$\boxed{\mathbf{B} = \nabla \times \mathbf{A}} \, . \tag{5.7}$$

This automatically satisfies eq. (5.2). Any given $\mathbf{A}$ uniquely determines $\mathbf{B}$. Conversely, there are many vector potentials for any given $\mathbf{B}$-field. Obviously,

$$\mathbf{A} \quad \text{and} \quad \mathbf{A}' = \mathbf{A} + \nabla\phi \tag{5.8}$$

produce the same $\mathbf{B}$-field. The reason is

$$\nabla \times \mathbf{A}' = \nabla \times \mathbf{A} + \nabla \times (\nabla\phi) = \nabla \times \mathbf{A} \, ,$$

where for any function ϕ the following identity holds

G. Lehner, *Electromagnetic Field Theory for Engineers and Physicists*,
DOI 10.1007/978-3-540-76306-2_5, © Springer-Verlag Berlin Heidelberg 2010

$$\nabla \times (\nabla \phi) = 0 \ .$$

Consequently, every **B**-field can be represented by an infinite number of vector potentials **A**. This allows one to impose additional restrictions on **A**. The proper choice of ϕ satisfies these restrictions, which can be used as a gauge. The transition from **A** to **A'** in eq. (5.8) is termed a *gauge transformation* of the vector potential. **B** remains unchanged and, thus, is said to be *gauge invariant*. The so-called *Coulomb gauge* is very useful for static problems.

$$\boxed{\nabla \bullet \mathbf{A} = 0} \ . \tag{5.9}$$

For time-dependent problems, the *Lorentz gauge* is usually used

$$\boxed{\nabla \bullet \mathbf{A} + \mu\varepsilon\frac{\partial \phi}{\partial t} = 0} \ . \tag{5.10}$$

Its importance will be made clear later.

From eqs. (5.1), (5.4). and (5.7) follows, that for currents in vacuum

$$\nabla \times \mathbf{H} = \nabla \times \frac{\mathbf{B}}{\mu_0} = \nabla \times \frac{1}{\mu_0}(\nabla \times \mathbf{A}) = \frac{1}{\mu_0}\nabla \times (\nabla \times \mathbf{A}) = \mathbf{g} \ .$$

Using the vector identity

$$\nabla \times (\nabla \times \mathbf{A}) = \nabla(\nabla \bullet \mathbf{A}) - \nabla^2 \mathbf{A} = \nabla(\nabla \bullet \mathbf{A}) - \Delta\mathbf{A} \tag{5.11}$$

and eq. (5.9) gives

$$\boxed{\nabla^2 \mathbf{A} = -\mu_0 \mathbf{g}} \ . \tag{5.12}$$

> The reader shall be cautioned in applying the Laplace operator to vectors, since only for Cartesian coordinates the result is simply an application of the double derivative to the individual components. This can easily be verified by applying the gradient, divergence, and curl in curvilinear coordinates (as we have discussed in Sections 3.1 through 3.3), and use eq. (5.11) to calculate $\nabla^2 \mathbf{A}$, *i.e.*,
>
> $$\nabla^2 \mathbf{A} = \nabla(\nabla \bullet \mathbf{A}) - \nabla \times (\nabla \times \mathbf{A}) \ .$$
>
> It is recommended to eliminate $\nabla^2 \mathbf{A}$ in this way when using curvilinear coordinates.

For Cartesian coordinates we find

$$\boxed{\begin{aligned} \nabla^2 A_x(\mathbf{r}) &= -\mu_0 g_x(\mathbf{r}) \\ \nabla^2 A_y(\mathbf{r}) &= -\mu_0 g_y(\mathbf{r}) \\ \nabla^2 A_z(\mathbf{r}) &= -\mu_0 g_z(\mathbf{r}) \end{aligned}} \ , \tag{5.13}$$

where ∇^2 represents the ordinary Laplacian or Laplace operator

$$\nabla^2 = \frac{\partial^2}{\partial x^2} + \frac{\partial^2}{\partial y^2} + \frac{\partial^2}{\partial z^2} \, .$$

The equation in cylindrical coordinates is

$$
\begin{array}{l}
\nabla^2 A_r - \dfrac{2}{r^2} \dfrac{\partial A_\varphi}{\partial \varphi} - \dfrac{A_r}{r^2} = -\mu_0 g_r \\[2ex]
\nabla^2 A_\varphi + \dfrac{2}{r^2} \dfrac{\partial A_r}{\partial \varphi} - \dfrac{A_\varphi}{r^2} = -\mu_0 g_\varphi \\[2ex]
\nabla^2 A_z \qquad\qquad\qquad = -\mu_0 g_z
\end{array}
\tag{5.14}
$$

where again ∇^2 represents the ordinary Laplace operator, whose form is given by eq. (3.33):

$$\nabla^2 = \frac{1}{r} \frac{\partial}{\partial r} r \frac{\partial}{\partial r} + \frac{1}{r^2} \frac{\partial^2}{\partial \varphi^2} + \frac{\partial^2}{\partial z^2}$$

Now we see that it would be wrong to attempt splitting eq. (5.12) in its cylindrical components and write it in the form

$$
\left.
\begin{array}{l}
\nabla^2 A_r = -\mu_0 g_r \\[1ex]
\nabla^2 A_\varphi = -\mu_0 g_\varphi \\[1ex]
\nabla^2 A_z = -\mu_0 g_z
\end{array}
\right\} \qquad \text{FALSE!}
$$

The reason for all this lies in the fact that the basis vectors for curvilinear coordinates are functions of the location, for instance in cylindrical coordinates $\mathbf{e}_r$ and $\mathbf{e}_\varphi$ (not $\mathbf{e}_z$, however), which results in additional terms when differentiating. In Cartesian coordinates we have

$$\nabla^2 \mathbf{A} = \nabla^2 (A_x \mathbf{e}_x + A_y \mathbf{e}_y + A_z \mathbf{e}_z) = \mathbf{e}_x \nabla^2 A_x + \mathbf{e}_y \nabla^2 A_y + \mathbf{e}_z \nabla^2 A_z \, ,$$

while for cylindrical coordinates we get

$$\nabla^2 \mathbf{A} = \nabla^2 (A_r \mathbf{e}_r + A_\varphi \mathbf{e}_\varphi + A_z \mathbf{e}_z) = \mathbf{e}_r (\nabla^2 A)_r + \mathbf{e}_\varphi (\nabla^2 A)_\varphi + \mathbf{e}_z (\nabla^2 A)_z \, ,$$

where

$$(\nabla^2 A)_r \neq \nabla^2 A_r \, , \qquad (\nabla^2 A)_\varphi \neq \nabla^2 A_\varphi \, .$$

To calculate the related fields when the currents $\mathbf{g}(\mathbf{r})$ are given, requires one to solve eq. (5.12), which in Cartesian coordinates is represented by the three scalar equations of (5.13), and for cylindrical coordinates by the three scalar equations given by (5.14). For formal reasons, we will initially restrict ourselves to Cartesian coordinates. We know already the solution of the three equations (5.13). These are three scalar Poisson equations. We have shown in electrostatics that the equation

$$\nabla^2 \varphi(\mathbf{r}) = -\frac{\rho(\mathbf{r})}{\varepsilon_0}$$

can be solved by the potential (see eq. (2.20))

$$\varphi(\mathbf{r}) = \frac{1}{4\pi\varepsilon_0}\int_V\frac{\rho(\mathbf{r'})}{|\mathbf{r}-\mathbf{r'}|}d\tau' \ .$$

Similarly, from eq. (5.13) follows:

$$\boxed{\begin{aligned} A_x(\mathbf{r}) &= \frac{\mu_0}{4\pi}\int_V\frac{g_x(\mathbf{r'})}{|\mathbf{r}-\mathbf{r'}|}d\tau' \\ A_y(\mathbf{r}) &= \frac{\mu_0}{4\pi}\int_V\frac{g_y(\mathbf{r'})}{|\mathbf{r}-\mathbf{r'}|}d\tau' \\ A_z(\mathbf{r}) &= \frac{\mu_0}{4\pi}\int_V\frac{g_z(\mathbf{r'})}{|\mathbf{r}-\mathbf{r'}|}d\tau' \end{aligned}} \ . \tag{5.15}$$

Of course we may combine these three equations into one vector equation

$$\boxed{\mathbf{A}(\mathbf{r}) = \frac{\mu_0}{4\pi}\int_V\frac{\mathbf{g}(\mathbf{r'})}{|\mathbf{r}-\mathbf{r'}|}d\tau'} \ . \tag{5.16}$$

However, important to remember is that *this equation is valid only in Cartesian coordinates.*

What remains to be proven is that the vector potential eq. (5.16) is indeed source-free, as our gauge (5.9) requires. We find

$$\nabla\bullet\mathbf{A}(\mathbf{r}) = \frac{\mu_0}{4\pi}\int_V\mathbf{g}(\mathbf{r'})\bullet\nabla_r\frac{1}{|\mathbf{r}-\mathbf{r'}|}d\tau' \ ,$$

having used the vector identity

$$\nabla\bullet(a f) = \mathbf{a}\bullet(\nabla f)+f(\nabla\bullet\mathbf{a})$$

and the fact that $\mathbf{g}(\mathbf{r'})$ is independent of the field point $\mathbf{r}$. Furthermore

$$\nabla\bullet\mathbf{A}(\mathbf{r}) = -\frac{\mu_0}{4\pi}\int_V\mathbf{g}(\mathbf{r'})\bullet\nabla_{r'}\frac{1}{|\mathbf{r}-\mathbf{r'}|}d\tau'$$

$$= \frac{-\mu_0}{4\pi}\int_V\left[(\nabla\bullet)_{r'}\frac{\mathbf{g}(\mathbf{r'})}{|\mathbf{r}-\mathbf{r'}|}-\frac{1}{|\mathbf{r}-\mathbf{r'}|}(\nabla\bullet)_{r'}\,\mathbf{g}(\mathbf{r'})\right]d\tau'$$

$$\nabla\bullet\mathbf{A}(\mathbf{r}) = \frac{\mu_0}{4\pi}\int_V(\nabla\bullet)_{r'}\frac{\mathbf{g}(\mathbf{r'})}{|\mathbf{r}-\mathbf{r'}|}d\tau' \ ,$$

because for stationary currents

$$\nabla\bullet\mathbf{g} = 0 \ .$$

And finally, because of Gauss' theorem

$$\nabla\bullet\mathbf{A}(\mathbf{r}) = -\frac{\mu_0}{4\pi}\oint\frac{\mathbf{g}(\mathbf{r'})}{|\mathbf{r}-\mathbf{r'}|}\bullet da' = 0 \ .$$

The reason is that the integral covers the entire space and there will be no currents crossing a sufficiently distant surface. Note that we use $d\mathbf{a}$ for the surface element here and whenever confusion with the vector potential $\mathbf{A}$ is possible.

One needs to beware of a false conclusion, here. We already saw that $\nabla \bullet \mathbf{A} = 0$ when $\nabla \bullet \mathbf{g} = 0$. However, this does not mandate that we select the Coulomb gauge when dealing with magnetostatic problems, where we have, of course, $\nabla \bullet \mathbf{g} = 0$. It only means that our approach does not bear any contradiction. Our assumption was the Coulomb gauge ($\nabla \bullet \mathbf{A} = 0$). Any different result would constitute a contradiction. On the other hand, if we choose current densities that are not source free ($\nabla \bullet \mathbf{g} \neq 0$), then we end up with a contradiction. Since always

$$\nabla \times (\nabla \times \mathbf{A}) = \mu_0 \mathbf{g} ,$$

the following must also be true

$$\nabla \bullet (\nabla \times (\nabla \times \mathbf{A})) = 0 = \mu_0 \nabla \bullet \mathbf{g}$$

$$\nabla \bullet \mathbf{g} = 0 \qquad .$$

Therefore, it should not be a surprise that for $\nabla \bullet \mathbf{g} \neq 0$, we obtain a wrong vector potential

$$\nabla \bullet \mathbf{A} \neq 0 .$$

When choosing a different gauge, then the vector potential calculated for a given current distribution satisfies this other gauge if and only if $\nabla \bullet \mathbf{g} = 0$.

In principle, this solves the calculation of magnetic fields due to any current distribution. In an actual case, this may still turn out to be difficult.

Instead of the vector potential, the field may be directly expressed by an integral. From eqs. (5.16) and (5.7) we obtain

$$\mathbf{B} = \nabla \times \mathbf{A} = \nabla \times \left[\frac{\mu_0}{4\pi} \int_V \frac{\mathbf{g}(\mathbf{r}')}{|\mathbf{r} - \mathbf{r}'|} d\tau' \right] ,$$

or

$$\mathbf{B} = \frac{\mu_0}{4\pi} \int_V (\nabla \times)_r \frac{\mathbf{g}(\mathbf{r}')}{|\mathbf{r} - \mathbf{r}'|} d\tau' .$$

We use

$$(\nabla \times)_r \frac{\mathbf{g}(\mathbf{r}')}{|\mathbf{r} - \mathbf{r}'|} = \frac{1}{|\mathbf{r} - \mathbf{r}'|} (\nabla \times)_r \, \mathbf{g}(\mathbf{r}') - \mathbf{g}(\mathbf{r}') \times \nabla_r \frac{1}{|\mathbf{r} - \mathbf{r}'|}$$

$$= -\mathbf{g}(\mathbf{r}') \times \left(-\frac{\mathbf{r} - \mathbf{r}'}{|\mathbf{r} - \mathbf{r}'|^3} \right)$$

$$= \mathbf{g}(\mathbf{r}') \times \left(\frac{\mathbf{r} - \mathbf{r}'}{|\mathbf{r} - \mathbf{r}'|^3} \right)$$

and therefore

$$\boxed{ \mathbf{B}(\mathbf{r}) = \frac{\mu_0}{4\pi} \int_V \frac{\mathbf{g}(\mathbf{r}') \times (\mathbf{r} - \mathbf{r}')}{|\mathbf{r} - \mathbf{r}'|^3} d\tau' } . \qquad (5.17)$$

This is the so-called *Biot-Savart's law* in its most general form. If there are only currents in relatively narrow conductors we may approximate (Fig. 5.1)

$$\mathbf{g}(\mathbf{r}') d\tau' = \mathbf{g}(\mathbf{r}') dA' ds' = I d\mathbf{s}' .$$

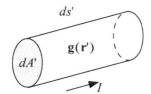

Fig. 5.1

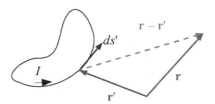

Fig. 5.2

This changes eqs. (5.16) and (5.17) to now read

$$\mathbf{A(r)} \; = \; \frac{\mu_0 I}{4\pi} \oint \frac{d\mathbf{s}'}{|\mathbf{r} - \mathbf{r}'|} \; ,$$
(5.18)

$$\mathbf{B(r)} \; = \; -\frac{\mu_0 I}{4\pi} \oint \frac{(\mathbf{r} - \mathbf{r}') \times d\mathbf{s}'}{|\mathbf{r} - \mathbf{r}'|^3} \; .$$
(5.19)

These integrals extend over the entire, closed current loop or all the closed current loops (Fig. 5.2). Frequently, it is said that the line element $d\mathbf{s}'$ creates the field

$$d\mathbf{B} \; = \; -\frac{\mu_0 I}{4\pi} \frac{(\mathbf{r} - \mathbf{r}') \times d\mathbf{s}'}{|\mathbf{r} - \mathbf{r}'|^3} \; .$$
(5.20)

Often, this is also called the Biot-Savart law. Unfortunately, this form is mistakable, we might almost say even wrong. Indeed, the field expressed by eq. (5.20) is a possible field, as it is source free. The related current density field is obtained from

$$\mathbf{g} \; = \; \nabla \times \mathbf{H} \; .$$

Taking the divergence on both sides reveals that it is always source free:

$$\nabla \bullet \mathbf{g} \; = \; \nabla \bullet (\nabla \times \mathbf{H}) \; = \; 0 \; .$$

However, the current I in the line element $d\mathbf{s}'$ is not source free. This constitutes an apparent contradiction. Nevertheless, the correct current density field can be calculated immediately. To simplify and make this easier to understand, just assume that the line element $d\mathbf{s}'$ is located at the origin and is oriented along the positive z-axis. When using spherical coordinates, $d\mathbf{B}$ and $d\mathbf{H}$ have only a φ-component

$$dH_\varphi = \frac{Ids'\sin\theta}{4\pi r^2} \ .$$

This makes

$$dg = \nabla \times d\mathbf{H} = \begin{cases} dg_r = \dfrac{Ids'}{4\pi r^3}\, 2\cos\theta \\[2ex] dg_\theta = \dfrac{Ids'}{4\pi r^3}\, \sin\theta \\[2ex] dg_\varphi = 0 \end{cases} \ .$$

We know the electrostatic analogue of this field very well. It represents the dipole field given by eq. (2.63). Thus, it describes the current I within a line element with the point-like isotropic current sources $+I$ at its upper end and $-I$ at its lower end, that is, a "dipole current density" which occupies the entire space. It is rotationally symmetric. Therefore, by using Ampere's law, one can show that this flux exactly causes the given magnetic field. It is furthermore apparent that all positive and negative point sources cancel each other when integrating over a closed contour, thereby leaving only the current I in the closed conductor. This explanation plausibly clarifies the integral result (5.19), which we had obtained in a purely formal manner.

To go beyond the current magnetostatic treatment of this problem and to regard it as time dependent problem is also possible. Currents with sources are now permissible, which however, requires time dependent volume charges, because of the charge conservation (continuity equation). Besides the magnetic field, one needs to consider also time dependent electric fields and thereby displacement current densities, *i.e.*, the current densities above are replaced by displacement current densities. We will not elaborate on this subject here.

To use the Biot-Savart law in its integral form (5.19) is advisable in magnetostatics. Nevertheless, the differential form can be used as well, if interpreted in the correct way, as just outlined.

Oftentimes, the vector potential is very useful to calculate the magnetic flux. Because of

$$\phi = \int_a \mathbf{B} \bullet d\mathbf{a} = \int_a (\nabla \times \mathbf{A}) \bullet d\mathbf{a},$$

and using Stokes' theorem we obtain:

$$\boxed{\phi = \oint \mathbf{A} \bullet d\mathbf{s}} \ . \tag{5.21}$$

The vector potential was introduced as an auxiliary quantity to calculate $\mathbf{B}$. At this point it is frequently said that only $\mathbf{B}$ has real meaning, while $\mathbf{A}$ has no significance beyond its role as an auxiliary quantity. This is not correct because in quantum mechanics, the field $\mathbf{A}$ is necessary and is a real field. The experiment by *Bohm* and *Aharonov*, interpreted in a quantum mechanical way shows *for example*, that the $\mathbf{A}$ field is important in certain regions (*e.g.*, outside of infinitely long coils,

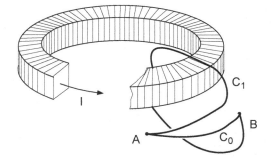

Fig. 5.3

see Section 5.2.3), where even when $\mathbf{B} = 0$, the $\mathbf{A}$ field remains $A \neq 0$. The Bohm-Aharonov experiment shall not be discussed here. Details can be found in Appendix A.3.

Besides the vector potential, there is also a *scalar magnetic potential*, whose usefulness is restricted to describe magnetic fields in regions that are current free. If in a region $\mathbf{g} = 0$, then

$$\nabla \times \mathbf{H} = 0 .$$

Therefore, $\mathbf{H}$ can be obtained from a scalar potential ψ by taking the gradient (note that ψ here is not related to the previously discussed flux function).

$$\boxed{\mathbf{H} = -\nabla \psi} \qquad (5.22)$$

ψ is a unique function in a simply connected region and shares most properties with the electric potential φ. When multiply connected regions enclose currents, ψ becomes ambiguous (Fig. 5.3). Consider a current carrying toroidal region. We pick the two points A and B in the current-free space and the two different paths C_0 and C_1, both starting at A and ending at B. Together C_0 and C_1 enclose the current carrying region. Ampere's law (5.6) states

$$\int_{B \atop (C_1)}^{A} \mathbf{H} \bullet d\mathbf{s} + \int_{A \atop (C_0)}^{B} \mathbf{H} \bullet d\mathbf{s} = I ,$$

that is

$$\int_{A \atop (C_1)}^{B} \mathbf{H} \bullet d\mathbf{s} = \int_{A \atop (C_0)}^{B} \mathbf{H} \bullet d\mathbf{s} - I .$$

More generally, we could be interested in a path that loops n times around the current I (Fig. 5.4). Then we get

$$\int_{A \atop (C_n)}^{B} \mathbf{H} \bullet d\mathbf{s} = \int_{A \atop (C_0)}^{B} \mathbf{H} \bullet d\mathbf{s} - nI . \qquad (5.23)$$

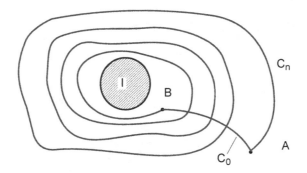

Fig. 5.4

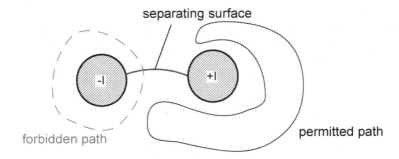

Fig. 5.5

If one defines the scalar potential

$$\psi(B) = \psi(A) - \int_{A}^{B} \mathbf{H} \bullet d\mathbf{s} ,$$ (5.24)

then ψ is determined but for whole number multiples of I. However, one can make ψ unique by introducing a cut in the space, so that a simply connected region emerges (Fig. 5.5). This cut represents a separating surface through which no integration path may go. By Ampere's law, along all permissible integration paths, this gives

$$\oint \mathbf{H} \bullet d\mathbf{s} = 0 .$$

This makes ψ unique. The scalar potential is important because it allows one to reduce many magnetostatic problems to problems whose solution is already known from electrostatics. The formal analogue between magnetostatics and electrostatics will be highlighted even more below. Specifically, Laplace's equation now applies:

$$\nabla \bullet \mathbf{B} = \nabla \bullet \mu_0 \mathbf{H} = \nabla \bullet (\mu_0 \nabla \psi) .$$

or

$$\boxed{\nabla^2 \psi = 0} \ . \tag{5.25}$$

Therefore, the formal methods (separation, conformal mapping) which we have discussed in detail in conjunction with electrostatics, are also of great consequence in magnetostatics.

5.2 Some Magnetic Fields

5.2.1 The Field of a Straight, Concentrated Current

Consider a current flowing along the z-axis of a Cartesian coordinate system, as shown in Fig. 5.6. Currents in magnetostatics are always source free. In this section, this is initially not the case. However, intuitively one can imagine the current to somehow form a closed loop at infinity, which would not provide any contributions to the magnetic field. To illustrate the methodology of the previous section, we will calculate the related magnetic field in three different ways: using the vector potential, by Biot-Savart's law, and by Ampere's law. The current density is

$$\mathbf{g} = \begin{cases} g_x = 0 \\ g_y = 0 \\ g_z = I\delta(x)\delta(y) \ . \end{cases} \tag{5.26}$$

First, we start with the vector potential. With (5.15) follows

$$A_x = A_y = 0$$

and

$$\begin{aligned} A_z &= \frac{\mu_0 I}{4\pi} \int_V \frac{\delta(x')\delta(y')\,dx'\,dy'\,dz'}{\sqrt{(x-x')^2 + (y-y')^2 + (z-z')^2}} \\ &= \frac{\mu_0 I}{4\pi} \int \frac{dz'}{\sqrt{x^2 + y^2 + (z-z')^2}} \ . \end{aligned}$$

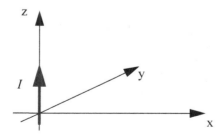

Fig. 5.6

The current along the z-axis could, for instance, be part of a circuit that includes the z-axis from $z = a$ to $z = b$. However, the current can as well flow from $-\infty$ to $+\infty$. In any case, one needs to calculate the integral

$$\int_a^b \frac{dz'}{\sqrt{x^2 + y^2 + (z - z')^2}}.$$

Substituting $z - z' = \zeta$ gives

$$-\int_{z-a}^{z-b} \frac{d\zeta}{\sqrt{x^2 + y^2 + \zeta^2}} = -\ln(\zeta + \sqrt{x^2 + y^2 + \zeta^2})\Big|_{z-a}^{z-b}$$

$$= \ln \frac{z - a + \sqrt{x^2 + y^2 + (z-a)^2}}{z - b + \sqrt{x^2 + y^2 + (z-b)^2}}.$$

Taking the limit $a \to -\infty$, $b \to +\infty$ gives

$$\lim_{\substack{a \to -\infty \\ b \to +\infty}} \ln \frac{z - a + \sqrt{x^2 + y^2 + (z-a)^2}}{z - b + \sqrt{x^2 + y^2 + (z-b)^2}} = \lim_{\substack{a \to -\infty \\ b \to +\infty}} \ln \frac{-a + |a|\sqrt{1 + \frac{x^2+y^2}{a^2}}}{-b + |b|\sqrt{1 + \frac{x^2+y^2}{b^2}}}$$

$$= \lim_{\substack{a \to -\infty \\ b \to +\infty}} \ln \frac{-a - a\sqrt{1 + \frac{x^2+y^2}{a^2}}}{-b + b\sqrt{1 + \frac{x^2+y^2}{b^2}}}$$

$$= \lim_{\substack{a \to -\infty \\ b \to +\infty}} \ln \frac{-2a}{-b + b + \frac{x^2+y^2}{2b}}$$

$$= \ln \frac{4(-a)b}{x^2 + y^2}.$$

Now, for arbitrary values of x_0 and y_0, one obtains for the vector potential

$$A_z = \frac{\mu_0 I}{4\pi} \ln \frac{4(-a)b}{x^2 + y^2} = \frac{\mu_0 I}{4\pi} \ln\left[\frac{4(-a)b}{x_0^2 + y_0^2}\right] - \frac{\mu_0 I}{4\pi} \ln\left[\frac{x^2 + y^2}{x_0^2 + y_0^2}\right].$$

Except for a very large, yet insignificant constant, the vector potential for the case of the "infinitely long current" is

$$A_z = -\frac{\mu_0 I}{4\pi} \ln\left[\frac{x^2 + y^2}{x_0^2 + y_0^2}\right].$$

Finally one obtains for the vector potential

$$\mathbf{A} = \left\langle 0, 0, -\frac{\mu_0 I}{4\pi} \ln\left[\frac{x^2 + y^2}{x_0^2 + y_0^2}\right]\right\rangle , \tag{5.27}$$

and for the magnetic field:

$$\begin{aligned}
B_x &= \frac{\partial A_z}{\partial y} - \frac{\partial A_y}{\partial z} = \frac{\partial A_z}{\partial y} = -\frac{\mu_0 I}{2\pi}\frac{y}{x^2 + y^2} \\[2mm]
B_y &= \frac{\partial A_x}{\partial z} - \frac{\partial A_z}{\partial x} = -\frac{\partial A_z}{\partial x} = +\frac{\mu_0 I}{2\pi}\frac{x}{x^2 + y^2} \\[2mm]
B_z &= \frac{\partial A_y}{\partial x} - \frac{\partial A_x}{\partial y} = 0
\end{aligned} \right\} . \tag{5.28}$$

The second method of solution is by means of the Biot-Savart law eq. (5.17). Here one starts with

$$\mathbf{g}(\mathbf{r}') \times (\mathbf{r} - \mathbf{r}') = \begin{vmatrix} \mathbf{e}_x & \mathbf{e}_y & \mathbf{e}_z \\ 0 & 0 & I\delta(x)\delta(y) \\ x - x' & y - y' & z - z' \end{vmatrix}$$

$$= \langle -(y - y'), (x - x'), 0\rangle \cdot I\delta(x)\delta(y) .$$

The magnetic field is

$$\begin{aligned}
B_x &= -\frac{\mu_0 I}{4\pi} \int_V \frac{(y - y')\delta(x')\delta(y')dx'\,dy'\,dz'}{\sqrt{(x - x')^2 + (y - y')^2 + (z - z')^2}^3} \\[2mm]
&= -\frac{\mu_0 I y}{4\pi} \int_{-\infty}^{+\infty} \frac{dz'}{\sqrt{x^2 + y^2 + (z - z')^2}^3} = -\frac{\mu_0 I y}{4\pi} \int_{-\infty}^{+\infty} \frac{d\zeta}{\sqrt{x^2 + y^2 + \zeta^2}^3} \\[2mm]
&= -\frac{\mu_0 I y}{4\pi} \left[\frac{\zeta}{(x^2 + y^2)\sqrt{x^2 + y^2 + \zeta^2}}\right]_{-\infty}^{+\infty} = -\frac{\mu_0 I y}{4\pi}\frac{2}{x^2 + y^2} .
\end{aligned}$$

Finally, as before one obtains

$$B_x = -\frac{\mu_0 I}{2\pi}\frac{y}{x^2 + y^2} .$$

In a similar way, calculating B_y leads to the previously obtained result.

One may also use cylindrical coordinates to calculate the field components

$$B_r = B_x\cos\varphi + B_y\sin\varphi$$

$$B_\varphi = -B_x\sin\varphi + B_y\cos\varphi$$

where

$$\cos\varphi = \frac{x}{r} = \frac{x}{\sqrt{x^2 + y^2}}$$

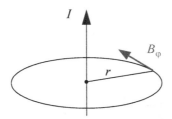

Fig. 5.7

and

$$\sin\varphi = \frac{y}{r} = \frac{y}{\sqrt{x^2 + y^2}}$$

Here, the components of the magnetic field are:

$$\left.\begin{array}{l} B_r = 0 \\ B_\varphi = \dfrac{\mu_0 I}{2\pi r} \\ B_z = 0 \end{array}\right\} . \tag{5.29}$$

Thus, the magnetic field **B** has only an azimuthal component, which has already been used in Chapter 1. Thirdly, one can now calculate the field for this problem again using Ampere's law. We already know based on the rotational symmetry, that there is only an azimuthal component (Fig. 5.7).

Then, B_φ can not depend on φ, which means that it has to be:

$$\mu_0 \oint \mathbf{H} \bullet d\mathbf{s} = \oint \mathbf{B} \bullet d\mathbf{s} = \mu_0 I$$

i.e.,

$$B_\varphi 2\pi r = \mu_0 I$$

and as already proven

$$B_\varphi = \frac{\mu_0 I}{2\pi r} .$$

Using

$$H_\varphi = \frac{I}{2\pi r}$$

and (5.24), one may describe the field by means of a scalar potential

$$\boxed{\psi = -\frac{I}{2\pi}(\varphi - \varphi_0)} . \tag{5.30}$$

This shows that the half-planes that originate at the z-axis and for which $\varphi = $ constant are equipotential surfaces. Conversely, from these equipotentials we find

$$H_\varphi = -\frac{1}{r}\frac{\partial \psi}{\partial \varphi} = \frac{I}{2\pi r} \ .$$

If one chooses a Cartesian coordinate system, such that

$$\tan(\varphi - \varphi_0) = \frac{y}{x} \ ,$$

then

$$\boxed{\psi = -\frac{I}{2\pi}\text{atan}\frac{y}{x}} \ , \qquad (5.31)$$

from which one finds the magnetic field:

$$B_x = \mu_0 H_x = -\mu_0\frac{\partial \psi}{\partial x} = -\frac{\mu_0 I}{2\pi}\frac{y}{x^2 + y^2}$$

$$B_y = \mu_0 H_y = -\mu_0\frac{\partial \psi}{\partial y} = +\frac{\mu_0 I}{2\pi}\frac{x}{x^2 + y^2}$$

$$B_z = 0 \ .$$

One might also write A_z in the form

$$A_z = -\frac{\mu_0 I}{4\pi}\ln\left(\frac{r}{r_0}\right)^2 = -\frac{\mu_0 I}{2\pi}\ln\left(\frac{r}{r_0}\right) \ . \qquad (5.32)$$

Notice that this tells us that A_z is constant on concentric circles ($r = $ const.) surrounding the current. These circles also represent the field lines. They are perpendicular to the equipotential surfaces ($\psi = $ const.). This allows one to view A_z as the flux function. The lines $A_z = $ const. and the lines $\psi = $ const. generate an orthogonal grid, parallel to the x-y plane. This is no coincidence, but a property of all "plane fields" and justifies the introduction of the complex potential, as well as the use of conformal mapping, similarly as we did for the electrostatic case. If one defines

$$w(z) = \frac{A_z}{\mu_0} + i\psi \ , \qquad (5.33)$$

then for the current case, one obtains

$$w(z) = -\frac{I}{2\pi}\ln\left(\frac{z}{z_0}\right) \ . \qquad (5.34)$$

This yields

$$w(z) = -\frac{I}{2\pi} \ln \frac{r \exp(i\varphi)}{r_0 \exp(i\varphi_0)} = -\frac{I}{2\pi} \ln \frac{r}{r_0} - \frac{I}{2\pi} \ln \exp[i(\varphi - \varphi_0)]$$

$$= -\frac{I}{2\pi} \ln \frac{r}{r_0} - i\frac{I}{2\pi}(\varphi - \varphi_0) \ .$$

Conversely, we could have introduced ψ as the real and A_z/μ_0 as the imaginary part of the complex potential.

The analogy between the complex potential (5.34) and the electric line charges (Sect. 3.12, Example 1) is obvious. However, notice that equipotential surfaces and field lines exchange their roles.

Of course, the fields of multiple currents can be superposed. For instance, consider the task to find the field of a current I_1 parallel to the z-axis, when I_1 passes through the x-y plane at $x = +d/2$, $y = 0$ and a second current I_2, also parallel to the z-axis, when I_2 passes through the x-y plane at $x = -d/2$, $y = 0$ (see Fig. 5.8). Then, by (5.28)

$$B_x = -\frac{\mu_0}{2\pi} \left[\frac{I_1 y}{\left(x - \frac{d}{2}\right)^2 + y^2} + \frac{I_2 y}{\left(x + \frac{d}{2}\right)^2 + y^2} \right]$$

$$B_y = +\frac{\mu_0}{2\pi} \left[\frac{I_1 \left(x - \frac{d}{2}\right)}{\left(x - \frac{d}{2}\right)^2 + y^2} + \frac{I_2 \left(x + \frac{d}{2}\right)}{\left(x + \frac{d}{2}\right)^2 + y^2} \right] \ .$$

As before in electrostatics, it is useful to find the stagnation points of the field, *i.e.*, the points where

$$B_x = B_y = 0 \ .$$

This scenario is independent of z. There is not only a stagnation point, but a whole "stagnation line" which consists entirely of stagnation points. Its coordinates are

$$x_s = \frac{d}{2} \frac{I_2 - I_1}{I_2 + I_1}$$

$$y_s = 0 \ .$$

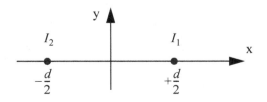

Fig. 5.8

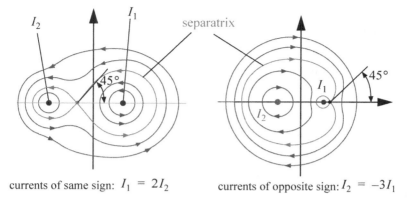

currents of same sign: $I_1 = 2I_2$ currents of opposite sign: $I_2 = -3I_1$

Fig. 5.9

If the currents have the same sign, then the stagnation point lies between the currents. For currents of opposite sign, it lies to the right or left of both currents (Fig. 5.9). The field is identical to that of two line charges, where again the field lines and the equipotentials exchange their role. The magnetic field lines in Fig. 5.9 correspond to the electric equipotentials of the line charges. A_z is constant along the magnetic field lines, $i.e.$, A_z may be regarded as flux function:

$$\mathbf{B} \bullet \nabla A_z = B_x \frac{\partial A_z}{\partial x} + B_y \frac{\partial A_z}{\partial y} = B_x(-B_y) + B_y B_x = 0 \ .$$

This produces the complex potential

$$w(z) = -\frac{I_1}{2\pi} \ln\left(\frac{z - d/2}{z_0}\right) - \frac{I_2}{2\pi} \ln\left(\frac{z + d/2}{z_0}\right) = 0 \ .$$

This result is the same as we would expect from superposing two potentials of the type (5.34), after shifting their center by $x = \pm d/2$, $y = 0$.

5.2.2 Field of a Rotationally Symmetric Current Distribution in Cylindrical Conductors

Consider a cylindrical conductor as shown in Fig. 5.10, with the current density

$$g_z = g_z(r) \ .$$

This produces a magnetic field that has only an azimuthal component which only depends on r. With Ampere's law one obtains (inside)

$$r \le r_0: \quad \mu_0 I(r) = \mu_0 \int_0^r g_z(r')2\pi r' \, dr'$$

$$= 2\pi r B_\varphi(r)$$

$$B_\varphi = \frac{\mu_0}{r}\int_0^r g_z(r')r'\,dr'$$ (5.35)

and for the field outside

$$r \geq r_0: \quad B_\varphi = \frac{\mu_0\int_0^{r_0} g_z(r')r'\,dr'}{r} = \frac{\mu_0 I}{2\pi r}.$$ (5.36)

$I(r)$ is the current flowing inside a cylinder of radius r. For a uniform current density inside the conductor one gets

$$r \leq r_0: \quad B_\varphi = \frac{\mu_0 g_z r^2}{r\ 2} = \frac{\mu_0 g_z r}{2} = \frac{\mu_0 I r}{2\pi r_0^2}$$ (5.37)

and for

$$r \geq r_0: \quad B_\varphi = \frac{\mu_0 I}{2\pi r}.$$ (5.38)

The field inside the conductor increases linearly with the radius and it falls outside as $1/r$ (see Fig. 5.11).

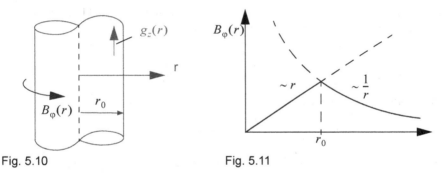

Fig. 5.10   Fig. 5.11

5.2.3   The Field of a simple Coil

The field of an infinitely long and ideal, densely wound coil (or solenoid) can easily be calculated using Ampere's law. Convince yourself that the field has to be parallel to the coil's axis and independent of both z and φ, i.e., only H_z does not vanish. The integral over the closed path C_1 of Fig. 5.12 is zero, i.e.,

$$\oint_{C_1} \mathbf{H} \bullet d\mathbf{s} = 0 = [H_{zo}(r_1) - H_{zo}(r_2)] \bullet d\mathbf{s}.$$

The field outside the solenoid has to be constant, i.e., independent of the radius. The same is true for the inside field when integrating along C_2. The inside field H_{zi} is independent of r. This requires that the outside field H_{zo} vanishes all together. The reasons is that H_{zo} has to vanish for $r \to \infty$. Finally, integrating along the closed loop C_3 gives.

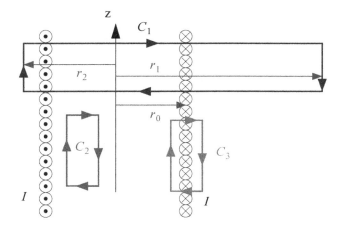

Fig. 5.12

$$H_{zi}ds = ndsI \ ,$$

where n represents the number of turns per unit length and I is the current per turn. Then

$$H_{zi} = nI \ , \hspace{4cm} (5.39)$$

The vector potential only has an azimuthal component A_φ, which is easily calculated from (5.21)

$$A_\varphi(r) = \frac{\phi(r)}{2\pi r} \ .$$

The flux inside the solenoid along a circular loop of radius r is

$$\phi(r) = nI\mu_0 r^2\pi$$

and therefore

$$A_\varphi(r) = \frac{\mu_0 nI}{2}r \ .$$

Outside the solenoid one has

$$\phi(r) = \mu_0 nI r_0^2\pi$$

and

$$A_\varphi(r) = \frac{\mu_0 nI r_0^2}{2r} \ .$$

The radial dependency of B_z and A_φ is shown in Fig. 5.13.

Notice that although $B_z = 0$ outside the solenoid, still A_φ is not zero. We have already mentioned that the outside field is significant and *for example*, has to be considered in Quantum Mechanics (see Appendix A.3).

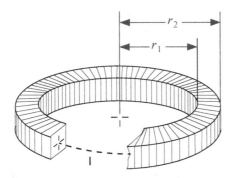

Fig. 5.13

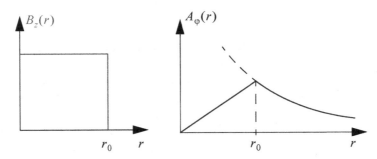

Fig. 5.14

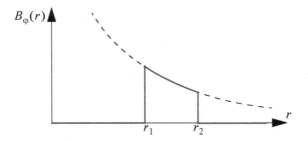

Fig. 5.15

The field of densely wound toroidal coils, as shown in Fig. 5.14, is also easy to write. The field for a circular, concentric path inside the coil is

$$\oint \mathbf{H} \bullet d\mathbf{s} = 2\pi r H_{\varphi i} = NI ,$$

or

$$H_{\varphi i} = \frac{NI}{2\pi r} ,$$
(5.40)

where N is the total number of turns and I the current in one turn. All fields outside the coil vanish if we make the idealized assumption that there is no current flows along the coil in azimuthal direction. Fig. 5.15 illustrates the field as a function of r. Note that all this does not depend on the shape of the coil's cross-sectional area.

5.2.4 The Field of a Circular Current and a Magnetic Dipole

Let an azimuthal current I flow in a circular loop as illustrated in Fig. 5.16. The loop is situated in the x-y plane. Its current density is

$$g_\varphi = I\delta(r - r_0)\delta(z) . \tag{5.41}$$

Cartesian coordinates allow the use of eq. (5.15) to calculate the vector potential.

$$g_x = -g_\varphi \sin\varphi$$
$$g_y = g_\varphi \cos\varphi . \tag{5.42}$$

Then, using eq. (5.15) gives

$$A_x(\mathbf{r}) = \frac{\mu_0 I}{4\pi} \int_V \frac{-\sin\varphi'\ \delta(r' - r_0)\delta(z')\,r'\,d\varphi'\,dr'\,dz'}{\sqrt{(r\cos\varphi - r'\cos\varphi')^2 + (r\sin\varphi - r'\sin\varphi')^2 + (z - z')^2}}$$

$$= \frac{\mu_0 I r_0}{4\pi} \int_0^{2\pi} \frac{-\sin\varphi'\ d\varphi'}{\sqrt{r^2 + r_0^2 + z^2 - 2rr_0\cos(\varphi - \varphi')}} .$$

Similarly the y-component becomes

$$A_y(\mathbf{r}) = \frac{\mu_0 I r_0}{4\pi} \int_0^{2\pi} \frac{\cos\varphi'\ d\varphi'}{\sqrt{r^2 + r_0^2 + z^2 - 2rr_0\cos(\varphi - \varphi')}} .$$

Returning to the components in cylindrical coordinates using the relation

$$A_r = A_x\cos\varphi + A_y\sin\varphi$$
$$A_\varphi = -A_x\sin\varphi + A_y\cos\varphi$$

one obtains

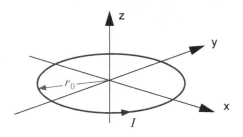

Fig. 5.16

$$A_r = \frac{\mu_0 I r_0}{4\pi} \int_0^{2\pi} \frac{\sin(\varphi - \varphi') \, d\varphi'}{\sqrt{r^2 + r_0^2 + z^2 - 2rr_0 \cos(\varphi - \varphi')}} = 0 \; , \tag{5.43}$$

$$A_\varphi = \frac{\mu_0 I r_0}{4\pi} \int_0^{2\pi} \frac{\cos(\varphi - \varphi') \, d\varphi'}{\sqrt{r^2 + r_0^2 + z^2 - 2rr_0 \cos(\varphi - \varphi')}} \; . \tag{5.44}$$

One might as well integrate from $-\pi$ through $+\pi$. A_r vanishes because of the odd integrand, A_φ does not, however. We will, once again, demonstrate the necessity of starting from Cartesian coordinates when using eq. (5.15). Had we tried to calculate A_φ and g_φ directly from (5.15), the result would not have been (5.44), but an incorrect one. This would have been the integral without the factor $\cos(\varphi - \varphi')$ in the numerator of the integrand. To find an analytic solution for A_φ of (5.44) is not possible. However, the integral is related to the total elliptic integrals. To see this, one has to resort to some mathematical tricks. First, we introduce the variable

$$\psi = \frac{\pi - (\varphi - \varphi')}{2}$$

or

$$\varphi - \varphi' = \pi - 2\psi \; .$$

This makes

$$\cos(\varphi - \varphi') = 2\sin^2\psi - 1 \; .$$

Furthermore, define the parameter

$$k^2 = \frac{4rr_0}{(r + r_0)^2 + z^2} \; . \tag{5.45}$$

Then, obtain

$$A_\varphi = \frac{\mu_0 I r_0}{4\pi\sqrt{(r + r_0)^2 + z^2}} \int_{\pi/2 - \varphi/2}^{3\pi/2 - \varphi/2} \frac{2\sin^2\psi - 1}{\sqrt{1 - k^2\sin^2\psi}} 2 \, d\psi \; .$$

Obviously, this is independent of φ because integration is always over a whole period of $\sin^2\psi$. Also the symmetry of the problem does not allow a φ-dependency of A_φ. This allows one to pick a convenient value, *for example*, $\varphi = 2\pi$. The integration is then from $\psi = -\pi/2$ through $\psi = +\pi/2$:

$$A_\varphi = \frac{\mu_0 I r_0}{2\pi\sqrt{(r + r_0)^2 + z^2}} \int_{-\pi/2}^{\pi/2} \frac{2\sin^2\psi - 1}{\sqrt{1 - k^2\sin^2\psi}} \, d\psi$$

$$= \frac{\mu_0 I r_0}{\pi\sqrt{(r + r_0)^2 + z^2}} \int_0^{\pi/2} \frac{2\sin^2\psi - 1}{\sqrt{1 - k^2\sin^2\psi}} \, d\psi$$

$$A_\varphi = \frac{\mu_0 I r_0}{\pi \sqrt{(r+r_0)^2 + z^2}} \int_0^{\pi/2} \frac{2\sin^2\psi - 1 + \left(\dfrac{2}{k^2} - \dfrac{2}{k^2}\right)}{\sqrt{1 - k^2\sin^2\psi}} d\psi$$

$$= \frac{\mu_0 I r_0}{\pi \sqrt{(r+r_0)^2 + z^2}} \left\{ -\frac{2}{k^2} \int_0^{\pi/2} \frac{1 - k^2\sin^2\psi}{\sqrt{1 - k^2\sin^2\psi}} d\psi + \left(\frac{2}{k^2} - 1\right) \int_0^{\pi/2} \frac{d\psi}{\sqrt{1 - k^2\sin^2\psi}} \right\}$$

$$= \frac{\mu_0 I r_0}{\pi \sqrt{(r+r_0)^2 + z^2}} \left\{ \left(\frac{2}{k^2} - 1\right) \int_0^{\pi/2} \frac{d\psi}{\sqrt{1 - k^2\sin^2\psi}} - \frac{2}{k^2} \int_0^{\pi/2} \sqrt{1 - k^2\sin^2\psi}\, d\psi \right\} .$$

These two integrals are called the *total elliptic integrals* of the first and the second kind, respectively:

$$K\left(\frac{\pi}{2}, k\right) = \int_0^{\pi/2} \frac{d\psi}{\sqrt{1 - k^2\sin^2\psi}} \quad , \tag{5.46}$$

$$E\left(\frac{\pi}{2}, k\right) = \int_0^{\pi/2} \sqrt{1 - k^2\sin^2\psi}\, d\psi \quad . \tag{5.47}$$

These enable one to write A_φ in the following way:

$$\boxed{A_\varphi = \frac{\mu_0 I}{2\pi r} \sqrt{(r+r_0)^2 + z^2} \left\{ \left(1 - \frac{k^2}{2}\right) K\left(\frac{\pi}{2}, k\right) - E\left(\frac{\pi}{2}, k\right) \right\}} . \tag{5.48}$$

In order to calculate A_φ, one needs to know the derivatives of K and E.

$$\left. \begin{array}{l} \dfrac{d}{dk} K\left(\dfrac{\pi}{2}, k\right) = \dfrac{E\left(\dfrac{\pi}{2}, k\right)}{k(1 - k^2)} - \dfrac{K\left(\dfrac{\pi}{2}, k\right)}{k} \\[4mm] \dfrac{d}{dk} E\left(\dfrac{\pi}{2}, k\right) = \dfrac{E\left(\dfrac{\pi}{2}, k\right) - K\left(\dfrac{\pi}{2}, k\right)}{k} \end{array} \right\} \tag{5.49}$$

We will limit ourselves to the case where the distance from the origin to the field point is much larger than r_0 :

$$\sqrt{r^2 + z^2} \gg r_0 .$$

A consequence of (5.45) is that $k \ll 1$ and both, K and E can be replaced by the first few terms of their respective series expansion

$$K\left(\frac{\pi}{2}, k\right) = \frac{\pi}{2}\left[1 + 2\frac{k^2}{8} + 9\left(\frac{k^2}{8}\right)^2 + \ldots\right]$$

$$E\left(\frac{\pi}{2}, k\right) = \frac{\pi}{2}\left[1 - 2\frac{k^2}{8} - 3\left(\frac{k^2}{8}\right)^2 - \ldots\right]$$

$$(5.50)$$

The correctness of the given terms of the two series can be proven by expanding the integrands of (5.46) and (5.47) into a series and then integrate them term by term. Using (5.50) in (5.48) for small values of k gives

$$A_\varphi = \frac{\mu_0 I}{2\pi r}\sqrt{(r + r_0)^2 + z^2} \cdot$$

$$\cdot \frac{\pi}{2}\left\{\left(1 - \frac{k^2}{2}\right)\left(1 + \frac{2}{8}k^2 + \frac{9}{64}k^4 + \ldots\right) - \left(1 - \frac{2}{8}k^2 - \frac{3}{64}k^4 - \ldots\right)\right\}$$

i.e.,

$$A_\varphi \approx \frac{\mu_0 I}{4r}\sqrt{(r + r_0)^2 + z^2}\left(\frac{k^4}{16}\right) \approx \frac{\mu_0 I}{4r}\frac{r^2 r_0^2}{(r^2 + z^2)^{3/2}}$$

Introducing spherical coordinates R, θ, φ (Fig. 5.17):

$$R = \sqrt{r^2 + z^2}$$

and

$$\sin\theta = \frac{r}{\sqrt{r^2 + z^2}} \cdot$$

This gives

$$A_\varphi \approx \frac{\mu_0 I r_0^2 \pi \sin\theta}{4\pi}\frac{\sin\theta}{R^2} \cdot$$

Using

$$\mathbf{B} = \nabla \times \mathbf{A}$$

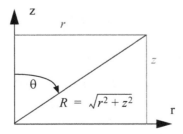

Fig. 5.17

gives

$$B_R = \frac{1}{R\sin\theta}\frac{\partial}{\partial\theta}(\sin\theta A_\varphi) = \frac{\mu_0 I r_0^2 \pi 2\cos\theta}{4\pi}\frac{}{R^3}$$

$$B_\theta = -\frac{1}{R\partial R}\frac{\partial}{}(RA_\varphi) = \frac{\mu_0 I r_0^2 \pi \sin\theta}{4\pi}\frac{}{R^3}$$

$$B_\varphi = 0 \ .$$

Now, we introduce the *magnetic dipole moment m*

$$\boxed{m = \mu_0 I r_0^2 \pi = \mu_0 I a} \tag{5.51}$$

and its vector representation

$$\boxed{\mathbf{m} = \mu_0 I \mathbf{a}} \ , \tag{5.52}$$

where the direction of the current and the direction of the area element are determined by a right handed system. This gives

$$\boxed{A_\varphi = \frac{m}{4\pi}\frac{\sin\theta}{R^2}} \tag{5.53}$$

and

$$\boxed{\begin{aligned} B_R &= \frac{2m}{4\pi}\frac{\cos\theta}{R^3} \\[4pt] B_\theta &= \frac{m}{4\pi}\frac{\sin\theta}{R^3} \\[4pt] B_\varphi &= 0 \ . \end{aligned}} \tag{5.54}$$

Unfortunately, the definition of $\mathbf{m}$ is not standardized and is oftentimes introduced without the factor μ_0. The field components in eq. (5.54) behave as functions of the location in the same manner as those of the electric dipole field (2.63). These two equations merge when exchanging p/ε_0 with m. The name magnetic dipole is based on this analogy.

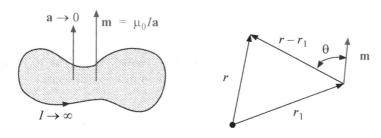

Fig. 5.18   Fig. 5.19

Just like the ideal electric dipole, the magnetic dipole has to be understood as the limit of an infinite current I while r_0^2 becomes infinitely small, leaving the

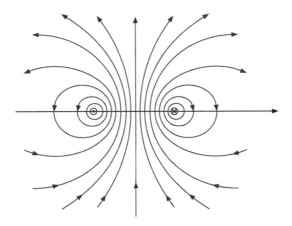

Fig. 5.20

product $\mu_0 I r_0^2 \pi$ finite. The area is not necessarily circular. In the limit of a vanishing area, its shape is immaterial for the field. What matters is the size of the area, which the definitions (5.51) and (5.52) already consider (Fig. 5.18).

A scalar potential was suitable to determine the field of an electric dipole (2.60). This is also the case for the magnetic dipole, for which we have

$$\psi = \frac{m \cos\theta}{4\pi\mu_0 |\mathbf{r} - \mathbf{r}_1|^2} = \frac{\mathbf{m} \bullet (\mathbf{r} - \mathbf{r}_1)}{4\pi\mu_0 |\mathbf{r} - \mathbf{r}_1|^3} , \tag{5.55}$$

with the quantities as shown in Fig. 5.19. When comparing (5.54) and (5.55), be aware that $\mathbf{H} = -\nabla\psi$ and $\mathbf{B} = \mu_0\mathbf{H}$. The scalar potential ψ is given in a form which is independent from a coordinate system. The vector potential (5.53) can also be written in a representation that is independent of a coordinate system:

$$\mathbf{A} = \frac{\mathbf{m} \times (\mathbf{r} - \mathbf{r}_1)}{4\pi |\mathbf{r} - \mathbf{r}_1|^3} = -\frac{\mathbf{m}}{4\pi} \times \nabla_r \frac{1}{|\mathbf{r} - \mathbf{r}_1|} = \frac{1}{4\pi} \nabla_r \times \frac{\mathbf{m}}{|\mathbf{r} - \mathbf{r}_1|} , \tag{5.56}$$

where we used the vector identity:

$$\nabla \times (\mathbf{b}\varphi) = \varphi(\nabla \times \mathbf{b}) - \mathbf{b} \times (\nabla\varphi) .$$

Using $\mathbf{m} = m\mathbf{e}_z$ and $r_1 = 0$ in (5.56), we find A_φ as of (5.53). The field produced by the current in a circular loop around a finite area is shown in Fig. 5.20. At a sufficiently large distance, it is indistinguishable from the field of two opposite charges of the same magnitude. The magnetic field of an ideal magnetic dipole corresponds entirely to the electric field of an ideal electric dipole (see Fig. 2.15).

The magnetic dipole is one of the most important concepts in magnetostatics. It will be the center piece of several of the following sections. Therefore, we will summarize the most important formulas, side by side with the electrostatic ones. Consider a dipole, oriented in the positive z-direction, located at the origin of a coordinate system. Then for spherical coordinates we have the following relations:

electric	magnetic
$E_R = \dfrac{2p\cos\theta}{4\pi\varepsilon_0 R^3}$	$H_R = \dfrac{2m\cos\theta}{4\pi\mu_0 R^3}$
$E_\theta = \dfrac{p\sin\theta}{4\pi\varepsilon_0 R^3}$	$H_\theta = \dfrac{m\sin\theta}{4\pi\mu_0 R^3}$
$E_\varphi = 0$	$H_\varphi = 0$
$\mathbf{D} = \varepsilon_0\mathbf{E}$	$\mathbf{B} = \mu_0\mathbf{H}$
$\mathbf{E} = -\nabla\varphi$	$\mathbf{H} = -\nabla\psi$
$\varphi = \dfrac{p\cos\theta}{4\pi\varepsilon_0 R^2}$	$\psi = \dfrac{m\cos\theta}{4\pi\mu_0 R^2}$
	$\mathbf{B} = \nabla\times\mathbf{A}$
	$\begin{aligned} A_R &= 0 \\ A_\theta &= 0 \\ A_\varphi &= \dfrac{m\sin\theta}{4\pi R^2} \end{aligned}$

Independent of any coordinate system, for a dipole ($\mathbf{p}$ or $\mathbf{m}$) at location $\mathbf{r}_1$ we have

electric	magnetic
$\varphi = \dfrac{\mathbf{p}\bullet(\mathbf{r}-\mathbf{r}_1)}{4\pi\varepsilon_0\lvert\mathbf{r}-\mathbf{r}_1\rvert^3}$	$\psi = \dfrac{\mathbf{m}\bullet(\mathbf{r}-\mathbf{r}_1)}{4\pi\mu_0\lvert\mathbf{r}-\mathbf{r}_1\rvert^3}$
	$\begin{aligned} \mathbf{A} &= \dfrac{\mathbf{m}\times(\mathbf{r}-\mathbf{r}_1)}{4\pi\lvert\mathbf{r}-\mathbf{r}_1\rvert^3} \\ &= -\dfrac{\mathbf{m}}{4\pi}\times\nabla_r\dfrac{1}{\lvert\mathbf{r}-\mathbf{r}_1\rvert} \\ &= \dfrac{1}{4\pi}\nabla_r\times\dfrac{\mathbf{m}}{\lvert\mathbf{r}-\mathbf{r}_1\rvert} \end{aligned}$

Anticipating the outcome of the ensuing discussion, we will give the result for a uniform dipole layer

electric	magnetic
$\varphi = \dfrac{\tau}{4\pi\varepsilon_0}\Omega$ $(\tau = \dfrac{dp}{da})$	$\psi = \dfrac{\dfrac{dm}{da}}{4\pi\mu_0}\Omega = \dfrac{I}{4\pi}\Omega$ (since $\dfrac{dm}{da} = I\mu_0$)

5.2.5   The Field of an Arbitrary Current Loop

An arbitrarily shaped current loop as illustrated in Fig. 5.21 can be thought of as being composed of many dipoles. If the currents in all loops are identical (I), then they cancel on the inside, even for infinitesimal sections. What remains is the current I on the boundary. Therefore, the current loop is equivalent to a *magnetic dipole layer*, *i.e.*, a layer of magnetic dipoles (in analogy to the electric dipole layer of Sect. 2.5.3). The area density of the dipole moment is

$$\frac{dm}{da} = \frac{\mu_0 I da}{da} = \mu_0 I, \tag{5.57}$$

i.e., a constant. This means that the dipole layer is uniform. In analogy to eq. (2.72) we obtain therefore

$$\psi = \frac{\mu_0 I}{4\pi\mu_0}\Omega = \frac{I}{4\pi}\Omega , \tag{5.58}$$

where Ω is the solid angle subtending the current loop as seen from the observation point. By eq. (2.73), the potential φ has a discontinuity of τ/ε_0 when passing through the electric dipole layer in the positive direction. Similarly, ψ has a discontinuity of I, when passing through a magnetic dipole layer in the positive direction. We have seen this result previously in a different form (refer to the

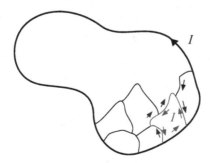

Fig. 5.21

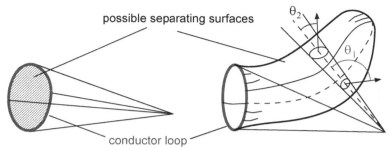

possible separating surfaces

θ_2

θ_1

conductor loop

Fig. 5.22

discussion on the scalar magnetic potential ψ of Sect. 5.1. The separating surface introduced then, now turns out to be a magnetic dipole layer).

Despite the perfect formal analogues, there is a major difference between electric and magnetic dipole layers. Electric dipole layers are a physical reality. By contrast, the magnetic dipole layer is a purely formal-fictitious construct and a substitute for finite current loops. This is obvious from the fact that they may be placed more or less arbitrarily since only the boundary matters (Fig. 5.22). The boundary defines the field in a unique way because the solid angle Ω does not depend on the choice of the separating surface. Positive and negative solid angles $d\Omega$ mutually cancel. Notice that $d\Omega$ has the same sign as $\cos\theta$.

Cylindrical dipole layers result when the currents only flow in z-direction (this is the analogue of cylindrical electric dipole layers, see Sect. 2.5.4). Fig. 5.23 allows to determine the angle α , which in analogy to eq. (2.85) gives

$$\psi = \frac{I}{2\pi}\alpha \ . \tag{5.59}$$

The same result follows from eq. (5.58) because for the situation of Fig. 5.23 we have

$$\Omega = 4\pi\frac{\alpha}{2\pi} = 2\alpha \ . \tag{5.60}$$

The fields of cylindric dipole layers are plane, *i.e.*, the fields are independent of z.

Let us calculate the field on the axis of a circular current loop (shown in Fig. 5.24) as an **example** on how to apply eq. (5.58). The solid angle Ω needed here, can be taken from the example of Sect. 2.5.3.

$$\Omega = 2\pi - \frac{2\pi z}{\sqrt{r_0^2 + z^2}} \qquad \text{for } 0 < z \ .$$

Then

$$\psi = \frac{I}{4\pi}\Omega = \frac{I}{2} - \frac{Iz}{2\sqrt{r_0^2 + z^2}}$$

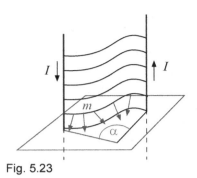

Fig. 5.23

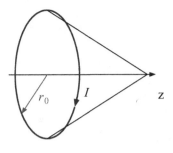

Fig. 5.24

and the field on the axis becomes

$$B_z = -\mu_0 \frac{\partial \psi}{\partial z} = \frac{\mu_0 I}{2} \frac{r_0^2}{\sqrt{r_0^2 + z^2}^3} . \qquad (5.61)$$

Of course, this result can also be derived from A_φ (5.48). The relations for r and k close to the axis are $r \ll r_0^2$ and $k^2 \ll 1$. Therefore

$$A_\varphi \approx \frac{\mu_0 I r_0^2}{4} \frac{r}{\sqrt{r_0^2 + z^2}}$$

and

$$B_z = \frac{1}{r}\frac{\partial}{\partial r}(rA_\varphi) = \frac{\mu_0 I r_0^2}{4r} \frac{2r}{\sqrt{r_0^2 + z^2}^3}$$

$$= \frac{\mu_0 I}{2} \frac{r_0^2}{\sqrt{r_0^2 + z^2}^3}$$

as before. In particular, the field at the center of the circle $r = 0$, $z = 0$ is

$$B_z = \frac{\mu_0 I}{2r_0} . \qquad (5.62)$$

5.2.6   The Field of a Conducting Loop in its Plane

The magnetic field in the plane of a plane conducting loop is a simple but useful application of Biot-Savart's law (Fig. 5.25). From

$$\mathbf{B} = -\frac{\mu_0 I}{4\pi} \oint \frac{(\mathbf{r} - \mathbf{r'}) \times d\mathbf{s'}}{|\mathbf{r} - \mathbf{r'}|^3}$$

it follows that $\mathbf{B}$ is perpendicular to the plane of the loop (= paper plane in Fig. 5.25). From a magnitude perspective, we have

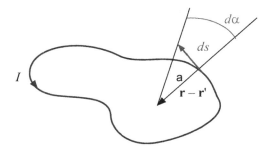

Fig. 5.25

$$|(\mathbf{r} - \mathbf{r}') \times d\mathbf{s}'| = |\mathbf{r} - \mathbf{r}'|^2 d\alpha .$$

Therefore, the magnitude of **B** is

$$B = \frac{\mu_0 I}{4\pi} \oint \frac{d\alpha}{|\mathbf{r} - \mathbf{r}'|} = \frac{\mu_0 I}{4\pi} \oint \frac{d\alpha}{a} , \tag{5.63}$$

where

$$a = |\mathbf{r} - \mathbf{r}'| .$$

If, for *example*, the loop consists of pieces of straight wire that form a polygon, then we need to add the contribution of each piece. When using

$$a\cos\alpha = a_s ,$$

the individual sections yield the following contribution (Fig. 5.26).

$$B = \frac{\mu_0 I}{4\pi} \int_{\alpha_a}^{\alpha_e} \frac{d\alpha}{a} = \frac{\mu_0 I}{4\pi} \int_{\alpha_a}^{\alpha_e} \frac{\cos\alpha \, d\alpha}{a_s}$$

$$= \frac{\mu_0 I}{4\pi a_s} \int_{\alpha_a}^{\alpha_e} \cos\alpha \, d\alpha = \frac{\mu_0 I}{4\pi a_s} [\sin\alpha]_{\alpha_a}^{\alpha_e} .$$

Finally:

$$B = \frac{\mu_0 I}{4\pi a_s} [\sin\alpha_e - \sin\alpha_a] . \tag{5.64}$$

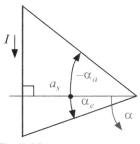

Fig. 5.26

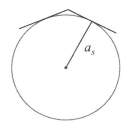

Fig. 5.27

If the conductor is infinitely long, then

$$\alpha_e = 90°, \qquad \alpha_a = -90°$$

and in accord with our previous result

$$B = \frac{\mu_0 I}{4\pi a_s} \cdot 2 = \frac{\mu_0 I}{2\pi a_s} .$$

The field at the center of a regular n-polygon (n sides) with the inside radius a_s is

$$B = n\frac{\mu_0 I}{4\pi a_s}2\sin\left(\frac{\pi}{n}\right) = \frac{n\mu_0 I}{2\pi a_s}\sin\left(\frac{\pi}{n}\right) .$$

For $n \gg 1$, this becomes

$$B = \frac{\mu_0 I}{2a_s} .$$

This is the same result as for a circle, which we already know: eq. (5.62) with $a_s = r_0$. It is also possible to write this result immediately:

$$B = \frac{\mu_0 I}{4\pi}\oint\frac{d\alpha}{a} = \frac{\mu_0 I}{4\pi r_0}\oint d\alpha = \frac{\mu_0 I}{4\pi r_0}2\pi = \frac{\mu_0 I}{2r_0} .$$

5.3   Magnetization

Several times already, we came to the conclusion that based on our current knowledge there are no "magnetic charges", which is the reason why **B** is source-free or solenoidal:

$$\nabla \cdot \mathbf{B} = 0 .$$

However, for convenience we will introduce the formal concept of fictitious magnetic charges. The reason is that it will simplify certain problems. Nevertheless, the physical source of static magnetic fields is always found in currents, *i.e.*, moving electric charges (disregarding the spin of particles). We have already found that all such fields can be thought of as being created by superposition of suitable dipole fields. The spin of elementary particles causes a magnetic moment, which has no explanation in classical physics. *Consequently, we can say that all static magnetic fields are ultimately caused by magnetic dipoles.* Magnetic dipoles are also fundamental in connection with the question about interaction between matter and magnetic fields. We will need to discuss this too. In doing so, we will again find a formal, broad analogy between electric and magnetic effects (see Sect. 5.5). First, we will deal with the field of a volume distribution of magnetic dipoles. It is convenient to define the *magnetization* as

$$\mathbf{M} = \frac{d\mathbf{m}}{d\tau} . \tag{5.65}$$

This represents the magnetic analogue to the electrostatic polarization **P**. Based on eq. (5.56), one concludes that a volume distribution of dipoles with the density **M(r)** creates the vector potential

$$\mathbf{A}(r) = \frac{1}{4\pi}\int_V \frac{\mathbf{M}(\mathbf{r}') \times (\mathbf{r} - \mathbf{r}')}{|\mathbf{r} - \mathbf{r}'|^3} d\tau' \ . \tag{5.66}$$

Written in a sightly different form gives

$$\mathbf{A}(r) = \frac{1}{4\pi}\int_V \mathbf{M}(\mathbf{r}') \times \nabla_{r'}\frac{1}{|\mathbf{r} - \mathbf{r}'|} \, d\tau' \ , \tag{5.67}$$

where the grad operator now operates on **r'**, which causes the sign change versus (5.56). Then we use the vector theorem

$$\nabla \times (\mathbf{b}\varphi) = \varphi(\nabla \times \mathbf{b}) - \mathbf{b} \times (\nabla\varphi) \ ,$$

which results in

$$\mathbf{A}(r) = -\frac{1}{4\pi}\int_V \nabla_{r'} \times \frac{\mathbf{M}(\mathbf{r}')}{|\mathbf{r} - \mathbf{r}'|} d\tau' + \frac{1}{4\pi}\int_V \frac{\nabla_{r'} \times \mathbf{M}(\mathbf{r}')}{|\mathbf{r} - \mathbf{r}'|} d\tau' \ .$$

The first of the two integrals can be further modified by applying the integral theorem

$$\int_V (\nabla \times \mathbf{c}) d\tau = -\oint \mathbf{c} \times d\mathbf{a} \ , \tag{5.68}$$

which is a variant of Gauss' theorem

$$\int_V (\nabla \bullet \mathbf{b}) d\tau = \oint b \bullet d\mathbf{a} \ .$$

and we obtain it from that by substituting

$$\mathbf{b} = \mathbf{c} \times \mathbf{d} \ ,$$

where the vector **c** depends on the location while **d** is location independent.

$$\int_V \nabla \bullet (\mathbf{c} \times \mathbf{d}) d\tau = \int_V \mathbf{d} \bullet (\nabla \times \mathbf{c}) d\tau - \int_V \mathbf{c} \bullet (\nabla \times \mathbf{d}) d\tau$$

$$= \mathbf{d} \bullet \int_V (\nabla \times \mathbf{c}) d\tau = \oint (\mathbf{c} \times \mathbf{d}) \bullet d\mathbf{a} \ .$$

$$\int_V \nabla \bullet (\mathbf{c} \times \mathbf{d}) d\tau = -\oint \mathbf{d} \bullet (\mathbf{c} \times d\mathbf{a}) = -\mathbf{d} \bullet \oint \mathbf{c} \times d\mathbf{a}$$

i.e., it holds for every vector **d** that

$$\mathbf{d} \bullet \int_V (\nabla \times \mathbf{c}) d\tau = -\mathbf{d} \bullet \oint \mathbf{c} \times d\mathbf{a} \ .$$

This proves (5.68). We obtain finally

$$\boxed{\mathbf{A}(\mathbf{r}) = \frac{1}{4\pi}\int_V \frac{\nabla_{r'} \times \mathbf{M}(\mathbf{r}')}{|\mathbf{r} - \mathbf{r}'|} d\tau' + \frac{1}{4\pi}\oint \frac{\mathbf{M}(\mathbf{r}') \times d\mathbf{a}'}{|\mathbf{r} - \mathbf{r}'|}} \ . \tag{5.69}$$

This is an important result. Comparing (5.69) and (5.16) reveals that a volume distribution of dipoles corresponds to a distribution of bound volume current densities

$$\boxed{\mathbf{g}_{mag}(\mathbf{r}) = \frac{1}{\mu_0}\nabla \times \mathbf{M}(\mathbf{r})} \tag{5.70}$$

and an additional distribution of bound surface current densities

$$\boxed{\mathbf{k}_{mag} = \frac{1}{\mu_0}\mathbf{M}(\mathbf{r}) \times \frac{d\mathbf{a}}{da} = \frac{1}{\mu_0}\mathbf{M}(\mathbf{r}) \times \mathbf{n}}.$$ (5.71)

The bound volume current density $(1/\mu_0)\nabla \times \mathbf{M}$, corresponding to the magnetization $\mathbf{M}$ is called *magnetization current density*. We refer to a surface current density when currents of infinite current density flow in a surface, where the current per unit length remains finite. This results plausibly form the limit illustrated in Fig. 5.28. Consider a current of density g flowing perpendicular to the paper plane within the thin layer of width d. In a section of length l, one finds the current

$$gld$$

or per unit length

$$gd .$$

If we let $g \to \infty$ while $d \to 0$, but leaving gd finite, then we obtain a surface current with the surface current density

$$k = gd , \quad [k] = \frac{A}{m} .$$

These formal results are justified as follows. Fig. 5.29 shows a "magnetized" volume, *i.e.*, a volume filled with dipoles. Let $\mathbf{M}$ be perpendicular to the upper and lower surface. If $\mathbf{M}$ is constant in the entire volume, then all internal currents cancel (Fig. 5.29b). All that remains is a surface current which has the direction of the cross product (vector product) of $\mathbf{M}$ and $\mathbf{n}$. If $\mathbf{M}$ is not uniform, then there are also bound volume currents inside the volume. These are connected to the circulation of $\mathbf{M}$ ($\nabla \times \mathbf{M}$).

A simple **example** is a circular cylinder of infinite length which is filled with uniform dipoles whose axes are oriented along the cylinder axis. There shall be no free currents inside. However, its surface shall carry surface currents. These are purely azimuthal. This problem is identical to the infinitely long coil of section 5.2.3, Fig. 5.12. With the magnetization $\mathbf{M}$, we get for the azimuthal component of the surface current density:

$$k_{\varphi mag} = \frac{M}{\mu_0}$$

and the field inside is

$$H_{zi} = k_{\varphi mag} = \frac{M}{\mu_0}$$

or

$$B_{zi} = \mu_0 H_{zi} = M .$$

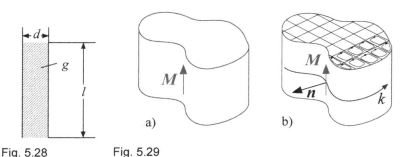

Fig. 5.28   Fig. 5.29

We can also calculate the scalar potential of a volume distribution of dipoles. By eq. (5.55) we have

$$\psi = \frac{1}{4\pi\mu_0}\int_V \frac{\mathbf{M(r')} \bullet (\mathbf{r}-\mathbf{r'})}{|\mathbf{r}-\mathbf{r'}|^3}d\tau' = \frac{1}{4\pi\mu_0}\int_V \mathbf{M(r')} \bullet \nabla_{r'}\frac{1}{|\mathbf{r}-\mathbf{r'}|}d\tau' \ ,$$

Because of the vector theorem

$$\nabla \bullet (\mathbf{b}\varphi) = \mathbf{b} \bullet (\nabla\varphi) + \varphi(\nabla \bullet \mathbf{b}) \ ,$$

this becomes

$$\psi = \frac{1}{4\pi\mu_0}\int_V \nabla_{r'} \bullet \frac{\mathbf{M(r')}}{|\mathbf{r}-\mathbf{r'}|}d\tau' - \frac{1}{4\pi\mu_0}\int_V \frac{1}{|\mathbf{r}-\mathbf{r'}|}\nabla_{r'} \bullet \mathbf{M(r')}d\tau' \ ,$$

and finally with Gauss' theorem

$$\psi = \frac{1}{4\pi\mu_0}\left[\oint \frac{\mathbf{M(r')} \bullet d\mathbf{a'}}{|\mathbf{r}-\mathbf{r'}|} - \int_V \frac{\nabla_{r'} \bullet \mathbf{M(r')}}{|\mathbf{r}-\mathbf{r'}|}d\tau'\right] \ . \tag{5.72}$$

From a formal perspective, this equation is entirely analogous to eq. (2.65). Therefore, it may be interpreted as being the potential of magnetic volume and surface charges. Of course, these are merely fictitious and only have a formal significance. In analogy to eqs. (2.66) and (2.67), one can define the (bound) magnetic volume charge density

$$\rho_{mag} = -\nabla \bullet \mathbf{M} \tag{5.73}$$

and the (bound) magnetic surface charge density

$$\sigma_{mag} = \mathbf{M} \bullet \frac{d\mathbf{a}}{da} = \mathbf{M} \bullet \mathbf{n} \ . \tag{5.74}$$

From a formal perspective and in total analogy to electrostatics, the entire magnetostatics can now be built on these concepts. We may now define magnetic charges

$$Q_{mag} = \int_V \rho_{mag}d\tau$$

and even write Coulomb's law for these

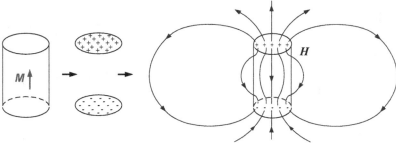

Fig. 5.30

$$F = \frac{Q_{mag\,1}Q_{mag2}}{4\pi r^2 \mu_0} \;,$$

or show that the force in the magnetic field is

$$\mathbf{F} = Q_{mag}\mathbf{H} \;,$$

and so forth. While eqs. (5.72) through (5.74) are valuable tools to calculate fields, the other analogues, *for example* Coulombs law, are not specifically relevant, not even from a formal perspective, thus, shall not be discussed any further.

As an **example**, consider a cylinder of finite length, uniformly filled with dipoles. It has no (bound) magnetic volume charges, however, there are (bound) magnetic surface charges at the top and bottom. They create an H field which compares to the electric field of two uniformly charged disks (Fig. 5.30). It only describes the field outside of the cylinder. We will need to discuss the inside field later.

In concluding this section, we summarize the most important results, side by side with the corresponding results from electrostatics in Table 5.1.

To minimize misunderstandings, the reader is cautioned that in literature these terms and quantities are not standardized. Frequently, the magnetic dipole moment in eq. (5.52) is defined without the factor μ_0. This also impacts the magnetization if we consider it to be the volume density of magnetic dipoles, *i.e.* it also lacks the factor μ_0. The quantity that is multiplied by μ_0 is then called magnetic polarization (it corresponds to the magnetization in this text). There is no need to distinguish between magnetic polarization and magnetization. This distinction relates to the historically understandable, nevertheless needless, distinction between an "elementary current theory" and a "bulk-magnetization theory" of magnetism. The former rests on eqs. (5.69) through (5.71), while the latter is based on eqs. (5.72) through (5.74). For us, these are not two different theories on magnetism, but just two equivalent formal consequences of the same theory, or the same thereby described physical reality. It is a question about the *equivalence of eddy ring and a dipole layer*, which we have met several times (for instance, in Sect. 5.2.5, where we thought of the field of a current loop as being the field of a dipole layer). This equivalence is what frequently allows to treat

electrostatic	magnetostatic
P	**M**
$\varphi = \dfrac{1}{4\pi\varepsilon_0}\oint\dfrac{\mathbf{P}(\mathbf{r}') \bullet \mathbf{n}'}{\lvert\mathbf{r}-\mathbf{r}'\rvert}\,da'$ $-\dfrac{1}{4\pi\varepsilon_0}\int\dfrac{\nabla_{r'} \bullet \mathbf{P}(\mathbf{r}')}{\lvert\mathbf{r}-\mathbf{r}'\rvert}\,d\tau'$	$\psi = \dfrac{1}{4\pi\mu_0}\oint\dfrac{\mathbf{M}(\mathbf{r}') \bullet \mathbf{n}'}{\lvert\mathbf{r}-\mathbf{r}'\rvert}\,da'$ $-\dfrac{1}{4\pi\mu_0}\int_V\dfrac{\nabla_{r'} \bullet \mathbf{M}(\mathbf{r}')}{\lvert\mathbf{r}-\mathbf{r}'\rvert}\,d\tau'$
$\rho_b = -\nabla\bullet P$ $\sigma_b = P \bullet \mathbf{n}$	$\rho_{mag} = -\nabla\bullet\mathbf{M}$ $\sigma_{mag} = \mathbf{M} \bullet \mathbf{n}$
	$\mathbf{g}_{mag} = \dfrac{1}{\mu_0}\nabla \times \mathbf{M}(\mathbf{r})$ $\mathbf{k}_{mag} = \dfrac{1}{\mu_0}\mathbf{M}(\mathbf{r}) \times \mathbf{n}$
	$\mathbf{A}(\mathbf{r}) = \dfrac{1}{4\pi}\int_V\dfrac{\nabla_{r'} \times \mathbf{M}(\mathbf{r}')}{\lvert\mathbf{r}-\mathbf{r}'\rvert}\,d\tau'$ $+\dfrac{1}{4\pi}\oint\dfrac{\mathbf{M}(\mathbf{r}') \times d\mathbf{a}'}{\lvert\mathbf{r}-\mathbf{r}'\rvert}$

Table 5.1

magnetostatic problems as if they were electrostatic ones and vice versa (notice that this also allows to treat an electric dipole layer as if it were a current loop where currents of fictitious magnetic charges flow).

5.4 Forces on Dipoles in Magnetic Fields

A moving charge in a magnetic field experiences the force
$$\mathbf{F} = Q\mathbf{v} \times \mathbf{B} .$$

If we observe the motion of a charge density distribution $\rho(\mathbf{r})$, then the force per unit volume, also called *force density*, is
$$\mathbf{f} = \rho\mathbf{v} \times \mathbf{B} .$$

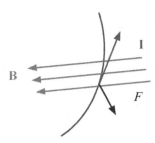

Fig. 5.31

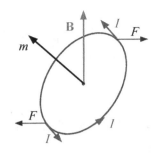

Fig. 5.32

Since

$$\rho\mathbf{v} = \mathbf{g}$$

is just the current density, we write

$$\mathbf{f} = \mathbf{g} \times \mathbf{B} .$$

Integrating this equation over the cross section of a current carrying wire gives the force per unit length at a location of the wire

$$\frac{\mathbf{F}(\mathbf{r})}{l} = \mathbf{I}(\mathbf{r}) \times \mathbf{B}(\mathbf{r}) ,$$

where **I** is a vector quantity pointing in the direction of a wire element, having the magnitude of the total current I (Fig. 5.31). First, consider a dipole, *i.e.*, a current loop within a uniform magnetic field (Fig. 5.32). Clearly, all forces cancel – there is no net force. What remains is a force pair with a torque, trying to orient **m** into the direction of **B**. It shall be noted without proof that this torque is

$$\frac{1}{\mu_0}\mathbf{m} \times \mathbf{B} . \tag{5.75}$$

If the magnetic field is not uniform, then there is a net force besides the torque. Fig. 5.33 illustrates that the net force points in the direction where the field increases when **m** is oriented parallel to **B**. If **m** is oriented anti-parallel to **B**, then the net force points in the direction of the decreasing field (Fig. 5.34). Again without proof, it shall be noted that the force on a dipole **m** in a field **B** where $\nabla \times \mathbf{B} = 0$ is

$$\mathbf{F}_{net} = \frac{(\mathbf{m} \bullet \nabla)\mathbf{B}}{\mu_0} = \frac{1}{\mu_0}\nabla(\mathbf{m} \bullet \mathbf{B}) . \tag{5.76}$$

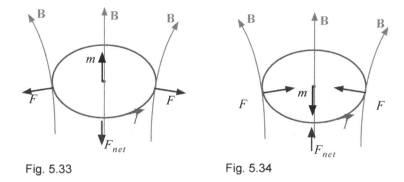

Fig. 5.33   Fig. 5.34

5.5   B and H in Magnetizable Media

So far, we discussed only fields in vacuum and how they are created by given currents or dipole distributions. If one brings any matter into an "external" magnetic field or applies a magnetic field to a space filled with matter, then this matter becomes "magnetized" and it thereby influences the net magnetic field. There are a number of concurrent effects, which results in a rather complicated overall picture. In the context of a phenomenological and macroscopic theory, these issues can only be touched briefly.

All matter consists of atoms, molecules, etc. and the shell electrons move around the nucleus according to certain laws (which can only be understood quantum mechanically). These moving charges cause currents and magnetic moments. Depending on the internal structure of a particular material, these magnetic moments may cancel each other or not, *i.e.*, a material might exhibit magnetic moments even without an applied field.

It shall be added, that besides the just discussed magnetic moments caused by circulating currents, there are other, elementary magnetic dipole moments which can only be explained in a quantum mechanical context (that is, they do not correspond to circulating currents, at least as far as we know today), just as the spin of elementary particles (electrons in particular), which also can only be explained quantum mechanically.

First, we consider a material that has no net magnetic dipole moment as long as there is no applied field. This kind of material is called *diamagnetic*. If we now apply a magnetic field that increases over time, then by the law of induction (see Sect. 1.11), a voltage is induced, which causes currents and thereby dipole moments. We then say that the medium is magnetized. Similarly as for the electric polarization (Sect. 2.8), in a first approximation we can assume a linear relationship between the magnetic field and the thereby caused magnetization. A consequence of the law of induction is that the induced **B** field weakens the applied field (Fig. 5.35). Formally, this is a consequence of the negative sign in the law of

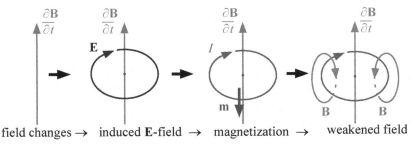

field changes $\rightarrow$ induced E-field $\rightarrow$ magnetization $\rightarrow$ weakened field

Fig. 5.35

induction (1.67), which is referred to as Lenz's law. Macroscopic, induced currents generally decay over time because of Ohmic resistance which any medium usually has. Superconductivity is an exception. The microscopic currents that cause magnetization (Ampere's molecular currents) do not decay. These flow without resistance and persist for as long as the external field remains. This is the reason why there is a unique relation between magnetization and magnetic field for which we may write

$$\mathbf{M} = \mu_0 \chi_m \mathbf{H} .$$ (5.77)

χ_m is called the *magnetic susceptibility*. Because of Lenz's law χ_m is negative. The specific formulation of (5.77) makes χ_m dimensionless. The magnetization $\mathbf{M}$ is oriented anti-parallel to $\mathbf{B}$. This allows to identify a diamagnetic material by the fact that in an inhomogeneous magnetic field, it experiences a force in the direction where the field decreases.

The molecules of a so-called *paramagnetic* material exhibit net magnetic moments even without an applied field. However, there is no preferred direction without an applied field and therefore no magnetization. The individual dipoles are statistically oriented in all directions and cancel each other on average (spatial and time average). An applied field causes a torque, which then tries to align the individual dipoles parallel to the external field. Still, temperature acts to mis-align the individual dipoles. This mis-alignment is more successful the higher the temperature. Nevertheless, partial orientation along the external field is achieved. The diamagnetic effect of induction applies simultaneously, in an attempt to decrease the magnetization. If paramagnetism is present in that material, then it prevails over diamagnetism and the same Ansatz as (5.77) can be made, now with positive χ_m. This is the reason why paramagnetic material is drawn toward the side where an inhomogeneous field increases. χ_m for paramagnetic materials is temperature dependent, while χ_m for diamagnetic materials does not depend on temperature.

There are many more magnetic phenomena. Particularly important is *ferromagnetism*, especially for electrical engineering. It is related to the spin of electrons and like paramagnetism, can qualitatively be described by the orientation of the related magnetic dipole moments. In contrast to paramagnetism, the magnetization due to ferromagnetism is several orders of magnitude higher and the relation between applied field and magnetization is not linear, not even unique (*i.e.*, it is not a single-valued function). This means that magnetization is not only a function of the external field but also of its previous states (*i.e.*, it depends on its history). Because of its non-linearity, a description of ferromagnetic materials by susceptibility depends on the current state which, from a theoretical perspective, does not necessarily make it useful. The susceptibility of diamagnetic and paramagnetic materials is very small ($|\chi_m| \ll 1$), while the numbers for ferromagnetism may reach some 10^4. Examples:

$$
\begin{array}{lll}
\text{Oxygen (O}_2) & \text{at } 18°\text{C (64°F)} & \chi_m = 1.8 \cdot 10^{-6} \\
\text{Palladium} & \text{at } 18°\text{C (64°F)} & \chi_m = 782 \cdot 10^{-6}
\end{array} \left.\vphantom{\begin{array}{l}1\\1\end{array}}\right\} \text{paramagnetic}
$$

$$
\begin{array}{ll}
\text{Nitrogen (N}_2) & \chi_m = -0.07 \cdot 10^{-6} \\
\text{Bismuth} & \chi_m = -160 \cdot 10^{-6}
\end{array} \left.\vphantom{\begin{array}{l}1\\1\end{array}}\right\} \text{diamagnetic}
$$

The reader is alerted about the non-standardized definitions of χ_m when comparing these numbers with other sources.

The fact that ferromagnetism can not be described by a linear relation has formal, far reaching consequences. The methods we have developed apply only to linear problems. There are hardly any mathematical methods available to treat non-linear problems analytically. Then, numerical methods need to be applied.

In the presence of magnetized media, it is possible to calculate the magnetic field as in vacuum, if one explicitly considers all currents, including those bound currents that originate from magnetization. We write

$$\nabla \times \mathbf{B} = \mu_0 \mathbf{g} \ ,$$

where

$$\mathbf{g} = \mathbf{g}_{free} + \mathbf{g}_{bound} = \mathbf{g}_{free} + \frac{1}{\mu_0} \nabla \times \mathbf{M} \ .$$

Therefore

$$\nabla \times \mathbf{B} = \mu_0 \mathbf{g}_f + \nabla \times \mathbf{M} \ ,$$

or

$$\boxed{\nabla \times \left(\frac{\mathbf{B} - \mathbf{M}}{\mu_0} \right) = \mathbf{g}_f} \ . \tag{5.78}$$

We define the magnetic field strength for magnetizable media

$$\boxed{\mathbf{H} = \frac{\mathbf{B} - \mathbf{M}}{\mu_0}} \ . \tag{5.79}$$

This results in our previous relation between **H** and **B** for vacuum if $\mathbf{M} = 0$. Conversely

$$\boxed{\mathbf{B} = \mu_0\mathbf{H} + \mathbf{M}} \ , \tag{5.80}$$

a relation which holds for any medium, *for example*, even for a permanent magnet which has a magnetization that persists even without an applied field. From (5.78) and (5.79) follows that

$$\boxed{\nabla \times \mathbf{H} = \mathbf{g}_f} \ . \tag{5.81}$$

This means that $\mathbf{H}$ is irrotational as long as there are only bound currents. However, $\mathbf{B}$ is not irrotational in this case, but rather

$$\nabla \times \mathbf{B} = \nabla \times \mathbf{M} \ .$$

For "linear media" follows from (5.80) that

$$\mathbf{B} = \mu_0 H + \mu_0\chi_m\mathbf{H} = \mu_0(1 + \chi_m)\mathbf{H} \ .$$

Defining the relative permeability

$$\boxed{\mu_r = 1 + \chi_m} \tag{5.82}$$

and the (absolute) permeability

$$\boxed{\mu = \mu_0\mu_r} \tag{5.83}$$

allows to write

$$\boxed{\mathbf{B} = \mu\mathbf{H}} \ . \tag{5.84}$$

When calculating $\mathbf{H}$, only free currents are relevant, while the bound currents are hidden in the relation between $\mathbf{B}$ and $\mathbf{H}$. This approach is similar to the one in electrostatics, where only the free charges are relevant when calculating $\mathbf{D}$ and the impact of bound charges is hidden in the relation between $\mathbf{D}$ and $\mathbf{E}$.

The $\mathbf{B}$ field is always source free and it is therefore always

$$\nabla\bullet\mathbf{B} = 0 \ .$$

The consequence is that $\mathbf{H}$ is not necessarily source free, namely if $\nabla\bullet\mathbf{M} \neq 0$, then:

$$\nabla\bullet\mathbf{H} = \nabla\bullet\frac{\mathbf{B} - \mathbf{M}}{\mu_0} = -\frac{1}{\mu_0}\nabla\bullet\mathbf{M} \ .$$

We have introduced $-\nabla\bullet\mathbf{M}$ as the fictitious magnetic charges eq. (5.73)

$$\rho_{mag} = -\nabla\bullet\mathbf{M} \ ,$$

which gives

$$\boxed{\nabla\bullet\mathbf{H} = \frac{1}{\mu_0}\rho_{mag}} \ . \tag{5.85}$$

The implication is that the $\mathbf{H}$ field originates from, or ends at bound magnetic charges. Because of

$$\mathbf{H} = -\nabla\psi \ ,$$

we can also write

$$\nabla\bullet(-\nabla\psi) = -\nabla^2\psi = +\frac{1}{\mu_0}\rho_{mag} \ ,$$

i.e.,

$$\boxed{\nabla^2\psi = -\frac{1}{\mu_0}\rho_{mag}}.$$

(5.86)

This represents the *magnetic Poisson equation.*

These properties of **B** and **H** can be confusing and are therefore summarized in Table 5.2 for three cases.

Table 5.2

Fields of free currents	Fields of magnetized matter	Fields of free currents and fields of magnetized matter
H, B both source free, but not curl free	**H** not source free, but curl free **B** source free, but not curl free	**H** neither source free nor curl free **B** source free, but not curl free
$\nabla\bullet\mathbf{B} = 0$ $\nabla\bullet\mathbf{H} = 0$ $\nabla\times\mathbf{B} = \mu_0\mathbf{g}_f$ $\nabla\times\mathbf{H} = \mathbf{g}_f$	$\nabla\bullet\mathbf{B} = 0$ $\nabla\bullet\mathbf{H} = \rho_{mag}/\mu_0$ $\nabla\times\mathbf{B} = \nabla\times\mathbf{M}$ $\nabla\times\mathbf{H} = 0$	$\nabla\bullet\mathbf{B} = 0$ $\nabla\bullet\mathbf{H} = \rho_{mag}/\mu_0$ $\nabla\times\mathbf{B} = \mu_0\mathbf{g}_f + \nabla\times\mathbf{M}$ $\nabla\times\mathbf{H} = \mathbf{g}_f$

The field of a cylindrical, uniformly magnetized permanent magnet shall serve as an **example**. The **H** field can be calculated like the electric field of two circular plates with a surface charge. This gives

$$\sigma_{mag} = \mathbf{M}\bullet\mathbf{n}.$$

The **B** field can be calculated like the field of a coil of finite length with the surface current

$$\mathbf{k} = k_\varphi\mathbf{e}_\varphi = \frac{1}{\mu_0}\mathbf{M}\times\mathbf{n}$$

or also as shown in Fig. 5.36 and Fig. 5.37, by calculating

$$\mathbf{B} = \mu_0\mathbf{H} + \mathbf{M}.$$

The field of **H** in Fig. 5.37a) is irrotational but has sources and sinks in the form of bound magnetic charges at the top and bottom surfaces, respectively. The field of **B** in Fig. 5.37b) is source free but has curl in the form of bound azimuthal currents in the cylinder wall. **M** in Fig. 5.37c) is neither source free nor curl free. Its curl is in the cylinder wall (as for **B**) and its sources are at the top and bottom surfaces (as for **H**). It shall be noted that Fig. 5.30 is not representing a magnetized material, but rather the vacuum field created by given dipoles, which can only be calculated outside of the cylinder. However, Fig. 5.37 is based on the now generalized

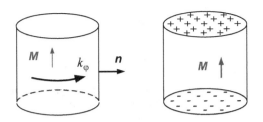

Fig. 5.36

definition of **H** as of eq. (5.79). Only this new definition defines **H** inside a magnetizable medium.

There are also anisotropic linear media for which χ_m and μ become a tensor. The relation in this case reads:

$$\left.\begin{aligned}
B_x &= \mu_{xx}H_x + \mu_{xy}H_y + \mu_{xz}H_z \\
B_y &= \mu_{yx}H_x + \mu_{yy}H_y + \mu_{yz}H_z \\
B_z &= \mu_{zx}H_x + \mu_{zy}H_y + \mu_{zz}H_z
\end{aligned}\right\} , \tag{5.87}$$

or in short

$$\mathbf{B} = \boldsymbol{\mu} \cdot \mathbf{H} . \tag{5.88}$$

The tensor $\boldsymbol{\mu}$ is symmetric, *i.e.*,

$$\mu_{ik} = \mu_{ki} . \tag{5.89}$$

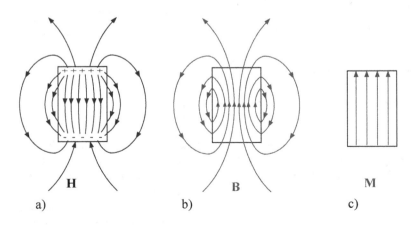

Fig. 5.37

5.6   Ferromagnetism

Ferromagnetism is a property of iron, cobalt, nickel, and certain alloys. The relationship between **M** and **H** or **B** and **H** is very complicated for these materials. As previously mentioned, it is neither linear nor unique. It depends on the history of the medium and is also rather different for the various kinds of iron. The exact relation has to be established by measurements. In principle, the measurement could be carried out as illustrated in Fig. 5.38. For simplicity reasons, it shall be assumed that the cross section of the coil is very small compared to its radius r. The primary current in the coil creates the field

$$H = \frac{NI}{2\pi r} \, ,$$

where N is the total number of turns the current path (primary loop) takes around the toroid. This creates an EMF in the induction loop (secondary loop) of magnitude

$$V_i = \dot\phi = \dot{B}A \ .$$

Integrating the EMF over time gives

$$\int V_i dt = BA \ .$$

Measuring the corresponding values of I and V_i provides the corresponding values of B and H. Charting these values gives the so-called hysteresis loop (Fig. 5.39).

If we expose a material that has not been magnetized to a H-field that increases, starting from zero, then B goes through the so-called *new curve* or *initial curve* until it reaches a region that is called *saturation*. Saturation is characterized by the fact that the related magnetization does not increase any more. The magnetization is shown in Fig. 5.40.

Now, letting H decrease, results not in a curve that is merely reversing direction, but it takes a different path. B or M decrease less than what they had increased when H was increased. The result is that even when H is zero, B still has a finite value (*remanence* or *remanent field*). To bring B to zero requires a negative

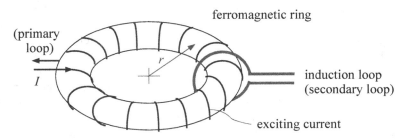

Fig. 5.38

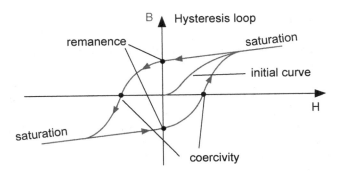

Fig. 5.39

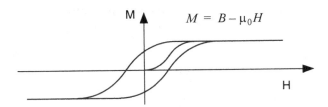

Fig. 5.40

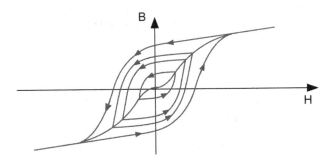

Fig. 5.41

field (the so-called *coercivity* or *coercive force*). Sufficiently large negative fields lead to saturation in the other direction. When increasing H thereafter, B will take the path through the other part of the hysteresis loop, eventually arriving again at the positive side of saturation. Smaller hysteresis loops are achieved when not going all the way to saturation (Fig. 5.41). The shape of the hysteresis loop depends on the material and can be wide or narrow, more or less tilted, or almost rectangular. There are materials with specific hysteresis loops that are more or less useful for any particular of the multitude of applications in electrical engineering. An important feature is the area enclosed when passing through an entire hysteresis loop, since this area represents the losses when changing magnetization. This is plausible: The work done per unit of time, *i.e.*, the power necessary to create the field given in Fig. 5.38 is

$$\frac{dW}{dt} = IV_e ,$$

where V_e is the voltage in the primary turns

$$V_e = N\dot{\phi} = NA\frac{dB}{dt} .$$

Therefore

$$\frac{dW}{dt} = INA\frac{dB}{dt} = \frac{IN}{2\pi r}A2\pi r\frac{dB}{dt} = H\tau\frac{dB}{dt} .$$

τ is the volume of the ferromagnetic ring. The necessary power needed per unit volume is

$$\frac{1}{\tau}\frac{dW}{dt} = H\frac{dB}{dt}$$

or

$$\frac{1}{\tau}dW = HdB \tag{5.90}$$

and

$$\frac{W}{\tau} = \int_{B_1}^{B_2} HdB .$$

W is the work necessary to establish the field B_2 starting at B_1. It corresponds per unit volume to the shaded area of Fig. 5.42.

The area of the integral for an entire loop is shown in Fig. 5.43

$$\frac{W}{\tau} = \oint HdB = \oint H(\mu_0 dH + dM) = \oint HdM . \tag{5.91}$$

This means that a wide hysteresis curve, which is characteristic of so-called "hard materials", leads to large losses during re-magnetization. Thus, hard materials are not suited for transformers, for which "soft materials", with a narrow hysteresis loop are better suited.

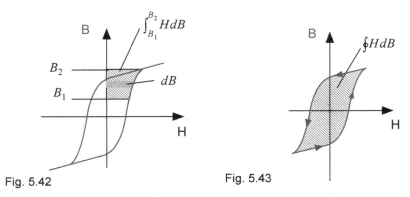

Fig. 5.42   Fig. 5.43

Although from a purely formal perspective, the relation between **B** and **H** is rather complicated, we can write the constitutive equation in this form

$$\mathbf{B} = \mu_0\mu_r\mathbf{H} = \mu\mathbf{H} ,$$

where μ_r or μ are now functions of **H** and its previous states. When subsequently using this terminology, it shall not be understood to suggest any kind of linearity. Rather a specific condition shall be characterized by a corresponding factor.

An electromagnet with a ferromagnetic core and a small air gap (Fig. 5.44) shall serve as an **example.** As long as the toroid is relatively slim and the gap sufficiently small, we may consider the fields H_1 inside the core and H_2 in the air-gap approximately uniform and can also neglect that the different force lines have different lengths. Then we may write

$$\oint \mathbf{H} \bullet d\mathbf{s} = H_1 l_1 + H_2 l_2 = NI$$

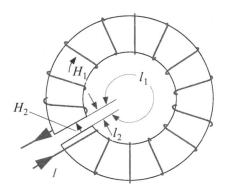

Fig. 5.44

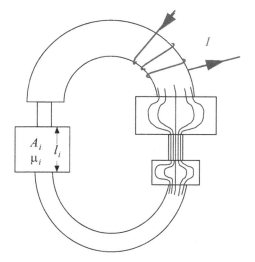

Fig. 5.45

and furthermore

$$B_1 = B_2 = B.$$

The reason is that the perpendicular components of **B** must be continuous at the boundary, which will be demonstrated in the next section. This leads to

$$\mu_1 H_1 = \mu_0 H_2 = B,$$

and furthermore

$$\frac{B}{\mu_1} l_1 + \frac{B}{\mu_0} l_2 = NI,$$

or

$$B = \frac{NI}{\dfrac{l_1}{\mu_1} + \dfrac{l_2}{\mu_0}}. \tag{5.92}$$

This represents a very simple example of a so-called *magnetic circuit*. In general, such a circuit may consist of several pieces of different length, different cross section, and different permeability l_i, A_i, μ_i (Fig. 5.45). We approximate that the flux is the same through each piece, which is equivalent to neglecting the flux leakage (fringe effects). One writes:

$$NI = \sum_{i=1}^{n} H_i l_i,$$

$$H_i = \frac{B_i}{\mu_i} = \frac{\phi}{A_i \mu_i},$$

and

$$NI = \sum_{i=1}^{n} \frac{\phi}{A_i \mu_i} l_i ,$$

or

$$\boxed{NI = \phi \sum_{i=1}^{n} \frac{l_i}{A_i \mu_i}} . \qquad (5.93)$$

For comparison, we write Ohm's law for a circuit of several resistors in series.

$$V = I \sum_{i=1}^{n} \frac{l_i}{A_i \kappa_i} . \qquad (5.94)$$

Notice the close formal analogy. NI replaces the voltage, which is why this relation is sometimes called the magnetomotive force. The flux ϕ replaces the current I. The sum

$$\sum_{i=1}^{n} \frac{l_i}{A_i \mu_i} ,$$

represents the resistance and is therefore called *magnetic resistance* R_{mag}. For an individual element we have now

$$R_{mag, i} = \frac{l_i}{A_i \mu_i} . \qquad (5.95)$$

Permeability is the formal equivalent of conductivity and may therefore be interpreted as *magnetic conductivity*. This analogy can be generalized to apply to entire networks with branching magnetic "currents". However, necessary is to remember that these are simply approximations, which are not necessarily accurate. To estimate accuracy may be difficult at times. Nevertheless, these kind of approximations may be acceptable when exact field theoretical calculations become too difficult.

5.7   Boundary Conditions for B and H, and the Refraction of Magnetic Force Lines

It always holds that

$$\nabla \times \mathbf{H} = \mathbf{g}_f ,$$

$$\nabla \cdot \mathbf{B} = 0 .$$

These relations allow one to derive the boundary conditions that $\mathbf{H}$ and $\mathbf{B}$ have to satisfy.

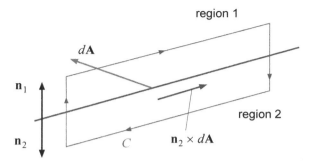

Fig. 5.46

Consider the area element $d\mathbf{A}$, where the vector $d\mathbf{A}$ lies in the boundary, *i.e.* the area element dA is perpendicular to the boundary (Fig. 5.46). One writes the integral

$$\int_A (\nabla \times \mathbf{H}) \cdot d\mathbf{A} = \oint_C \mathbf{H} \cdot d\mathbf{s}$$

$$= \mathbf{H}_2 \cdot \frac{\mathbf{n}_2 \times d\mathbf{A}}{dA} ds - \mathbf{H}_1 \cdot \frac{\mathbf{n}_2 \times d\mathbf{A}}{dA} ds$$

$$= \int_A \mathbf{g}_f \cdot d\mathbf{A} = \mathbf{k}_f \cdot \frac{d\mathbf{A}}{dA} ds$$

where $\mathbf{k}_f$ is the surface current density in the boundary (note that $\mathbf{k}_f$ does not contain any magnetization currents). Therefore

$$(\mathbf{H}_2 - \mathbf{H}_1) \cdot \frac{\mathbf{n}_2 \times d\mathbf{A}}{dA} ds = \mathbf{k}_f \cdot \frac{d\mathbf{A}}{dA} ds$$

or

$$(\mathbf{H}_2 - \mathbf{H}_1) \times \mathbf{n}_2 \cdot d\mathbf{A} = \mathbf{k}_f \cdot d\mathbf{A} .$$

This is true for every surface element on the boundary and therefore

$$\boxed{(\mathbf{H}_2 - \mathbf{H}_1) \times \mathbf{n}_2 = \mathbf{k}_f} . \qquad (5.96)$$

If there is no free current in the surface $\mathbf{k}_f = 0$ this reduces to

$$(\mathbf{H}_2 - \mathbf{H}_1) \times \mathbf{n}_2 = 0$$

or

$$\boxed{H_{2t} = H_{1t}} , \qquad (5.97)$$

that is, the tangential components of $\mathbf{H}$ are continuous. The presence of a surface current density causes a discontinuity of the tangential components.

Now, consider the small slice of volume as illustrated in Fig. 5.47. Here, we have

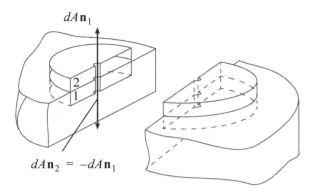

Fig. 5.47

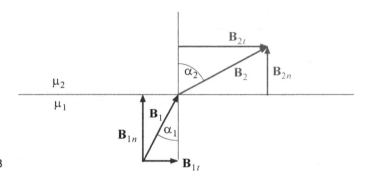

Fig. 5.48

$$\int (\nabla \bullet \mathbf{B})\,d\tau \;=\; \oint B \bullet d\mathbf{A} \;=\; (\mathbf{B}_2 - \mathbf{B}_1) \bullet \mathbf{n}_1\, dA \;=\; 0$$

i.e.,

$$(\mathbf{B}_2 - \mathbf{B}_1) \bullet \mathbf{n}_1 \;=\; 0$$

or

$$\boxed{B_{2n} \;=\; B_{1n}}\,. \tag{5.98}$$

The normal components of **B** are always continuous.

From eqs. (5.97) and (5.98) follows the law of refraction for magnetic field lines. For simplicity reasons, we assume no free currents in the surface. Using Fig. 5.48, we find

$$\frac{\tan \alpha_1}{\tan \alpha_2} \;=\; \frac{\dfrac{B_{1t}}{B_{1n}}}{\dfrac{B_{2t}}{B_{2n}}} \;=\; \frac{B_{1t}}{B_{2t}} \;=\; \frac{\mu_1 H_{1t}}{\mu_2 H_{2t}}$$

that is

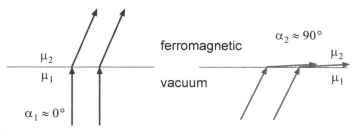

Fig. 5.49

$$\boxed{\frac{\tan\alpha_1}{\tan\alpha_2} = \frac{\mu_1}{\mu_2}} \ . \tag{5.99}$$

If $\mu_1 \ll \mu_2$ (*for example* at the boundary between a ferromagnetic material and vacuum) then we have either $\alpha_2 \approx 90°$ or $\alpha_1 \approx 0°$. This means that the force lines either enter the ferromagnetic material perpendicularly, or they are tangential to its surface (Fig. 5.49).

The formulas derived above do not consider possible effects of dipole layers. As before in the electrostatic case, dipole layers modify the boundary conditions.

Sometimes, the boundary conditions for **A** are needed. Because of

$$\nabla \cdot \mathbf{A} = 0$$

and

$$\nabla \times \mathbf{A} = \mathbf{B} \ ,$$

similar considerations as before lead to the result that both, the tangential and the normal components of **A** have to be continuous at the boundary:

$$\boxed{\begin{aligned} A_{1n} &= A_{2n} \\ A_{1t} &= A_{2t} \end{aligned}} \ . \tag{5.100}$$

The following sections are dedicated to illustrate the boundary conditions by some examples.

5.8 Plate, Sphere, Hollow Sphere in a Uniform Magnetic Field

5.8.1 The Planar Plate

Consider a planar plate of magnetic material (μ) be placed inside a uniform field $\mathbf{H}_a$ which is perpendicular to the plates (Fig. 5.50). This causes a uniform magnetization inside the plate

$$\mathbf{M} = \mu_0 \chi_m \mathbf{H}_i \ .$$

The magnetization $\mathbf{M}$ causes an opposing field $\mathbf{H}_g$ (more precisely, this field weakens the applied field in ferro- and paramagnetic materials, while it amplifies the applied field $\mathbf{H}_a$ in diamagnetic materials). The resulting net field inside is

$$\mathbf{H}_i = \mathbf{H}_a + \mathbf{H}_g \ .$$

And so for the magnetization:

$$\mathbf{M} = \mu_0 \chi_m \mathbf{H}_a + \mu_0 \chi_m \mathbf{H}_g \ .$$

Now we determine the field $\mathbf{H}_g$ caused by $\mathbf{M}$. At the surface, $\mathbf{M}$ causes fictitious magnetic charges $\pm\rho_{mag}$ (*for example* for paramagnetism + on top and - at the bottom). The result for paramagnetism is a field pointing downward.

$$H_g = \frac{M}{\mu_0} \ .$$

Any case, it is always true

$$\mathbf{H}_g = -\frac{\mathbf{M}}{\mu_0} \ . \tag{5.101}$$

Therefore

$$\mathbf{M} = \mu_0 \chi_m \mathbf{H}_a - \chi_m \mathbf{M}$$

i.e.,

$$\mathbf{M} = \frac{\mu_0 \chi_m \mathbf{H}_a}{1 + \chi_m} \ , \tag{5.102}$$

and

$$\mathbf{H}_i = \mathbf{H}_a - \frac{\mathbf{M}}{\mu_0} \ . \tag{5.103}$$

Analogous to the definition of the de-electrification factor in Sect. 2.12.4, eq. (2.141), one may now define the *de-magnetization* factor of the plate ($1/\mu_0$). With eq. (5.103) one obtains

$$\mathbf{H}_i = \mathbf{H}_a - \frac{\mathbf{M}}{\mu_0} = \mathbf{H}_a - \frac{\chi_m \mathbf{H}_a}{1 + \chi_m} = \frac{\mathbf{H}_a}{1 + \chi_m} \ .$$

i.e.,

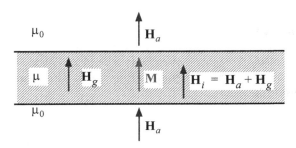

Fig. 5.50

$$\mathbf{H}_i = \frac{\mathbf{H}_a}{\mu_r}.$$

(5.104)

This shows that for a paramagnetic plate ($\mu_r > 1$) $H_i < H_a$ and for a diamagnetic plate ($\mu_r < 1$) $H_i > H_a$, as mentioned above already. Furthermore,

$$B_i = \mu H_i = \mu_r \mu_0 H_i = \mu_r \mu_0 \frac{H_a}{\mu_r} = \mu_0 H_a = B_a,$$

that is, B is continuous, as necessary. We could have used this fact to directly obtain eq. (5.104).

5.8.2   The Sphere

Consider a sphere (of μ_i) in a space (of μ_a) and placed in an applied uniform field $\mathbf{H}_{a,\infty}$ extending to infinity (Fig. 5.51). As for the electrostatic case, we can solve this problem by superposition of a dipole field on the outside and a uniform field inside. This solution is also the only one possible. We try the following Ansatz

$$\left.\begin{array}{l} \psi_a = -H_{a,\infty} r \cos\theta + \dfrac{C\cos\theta}{r^2} \\[2mm] \psi_i = -H_i r \cos\theta \end{array}\right\}.$$

(5.105)

Two constants H_i and C need to be determined. They result from the boundary conditions at the sphere's surface $r = r_s$, where B_r and H_θ have to be continuous. We start with

$$H_\theta = -\frac{1}{r}\frac{\partial\psi}{\partial\theta},$$

i.e.,

$$\left.\begin{array}{l} H_{\theta a} = -H_{a,\infty}\sin\theta + \dfrac{C\sin\theta}{r^3} \\[2mm] H_{\theta i} = -H_i\sin\theta \end{array}\right\}$$

and

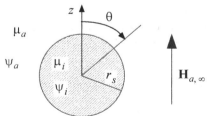

Fig. 5.51

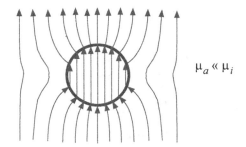

$\mu_a \ll \mu_i$

Fig. 5.52

$$B_r = -\mu \frac{\partial \psi}{\partial r} \; ,$$

i.e., the *B* field outside and inside is

$$\left. \begin{array}{l} B_{ra} = +\mu_a H_{a,\,\infty} \cos\theta + 2\mu_a \dfrac{C\cos\theta}{r^3} \\[2mm] B_{ri} = +\mu_i H_i \cos\theta \end{array} \right\} .$$

This determines the boundary conditions to be

$$\left. \begin{array}{l} -H_{a,\,\infty} + \dfrac{C}{r_s^3} = -H_i \\[3mm] +\mu_a H_{a,\,\infty} + 2\mu_a \dfrac{C}{r_s^3} = +\mu_i H_i \end{array} \right\} .$$

The result for H_i after eliminating C becomes

$$H_i = \frac{3\mu_a}{2\mu_a + \mu_i} H_{a,\,\infty} \tag{5.106}$$

and

$$B_i = \frac{3\mu_i}{2\mu_a + \mu_i} B_{a,\,\infty} \; . \tag{5.107}$$

Those results are, again, entirely analogous to those obtained in electrostatics. Only replacing ε by μ gives the previous results from electrostatics. The field inside is uniform, which is only the case for ellipsoids and their limiting cases.

If $\mu_i \gg \mu_a$ (ferromagnetic sphere in vacuum, see Fig. 5.52), then

$$B_i \approx 3 B_{a,\,\infty} \; ,$$

i.e., the field is magnified by a factor of 3. However, H_i becomes very small

$$H_i \approx 3\frac{\mu_a}{\mu_i}H_{a,\,\infty} \ .$$

The force lines outside are perpendicular to the surface. The field is equivalent to the electric field of a conductive sphere in a uniform external field.

For a sphere in vacuum the relation is $\mu_a = \mu_0$ and therefore, we can write H_i in the following form:

$$H_i = H_{a,\,\infty} - \frac{1}{3\mu_0}M \tag{5.108}$$

where, of course,

$$M = \mu_0 \chi_{mi} H_i = \mu_0(\mu_{ri} - 1)H_i = (\mu_i - \mu_0)H_i \ .$$

The de-magnetization factor of the sphere is therefore $1/3\mu_0$. This relates to the fact that the field of a uniformly polarized sphere (**M**) creates the field inside

$$\mathbf{H} = \frac{-\mathbf{M}}{3\mu_0} \ .$$

5.8.3   The Hollow Sphere

The problem of a hollow sphere in a uniform, applied field $\mathbf{H}_{a,\,\infty}$ can be solved in a similar manner (Fig. 5.53). There are three regions with the permeabilities μ_a, μ_m, μ_i. For the potentials, we try the Ansatz

$$\left.\begin{aligned}
\psi_a &= -H_{a,\,\infty}r\cos\theta + \frac{C\cos\theta}{r^2} \\[6pt]
\psi_m &= -H_m r\cos\theta + \frac{D\cos\theta}{r^2} \\[6pt]
\psi_i &= -H_i r\cos\theta
\end{aligned}\right\} \ . \tag{5.109}$$

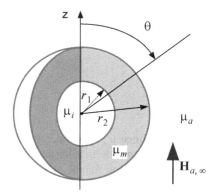

Fig. 5.53

The four constants C, D, H_m, H_i are determined in the familiar way by two boundary conditions at $r = r_1$ and two at $r = r_2$. They result in these equations:

$$
\left.
\begin{aligned}
-H_{a,\infty} + \frac{C}{r_2^3} &= -H_m + \frac{D}{r_2^3} \\[2mm]
-H_m + \frac{D}{r_1^3} &= -H_i \\[2mm]
\mu_a H_{a,\infty} + 2\mu_a \frac{C}{r_2^3} &= \mu_m H_m + 2\mu_m \frac{D}{r_2^3} \\[2mm]
\mu_m H_m + 2\mu_m \frac{D}{r_1^3} &= \mu_i H_i
\end{aligned}
\right\}
\qquad (5.110)
$$

Cramer's rule may be used to solve for H_i. We skip the laborious, but trivial algebra. The final result is

$$
H_i = H_{a,\infty} \frac{1}{\dfrac{2}{3} + \dfrac{\mu_i}{3\mu_a} + \dfrac{2(\mu_m - \mu_i)(\mu_m - \mu_a)}{9\mu_m\mu_a}\left[1 - \left(\dfrac{r_1}{r_2}\right)^3\right]} . \qquad (5.111)
$$

The previously calculated case of the sphere is a special case of this. Letting $r_1 = r_2$ gives again eq. (5.106). Another interesting case is for $\mu_i = \mu_a = \mu_0$ (vacuum), $\mu_m \neq \mu_0$. Now the field is

$$
H_i = H_{a,\infty} \frac{1}{1 + \dfrac{2\left(\dfrac{\mu_m}{\mu_0} - 1\right)^2}{9\dfrac{\mu_m}{\mu_0}}\left[1 - \left(\dfrac{r_1}{r_2}\right)^3\right]} .
$$

For a uniform ferromagnetic hollow sphere we have $\mu_m \gg \mu_0$ and therefore

$$
H_i = \frac{9 H_{a,\infty}}{2\dfrac{\mu_m}{\mu_0}\left[1 - \left(\dfrac{r_1}{r_2}\right)^3\right]} .
$$

If also $(r_1/r_2)^3 \ll 1$, then

$$
H_i \approx \frac{9 H_{a,\infty}}{2\dfrac{\mu_m}{\mu_0}} .
$$

This is an important and handy result. Since μ_m may take on values of the order of $10^4 \mu_0$, i.e., H_i is by 3 to 4 orders of magnitude smaller than the outside field. This means that highly permeable materials allow for the shielding of external fields (Fig. 5.54). Of course, the B field is:

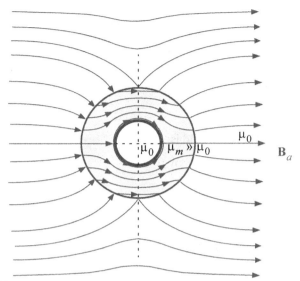

Fig. 5.54

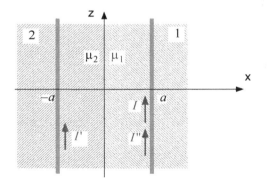

Fig. 5.55

$$B_i \approx \frac{9 B_{a,\infty}}{2 \dfrac{\mu_m}{\mu_0}} \; .$$

5.9 Imaging at a Plane

The next problem considered deals with the magnetic field of an infinitely long, straight, current carrying wire. The wire shall be parallel to a boundary that separates two materials of different permeability (Fig. 5.55). If we let ourselves be guided by the broad analogy between electrostatics and magnetostatics above, then we might suppose that the problem can be solved using *image currents* I' and I'' (compare sects. 2.11.2 and 4.5.1.). Therefore, we try (and also show that this is correct) to express the field in region 1 as the superposition of the fields of I and I', and in region 2 as the field of I''. According to eq. (5.28), the field of a current $I(x = a, y = 0)$ has the field

$$\mathbf{H} = \frac{I}{2\pi[(x-a)^2 + y^2]}\langle -y, x-a, 0\rangle \,,$$

for the current $I'(x = -a, y = 0)$, the field is

$$\mathbf{H'} = \frac{I'}{2\pi[(x+a)^2 + y^2]}\langle -y, x+a, 0\rangle \,,$$

and for the current $I''(x = a, y = 0)$, the field is

$$\mathbf{H''} = \frac{I''}{2\pi[(x-a)^2 + y^2]}\langle -y, x-a, 0\rangle.$$

Therefore, we write for the field in region 1:

$$\left.\begin{aligned}
H_{1x} &= -\frac{Iy}{2\pi[(x-a)^2 + y^2]} - \frac{I'y}{2\pi[(x+a)^2 + y^2]} \\
H_{1y} &= +\frac{I(x-a)}{2\pi[(x-a)^2 + y^2]} + \frac{I'(x+a)}{2\pi[(x+a)^2 + y^2]}
\end{aligned}\right\} \,.$$

(5.112)

and for region 2:

$$\left.\begin{aligned}
H_{2x} &= -\frac{I''y}{2\pi[(x-a)^2 + y^2]} \\
H_{2y} &= +\frac{I''(x-a)}{2\pi[(x-a)^2 + y^2]}
\end{aligned}\right\} \,.$$

(5.113)

Because for $x = 0$, both H_y and $B_x = \mu H_x$ have to be continuous, it has to be

$$\left.\begin{aligned}
-\frac{Ia}{2\pi[a^2 + y^2]} + \frac{I'a}{2\pi[a^2 + y^2]} &= -\frac{I''a}{2\pi[a^2 + y^2]} \\
-\frac{Iy\mu_1}{2\pi[a^2 + y^2]} - \frac{I'y\mu_1}{2\pi[a^2 + y^2]} &= -\frac{I''y\mu_2}{2\pi[a^2 + y^2]}
\end{aligned}\right\} \,.$$

When cancelling common terms we get

$$\left.\begin{aligned}
I'' &= I - I' \\
\mu_2 I'' &= \mu_1(I + I')
\end{aligned}\right\} \,.$$

(5.114)

Solving for I' and I'' gives

$$I' = \frac{\mu_2 - \mu_1}{\mu_1 + \mu_2} I \left.\vphantom{\frac{\mu_2 - \mu_1}{\mu_1 + \mu_2}}\right\}$$
$$I'' = \frac{2\mu_1}{\mu_1 + \mu_2} I \left.\vphantom{\frac{2\mu_1}{\mu_1 + \mu_2}}\right\} \quad . \tag{5.115}$$

Notice when comparing these results with the corresponding electrostatic problems of Sect. 2.11.2, that the relations become the same when replacing ε by $1/\mu$ (not μ).

The field in region 2 is the result of two currents, in the same manner as it was discussed in Sect. 5.2.1, Fig. 5.9. The currents I and I' are parallel when $\mu_1 < \mu_2$ and they are anti-parallel when $\mu_1 > \mu_2$. The field lines in region 2 are concentric circles. The fields are planar, *i.e.*, they are independent of z and the fieldlines are in planes parallel to $z = 0$. Figs. 5.56 through 5.58 show examples of such fields. Shown are the lines of **B,** which are source-free at the boundary.

We have solved the problem with the Ansatz (5.112) and (5.113) by formally applying the boundary conditions. This disguises what really happens. The current I magnetizes both materials. The result are magnetization currents at the boundary at $x = 0$ and also in the vicinity of the current I.

We start with the boundary between the two materials at $x = 0$. Using eq. (5.77) and with the fields described in (5.112) and (5.113), we find for the boundary

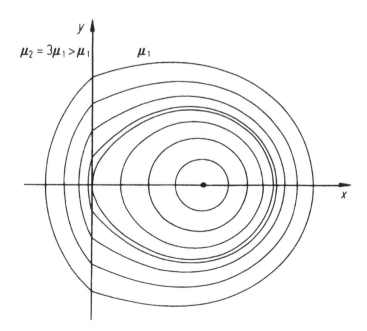

Fig. 5.56

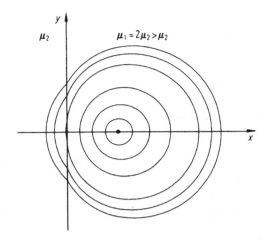

Fig. 5.57

$$M_{1x} = -\frac{(\mu_1 - \mu_0)y}{2\pi(a^2 + y^2)}(I + I')$$

$$M_{1y} = -\frac{(\mu_1 - \mu_0)a}{2\pi(a^2 + y^2)}(I - I')\bigg\}$$ (5.116)

$$M_{1z} = 0$$

and

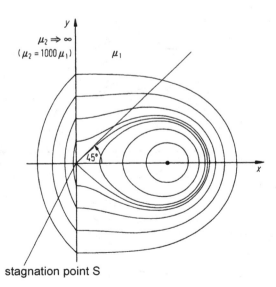

stagnation point S

Fig. 5.58

$$M_{2x} = -\frac{(\mu_2 - \mu_0)y}{2\pi(a^2 + y^2)}I'$$

$$M_{2y} = -\frac{(\mu_2 - \mu_0)a}{2\pi(a^2 + y^2)}I' \quad . \qquad (5.117)$$

$$M_{2z} = 0$$

The surface current density of the magnetization current in the surface results from (5.71)

$$\mathbf{k} = \frac{1}{\mu_0}(\mathbf{M}_1 \times \mathbf{n}_1 + \mathbf{M}_2 \times \mathbf{n}_2) \quad .$$

where

$$\mathbf{n}_1 = \langle -1, 0, 0 \rangle$$

and

$$\mathbf{n}_2 = \langle +1, 0, 0 \rangle \quad .$$

Thus, $\mathbf{k}$ has only a z-component, namely

$$k_z(y) = \frac{\mu_1 a I'}{\mu_0 \pi(a^2 + y^2)} \quad . \qquad (5.118)$$

Now, we look at the neighborhood of the current itself. It flows inside the center of a small cylindrical vacuum, which is removed from the surrounding medium (μ_1) (see Fig. 5.59). At the surface of the medium one has

$$H_\varphi = \frac{I}{2\pi\rho} \quad .$$

or

$$M_\varphi = \frac{(\mu_1 - \mu_0)I}{2\pi\rho} \quad .$$

This results in the surface current density

$$k_z = \frac{(\mu_1 - \mu_0)}{\mu_0 2\pi\rho}I$$

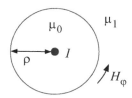

Fig. 5.59

and the magnetization current

$$k_z 2\pi\rho = \left(\frac{\mu_1}{\mu_0} - 1\right)I.$$

The total or effective current, including I is now

$$I_{eff} = \frac{\mu_1}{\mu_0}I = \mu_r I. \tag{5.119}$$

This does not change, even if we let the radius ρ approach zero. Compare this result with the effective charge $Q' = Q_{eff} = Q/\varepsilon_r$ of Sect. 2.11.1. Again, μ replaces $1/\varepsilon$.

Eqs. (5.118) and (5.119) define all currents. It is possible to calculate their magnetic field **B**, which will manifest itself in the field given by eqs. (5.112), (5.113), and (5.115). First, we calculate the field caused by the surface current k_z. By (5.28) and (5.118), we write

$$\left.\begin{aligned}
B_x &= \frac{\mu_1 aI'}{2\pi^2}\int_{-\infty}^{+\infty} \frac{-(y-y')dy'}{[x^2 + (y'-y)^2][a^2 + y'^2]}\\
B_y &= \frac{\mu_1 aI'x}{2\pi^2}\int_{-\infty}^{+\infty} \frac{dy'}{[x^2 + (y'-y)^2][a^2 + y'^2]}
\end{aligned}\right\}. \tag{5.120}$$

These two integrals can be solved with the usual methods of calculus. After some cumbersome algebra we obtain

$$\int_{-\infty}^{+\infty} \frac{(y'-y)dy'}{[x^2 + (y'-y)^2][a^2 + y'^2]} = -\frac{\pi y}{a[y^2 + (a+|x|)^2]} \tag{5.121}$$

$$\int_{-\infty}^{+\infty} \frac{dy'}{[x^2 + (y'-y)^2][a^2 + y'^2]} = \frac{\pi(a+|x|)}{a|x|[y^2 + (a+|x|)^2]}. \tag{5.122}$$

Finally for $x > 0$, this results in

$$\left.\begin{aligned}
B_x &= -\frac{\mu_1 I'}{2\pi}\frac{y}{y^2 + (x+a)^2}\\
B_y &= +\frac{\mu_1 I'}{2\pi}\frac{x+a}{y^2 + (x+a)^2}
\end{aligned}\right\} \tag{5.123}$$

and for $x < 0$

$$\left.\begin{aligned}
B_x &= -\frac{\mu_1 I'}{2\pi}\frac{y}{y^2 + (x-a)^2}\\
B_y &= +\frac{\mu_1 I'}{2\pi}\frac{x-a}{y^2 + (x-a)^2}
\end{aligned}\right\}. \tag{5.124}$$

The field of the current $(\mu_1/\mu_0)I$ at $x = a, y = 0$ needs to be added. By (5.28)

$$
\begin{aligned}
B_x &= -\frac{\mu_1 I y}{2\pi[y^2+(x-a)^2]} \\[2mm]
B_y &= +\frac{\mu_1 I(x-a)}{2\pi[y^2+(x-a)^2]}
\end{aligned}\Bigg\} .
\tag{5.125}
$$

The total field in region 1 $(x>0)$ is finally

$$
\begin{aligned}
B_{1x} &= -\mu_1\left\{\frac{I'y}{2\pi[y^2+(x+a)^2]}+\frac{Iy}{2\pi[y^2+(x-a)^2]}\right\} = \mu_1 H_{1x} \\[2mm]
B_{1y} &= +\mu_1\left\{\frac{I'(x+a)}{2\pi[y^2+(x+a)^2]}+\frac{I(x-a)}{2\pi[y^2+(x-a)^2]}\right\} = \mu_1 H_{1y}
\end{aligned}\Bigg\}
$$

and in region 2 $(x<0)$, when using (5.114)

$$
\begin{aligned}
B_{2x} &= -\frac{\mu_1(I'+I)y}{2\pi[y^2+(x-a)^2]} = -\frac{\mu_2 I''y}{2\pi[y^2+(x-a)^2]} = \mu_2 H_{2x} \\[2mm]
B_{2y} &= +\frac{\mu_1(I'+I)(x-a)}{2\pi[y^2+(x-a)^2]} = +\frac{\mu_2 I''(x-a)}{2\pi[y^2+(x-a)^2]} = \mu_2 H_{2y}
\end{aligned}\Bigg\} .
$$

As expected, this is exactly our previous result of (5.112), (5.113), and (5.115).

Formally, one can utilize the fictitious magnetic charges which are caused by magnetization. One can imagine that the entire field is created by the current I on one hand, and the fictitious magnetic charges on the other. The fictitious magnetic charges are located exclusively on the boundary $x = 0$. According to (5.74), these surface charges are

$$
\sigma_{mag} = \mathbf{M}_1 \bullet \mathbf{n}_1 + \mathbf{M}_2 \bullet \mathbf{n}_2 = -\mu_0 \frac{I'y}{\pi(a^2+y^2)} .
\tag{5.126}
$$

These can be thought of as parallel line charges in z-direction. An individual line charge $\sigma_{mag}(y')dy'$ causes the field

$$
\begin{aligned}
dH_x &= \frac{\sigma_{mag}(y')dy'}{2\pi\mu_0}\frac{x}{[x^2+(y-y')^2]} \\[2mm]
dH_y &= \frac{\sigma_{mag}(y')dy'}{2\pi\mu_0}\frac{(y-y')}{[x^2+(y-y')^2]}
\end{aligned}\Bigg\} ,
\tag{5.127}
$$

which can be shown by analogy to electrostatics using (5.85) and symmetry arguments. Therefore, the field is

$$H_x = -\frac{I'x}{2\pi^2}\int_{-\infty}^{+\infty}\frac{y'\,dy'}{[x^2+(y-y')^2][a^2+y'^2]}\Bigg|$$

$$H_y = -\frac{I'}{2\pi^2}\int_{-\infty}^{+\infty}\frac{(y-y')y'\,dy'}{[x^2+(y-y')^2][a^2+y'^2]}\Bigg|$$

(5.128)

These integrals can, for the most part, be reduced to the ones we had before (5.121) and (5.122).

$$\int_{-\infty}^{+\infty}\frac{y'\,dy'}{[x^2+(y'-y)^2][a^2+y'^2]} = \frac{\pi y}{|x|[y^2+(a+|x|)^2]}$$

(5.129)

$$\int_{-\infty}^{+\infty}\frac{y'(y-y')\,dy'}{[x^2+(y'-y)^2][a^2+y'^2]} = \frac{-\pi(a+|x|)}{[y^2+(a+|x|)^2]}.$$

(5.130)

Then, the field of the fictitious magnetic charges alone for $x > 0$ is

$$H_x = -\frac{I'y}{2\pi[y^2+(x+a)^2]}\Bigg|$$

$$H_y = +\frac{I'(x+a)}{2\pi[y^2+(x+a)^2]}\Bigg|$$

(5.131)

and for $x < 0$

$$H_x = +\frac{I'y}{2\pi[y^2+(x-a)^2]}\Bigg|$$

$$H_y = -\frac{I'(x-a)}{2\pi[y^2+(x-a)^2]}\Bigg|$$

(5.132)

The field of current I is

$$H_x = -\frac{Iy}{2\pi[y^2+(x-a)^2]}$$

$$H_y = +\frac{I(x-a)}{2\pi[y^2+(x-a)^2]}.$$

Combining all these fields gives the field of eqs. (5.112), (5.113), and (5.115).

With this example, and in harmony with the general theory, we have shown that the impact of a magnetized medium can be calculated by magnetization currents or by fictitious magnetic charges, both rendering the same result.

It shall be noted that the case of $\mu_2 = 0$ allows, at least formally, for an interesting interpretation. In anticipation of the discussion on the *skin effect* in Chapter 6, consider an ideal conductor in region 2. Then no **B**-field can penetrate this medium from outside (there could be pre-existing fields inside, however). Suddenly applying a current I outside (region 1), causes induced currents at the surface $x = 0$ (of an infinitely conductive medium in region 2) which are such, that they exactly cancel all fields that this current otherwise would have created internally (in region 2). In case of finite conductivity, these currents decay gradually, which allows the external field to gradually penetrate the conductor. However, in case of infinite conductivity, these currents do not decay and the fields remain excluded. The proof of this will be provided in Chapter 6. Nevertheless, we can discuss the corresponding image problem already. Although a material with infinite conductivity may posses any permeability, we may still formally satisfy the condition that all **B**-fields vanish in region 2, by letting $\mu_2 = 0$ in above result. This yields from (5.115)

$$I' = -I, \tag{5.133}$$

and from (5.118) with $\mu_1 = \mu_0$ we get

$$k_z(y) = -\frac{aI}{\pi(a^2 + y^2)} \tag{5.134}$$

where

$$\int_{-\infty}^{+\infty} k_z(y)\,dy = -\frac{aI}{\pi}\int_{-\infty}^{+\infty}\frac{dy}{a^2 + y^2} = -I. \tag{5.135}$$

Of course in reality, there is no current in region 2. There is only a surface current (5.134) at the surface at $x = 0$, which according to (5.135) just totals the image current $-I$. Calculating the field of the current $k_z(y)$ in region 2, using (5.124), (5.133), and $\mu_1 = \mu_0$ gives

$$B_x = +\frac{\mu_0 I y}{2\pi[y^2 + (x-a)^2]}$$

$$B_y = -\frac{\mu_0 I (x-a)}{2\pi[y^2 + (x-a)^2]},$$

and the field of the current I

$$B_x = -\frac{\mu_0 I y}{2\pi[y^2 + (x-a)^2]}$$

$$B_y = +\frac{\mu_0 I (x-a)}{2\pi[y^2 + (x-a)^2]},$$

is, in deed, exactly canceled.

5.10 Planar Problems

For an arbitrary distribution of currents in z-direction

$$\mathbf{g} = \langle 0, 0, g_z(x,y)\rangle, \tag{5.136}$$

the vector potential is by (5.15)

$$\mathbf{A} = \langle 0, 0, A_z(x,y)\rangle. \tag{5.137}$$

The corresponding magnetic field **B** is

$$\mathbf{B} = \nabla\times\mathbf{A} = \langle \frac{\partial A_z}{\partial y}, -\frac{\partial A_z}{\partial x}, 0\rangle. \tag{5.138}$$

A_z is constant along a force line, *i.e.*, A_z can take the role of the flux function:

$$\mathbf{B} \bullet \nabla A_z = B_x \frac{\partial A_z}{\partial x} + B_y \frac{\partial A_z}{\partial y} + B_z \frac{\partial A_z}{\partial z}$$

$$= \frac{\partial A_z}{\partial y} \frac{\partial A_z}{\partial x} - \frac{\partial A_z}{\partial x} \frac{\partial A_z}{\partial y} + 0 = 0$$

(5.139)

On the other hand, the situation outside of the current carrying region it is

$$\mathbf{H} = -\nabla \psi \ . \tag{5.140}$$

where

$$\psi = \psi(x, y) \ . \tag{5.141}$$

The lines ψ = const. are perpendicular to the field lines. Therefore, the lines ψ = const. and the lines A_z = const. establish an orthogonal grid (see Sect. 3.10 through 3.12). We obtain

$$\left.\begin{array}{l} H_x = \dfrac{1}{\mu_0}\dfrac{\partial A_z}{\partial y} = -\dfrac{\partial \psi}{\partial x} \\[3mm] H_y = -\dfrac{1}{\mu_0}\dfrac{\partial A_z}{\partial x} = -\dfrac{\partial \psi}{\partial y} \end{array}\right\} \ . \tag{5.142}$$

Therefore, the functions $(1/\mu_0)A_z(x, y)$ and $\psi(x, y)$ satisfy the Cauchy-Riemann equations (3.380) and thus, may be considered as the real and imaginary part of a complex potential $w(z)$ (see also (5.33)):

$$w(z) = \frac{A_z(x, y)}{\mu_0} + i\psi(x, y) \ . \tag{5.143}$$

The consequence is that the methods of conformal mapping apply to magnetostatics as well.

We have already met the example of the complex potential of an infinitely long, straight wire in Sect. 5.2.1, eq. (5.34).

5.11 Cylindrical Boundary Value Problems

5.11.1 Separation of Variables

The separation of variables method is also important in solving magnetostatic problems. It will suffice to discus a few examples in cylindrical coordinates.

In Sect. 5.1, we have restricted ourselves to Cartesian coordinates, which enabled us to use the known solution of the scalar Poisson equation to solve Poisson's vector equation (5.12). It is also possible to use curvilinear coordinates, *for example*, cylindrical coordinates. Then one needs to solve the set of equations given in (5.14). For simplicity reasons, we restrict ourselves to rotationally symmetric fields, and assume that there are only azimuthal currents.

$$\mathbf{g} = \langle 0, g_\varphi(r, z), 0 \rangle \ . \tag{5.144}$$

Based on Sect. 5.2.4, one concludes that $\mathbf{A}$ also has only an azimuthal component

$$\mathbf{A} = \langle 0, \mathbf{A}_\varphi(r, z), 0 \rangle \ . \tag{5.145}$$

According to (5.14) it is for $\mathbf{A}_\varphi$:

$$\nabla^2 \mathbf{A}_\varphi(r, z) - \frac{\mathbf{A}_\varphi(r, z)}{r^2} = -\mu_0 g_\varphi(r, z) \ .$$

In particular for the current-free space

$$\nabla^2 \mathbf{A}_\varphi - \frac{\mathbf{A}_\varphi}{r^2} = 0 \ , \tag{5.146}$$

or written more explicitly using (3.33):

$$\frac{1}{r}\frac{\partial}{\partial r} r \frac{\partial}{\partial r} \mathbf{A}_\varphi(r, z) + \frac{\partial^2}{\partial z^2} \mathbf{A}_\varphi(r, z) - \frac{\mathbf{A}_\varphi(r, z)}{r^2} = 0 \ . \tag{5.147}$$

In order to solve this equation by separation of variables, we write the Ansatz

$$\mathbf{A}_\varphi(r, z) = R(r)Z(z) \ , \tag{5.148}$$

as before in Sect. 3.7, and obtain

$$\frac{\partial^2}{\partial z^2} Z(z) = k^2 Z(z) \tag{5.149}$$

and

$$\frac{1}{r}\frac{\partial}{\partial r} r \frac{\partial}{\partial r} R(r) + \left(k^2 - \frac{1}{r^2} \right) R(r) = 0 \ , \tag{5.150}$$

with the solution

$$Z(z) = A_1 \cosh(kz) + A_2 \sinh(kz) \tag{5.151}$$

or

$$Z(z) = \tilde{A}_1 \exp(kz) + \tilde{A}_2 \exp(-kz) \tag{5.152}$$

and

$$R(r) = C_1 J_1(kr) + C_2 N_1(kr) \ . \tag{5.153}$$

Alternatively, one may use

$$\frac{\partial^2}{\partial z^2} Z(z) = -k^2 Z(z) \tag{5.154}$$

to obtain

$$Z(z) = A_1 \cos(kz) + A_2 \sin(kz) \ . \tag{5.155}$$

and

$$R(r) = C_1 I_1(kr) + C_2 K_1(kr) \ . \tag{5.156}$$

To calculate the scalar potential ψ is possible. The relation for the current-free space is then

$$\nabla^2 \psi = 0 \ , \tag{5.157}$$

or

$$\frac{1}{r}\frac{\partial}{\partial r} r \frac{\partial}{\partial r} \psi + \frac{\partial^2}{\partial z^2} \psi = 0 \ . \tag{5.158}$$

Using the separation Ansatz

$$\psi(r, z) = R(r)Z(z) \tag{5.159}$$

gives the same result as before, except that for all cylinder functions, the index 1 is replaced by 0.

The choice here is to either pick Z as of (5.151) or (5.152) with

$$R(r) = C_1 J_0(kr) + C_2 N_0(kr) \tag{5.160}$$

or pick Z as of (5.155) with

$$R(r) = C_1 I_0(kr) + C_2 K_0(kr) \ . \tag{5.161}$$

Before discussing examples of boundary value problems in detail, we shall provide some insight into rotationally symmetric fields.

5.11.2 Structure of Rotationally Symmetric Magnetic Fields

The function $r A_\varphi(r, z)$ is constant along field lines. From

$$\mathbf{B} = \nabla \times \mathbf{A}$$

follows with (5.145) that

$$\mathbf{B} = \langle -\frac{\partial}{\partial z} A_\varphi , 0, \frac{1}{r}\frac{\partial}{\partial r}(r A_\varphi) \rangle \ .$$

Therefore, for rotational symmetry $\partial / \partial \varphi = 0$, it is:

$$\mathbf{B} \bullet \nabla(r A_\varphi) = B_r \frac{\partial}{\partial r}(r A_\varphi) + B_z \frac{\partial}{\partial z}(r A_\varphi)$$

$$= -\frac{\partial}{\partial z} A_\varphi \frac{\partial}{\partial r}(r A_\varphi) + \frac{1}{r}\frac{\partial}{\partial r}(r A_\varphi)\frac{\partial}{\partial z}(r A_\varphi) = 0 \ .$$

This means that $r A_\varphi(r, z)$ is the flux function of the rotationally symmetric field. The lines $r A_\varphi(r, z) = $ const. are the field lines laying in the r-z plane ($\varphi = $ const.)

The field used initially based on $\mathbf{g}$ as of eq. (5.144) is not the most general rotationally symmetric field, which is rather

$$\mathbf{g} = \langle g_r(r, z), g_\varphi(r, z), g_z(r, z) \rangle \ ,$$

which yields

$$\mathbf{B} = \langle B_r(r, z), B_\varphi(r, z), B_z(r, z) \rangle \; .$$

Both vector fields are source-free:

$$\nabla \bullet \mathbf{g} = \frac{1}{r}\frac{\partial}{\partial r}(rg_r) + \frac{\partial}{\partial z}(g_z) = 0$$

$$\nabla \bullet \mathbf{B} = \frac{1}{r}\frac{\partial}{\partial r}rB_r + \frac{\partial}{\partial z}(B_z) = 0 \; .$$

Both conditions are satisfied by two arbitrary functions $F(r, z)$ and $G(r, z)$ when calculating the r- and z-components of $\mathbf{g}$ and $\mathbf{B}$ as follows:

$$\left.\begin{array}{ll} g_r = -\dfrac{1}{r}\dfrac{\partial F}{\partial z}, & g_z = \dfrac{1}{r}\dfrac{\partial F}{\partial r} \\[2ex] B_r = -\dfrac{1}{r}\dfrac{\partial G}{\partial z}, & B_z = \dfrac{1}{r}\dfrac{\partial G}{\partial r} \end{array}\right\} \; . \tag{5.162}$$

This allows one to calculate:

$$\left.\begin{array}{l} \mathbf{g} \bullet \nabla F = -\dfrac{1}{r}\dfrac{\partial F}{\partial z}\dfrac{\partial F}{\partial r} + g_\varphi \cdot 0 + \dfrac{1}{r}\dfrac{\partial F}{\partial r}\dfrac{\partial F}{\partial z} = 0 \\[2ex] \mathbf{B} \bullet \nabla G = -\dfrac{1}{r}\dfrac{\partial G}{\partial z}\dfrac{\partial G}{\partial r} + B_\varphi \cdot 0 + \dfrac{1}{r}\dfrac{\partial G}{\partial r}\dfrac{\partial G}{\partial z} = 0 \end{array}\right\} \; .$$

Thereby, G is constant along the field lines and F along the current density lines. Furthermore, from

$$\mathbf{B} = \nabla \times \mathbf{A}$$

$$\mathbf{g} = \frac{1}{\mu_0}\nabla \times \mathbf{B}$$

follows that

$$G(r, z) = rA_\varphi(r, z)$$

$$F(r, z) = \frac{1}{\mu_0}rB_\varphi(r, z) \; . \tag{5.163}$$

The term $2\pi G$ can be regarded as the magnetic flux through a disk of radius r, oriented perpendicular to the z-axis and the term $2\pi F$ as the current through this region.

$$\left.\begin{array}{l} \phi = \int_A B_z dA = \int_0^r B_z 2\pi r' dr' = \int_0^r 2\pi r'\dfrac{1}{r'}\dfrac{\partial G}{\partial r'}dr' = 2\pi G(r, z) \\[2ex] I = \int_A g_z dA = \int_0^r g_z 2\pi r' dr' = \int_0^r 2\pi r'\dfrac{1}{r'}\dfrac{\partial F}{\partial r'}dr' = 2\pi F(r, z) \end{array}\right\} \; .$$

Therefore, by Ampere's law

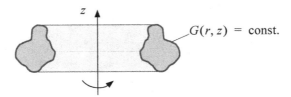

$G(r, z)$ = const.

Fig. 5.60

$$B_\varphi = \frac{\mu_0 I}{2\pi r} = \frac{\mu_0 2\pi F}{2\pi r} = \frac{\mu_0 F}{r}$$

and because of eq. (5.21)

$$A_\varphi = \frac{\phi}{2\pi r} = \frac{2\pi G}{2\pi r} = \frac{G}{r}$$

which makes the eqs. (5.163) plausible.

The fields lines run entirely on the surface $G(r, z)$ = const., *i.e.*, on toroidal, rotationally symmetric surfaces with a cross section *for example*, as shown in Fig. 5.60. Thus, a given force line never leaves this surface. There is no azimuthal field if $F(r, z) = 0$ and then, all field lines lie on the surface φ = const. When superposing azimuthal fields $F \neq 0$, then the field lines spiral around the toroidal surfaces $G(r, z)$ = const. This makes it possible that the field lines close themselves after a number of loops. However, that has to be regarded as the exception. *We emphasize this here because of the frequent misconception that in order for* **B** *to be source free, the* **B** *lines have to either close or go to infinity.* This is wrong. Field lines can remain in the finite space and still never close up, *for example*, on a toroidal surface as in Fig. 5.60 whereby filling the surface arbitrarily dense. We could rightfully state that a field line, when it does not close, creates the toroidal surface (the so-called magnetic surface). Of course, if we were to trace a field line from a starting point through sufficiently many loops, then we would find that this line will come arbitrarily close to the starting point and thus, the line "nearly" closes.

5.11.3 Examples

5.11.3.1 Cylinder with an Azimuthal Surface Current

Consider a cylinder of radius r_0 carrying the azimuthal surface current $k_\varphi(z)$, where

$$k_\varphi(z) = k_\varphi(-z) . \tag{5.164}$$

Then one writes for A_φ in the regions 1 and 2 (shown in Fig. 5.61)

$$A_\varphi^{(1)} = \int_0^\infty f_1(k) I_1(kr) \cos(kz)\,dk \Bigg\} .$$
$$A_\varphi^{(2)} = \int_0^\infty f_2(k) K_1(kr) \cos(kz)\,dk \Bigg\}$$

(5.165)

Arguments of the kind used in Sect. 3.7 serve to justify this statement. Use of $\cos(kz)$ alone, without $\sin(kz)$, is a result of the symmetry (5.164). Furthermore, notice that K_1 diverges at the origin, while I_1 diverges at infinity. The functions f_1 and f_2 are determined by the boundary conditions at $r = r_0$, where

$$B_r = -\frac{\partial A_\varphi}{\partial z}$$

has to be continuous and the z-components of the field have to fit the current, that is, according to (5.96) it must be

$$\left(\frac{1}{r}\frac{\partial}{\partial r} r A_\varphi^{(1)} - \frac{1}{r}\frac{\partial}{\partial r} r A_\varphi^{(2)} \right)_{r=r_0} = B_{z1} - B_{z2} = \mu_0 k_\varphi(z).$$

(5.166)

Continuity of B_r results in

$$\int_0^\infty [f_1(k) I_1(kr_0) - f_2(k) K_1(kr_0)] k \sin(kz)\,dk = 0$$

i.e.,

$$f_1(k) I_1(kr_0) = f_2(k) K_1(kr_0) .$$

(5.167)

Using (5.166) and

$$\frac{d}{dz}(z I_1(z)) = z I_0(z)$$

(5.168)

$$\frac{d}{dz}(z K_1(z)) = -z K_0(z)$$

(5.169)

gives

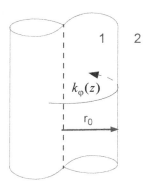

Fig. 5.61

$$\int_0^\infty [f_1(k)I_0(kr_0) + f_2(k)K_0(kr_0)]k\cos(kz)\,dk$$

$$= \mu_0 k_\varphi(z) = \mu_0 \int_0^\infty \tilde{k}_\varphi(k)\cos(kz)\,dk$$

where $\tilde{k}_\varphi(k)$ is the Fourier transform of $k_\varphi(z)$, and then

$$f_1(k)I_0(kr_0) + f_2(k)K_0(kr_0) = \frac{\mu_0 \tilde{k}_\varphi(k)}{k}. \tag{5.170}$$

Eqs (5.167) and (5.170) allow to calculate the two functions $f_1(k)$ and $f_2(k)$. Specifically, if

$$k_\varphi(z) = I\delta(z) = \frac{I}{\pi}\int_0^\infty \cos(kz)\,dk \tag{5.171}$$

(see eq. (3.198)), then

$$\tilde{k}_\varphi(k) = \frac{I}{\pi}. \tag{5.172}$$

When applying

$$I_0(z)K_1(z) + K_0(z)I_1(z) = \frac{1}{z}$$

this gives f_1 and f_2

$$f_1 = \frac{\mu_0 I r_0}{\pi}K_1(kr_0) \quad \text{and} \quad f_2 = \frac{\mu_0 I r_0}{\pi}I_1(kr_0),$$

and for the vector potential

$$\left.\begin{aligned} A_\varphi^{(1)} &= \frac{\mu_0 I r_0}{\pi}\int_0^\infty K_1(kr_0)I_1(kr)\cos(kz)\,dk \\[2mm] A_\varphi^{(2)} &= \frac{\mu_0 I r_0}{\pi}\int_0^\infty I_1(kr_0)K_1(kr)\cos(kz)\,dk \end{aligned}\right\} . \tag{5.173}$$

Similar considerations allow to calculate the scalar potential. The calculation is left for the reader as an exercise. The scalar potential for the loop current (5.171) is

$$\left.\begin{aligned} \psi^{(1)} &= -\frac{I r_0}{\pi}\int_0^\infty K_1(kr_0)I_0(kr)\sin(kz)\,dk \\[2mm] \psi^{(2)} &= +\frac{I r_0}{\pi}\int_0^\infty I_1(kr_0)K_0(kr)\sin(kz)\,dk \end{aligned}\right\} . \tag{5.174}$$

Compatibility of the two results is easy to verify. It has to be

$$\left.\begin{array}{l} B_r = -\dfrac{\partial A_\varphi}{\partial z} = -\mu_0\dfrac{\partial\psi}{\partial r} \\[4mm] B_z = \dfrac{1}{r}\dfrac{\partial}{\partial r}rA_\varphi = -\mu_0\dfrac{\partial\psi}{\partial z} \end{array}\right\} . \tag{5.175}$$

Indeed, both A_φ and ψ result in the same field

$$\left.\begin{array}{l} B_r^{(1)} = +\dfrac{\mu_0 I}{\pi}\displaystyle\int_0^\infty K_1(kr_0)I_1(kr)kr_0\sin(kz)\,dk \\[4mm] B_r^{(2)} = +\dfrac{\mu_0 I}{\pi}\displaystyle\int_0^\infty I_1(kr_0)K_1(kr)kr_0\sin(kz)\,dk \\[4mm] B_z^{(1)} = +\dfrac{\mu_0 I}{\pi}\displaystyle\int_0^\infty K_1(kr_0)I_0(kr)kr_0\cos(kz)\,dk \\[4mm] B_z^{(2)} = -\dfrac{\mu_0 I}{\pi}\displaystyle\int_0^\infty I_1(kr_0)K_0(kr)kr_0\cos(kz)\,dk \end{array}\right\} , \tag{5.176}$$

where besides eqs. (5.168) and (5.169), the following relations are also needed

$$\frac{\mathrm{d}}{\mathrm{d}z}I_0 = I_1(z), \tag{5.177}$$

$$\frac{\mathrm{d}}{\mathrm{d}z}K_0 = -K_1(z). \tag{5.178}$$

Obviously, from (5.176) it follows that these fields exhibit the correct symmetry relative to z. The radial components are anti-symmetric in the case of symmetric current distributions and the longitudinal components are symmetric. This behavior is a result of (5.175) if A_φ is stated as a symmetric and ψ as an anti-symmetric function.

Of course, this special case (5.171) corresponds to the current loop, discussed previously in detail (Sect. 5.2.4). The vector potential (5.173) is initially just a rather complicated Fourier series of the previously found vector potential (5.48). Both equations are equivalent. Nevertheless, the Fourier series (5.173) is much more useful when solving boundary value problems than the other, much handier equation (5.48). An example shall serve to illustrate this (Sect. 5.11.3.3). However, in the following example (Sect. 5.11.3.2), we will first derive another form of the same potential.

5.11.3.2 Azimuthal Surface Currents in the x-y-Plane

Let an azimuthal surface current flow in the plane $z = 0$ and let it depend only on r (Fig. 5.62). This is an opportunity to use the other solution (5.151) through (5.153) and (5.155), (5.156), respectively. The example of Sect. 5.11.3.1 required one to use the second form, since a function of z was given on a cylinder, which required a Fourier transform with respect to z and thereby the Ansatz with the trigonometric functions. In the current example, a function of r is given. This requires a Hankel transform, which is possible with the first of the two trial functions. To satisfy the boundary conditions for $z \to \pm\infty$, where the fields have to vanish, we write

$$\left. \begin{aligned} A_\varphi^{(1)} &= \int_0^\infty f_1(k)J_1(kr)\exp(-kz)dk \\ A_\varphi^{(2)} &= \int_0^\infty f_2(k)J_1(kr)\exp(+kz)dk \end{aligned} \right\} , \tag{5.179}$$

where the index 1 addresses the region $z \geq 0$ and the index 2 the region $z \leq 0$. First, for $z = 0$ is must be

$$B_z^{(1)} = B_z^{(2)} ,$$

from which results

$$f_1(k) = f_2(k) = f(k) . \tag{5.180}$$

This allows to combine the two trial functions of (5.179) into

$$A_\varphi = \int_0^\infty f(k)J_1(kr)\exp(-k|z|)dk . \tag{5.181}$$

The task is now to find $f(k)$ for $z = 0$ such that (see (5.96))

$$B_r^{(1)} - B_r^{(2)} = \mu_0 k_\varphi(r). \tag{5.182}$$

Based on (3.224), we introduce the Hankel transformed function

$$\tilde{k}_\varphi(k) = \int_0^\infty r k_\varphi(r)J_1(kr)dr \tag{5.183}$$

where by (3.225)

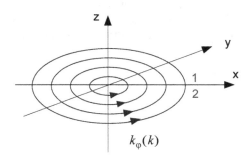

Fig. 5.62

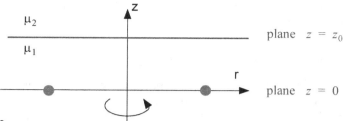

Fig. 5.63

$$k_\varphi(r) = \int_0^\infty k\tilde{k}_\varphi(k)J_1(kr)\,dk .$$

(5.184)

Then for $z = 0$

$$B_r^{(1)} - B_r^{(2)} = \mu_0\int_0^\infty k\tilde{k}_\varphi(k)J_1(kr)\,dk .$$

(5.185)

Using

$$B_r = -\frac{\partial A_\varphi}{\partial z}$$

gives for $z = 0$

$$\int_0^\infty [f(k)J_1(kr)k + f(k)J_1(kr)k]\,dk = \mu_0\int_0^\infty k\tilde{k}_\varphi(k)J_1(kr)\,dk .$$

or

$$2f(k) = \mu_0\tilde{k}_\varphi(k) .$$

(5.186)

In particular, consider the current loop I at $r = r_0$

$$k_\varphi = I\delta(r - r_0) .$$

(5.187)

Then by (5.183)

$$\tilde{k}_\varphi(k) = \int_0^\infty rI\delta(r - r_0)J_1(kr)\,dr = Ir_0 J_1(kr_0)$$

and in our case

$$f(k) = \frac{\mu_0 Ir_0 J_1(kr_0)}{2} ,$$

(5.188)

i.e.,

$$A_\varphi = \frac{\mu_0 Ir_0}{2}\int_0^\infty J_1(kr_0)J_1(kr)\exp(-k|z|)\,dk .$$

(5.189)

This is yet another representation of the potential of a current loop. It is equivalent to the two previous ones, (5.48) and (5.173). Certain boundary value problems require this form. Such a problem would, for instance, be that of a current loop located in the plane at $z = 0$, opposite to a plane boundary at $z = z_0$ between two media of different permeability (μ_1, μ_2), as illustrated in Fig. 5.63. We will skip this problem here, except to note that it can be solved in a similar way as the problem in the next example.

5.11.3.3 Current Loop and Magnetizable Cylinder

Let the space contain two media of different permeability. Permeability in region 1 ($r > r_1$) is μ_1 and μ_2 for region 2 ($r < r_1$). Region 1 hosts a circular current loop with the current I. Its radius is r_0 (Fig. 5.64). We want to find the behavior of the fields in the two regions.

There are no currents in region 2. This is equivalent to a vacuum field. We try

$$A_\varphi^{(2)} = \frac{\mu_1 I r_0}{\pi} \int_0^\infty [f_2(k) + K_1(kr_0)] I_1(kr) \cos(kz) dk , \tag{5.190}$$

The current I needs to be considered in region 1, *i.e.*, the inhomogeneous equation (5.146) needs to be solved. We already know the special solution for the ring current I. The solution for the general solution is just the superposition of that particular solution, and the solution of the homogeneous equation, *i.e.*, the solution for the vacuum. Therefore for $r_1 \leq r \leq r_0$:

$$A_\varphi^{(1)} = \frac{\mu_1 I r_0}{\pi} \int_0^\infty [f_1(k)K_1(kr) + K_1(kr_0)I_1(kr)] \cos(kz) dk \tag{5.191}$$

and for $r_0 \leq r$

$$A_\varphi^{(1)} = \frac{\mu_1 I r_0}{\pi} \int_0^\infty [f_1(k)K_1(kr) + I_1(kr_0)K_1(kr)] \cos(kz) dk . \tag{5.192}$$

For $f_1 = f_2 = 0$, this gives exactly the field of the current I alone, *i.e.*, f_1 and f_2 describe the effect caused by the existence of the second medium (μ_2). In other words: f_1 and f_2 express the magnetization currents on the boundary at $r = r_1$. f_1 and f_2 are determined by the boundary conditions at $r = r_1$, which are:

$$B_r^{(1)} = B_r^{(2)}, \tag{5.193}$$

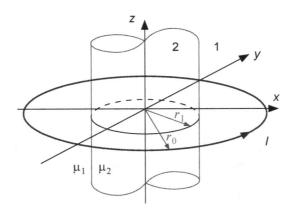

Fig. 5.64

and

$$\frac{1}{\mu_1}B_z^{(1)} = \frac{1}{\mu_2}B_z^{(2)} .$$
(5.194)

This gives

$$f_2(k)I_1(kr_1) + K_1(kr_0)I_1(kr_1) = f_1(k)K_1(kr_1) + K_1(kr_0)I_1(kr_1)$$
(5.195)

and

$$\frac{f_2(k)I_0(kr_1) + K_1(kr_0)I_0(kr_1)}{\mu_2} = \frac{K_1(kr_0)I_0(kr_1) - f_1(k)K_0(kr_1)}{\mu_1} .$$
(5.196)

Solving for f_1 and f_2 gives

$$f_1(k) = \frac{\left(\frac{\mu_2}{\mu_1} - 1\right)K_1(kr_0)I_0(kr_1)I_1(kr_1)}{K_1(kr_1)I_0(kr_1) + \frac{\mu_2}{\mu_1}K_0(kr_1)I_1(kr_1)}$$
(5.197)

$$f_2(k) = \frac{\left(\frac{\mu_2}{\mu_1} - 1\right)K_1(kr_0)I_0(kr_1)K_1(kr_1)}{K_1(kr_1)I_0(kr_1) + \frac{\mu_2}{\mu_1}K_0(kr_1)I_1(kr_1)} .$$
(5.198)

This basically solves the problem. Of course, the special case of $\mu_1 = \mu_2$ with $f_1 = f_2 = 0$ is just the field of a current loop in a uniform medium.

As before in Sect. 5.9, it is possible to calculate the magnetization currents in the surface $r = r_1$ and show that these currents just represent the additional field created by f_1 and f_2. First

$$k_\varphi(z) = \left(\frac{M_z^{(2)} - M_z^{(1)}}{\mu_0}\right)_{r=r_1} = \left(\frac{\mu_2 - \mu_0}{\mu_2\mu_0}B_z^{(2)} - \frac{\mu_1 - \mu_0}{\mu_1\mu_0}B_z^{(1)}\right)_{r=r_1}$$

$$= \left(\frac{1}{\mu_0}[B_z^{(2)} - B_z^{(1)}]\right)_{r=r_1} ,$$

where with

$$B_z = \frac{1}{r}\frac{\partial}{\partial r}(rA_\varphi)$$
(5.199)

and applying the already frequently used formula

$$K_1(z)I_0(z) + K_0(z)I_1(z) = \frac{1}{z}$$
(5.200)

yields

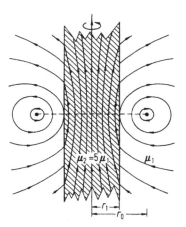

Fig. 5.65

$$k_\varphi(z) = \frac{(\mu_2 - \mu_1)Ir_0}{\pi\mu_0 r_1} \int_0^\infty \frac{K_1(kr_0)I_0(kr_1)\cos(kz)}{K_1(kr_1)I_0(kr_1) + \dfrac{\mu_2}{\mu_1}K_0(kr_1)I_1(kr_1)} \, dk \ . \qquad (5.201)$$

One can show that the integral is positive. Therefore, k_φ is positive when $\mu_2 > \mu_1$, and is negative when $\mu_2 < \mu_1$ (assuming a positive current I). The former case results in attracting forces and the latter in repelling ones.

If the current loop (in region 1) is in vacuum ($\mu_1 = \mu_0$), then it is attracted by a paramagnetic cylinder, and repelled from a diamagnetic one. This is true also for differently shaped bodies and current loops. The total magnetization current is

$$\int_{-\infty}^{+\infty} k_\varphi(z)\,dk = \frac{(\mu_2-\mu_1)Ir_0}{\pi\mu_0 r_1} \int_0^\infty \frac{K_1(kr_0)I_0(kr_1)\int_{-\infty}^{+\infty}\cos(kz)\,dz}{K_1(kr_1)I_0(kr_1) + \dfrac{\mu_2}{\mu_1}K_0(kr_1)I_1(kr_1)}\,dk$$

$$\int_{-\infty}^{+\infty} k_\varphi(z)\,dk = \frac{(\mu_2-\mu_1)Ir_0}{\pi\mu_0 r_1} \int_0^\infty \frac{K_1(kr_0)I_0(kr_1)2\pi\delta(k)}{K_1(kr_1)I_0(kr_1) + \dfrac{\mu_2}{\mu_1}K_0(kr_1)I_1(kr_1)}\,dk$$

$$\int_{-\infty}^{+\infty} k_\varphi(z)\,dk = \frac{(\mu_2-\mu_1)Ir_0}{\mu_0 r_1} \lim_{k\to 0} \frac{K_1(kr_0)I_0(kr_1)}{K_1(kr_1)I_0(kr_1) + \dfrac{\mu_2}{\mu_1}K_0(kr_1)I_1(kr_1)} \ .$$

According to eqs. (3.178) through (3.180), this limit is r_1/r_0 and therefore

$$\int_{-\infty}^{+\infty} k_\varphi(z)\,dk = \frac{(\mu_2-\mu_1)I}{\mu_0} \ . \qquad (5.202)$$

The case when an external field shall be shielded by a cylinder with infinite conductivity can formally be described by letting $\mu_2 = 0$. With $\mu_1 = \mu_0$ one obtains

$$k_{\varphi}(z) = -\frac{Ir_0}{\pi r_1}\int_0^{\infty}\frac{K_1(kr_0)}{K_1(kr_1)}\cos(kz)dk \tag{5.203}$$

and the total current is

$$\int_{-\infty}^{+\infty}k_{\varphi}(z)dz = -I . \tag{5.204}$$

The entire surface current in this case totals to exactly -I, *i.e.*, the same amount as the current in the loop, anti-parallel, however. This is necessary. Otherwise, there had to be a magnetic field in the cylinder due to Ampere's law.
Fig. 5.65 provides an example for the field lines when $\mu_2 = 5\mu_1$.

5.12   Magnetic Energy, Magnetic Flux and Inductance Coefficients

5.12.1   Magnetic Energy

We have discovered already in Sect. 2.14.1 that the potential energy density stored in a magnetic field is $(1/2)\mu H^2 = (1/2)\mathbf{H} \bullet \mathbf{B}$. The total energy is therefore

$$W = \frac{1}{2}\int_V \mathbf{H} \bullet \mathbf{B}d\tau .$$

Using (5.7) yields

$$W = \frac{1}{2}\int_V \mathbf{H} \bullet (\nabla \times \mathbf{A})d\tau .$$

With the vector identify

$$\nabla \bullet (\mathbf{A} \times \mathbf{H}) = \mathbf{H} \bullet (\nabla \times \mathbf{A}) - \mathbf{A} \bullet (\nabla \times \mathbf{H})$$

this gives

$$W = \frac{1}{2}\int_V \nabla \bullet (\mathbf{A} \times \mathbf{H})d\tau + \frac{1}{2}\int_V \mathbf{A} \bullet (\nabla \times \mathbf{H})d\tau$$

or

$$W = \frac{1}{2}\oint(\mathbf{A} \times \mathbf{H}) \bullet da + \frac{1}{2}\int_V \mathbf{A} \bullet (\nabla \times \mathbf{H})d\tau .$$

The surface integral has to include all possible boundaries. A sphere at infinity does not provide any contributions because of the sufficiently fast decrease of $\mathbf{A} \times \mathbf{H}$ as R increases (the product is proportional to R^{-3}). This can be seen from eqs. (5.16) and (5.17). Inside surfaces – they need to be considered from both sides – do not provide any contribution as long as $(\mathbf{A} \times \mathbf{H}) \bullet da$ is continuous. As we will see shortly, this is the case, as long as there are no surface currents in the boundary. The energy is then:

$$\boxed{W = \frac{1}{2}\int_V \mathbf{A} \bullet \mathbf{g} \, d\tau} . \tag{5.205}$$

This interesting result should be compared to a similar equation obtained for the electrostatic energy (2.171):

$$W = \frac{1}{2}\int_V \rho\varphi\, d\tau \ .$$

If there are surface currents in the boundary, then the surface integral contributes to the energy, namely

$$W_a = \frac{1}{2}\oint(\mathbf{A}\times\mathbf{H})\bullet d\mathbf{a} = \frac{1}{2}\sum_i\int_{a_i}[\mathbf{A}\times(\mathbf{H}_2-\mathbf{H}_1)]\bullet\mathbf{n}_2 da_i$$

$$= \frac{1}{2}\sum_i\int_{a_i}\mathbf{A}\bullet[(\mathbf{H}_2-\mathbf{H}_1)\times\mathbf{n}_2]da_i \ .$$

By eq. (5.96), this gives

$$W_a = \frac{1}{2}\sum_i\int_{a_i}\mathbf{A}\bullet\mathbf{k}\, da_i \ .$$

These contributions, originating from surface currents, can be thought of as being contained in eq. (5.205). Now we have to consider all currents in this equation, even surface currents, for which, indeed, $\mathbf{g}$ is infinitely large, but $\mathbf{g}\, d\tau$ remains finite.

By (5.16) and (5.205) it is finally:

$$W = \frac{\mu_0}{8\pi}\int_V\int_{V'}\frac{\mathbf{g}(\mathbf{r})\bullet\mathbf{g}(\mathbf{r}')d\tau\, d\tau'}{|\mathbf{r}-\mathbf{r}'|} \ . \tag{5.206}$$

Now consider a system consisting of n closed conductors with the currents $\mathbf{g}_i(\mathbf{r})$, $(i = 1, 2, ..., n)$. Then

$$\mathbf{g}(\mathbf{r}) = \sum_{i=1}^n\mathbf{g}_i(\mathbf{r}) \ ,$$

and

$$W = \frac{\mu_0}{8\pi}\sum_{i,j=1}^n\int_V\int_{V'}\frac{\mathbf{g}_i(\mathbf{r})\bullet\mathbf{g}_j(\mathbf{r}')d\tau\, d\tau'}{|\mathbf{r}-\mathbf{r}'|}$$

$$= \frac{\mu_0}{8\pi}\sum_{i,j=1}^n I_iI_j\int_V\int_{V'}\frac{\mathbf{g}_i(\mathbf{r})\bullet\mathbf{g}_j(\mathbf{r}')d\tau\, d\tau'}{I_iI_j|\mathbf{r}-\mathbf{r}'|} \ .$$

i.e.,

$$W = \frac{1}{2}\sum_{i,j=1}^n L_{ij}I_iI_j \ , \tag{5.207}$$

having defined the so-called *inductance coefficients* L_{ij} in the following way

$$L_{ij} = \frac{\mu_0}{4\pi I_iI_j}\int_V\int_{V'}\frac{\mathbf{g}_i(\mathbf{r})\bullet\mathbf{g}_j(\mathbf{r}')d\tau\, d\tau'}{|\mathbf{r}-\mathbf{r}'|} \ . \tag{5.208}$$

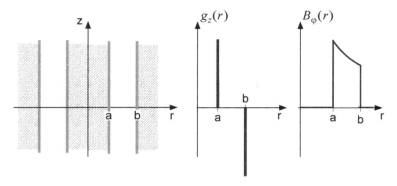

Fig. 5.66

The reciprocity (symmetry) of these coefficients

$$\boxed{L_{ij} = L_{ji}}$$

(5.209)

is an immediate consequence of definition (5.208). When $i = j$, these coefficients are called *self-inductance coefficients*.

There are two ways to calculate L_{ij}: either use (5.208) or calculate W and then compare it with (5.207). The latter is usually the more convenient approach. Two simple examples shall serve to illustrate this.

1) The Coaxial Cable (Fig. 5.66)

If the conductors in the cable are very thin (or they have infinite conductivity), then the situation is as sketched in Fig. 5.66. The field is

$$B_\varphi(r) = \frac{\mu_0 I}{2\pi r}$$

$$H_\varphi(r) = \frac{I}{2\pi r}$$

$$\frac{1}{2}\mathbf{B} \bullet \mathbf{H} = \frac{\mu_0 I^2}{8\pi^2 r^2} \;.$$

Total energy becomes infinite if the length is infinite. Per unit length, however, it remains finite.

$$\frac{W}{l} = \frac{\mu_0}{8\pi^2}\int_a^b \frac{I^2}{r^2} 2\pi r\, dr = \frac{\mu_0 I^2}{4\pi}\int_a^b \frac{dr}{r}$$

$$= \frac{\mu_0 I^2}{4\pi}\ln\frac{b}{a} = \frac{1}{l}\frac{1}{2}L_{11}I^2 \;,$$

which yields the self-inductance coefficients per unit length

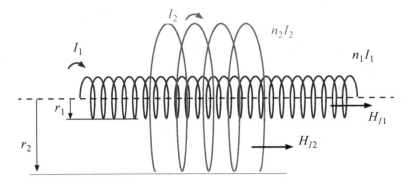

Fig. 5.67

$$\frac{L}{l} = \frac{L_{11}}{l} = \frac{\mu_0}{2\pi} \ln \frac{b}{a} \, , \qquad\qquad\qquad (5.210)$$

a frequently used, and in practice important result.

2) Two infinitely long Coils (Fig. 5.67)

Consider the currents I_1 and I_2 flowing in the two coils with n_1 and n_2 turns per unit length, respectively. Inside coil 1, this creates the field

$$H_{z1} = n_1 I_1 + n_2 I_2$$

and between coil 2 and coil 1 the field

$$H_{z2} = n_2 I_2 \, .$$

Then

$$\begin{aligned}
\frac{W}{l} &= \frac{\mu_0 (n_1 I_1 + n_2 I_2)^2}{2} r_1^2 \pi + \frac{\mu_0 (n_2 I_2)^2}{2} (r_2^2 - r_1^2) \pi \\
&= \frac{\mu_0 \pi}{2} [n_1^2 I_1^2 r_1^2 + 2 n_1 n_2 I_1 I_2 r_1^2 + n_2^2 I_2^2 r_2^2] \\
&= \frac{1}{2} \frac{L_{11}}{l} I_1^2 + \frac{1}{2} \frac{L_{12}}{l} I_1 I_2 + \frac{1}{2} \frac{L_{21}}{l} I_1 I_2 + \frac{1}{2} \frac{L_{22}}{l} I_2^2 \\
&= \frac{1}{2} \frac{L_{11}}{l} I_1^2 + \frac{L_{12}}{l} I_1 I_2 + \frac{1}{2} \frac{L_{22}}{l} I_2^2
\end{aligned}$$

and by comparing coefficients

$$\left.\begin{array}{rcl} \frac{1}{I}L_{11} &=& \mu_0 \pi r_1^2 n_1^2 \\[2mm] \frac{1}{I}L_{12} = \frac{1}{I}L_{21} &=& \mu_0 \pi r_1^2 n_1 n_2 \\[2mm] \frac{1}{I}L_{22} &=& \mu_0 \pi r_2^2 n_2^2 \end{array}\right\} . \qquad (5.211)$$

Notice that the currents $I_{1,2}$ carry a sign and the mixed terms (here $\sim I_1 I_2$) may be both, positive or negative.

5.12.2 Magnetic Flux

Consider one or more circuits with the magnetic energy

$$W = \frac{\mu_0}{2} \int_V H^2 d\tau .$$

In the outside space, *i.e.*, outside the current carrying conductor, the field may be written by means of the scalar potential ψ

$$\mathbf{H} = -\nabla \psi .$$

We now assume that with respect to the extent of the outside space, the width of the conductor is negligible. Consequently, we only need to take the volume integral over the outside space and the inside contribution, including that of the magnetic energy inside is negligible. This gives

$$W = \frac{\mu_0}{2} \int_V (\nabla \psi)^2 d\tau .$$

Because of the vector identity

$$\nabla \bullet (\psi \nabla \psi) = \psi \nabla^2 \psi + (\nabla \psi)^2$$

this becomes

$$W = \frac{\mu_0}{2} \int_V [\nabla \bullet (\psi \nabla \psi) - \psi \nabla^2 \psi] d\tau .$$

Using

$$\nabla^2 \psi = 0$$

and applying Gauss' integral theorem gives

$$W = \frac{\mu_0}{2} \int_A [\psi \nabla \psi] \bullet d\mathbf{A} .$$

If we initially concentrate on one conductor (shown in Fig. 5.68) we get

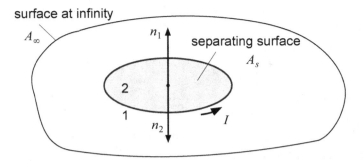

Fig. 5.68

$$W = \frac{\mu_0}{2}\int_{A_\infty}[\psi\nabla\psi]\bullet d\mathbf{A} + \frac{\mu_0}{2}\int_{A_s}[\psi\nabla\psi]\bullet d\mathbf{A}$$

$$= 0 + \frac{\mu_0}{2}\int_{A_s}[\psi\nabla\psi]\bullet d\mathbf{A}$$

$$= \frac{\mu_0}{2}\int_{A_s}\left[\psi_1\left(\frac{\partial\psi}{\partial x}\right)_{1n} - \psi_2\left(\frac{\partial\psi}{\partial x}\right)_{2n}\right]\bullet dA_1 \quad .$$

Because

$$\mu_0\frac{\partial\psi}{\partial x} = -B_n$$

is continuous, that is,

$$\mu_0\left(\frac{\partial\psi}{\partial x}\right)_{1n} = \mu_0\left(\frac{\partial\psi}{\partial x}\right)_{2n} = \mu_0\left(\frac{\partial\psi}{\partial x}\right)_n = -B_n$$

and therefore the energy becomes

$$W = \frac{\mu_0}{2}\int_{A_s}(\psi_1 - \psi_2)\left(\frac{\partial\psi}{\partial x}\right)_n dA_1 = \frac{1}{2}\int_{A_s}(\psi_2 - \psi_1)B_n dA_1 \quad .$$

By (5.58) and in analogy to the electrostatic dipole layer (see Sect. 2.5.3), $\psi_2 - \psi_1$ is constant along the entire separating surface

$$\psi_2 - \psi_1 = I \quad .$$

This makes

$$W = \frac{1}{2}I\phi \quad , \tag{5.212}$$

having used the fact that

$$\int_{A_s}B_n dA = \phi$$

is the flux through the separating surface. On the other hand, since

$$W = \frac{1}{2}L_{11}I^2 = \frac{1}{2}(L_{11}I)I \ ,$$

comparison shows that consequently

$$\phi = L_{11}I \ . \tag{5.213}$$

This result may be generalized to arbitrarily many conductors:

$$\boxed{W = \frac{1}{2}\sum_{i=1}^{n} I_i \phi_i} \ . \tag{5.214}$$

Conversely, because of

$$W = \frac{1}{2}\sum_{i,j=1}^{n} L_{ij}I_iI_j = \frac{1}{2}\sum_{i=1}^{n} I_i \sum_{j=1}^{n} L_{ij}I_j \ ,$$

it is also

$$\boxed{\phi_i = \sum_{j=1}^{n} L_{ij}I_j} \ . \tag{5.215}$$

One concludes that the flux through a conductor loop is a linear function of all currents

$$\phi_1 = L_{11}I_1 + L_{12}I_2 + L_{13}I_3 + \dots$$

$$\phi_2 = L_{21}I_1 + L_{22}I_2 + L_{23}I_3 + \dots$$

$$\dots\dots\dots\dots\dots\dots\dots\dots\dots\dots$$

If the currents are time dependent, then the magnetic flux is also time dependent. The law of electromagnetic induction states, except for the sign, that the temporal change of $\partial\phi_i/\partial t$ equals the EMF induced in the conductor loop. That is the reason why the inductance coefficients are important for the induced voltages in a network, which also explains their name.

Eq. (5.215) provides another avenue to calculate the inductance coefficients. Consider *for example*, the two coils calculated above. For these one has:

$$\frac{1}{I}\phi_1 = n_1(r_1^2\pi\mu_0 n_1 I_1 + r_1^2\pi\mu_0 n_2 I_2)$$

$$\frac{1}{I}\phi_2 = n_2(r_1^2\pi\mu_0 n_1 I_1 + r_2^2\pi\mu_0 n_2 I_2) \ ,$$

which, in accordance with (5.211) gives

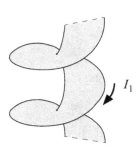

Fig. 5.69

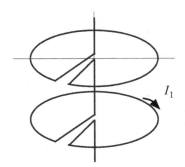

Fig. 5.70

$$\frac{1}{l}L_{11} = \mu_0 \pi r_1^2 n_1^2$$

$$\frac{1}{l}L_{12} = \frac{1}{l}L_{21} = \mu_0 \pi r_1^2 n_1 n_2$$

$$\frac{1}{l}L_{22} = \mu_0 \pi r_2^2 n_2^2 \ .$$

The fact that L_{11} is proportional to n_1^2 shall be explained more plausibly. One factor n_1 is a result of the magnetic inductance by the number of turns per unit length. The second factor n_1 results from the surface, shown in Fig. 5.69, that borders the coil and which connects the individual turns in a helical way. The projection of this surface onto a plane perpendicular to the coil's axis gives the n_1 multiple of the coil's cross section. Fig. 5.70 gives a schematic representation of this.

6 Time Dependent Problems I (Quasi Stationary Approximation)

Maxwell's equations were introduced in Chapter 1, but with a few exceptions, we have not discussed them in their complete form. So far, we have focused on time independent problems, specifically electrostatics in Chapter 2 and 3, stationary electric currents in Chapter 4 and magnetostatics in Chapter 5. Now, we turn to time dependent problems. We will do this in two steps. The time dependent Maxwell equations differ from the stationary ones by two terms, the displacement current $\partial \mathbf{D}/\partial t$ in the first and the magnetic induction $\partial \mathbf{B}/\partial t$ in the second Maxwell equation. We will account for this fact in our proceedings. Specifically, the displacement current may be neglected in a first approximation, while the law of induction needs to be considered. This approximation does not allow us to describe electromagnetic waves, since the displacement current is vital for these. On the other hand, skin effect, eddy currents, and similar effects can be described without the displacement current term. The applicability of this so-called quasi stationary approximation is limited to cases where the temporal changes do not occur too rapidly. We will postpone the discussion of the complete version of Maxwell's equations to the next chapter, (Chapter 7), where we will study electromagnetic waves by including the displacement current term.

6.1 Faraday's Law of Magnetic Induction

6.1.1 Induction by a Temporal Change of B

Consider a time dependent magnetic field described by the magnetic induction or also called the magnetic flux density $\mathbf{B}(\mathbf{r}, t)$ and a contour C, fixed in space. The contour may be implemented by an infinitely thin, conducting material. If A is the area that C circumscribes, then by our definition, eq. (1.66), the magnetic flux penetrating this area is

$$\phi = \int_A \mathbf{B}(\mathbf{r}, t) \bullet d\mathbf{A} \; . \tag{6.1}$$

The time dependent magnetic field $\mathbf{B}(\mathbf{r}, t)$ induces an electric field $\mathbf{E}(\mathbf{r}, t)$, which can be described using eq. (1.68)

$$\nabla \times \mathbf{E}(\mathbf{r}, t) = -\frac{\partial}{\partial t} \mathbf{B}(\mathbf{r}, t) \; . \tag{6.2}$$

The line integral of the electric field is given by the time derivative of the magnetic flux:

$$\oint_C \mathbf{E} \bullet d\mathbf{s} = \int_A (\nabla \times \mathbf{E}) \bullet d\mathbf{A} = -\int_A \frac{\partial \mathbf{B}}{\partial t} d\mathbf{A} = -\frac{\partial}{\partial t} \int_A \mathbf{B} \bullet d\mathbf{A} = -\frac{\partial \phi}{\partial t} \; . \tag{6.3}$$

Expressed in a different form, one writes

G. Lehner, *Electromagnetic Field Theory for Engineers and Physicists*,
DOI 10.1007/978-3-540-76306-2_6, © Springer-Verlag Berlin Heidelberg 2010

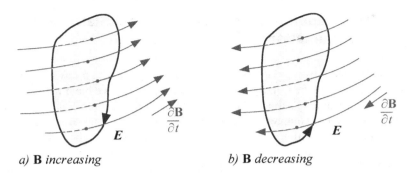

a) **B** *increasing*   *b)* **B** *decreasing*

Fig. 6.1

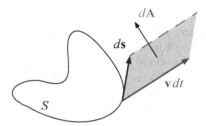

Fig. 6.2

$$V_i = \oint_C \mathbf{E} \cdot d\mathbf{s} = -\frac{\partial \phi}{\partial t} \ , \qquad (6.4)$$

where V_i is the voltage (EMF), induced within the closed loop. Figure 6.1 shows the direction of the induced field for an increasing and a decreasing **B**-field, respectively.

6.1.2   Induction through the Motion of the Conductor

Inside a magnetic field **B**, the *Lorentz force*

$$\mathbf{F} = Q(\mathbf{v} \times \mathbf{B}) \qquad (6.5)$$

acts on a particle with charge Q, and velocity **v**.
One can also explain this force as being the effect of an electric field **E** on a particle within a moving reference frame, where

$$\mathbf{E} = \mathbf{v} \times \mathbf{B} \ . \qquad (6.6)$$

If one now moves a closed conductor loop (made of infinitely thin wire) within a magnetic field that is constant in time, then the line integral along the path S (Fig. 6.2) is

$$\oint_S \mathbf{E} \bullet d\mathbf{s} = \oint_S (\mathbf{v} \times \mathbf{B}) \bullet d\mathbf{s} = -\oint_S \mathbf{B} \bullet (\mathbf{v} \times d\mathbf{s}) \ .$$

While the conductor loop moves during the time period dt, the path element $d\mathbf{s}$ covers an area element $d\mathbf{A}$ (Fig. 6.2):

$$\mathbf{v} \times d\mathbf{s}\,dt = d\mathbf{A} \ .$$

Therefore

$$V_i = \oint_S \mathbf{E} \bullet d\mathbf{s} = -\oint_S \mathbf{B} \bullet \frac{d\mathbf{A}}{dt} = -\frac{d}{dt}\int_A \mathbf{B} \bullet d\mathbf{A} = -\frac{d\phi}{dt} \ , \tag{6.7}$$

where we have only considered the flux change based on motion of the loop within a constant $\mathbf{B}$ field. Temporal changes of $\mathbf{B}$ itself are currently excluded, as outlined above. Consequently, as a result of the Lorentz force, the temporal change of the magnetic flux ϕ in a closed loop is given by line integral $\oint \mathbf{E} \bullet d\mathbf{s}$ (compare (6.3) with (6.7)).

The loop of a conductor does not have to be closed. It is possible *for example*, to move a piece of wire perpendicular to a magnetic field (Fig. 6.3) This causes an induced field $\mathbf{E} = \mathbf{v} \times \mathbf{B}$ inside the wire of magnitude

$$E = vB \ , \tag{6.8}$$

An electric current flows inside the conductor as a result of this field. This continues until E becomes zero ($\mathbf{E} = 0$). Charges are moved to the surface of the conductor and the resulting electrostatic field gradually cancels the induced field inside the conductor.

The final state is an electric field between the ends of the conductor that creates a field outside, whose potential difference is exactly the EMF which was initially induced. Its magnitude is

$$|V| = vBl \ . \tag{6.9}$$

No EMF is induced when moving an entire conductor loop transverse to a uniform magnetic field (Fig. 6.4). The reason is that the partial EMFs mutually cancel. The flux through the loop remains unchanged. When moving just one part of the loop, while maintaining contact with the rest of the loop, then there will be an EMF of

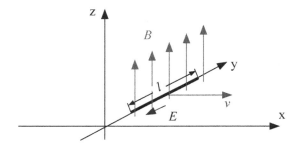

Fig. 6.3

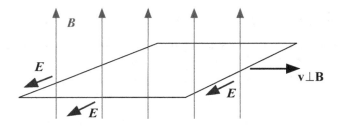

Fig. 6.4

magnitude vBl (Fig. 6.5). This results from (6.9) as before, and correlates to the temporal flux change because of

$$\phi = Bla(t) = Bl(a_0 + vt) \ ,$$

where

$$V_i = -\frac{\partial \phi}{\partial t} = -vBl \ . \tag{6.10}$$

Faraday's law of electromagnetic induction is of great importance for many technological applications. Most generators, *i.e.*, current producing machines, rely on the EMF generated in a conductor loop when rotating within a magnetic field. During this important process of energy transformation, mechanical energy is transformed into electrical energy (Fig. 6.6).

If ω is the angular velocity, then the flux encompassed during the time period t is

$$\phi = Bal\cos\omega t \ , \tag{6.11}$$

which lets the induced EMF become

$$V_i = \oint \mathbf{E} \bullet d\mathbf{s} = -\dot{\phi} = Bal\omega\sin\omega t \ . \tag{6.12}$$

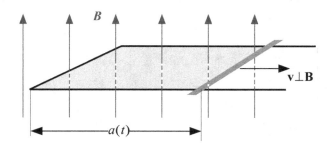

Fig. 6.5

This is the principle of alternating current generators. Direct current can be produced by means of commutators, a subject which we will leave untouched.

6.1.3   Induction by simultaneous temporal Change of *B* and Motion of the Conductor

The two effects discussed in sects. 6.1.1 and 6.1.2 can also occur simultaneously and in such case need to be added. The equations read

$$\oint \mathbf{E} \bullet d\mathbf{s} \ = \ -\frac{\partial}{\partial t}\int_A \mathbf{B} \bullet d\mathbf{A} - \oint \mathbf{B} \bullet (\mathbf{v} \times d\mathbf{s}) \ ,$$

$$\boxed{\oint \mathbf{E} \bullet d\mathbf{s} \ = \ -\frac{d\phi}{dt}} \ . \tag{6.13}$$

Notice the "total time derivative" in the operator d/dt, which indicates that the time derivative of the total flux applies, regardless of the origin of its change by either the change of the magnetic field or by motion of the conductor. When using the vector potential **A** (to avoid ambiguity, we will now change the symbol for the area element to d**a** and the area to **a**), to obtain

$$\oint \mathbf{E} \bullet d\mathbf{s} \ = \ -\frac{\partial}{\partial t}\int_a \nabla\times\mathbf{A} \bullet d\mathbf{a} + \oint (\mathbf{v} \times \mathbf{B}) \bullet d\mathbf{s}$$

$$= \ -\frac{\partial}{\partial t}\int_a \nabla\times\mathbf{A} \bullet d\mathbf{a} + \int_a \nabla\times(\mathbf{v} \times \mathbf{B}) \bullet d\mathbf{a}$$

or

$$\int_a \nabla\times\left(\mathbf{E} +\frac{\partial \mathbf{A}}{\partial t} - \mathbf{v} \times \mathbf{B}\right) \bullet d\mathbf{a} \ = \ 0 \ .$$

Since this is true for any surface, this has to be true for the integrand

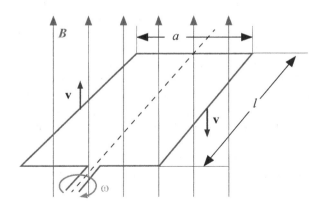

Fig. 6.6

Fig. 6.7

$$\nabla \times \left(\mathbf{E} + \frac{\partial \mathbf{A}}{\partial t} - \mathbf{v} \times \mathbf{B} \right) = 0 \ .$$

Introducing a suitable scalar function φ, we may write

$$\mathbf{E} + \frac{\partial \mathbf{A}}{\partial t} - \mathbf{v} \times \mathbf{B} = -\nabla \varphi \ . \tag{6.14}$$

The field $\mathbf{E}$ here is the one which an observer would "see" when moving with the conductor. An observer at rest then sees the field

$$\mathbf{E} = -\nabla \varphi - \frac{\partial \mathbf{A}}{\partial t} \ , \tag{6.15}$$

a relation which is a generalization of eq. (1.47) for time dependent problems, and will occupy us some more later on.

Equation (6.13) expresses that the EMF induced in a conducting loop is always given by the total time derivative of the overall magnetic flux, regardless of whether or not the loop is moving or flexible. Notice that this peculiar phenomenon, that two such apparently different effects – temporal change of a field and motion or change in shape of a closed conductor loop – are related in such a simple manner. This can be made plausible by imagining the individual force lines. Then, an increasing field can be thought of as adding lines of force to the existing field inside the closed loop. Conversely, when field lines move out of the loop, then the overall field decreases over time. The lines added or the lines removed are in motion relative to the conductor, which one considers as being at rest initially. The effect is the same, whether a line of force moves in a certain direction through the conductor or if the conductor moves in the opposite direction relative to the force line. This intuitive understanding can also be used for quantitative analysis and allows to unify understanding of these initially different effects.

Often times, the task is to analyze not closed, but open loops. For a path that goes in part through the conductor and in part through vacuum (Fig. 6.7), the integral becomes

$$\oint \mathbf{E} \bullet d\mathbf{s} = -\oint \nabla \varphi \bullet d\mathbf{s} - \frac{d\phi}{dt} \ .$$

This considers already that volume charges caused by currents (or other preexisting volume charges) superimpose an electrostatic field $\mathbf{E}_s$ over the induced electric field $\mathbf{E}_i$, where the electrostatic field $\mathbf{E}_s$ is derived from the potential φ. The latter is, of course, irrotational. The field is

$$\mathbf{E} = \mathbf{E}_s + \mathbf{E}_i \ ,$$

where

$$\nabla \times \mathbf{E}_s = 0$$

or

$$\oint \mathbf{E}_s \bullet d\mathbf{s} = -\oint \nabla \varphi \bullet d\mathbf{s} = 0 \ .$$

Therefore, it is always

$$\oint \mathbf{E} \bullet d\mathbf{s} = -\frac{d\phi}{dt} \ .$$

Eq. (6.13) derived above also holds in the case of superpositioning of electrostatic fields, regardless of whether these fields are a result of induced currents or not.

The reader is alerted to a possible miss-interpretation of eq. (6.13). This is best discussed by means of two examples. For this purpose, the unipolar machine is described and interpreted first, followed by Hering's experiment.

6.1.4   Unipolar Machine

Consider the wheel made of conductive material as shown in Fig. 6.8, rotating with a constant ω in a magnetic field. For simplicity reasons, we assume that the

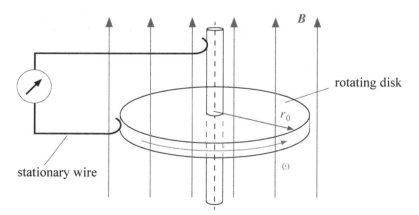

Fig. 6.8

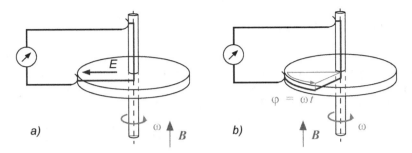

Fig. 6.9

magnetic field is uniform and oriented parallel to the axis of rotation. Two brushes provide connectivity between the stationary outside part of the circuit that includes a voltmeter and the rotating part inside. The connection to the rotating part of the circuit is located at the rotating axis on one hand and the edge of the wheel on the other. The result of this closed circuit is an induced EMF that can be calculated as follows: Velocity is a function of the distance to the center

$$v_\varphi = \omega r \; . \tag{6.16}$$

Eq. (6.6) allows to calculate the electric field

$$E_r = \omega B r \; . \tag{6.17}$$

and

$$V_i = \oint \mathbf{E}(r) \bullet d\mathbf{s} = \oint_{r_0}^{0} E_r(r) \bullet dr = -\frac{\omega B r_0^2}{2} \; . \tag{6.18}$$

The problem can be solved in a rather simple manner. However, if we start with the magnetic flux through the circuit, we encounter difficulties. First, notice how unclear the quantitative description of this flux actually is. One could state that the flux is always, identically zero $\phi = 0$ (Fig. 6.9a). Then, the induced EMF would have to vanish, an obviously incorrect conclusion. However, the circuit as shown in Fig. 6.9b is also a possibility. This yields

$$\phi = \frac{r_0^2 \varphi}{2} B = \frac{r_0^2 \omega t}{2} B \; ,$$

which is the correct result for the induced EMF

$$V_i = -\frac{d\phi}{dt} = -\frac{r_0^2 \omega}{2} B \; .$$

However, this explanation is not very persuasive. While there are a number of possible ways to define ϕ, it is not unique. The definition in Fig. 6.9b was designed to provide the correct result, but that does not prove anything in and by itself. What is always true is

$$\mathbf{E} = \mathbf{v} \times \mathbf{B} \, ,$$

and, is therefore a useful basis to calculate the EMF. Using the flux in this and other similar situation makes no sense.

6.1.5   Hering's experiment

Fig. 6.10 shows a magnet for which one assumes that there is no leakage of its flux. It has an air gap that is penetrated by a magnetic field. Outside the magnet, there is a conducting loop with elastic contacts that ensure a closed circuit at all times. The circuit also includes a voltmeter. Now, we move the conducting loop from its position Fig. 6.10a to position Fig. 6.10b. This causes a voltage impulse.

$$\int V dt = -\Delta \phi \, ,$$

in the voltmeter. The spring loaded contacts ensure that the circuit remains closed. The loop is further moved to position Fig. 6.10c, whereby nothing happens, *i.e.*, there is no induced EMF. Finally, the loop is pulled from the magnet to the outside, as illustrated in Fig. 6.10d. The spring contacts open, but stay in contact with the magnet, so the circuit is still closed at all times.

The last phase is shown in more detail in Fig. 6.11. The question is now, what EMF is induced during the process of going from state Fig. 6.11a to Fig. 6.11c. Undoubtedly, there is an initial flux $\phi \neq 0$, while there is no flux in the final state. Nevertheless – and as can be verified by experiment – there is no induced EMF. This may appear like a paradox which represents a contradiction to eq. (6.13). Essential is, however, that the mentioned flux change is not related to the motion of the conductor. If we start from the safe grounds of the relation for the locally induced electrical field

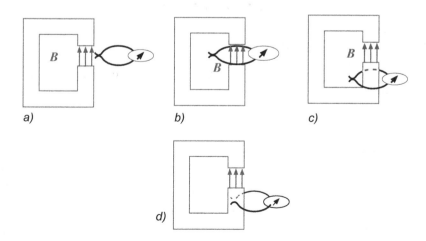

Fig. 6.10

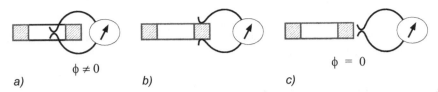

a) $\phi \neq 0$ b)   c) $\phi = 0$

Fig. 6.11

$$\mathbf{E} = \mathbf{v} \times \mathbf{B} ,$$

then we immediately see that it vanishes everywhere. From Fig. 6.11b, it becomes clear that during this phase of the process, inside of the magnet $\mathbf{B} \neq 0$ but $\mathbf{v} = 0$, while outside, $v \neq 0$ but $\mathbf{B} = 0$ (because of the ideal assumption that there is no leakage of B). Since $\mathbf{E}$ vanishes everywhere, so has to vanish the integral $\oint \mathbf{E} \bullet d\mathbf{s}$. Thereby, the result of Hering's experiment which initially appeared to be paradoxical, now becomes downright self-evident. The seeming paradox is simply a consequence of using the law of induction in a form (6.13) which is not legitimate for this particular case.

Eq. (6.13) only applies in situations when the loop during its motion or deformations maintains its material identity and is penetrated by a uniquely identifiable flux. This is neither the case for the Unipolar machine nor Hering's experiment. Looking back, we could have supposed this because of the spring contacts, which may have seemed minor. Brushes and sliding contacts require extra caution. In case of doubt, it is best to go back to the fundamental laws. This is good advice, not only in the realm of induction, as applying simple recipes and summary results to complex problems may lead to errors and conflicts in any subject matter.

6.2   Diffusion of Electromagnetic Fields

6.2.1   Equations for *E, g, B, A*

When neglecting the displacement current but considering induction, then Maxwells equations are

$$\nabla \times \mathbf{E} = -\frac{\partial \mathbf{B}}{\partial t} \tag{6.19}$$

$$\nabla \times \mathbf{H} = \mathbf{g} \tag{6.20}$$

$$\nabla \bullet \mathbf{B} = 0 \tag{6.21}$$

$$\nabla \bullet \mathbf{D} = \rho \tag{6.22}$$

To solve these equations requires the following relations

$$\mathbf{D} = \varepsilon \mathbf{E} \tag{6.23}$$

$$\mathbf{B} = \mu \mathbf{H} \tag{6.24}$$

$$\mathbf{g} = \kappa \mathbf{E} \tag{6.25}$$

ε, μ, κ shall be location independent factors. Substituting eq. (6.23) in (6.22) while letting $\rho = 0$ gives

$$\nabla \bullet \mathbf{E} = 0 . \tag{6.26}$$

Using Eqs. (6.19), (6.24), (6.20), (6.25) one finds

$$\nabla \times \nabla \times \mathbf{E} = -\frac{\partial}{\partial t} \nabla \times \mathbf{B} = -\frac{\partial}{\partial t} \mu \nabla \times \mathbf{H} = -\frac{\partial}{\partial t} \mu \mathbf{g} = -\frac{\partial}{\partial t} \mu \kappa \mathbf{E} .$$

Applying (6.26) and the vector identity (5.11) gives

$$\nabla \times \nabla \times \mathbf{E} = \nabla(\nabla \bullet \mathbf{E}) - \nabla \bullet (\nabla \mathbf{E}) = \nabla(\nabla \bullet \mathbf{E}) - \nabla^2 \mathbf{E}$$
$$= -\nabla^2 \mathbf{E} ,$$

that is

$$\boxed{\nabla^2 \mathbf{E} = \mu \kappa \frac{\partial \mathbf{E}}{\partial t}} . \tag{6.27}$$

Using (6.25) gives

$$\boxed{\nabla^2 \mathbf{g} = \mu \kappa \frac{\partial \mathbf{g}}{\partial t}} . \tag{6.28}$$

In similar manner, for $\mathbf{B}$ we obtain from eqs. (6.20), (6.25), (6.19), (6.24), and (6.21)

$$\nabla \times \nabla \times \mathbf{H} = \nabla \times \mathbf{g} = \nabla \times \kappa \mathbf{E} = \kappa \nabla \times \mathbf{E}$$

$$= -\kappa \frac{\partial \mathbf{B}}{\partial t} = +\frac{1}{\mu} \nabla \times \nabla \times \mathbf{B}$$

$$= -\frac{1}{\mu} \nabla \bullet (\nabla \mathbf{B}) = -\frac{1}{\mu} \nabla^2 \mathbf{B}$$

that is

$$\boxed{\nabla^2 \mathbf{B} = \mu \kappa \frac{\partial \mathbf{B}}{\partial t}} . \tag{6.29}$$

Since

$$\mathbf{B} = \nabla \times \mathbf{A}$$

and with proper gauge choice for $\mathbf{A}$, it follows from (6.29) that

$$\boxed{\nabla^2 \mathbf{A} = \mu \kappa \frac{\partial \mathbf{A}}{\partial t}} . \tag{6.30}$$

Applying the curl on both sides of (6.30) yields again (6.29).

Notice that we obtain the same type of equation for all quantities $\mathbf{E}$, $\mathbf{g}$, $\mathbf{B}$, $\mathbf{A}$. Of course these quantities are still distinct according to their physical meaning. Their distinctiveness is, from a formal perspective, not expressed in the equations themselves, but in their different initial and boundary conditions which are necessary to solve the equations.

6.2.2   The Physics of these Equations

In eqs. (6.27) through (6.30), we have found a type of equation that has the same typical structure for **E**, **g**, **B**, **A**. This equation is the so-called *diffusion equation* and is of paramount importance in physics. It received its name from the fact that (as a scalar equation), it describes the diffusion of particles. From a thermodynamics perspective, this is a typical behavior of irreversible (entropy increasing) processes. The formula for heat conduction is of this type. The only difference between the scalar diffusion equation or the heat equation and eqs. (6.27) through (6.30) is simply that the current equations are not scalar, but rather *vector diffusion equations*. In this context and from a purely formal perspective, the reader shall be reminded on the discussion about the application of the ∇^2-operator (Sect. 5.1). Also note that here, we deal with an irreversible process, too. Consider the path of a conductor forming a closed loop, where at time $t = 0$, the conductor shall carry a current. Its conductivity shall be κ. There shall be no external voltage source. The initial current may be the result of *for example*, an induction or an external voltage source that was shorted after creating the current (Fig. 6.12). When leaving this circuit all by itself, the current decays over time. Responsible for this is the resistance (R) of the conductor, which produces irreversible heat (RI^2 per unit of time), as long as the current flows. Because of the law of conservation of energy, this energy has to be taken from another reservoir, whose energy content is thereby depleted. This energy is that of the magnetic field. Thus, current and magnetic field have to decrease. This process ceases when all available energy was transformed into heat.

The energy principle should allow for the reverse process, as well. By only considering the energy balance, one can imagine that a current emerges in a closed conductor loop, while using the required energy from the energy stored in the temperature of the conductor, which would cool down. This has never been observed. Besides the first law of thermodynamics (the energy principle, or the law of the impossibility of a perpetuum mobile of the first kind), there is the second law of thermodynamics (entropy theorem, or the law of the impossibility of a

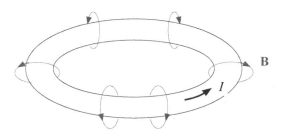

Fig. 6.12

perpetuum mobile of the second kind). It is this second law that does not entirely prohibit the described process, but declares it as very improbable, so that one can not expect to actually witness it. The basis of this are probability evaluations of micro states. In our specific case, it is very improbable that the charges in a conductor, by pure chance, move in such a way that a macroscopic current results (by chance shall mean that the motion is not caused by an external field). Much more likely is that the charges move disorderly with different velocities in all possible directions, mutually cancelling each other in a spatial and temporal average. Nevertheless, taking sufficiently accurate measurements (with sufficient spatial and temporal resolution), will reveal small, fluctuating currents, which are responsible for the permanent background *noise* of the macroscopic events. The problems arising from noise are not only theoretically interesting but also of great practical importance (*for example*, because they define the limits of accuracy for very exact measurements), but this is not the topic of this text. Here, we merely want to note that the processes described in eqs. (6.27) through (6.30) are macroscopically irreversible. This irreversibility is formally expressed by the simple time derivative. Replacing *t* by *-t* changes the equation, it is – as it is known – not *invariant against time reversal*. In other words, there is a difference whether time increases or decreases. Therefore, the process can not just run backwards.

The *wave equation*, which covers electromagnetic waves, will be discussed in Chapter. 7. Its form for **E** is

$$\nabla^2 \mathbf{E} = \mu\varepsilon\frac{\partial^2 \mathbf{E}}{\partial t^2} \ . \tag{6.31}$$

The only significant difference to eq. (6.27) is that it contains the second time derivative. This makes it invariant against time reversal. As we will see, it describes processes (waves), which may proceed both forward and backward in time.

To envision the difference, one might imagine taking a movie of such irreversible processes (*for example* diffusion) or reversible processes (*for example* waves). Then play those movies backwards. In case of a wave (more precisely: not attenuated wave, *i.e.*, one that does not irreversibly loose energy) the movie played backwards describes the same natural situation as the one played forward. On the other hand, the movie of the irreversible process played backwards would seem unnatural and very puzzling.

Another remark on the formality of the underlaying mathematics shall be made: One distinguishes three types of partial differential equations of second order. They are called *elliptic*, *parabolic*, and *hyperbolic*. All three types are very important in science, and the formal differences also manifest themselves in practical significance. That is, these three types of equations describe three significantly different phenomena. When just considering two independent variables (*x*, *y* or *x*, *t*), then the equation

$$\frac{\partial^2 \varphi}{\partial x^2} + \frac{\partial^2 \varphi}{\partial y^2} = -\frac{\rho}{\varepsilon_0} \tag{6.32}$$

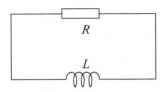

Fig. 6.13

is an elliptic equation. From an application perspective, this is a potential equation (Poisson equation). The equation

$$\frac{\partial^2 \varphi}{\partial x^2} = \frac{\partial \varphi}{\partial t} \qquad (6.33)$$

is a parabolic equation. We have called it a (scalar) diffusion equation. The equation

$$\frac{\partial^2 \varphi}{\partial x^2} = \frac{\partial^2 \varphi}{\partial t^2} \qquad (6.34)$$

is a hyperbolic equation. We recognize in it the just mentioned type of the wave equation. Summarizing in a table:

Equation	mathematical term	refers to application
(6.32)	elliptic equation	potential equation
(6.33)	parabolic equation.	diffusion equation
(6.34)	hyperbolic equation.	wave equation

When neglecting field theoretical details, the above discussed example of the conductor with a decaying current (Fig. 6.12) can be approximated in a summary description by a simple RL circuit (Fig. 6.13), where

$$RI(t) + L\frac{\mathrm{d}}{\mathrm{d}t}I(t) = 0 \qquad (6.35)$$

with the general solution

$$I(t) = C \exp\left(-\frac{R}{L}t\right)$$

If $I = I_0$ when $t = 0$, then

$$C = I_0$$

and

$$I(t) = I_0 \exp\left(-\frac{R}{L}t\right) .$$

(6.36)

This results in a decaying current, as the 2nd law of thermodynamics mandates. Multiplying (6.35) by I gives

$$RI^2(t) + LI\frac{\mathrm{d}}{\mathrm{d}t}I(t) = 0 .$$

Integrating this equation over time yields

$$\int_0^t RI^2(t)\,dt + \frac{1}{2}LI^2(t) = \frac{1}{2}LI_0^2 .$$

(6.37)

This is nothing else than the first law of thermodynamics (energy principle), applied to this problem. Eq. (6.37) states that the heat produced by the current during the time period 0 through t is

$$\int_0^t RI^2(t)\,dt ,$$

which is drawn from the magnetic energy reservoir. While there was the initial energy $(1/2)LI_0^2$, at time t only the magnetic energy

$$\frac{1}{2}LI^2 = \frac{1}{2}LI_0^2 - \int_0^t RI^2(t)\,dt$$

remains. The magnetic field, which is produced by the current, decays with the current.

To describe in detail the process of field diffusion, as it is also referred to because of the formal analogy to the diffusion process, requires one to solve the equation

$$\nabla^2 \mathbf{B} = \mu\kappa\frac{\partial \mathbf{B}}{\partial t}$$

with its corresponding boundary conditions and initial values. This is significantly more difficult than solving eq. (6.35). The following sections will provide examples of such boundary and initial value problems (*skin effect, eddy currents*). A major mathematical tool is the *Laplace transform*. A few formulas and theorems shall be compiled in Sect. 6.3.

6.2.3 Approximations and Similarity Theorems

Before delving into that subject, here we show how a rough, still useful approximation of diffusion problems can be obtained with hardly any calculation at all. For instance, insert a conductor suddenly into a magnetic field. Its inside shall initially be without a field. Once inside the field, it gradually penetrates, diffuses into the conductor. One would like to approximate how long it will take for the field to penetrate the conductor (Fig. 6.14).

The typical length of the conductor (which may be cube shaped) shall be of the order of magnitude l. The diffusion time shall be approximately t_0. This allows for a rough approximation of eq. (6.29) in the form

$$\frac{\mathbf{B}}{l^2} \approx \mu\kappa\frac{\mathbf{B}}{t_0} \; . \tag{6.38}$$

This is justified because it is approximately

$$\nabla^2\mathbf{B} \approx \frac{\mathbf{B}}{l^2} \; .$$

and

$$\frac{\partial\mathbf{B}}{\partial t} \approx \frac{\mathbf{B}}{t_0} \; .$$

Consequently

$$\boxed{t_0 \approx \mu\kappa l^2} \; . \tag{6.39}$$

Rather typical for diffusion processes is, that the time is not proportional to l, but to l^2. This is a consequence of the fact that these processes are a result of random behavior (stochastic processes). In the case of field diffusion, it is the resistance of the conductor, which depends on the statistical behavior of charges, $i.e.$, the collisions which occur statistically.

For comparison, a similar approximation of the wave equation shall be provided. It will be shown that the wave equation for $\mathbf{B}$ is of the form

$$\nabla^2\mathbf{B} = \varepsilon\mu\frac{\partial^2\mathbf{B}}{\partial t^2} \; . \tag{6.40}$$

From this follows that

$$\frac{\mathbf{B}}{l^2} \approx \varepsilon\mu\frac{\mathbf{B}}{t_0^2}$$

and

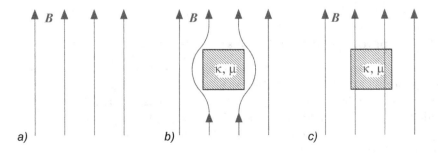

Fig. 6.14

$$\boxed{t_0 \approx \sqrt{\varepsilon \mu}\, l}\ . \tag{6.41}$$

This expresses a linear relation between l and t_0, as it is typical for orderly motion. In this context, the velocity corresponds to

$$\frac{l}{t_0} \approx \frac{1}{\sqrt{\varepsilon \mu}}\ , \tag{6.42}$$

which we will later recognize as the propagation velocity (phase velocity) of electromagnetic waves. The difference between the two relations (6.39) and (6.41), again, is formally a result of the first and second time derivative.

One may interpret (6.39) also with respect to the skin effect. For instance, if a magnetic field is applied to the surface of a conductor for a certain time t_0, then within this time the field can penetrate into the conductor to a depth l. This is the skin depth which is by (6.39)

$$l \approx \sqrt{\frac{t_0}{\mu \kappa}}\ . \tag{6.43}$$

If the applied field is alternating with the frequency f, then we have

$$t_0 \approx \frac{1}{f} = \frac{2\pi}{\omega}$$

i.e., disregarding purely numeric factors, we obtain

$$l \approx \frac{1}{\sqrt{\mu \kappa \omega}}\ . \tag{6.44}$$

This formula was obtained in a rather simple manner. The difference with exact results which are based on detailed calculations and consider boundary conditions as well as initial values, manifests itself, however, merely in numerical factors, which are usually unimportant in a first approximation. Nonetheless, the dependence of the skin depth on μ, κ, and ω is described exactly by above formula.

To make this transparent and to avoid the need to carry along all these many coefficients when solving, *for example*, the diffusion equation for the magnetic field

$$\left(\frac{\partial^2}{\partial x^2} + \frac{\partial^2}{\partial y^2} + \frac{\partial^2}{\partial z^2} \right) \mathbf{B}(x, y, z, t) = \mu \kappa \frac{\partial}{\partial t} \mathbf{B}(x, y, z, t)\ , \tag{6.45}$$

one introduces dimensionless variables, which is also advantageous for many other problems. We take the liberty to introduce a more or less arbitrary standard length l in the following manner:

$$\left. \begin{array}{l} \xi = \dfrac{x}{l} \\[2mm] \eta = \dfrac{y}{l} \\[2mm] \zeta = \dfrac{z}{l} \\[2mm] \tau = \dfrac{t}{t_0} = \dfrac{t}{\mu\kappa l^2} \end{array} \right\} \ . \tag{6.46}$$

This simplifies (6.45) to an equation without the parameters μ and κ.

$$\left(\frac{\partial^2}{\partial\xi^2} + \frac{\partial^2}{\partial\eta^2} + \frac{\partial^2}{\partial\zeta^2} \right) \mathbf{B} = \frac{\partial \mathbf{B}}{\partial\tau} \ . \tag{6.47}$$

The advantage of introducing dimensionless variables by eqs. (6.46) is that once a problem of a certain kind was solved for specific values of μ and κ, then by similarity transformation (changing the scale factor) in conjunction with (6.46), allows one to solve problems with different values of these parameters. It is said that the results are scalable, or that it is possible to provide *similarity laws*. Because of the freedom to choose the length l, we may reduce the behavior of fields in similar conductors of different expansion by similarity transformation onto each other.

6.3   Laplace Transform

The Laplace transform is oftentimes a very helpful tool to solve initial value problems. In the theory of circuits, this approach is used to reduce ordinary differential equations, whose independent variable is the time, to algebraic equations. In field theory, the equations are partial differential equations, *for example* in x, y, z, and t. Applying the Laplace transform results in a new partial differential equation in x, y, and z, where the initial values are automatically considered. This reduces the problem to a spatial boundary value problem, which can be solved using the methods introduced in Chapter 3. In the simplest case, the spatial problem is one dimensional, *for example*, x-dependent (planar) or r-dependent (cylindrical). In this case, the Laplace transform reduced the original partial differential equation to an ordinary differential equation in x or r.

Now, we will compile a list of important relations about the Laplace transform, whereby no explanation or derivation is provided. More over, no attempt for completeness will be made.

The (one sided) Laplace transform of $\tilde{f}(p)$ for a function $f(t)$ is defined by

$$\boxed{\tilde{f}(p) = \int_0^\infty f(t)\exp(-pt)dt} \ . \tag{6.48}$$

$\mathcal{L}$ shall serve as the symbol to represent the Laplace transform of a time dependent function and $\mathcal{L}^{-1}$ shall represent the inverse Laplace transformation. Thus, we will frequently write

$$\mathcal{L}\{f(t)\} = \tilde{f}(p) ,\qquad\qquad (6.49)$$

or

$$\mathcal{L}^{-1}\{\tilde{f}(p)\} = f(t) .\qquad\qquad (6.50)$$

p is a complex number. Of course, all this depends on the existence of the integral on the right side of (6.48), which shall not be our concern here, however.

A few pairs of related functions of $\tilde{f}(p)$ and $f(t)$ are listed in table 6.1. Some of these can be found immediately by applying (6.48), some will be discussed in conjunction with solved problems in subsequent sections. There are also a number of tables for the Laplace transform. A particularly detailed table for both the Laplace transform and its inverse can be found in [5] volume 1.

Particularly important in the context of initial value problems is, that for the n-th time derivative of a function $f(t)$ the following relation holds:

Table 6.1

$f(t) = \mathcal{L}^{-1}\{\tilde{f}(p)\}$	$\tilde{f}(p) = \mathcal{L}\{f(t)\}$	Convergence region
t^{α}	$\dfrac{\alpha!}{p^{\alpha+1}}$	$\mathfrak{Re}\{p\} > 0$
$\delta(t)$	1	
$\sin(\omega t)$	$\dfrac{\omega}{p^2 + \omega^2}$	$\mathfrak{Re}\{p\} > \lvert\mathfrak{Im}\{\omega\}\rvert$
$\cos(\omega t)$	$\dfrac{p}{p^2 + \omega^2}$	$\mathfrak{Re}\{p\} > \lvert\mathfrak{Im}\{\omega\}\rvert$
$\exp(\alpha t)$	$\dfrac{1}{p - \alpha}$	$\mathfrak{Re}\{p\} > \mathfrak{Re}\{\alpha\}$
$\dfrac{1}{\sqrt{4\pi t}}\exp\!\left(-\dfrac{x^2}{4t}\right)$	$\dfrac{1}{2\sqrt{p}}\exp(-\lvert x\rvert\sqrt{p})$	$\mathfrak{Re}\{p\} > 0$
$\dfrac{x}{\sqrt{4\pi t^3}}\exp\!\left(-\dfrac{x^2}{4t}\right)$	$\exp(-\lvert x\rvert\sqrt{p})$	$\mathfrak{Re}\{p\} \geq 0$
$f(t) = \begin{cases} 0 & \text{for } (t < t') \\ g(t-t') & \text{for } (t > t') \end{cases}$	$\tilde{g}(p)\exp(-pt')$	$\mathfrak{Re}\{p\} \geq 0$

time shifting theorem

$$\mathcal{L}\left\{\frac{d^n}{dt^n}f(t)\right\} = p^n\tilde{f}(p) - p^{n-1}f(0) - p^{n-2}f'(0) - p^{n-3}f''(0)$$ (6.51)

$$- \dots - p\,f^{(n-2)}(0) - f^{(n-1)}(0) \,.$$

Here, $f'(0)$ is the first, $f''(0)$, the second, $f^{(n)}(0)$, the n-th time derivative of $f(t)$ for $t = 0$. This can be proven by continuously integrating (6.48) by parts. This still holds for the following relation,

$$\mathcal{L}\left\{\int_0^t dt_n \int_0^{t_n} dt_{n-1} \int_0^{t_{n-1}} dt_{n-2} \dots \int_0^{t_3} dt_2 \int_0^{t_2} dt_1 f(t_1)\right\} = \frac{\tilde{f}(p)}{p^n}\,,$$ (6.52)

which is used to transform multiple time integrals. Apart from the terms $f(0)$, $f'(0)$, etc., which represent the initial values, every differentiation creates another factor p and every integration a factor $1/p$. This highlights in a striking expression the fact, that integration and differentiation are mutually inverse operations. It also illustrates that, or how, the Laplace transform under certain conditions reduces differential and integral equations to algebraic ones.

The *convolution* or *faltungs theorem* shall prove itself useful shortly. Consider two time dependent functions $f_1(t)$ and $f_2(t)$ with their Laplace transform $\tilde{f}_1(p)$ and $\tilde{f}_2(p)$. The integral

$$F(t) = \int_0^t f_1(t_0)f_2(t-t_0)dt_0 = \int_0^t f_2(t_0)f_1(t-t_0)dt_0$$ (6.53)

is the so-called *convolution integral* of the two functions $f_1(t)$ and $f_2(t)$. Its Laplace transform can be shown to be

$$\mathcal{L}\{F(t)\} = \tilde{F}(p) = \tilde{f}_1(p) \cdot \tilde{f}_2(p) \,.$$ (6.54)

When solving a problem by means of the Laplace transform, we obtain a result in the p-domain. To get the solution in the time domain, an inverse transformation is necessary. In simple cases, this can be achieved by finding the transform in a table. In general, the inverse needs to be calculated. The formula for inversion is an integral in the complex p-plane, and one can prove by means of equations known from the Fourier transform that:

$$\boxed{f(t) = \frac{1}{2\pi i} \int_{\sigma - i\infty}^{\sigma + i\infty} f(p)\exp(pt)dp}\,.$$ (6.55)

This is called the *Fourier-Mellin theorem*. Fig. 6.15 shows the complex p-plane with the integration path from $\sigma - i\infty$ to $\sigma + i\infty$, which runs at a distance σ parallel to the imaginary axis. The path has to be chosen such that it is to the right of all poles of $\tilde{f}(p)$. This means that there is not a complete freedom in the choice of σ. Function theory is frequently used to evaluate the inversion integral (6.55). If the integrand vanishes at infinity, then the integral (6.55) can be replaced by one around a closed loop as shown in Fig. 6.16. The path closes at infinity but does not add to the integral.

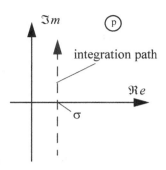

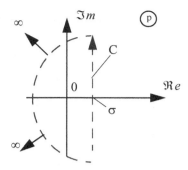

Fig. 6.15   Fig. 6.16

The integral is then:

$$f(t) = \frac{1}{2\pi i} \oint_C \tilde{f}(p) \exp(pt) dp .$$ (6.56)

This integral can be evaluated with complex analysis, by accounting for all poles inside the closed contour C (Fig. 6.16). One way is to add all residues of the integrand. This requires explanation of a few terms from function theory, complex analysis in particular. Consider an annular domain in the complex z-plane, centered around the point z_0. If a function is analytic in that domain, then it has a series representation of the form

$$f(z) = \sum_{n=-\infty}^{+\infty} a_n (z - z_0)^n ,$$ (6.57)

which is the so-called *Laurent series*. If there are terms with negative exponents of n (*i.e.*, $n<0$), then $f(z)$ diverges for $z = z_0$ and $f(z)$ is said to have a singularity at $z = z_0$. If there are an infinite number of such terms, then the singularity is said to be an *essential singularity*. If the series ends with the term $a_{-m}(z-z_0)^{-m}$ (*i.e.*, $a_{-m} \neq 0$ and $a_{-m-\nu} = 0$ for all $\nu \geq 1$), then the singularity is called a *pole of order m*. The coefficient a_{-1} of this Laurent series is very special. It is called the *residue* of the function $f(z)$ at the point $z = z_0$. The significance of the residue a_{-1} (its name suggests that it remains) stems from the fact that the integral for any positively oriented closed contour is

$$\oint_C f(z) dz = 2\pi i \cdot a_{-1} ,$$ (6.58)

as long as the path encloses z_0. If a path includes several poles z_k, then the values of all residues need to be added:

$$\oint_C f(z) dz = 2\pi i \sum_k a_{-1}^{(k)} .$$ (6.59)

This allows one to reduce the inverse Laplace transform to finding all residues of all poles and essential singularities. With (6.56) and (6.59) one can write

$$f(t) = \sum \text{sum of all residues of } [\tilde{f}(p)\exp(pt)] . \tag{6.60}$$

This requires to find all singularities and their residues. The Laurent series (6.57) shows that a residue of $f(z)$ for a pole of order 1 can be written as

$$a_{-1} = \lim_{z \to z_0} f(z)(z - z_0) . \tag{6.61}$$

For a pole of order m, the Laurent series tells us that

$$a_{-1} = \frac{1}{(m-1)!} \lim_{z \to z_0} \frac{d^{m-1}}{dz^{m-1}} [f(z)(z - z_0)^m] . \tag{6.62}$$

In case of an essential singularity, $i.e.$, when m is infinitely large, then eq. (6.62) does not help. In this case we have to go back to the Laurent series itself. Frequently, if $f(a) = g(a) = 0$, but $g'(a) \neq 0$ or $f'(a) \neq 0$, then l'Hospital's rule

$$\lim_{x \to a} \frac{f(x)}{g(x)} = \frac{f'(a)}{g'(a)} , \tag{6.63}$$

is a useful tool to determine the necessary limits to find the residues according to (6.61) and (6.62). However, if $g'(a) = 0$ and $f'(a) = 0$, then the procedure may be repeated:

$$\lim_{x \to a} \frac{f(x)}{g(x)} = \lim_{x \to a} \frac{f'(x)}{g'(x)} = \frac{f''(x)}{g''(x)} , \tag{6.64}$$

etc.

6.4 Field Diffusion in the Two-sided Infinite Space

We will investigate the behavior of a magnetic field $B_z(x, t)$ in a uniformly conductive material (κ, μ). It satisfies the diffusion equation

$$\frac{\partial^2}{\partial x^2} \mathbf{B}(x, t) = \mu\kappa \frac{\partial}{\partial t} \mathbf{B}_z(x, t) . \tag{6.65}$$

Using the dimensionless time

$$\tau = \frac{t}{t_0} = \frac{t}{\mu\kappa l^2} \tag{6.66}$$

and the dimensionless spatial coordinate

$$\xi = \frac{x}{l} , \tag{6.67}$$

where as before, l is an arbitrary length, then

$$\frac{\partial^2}{\partial \xi^2} B_z(\xi, \tau) = \frac{\partial}{\partial \tau} B_z(\xi, \tau) . \tag{6.68}$$

A solution to this equation is *for example,*

$$B_z(\xi, \tau) = B_0 \frac{\exp\left[-\dfrac{(\xi - \xi')^2}{4\tau}\right]}{\sqrt{4\pi\tau}} , \tag{6.69}$$

which can easily be verified by substitution. The solution also makes physically sense because it remains finite for both $\xi \to -\infty$, as well as for $\xi \to +\infty$ (it vanishes). This is a Gaussian curve. Its width depends on the elapsed time, more precisely, it widens for increasing times. For small times, the curve is narrow and very high. Its integral from $\xi \to -\infty$ to $\xi \to +\infty$ is constant (B_0) for all times

$$B_0 \int_{-\infty}^{+\infty} \frac{1}{\sqrt{4\pi\tau}} \exp\left[-\frac{(\xi - \xi')^2}{4\tau}\right] d\xi = B_0 . \tag{6.70}$$

Disregarding the factor B_0, for $\tau \to 0$ this expresses a δ-function, which can be defined as the limit of a Gaussian curve (see Sect. 3.4.5):

$$\lim_{\tau \to 0} \frac{\exp\left[-\dfrac{(\xi - \xi')^2}{4\tau}\right]}{\sqrt{4\pi\tau}} = \delta(\xi - \xi') , \tag{6.71}$$

and therefore

$$B_z(\xi, 0) = B_0 \delta(\xi - \xi') . \tag{6.72}$$

Consequently, according to (6.69), $B_z(\xi, \tau)$ is the solution to the diffusion problem in the infinite space with the initial condition given by (6.72). The initially closely localized field flows gradually more and more apart (Fig. 6.17). From a formal perspective, this behavior is entirely analogous to the theory of heat conductivity. The special solution given here gives us – and this makes it so significant – the solution to much more general problems. If the initial field is given in an arbitrary shape

$$B_z(\xi, 0) = h(\xi) , \tag{6.73}$$

then we can find its solution by the superposition of many δ-functions:

$$B_z(\xi, 0) = h(\xi) = \int_{-\infty}^{+\infty} h(\xi_0)\delta(\xi - \xi_0)d\xi_0 . \tag{6.74}$$

That is, the initial field consists of pieces of the form

$$dB_z(\xi, 0) = \delta(\xi - \xi_0)h(\xi_0)d\xi_0 . \tag{6.75}$$

The contribution of such a piece at a later time is

$$dB_z(\xi, \tau) = h(\xi_0)\frac{\exp\left[-\dfrac{(\xi - \xi_0)^2}{4\tau}\right]}{\sqrt{4\pi\tau}} d\xi_0 . \tag{6.76}$$

The overall field results when superposing all the contributions, *i.e.*, the integral

$$B_z(\xi, \tau) = \int\limits_{-\infty}^{+\infty} h(\xi_0) \frac{\exp\left[-\dfrac{(\xi - \xi_0)^2}{4\tau}\right]}{\sqrt{4\pi\tau}} d\xi_0 ,$$ (6.77)

or when returning to the quantities with dimensions x and t:

$$B_z(x, t) = \int\limits_{-\infty}^{+\infty} h(x_0) \frac{\exp\left[-\dfrac{(x - x_0)^2 \mu\kappa}{4t}\right]}{\sqrt{4\pi t}} \sqrt{\mu\kappa}\, dx_0 .$$ (6.78)

The relation, expressed in the form of (6.77) or (6.78) is of paramount interest because it describes the problem in full generality for every initial condition. $B_z(\xi, \tau)$ is obtained, so to speak, out of $B_z(\xi, 0)$ by an integral transform, where the solution belonging to the δ-function as initial condition (6.69), serves as the integral kernel. It is *Green's function* for this problem.

Equation (6.74) can be regarded as the expansion of the function $h(\xi)$ by the complete orthogonal and normalized systems of functions $\delta(\xi - \xi_0)$. The orthogonality relation is

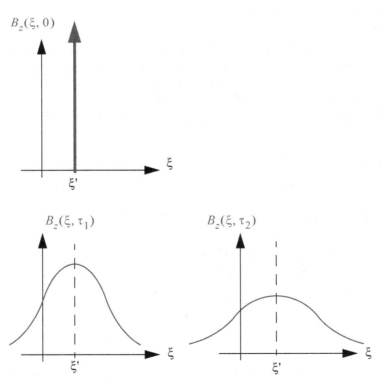

Fig. 6.17

$$\int_{-\infty}^{+\infty} \delta(\xi - \xi_0)\delta(\xi - \xi'_0)d\xi = \delta(\xi_0 - \xi'_0) .$$

Applying the Ansatz

$$h(\xi) = \int_{-\infty}^{+\infty} g(\xi_0)\delta(\xi - \xi_0)d\xi_0$$

yields

$$g(\xi_0) = h(\xi_0) ,$$

i.e., the coefficient function results in the expanded function itself, which is nothing more than an unfamiliar, however useful, interpretation of the defining property of the δ-function. If the problem is solved for a basis function, then it is solved for any function that can be expanded as a series of this base. This is the reason why the δ-function allows one to find the Green's function for the problem.

There is another approach, that is more systematic. We just gave the specific solution (6.69), which is not very satisfying. Now, we want to derive a solution starting from initial values and boundary conditions. For that purpose, we now analyze $\tilde{B}_z(\xi, p)$, rather than $B_z(\xi, \tau)$. Then we use (6.51) and (6.68) to obtain

$$\frac{\partial^2}{\partial \xi^2}\tilde{B}_z(\xi, p) = p\tilde{B}_z(\xi, p) - h(\xi) , \tag{6.79}$$

i.e., the partial differential equation in ξ and τ becomes an ordinary differential equation in ξ and the initial condition is already included. The task is to find the solution that remains finite for both $\xi \to -\infty$ and $\xi \to +\infty$. Those are the necessary boundary conditions (which are oftentimes implicitly assumed) to make the solution unique. The general solution to (6.79) results from adding the solution of the related homogeneous equation to a specific solution of the inhomogeneous equation.

The special solution is derived from the solution of the homogeneous equation by means of method variation of parameters. Leaving out the details, the general solution of (6.79) is

$$\tilde{B}_z(\xi, p) = A_1 \exp[-\sqrt{p}\xi] + A_2 \exp[+\sqrt{p}\xi] - \int_{-\infty}^{\xi} h(\xi_0)\frac{\sinh\sqrt{p}(\xi - \xi_0)}{\sqrt{p}}d\xi_0 \tag{6.80}$$

The integral is a special solution of the inhomogeneous equation (6.79). The lower boundary may be replaced by an arbitrary constant. Such a new integral is, again, a solution of the inhomogeneous equation. On the other hand, the difference between the two integrals is a solution of the homogeneous ordinary differential equation. This means that it can be expressed by a suitable superposition of the two integrals. In other words, the solution (6.80) remains unchanged when changing the lower boundary of the integral, when simultaneously changing the coefficients A_1 and A_2 in a suitable manner.

In particular, if

$$h(\xi) = B_0 \delta(\xi - \xi') ,$$ (6.81)

then letting $C_{1,2} = A_{1,2}/B_0$ gives

$$\frac{\tilde{B}_z(\xi, p)}{B_0} = \begin{cases} C_1 \exp[-\sqrt{p}\xi] + C_2 \exp[+\sqrt{p}\xi] & \text{for } \xi < \xi' \\[2ex] C_1 \exp[-\sqrt{p}\xi] + C_2 \exp[+\sqrt{p}\xi] - \dfrac{\sinh[\sqrt{p}(\xi - \xi')]}{\sqrt{p}} & \text{for } \xi > \xi'. \end{cases}$$ (6.82)

To prevent that $\tilde{B}_z$ diverges for $\xi \to -\infty$ requires that

$$C_1 = 0 .$$ (6.83)

For very large values of ξ we have

$$\frac{\tilde{B}_z(\xi, p)}{B_0} \approx C_2 \exp[+\sqrt{p}\xi] - \frac{1}{2\sqrt{p}} \exp[\sqrt{p}(\xi - \xi')] .$$ (6.84)

To prevent that $\tilde{B}_z$ diverges for $\xi \to +\infty$ requires that

$$C_2 = \frac{1}{2\sqrt{p}} \exp[-\sqrt{p}\xi'] .$$ (6.85)

With (6.82), (6.83), and (6.85), we finally obtain

$$\frac{\tilde{B}_z(\xi, p)}{B_0} = \frac{1}{2\sqrt{p}} \exp[-\sqrt{p}|\xi - \xi'|] .$$ (6.86)

Writing the absolute value of the difference $\xi - \xi' \Rightarrow |\xi - \xi'|$ allows one to omit the distinction of the two equations that initially result from using (6.82), where

$$|\xi - \xi'| = \begin{cases} \xi - \xi' & \text{for } \xi > \xi' \\ \xi' - \xi & \text{for } \xi < \xi' . \end{cases}$$ (6.87)

For symmetry reasons, a result of the form as of (6.86) can be expected because the field has to behave in the same way left and right of $\xi = \xi'$. Therefore, the result may only depend on $|\xi - \xi'|$. Returning to the time domain, when using (6.86) gives

$$B_z(\xi, \tau) = B_0 \frac{\exp\left[-\dfrac{(\xi - \xi')^2}{4\tau}\right]}{\sqrt{4\pi\tau}} .$$ (6.88)

The reason is that

$$\mathcal{L}^{-1}\left\{ \frac{1}{2\sqrt{p}} \exp[-\sqrt{p}|\xi - \xi'|] \right\} = \frac{\exp\left[-\dfrac{(\xi - \xi')^2}{4\tau}\right]}{\sqrt{4\pi\tau}} .$$ (6.89)

The result (6.88) is consistent with our previous result (6.69), which we now have derived in a systematic manner. The general solution (6.77) follows from (6.88) as before.

6.5   Diffusion of a Field in a Half-Space

6.5.1   General solution

The next problem we shall discus, is the field diffusion in the half-space $\xi > 0$ (Fig. 6.18). Again, this is given by the solution of the equation

$$\frac{\partial^2}{\partial \xi^2} B_z(\xi, \tau) = \frac{\partial}{\partial \tau} B_z(\xi, \tau) \ , \tag{6.90}$$

where the boundary conditions are now

$$[B_z(\xi, \tau)]_{\xi \, = \, 0^+} = B_z(0, \tau) = f(\tau) \ . \tag{6.91}$$

$$[B_z(\xi, \tau)]_{\xi \, \to \, \infty} = B_z(\infty, \tau) = \text{finite} \ , \tag{6.92}$$

and the initial condition is

$$[B_z(\xi, \tau)]_{\tau \, = \, 0} = B_z(\xi, 0) = h(\xi) \ . \tag{6.93}$$

As before, for $B_z(\xi, p)$ we use (6.79)

$$\frac{\partial^2}{\partial \xi^2} \tilde{B}_z(\xi, p) = p \tilde{B}_z(\xi, p) - h(\xi) \ , \tag{6.94}$$

with the general solution similar to (6.80),

$$\tilde{B}_z(\xi, p) = A_1 \exp[-\sqrt{p}\xi] + A_2 \exp[+\sqrt{p}\xi] - \int_0^\xi h(\xi_0) \frac{\sinh \sqrt{p}(\xi - \xi_0)}{\sqrt{p}} d\xi_0 \tag{6.95}$$

Only the lower boundary of the integral (6.80) was changed from $\xi_0 = -\infty$ to $\xi_0 = 0$. The reason is that this problem only deals with the positive half-space. Again, we choose

$$h(\xi) = B_0 \delta(\xi - \xi') \ , \tag{6.96}$$

and in analogy to (6.82) one obtains

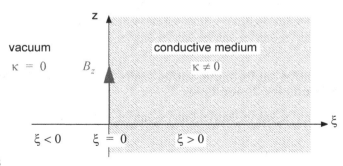

Fig. 6.18

$$\frac{\tilde{B}_z(\xi,p)}{B_0} = \begin{cases} C_1 \exp[-\sqrt{p}\,\xi] + C_2 \exp[\sqrt{p}\,\xi] & \text{for } 0 \leq \xi < \xi' \\ C_1 \exp[-\sqrt{p}\,\xi] + C_2 \exp[\sqrt{p}\,\xi] - \dfrac{\sinh[\sqrt{p}(\xi-\xi')]}{\sqrt{p}} & \text{for } 0 \leq \xi' < \xi. \end{cases}$$

(6.97)

From this and the two boundary conditions (6.91) and (6.92), it follows that

$$B_0(C_1 + C_2) = \tilde{f}(p)$$

(6.98)

and

$$C_2 = \frac{1}{2\sqrt{p}}\exp[-\sqrt{p}\,\xi'] \,,$$

(6.99)

entirely in accordance with (6.85). Eliminating C_2 from (6.98) yields

$$C_1 = \frac{\tilde{f}(p)}{B_0} - \frac{1}{2\sqrt{p}}\exp[-\sqrt{p}\,\xi'] \,.$$

(6.100)

Finally, with (6.97), (6.99), and (6.100), one obtains the solution to our problem in the p-domain:

$$\tilde{B}_z(\xi,p) = \tilde{f}(p)\exp[-\sqrt{p}\,\xi] + \frac{B_0}{2\sqrt{p}}\{-\exp[-\sqrt{p}(\xi+\xi')] + \exp[-\sqrt{p}|\xi-\xi'|]\}.$$

(6.101)

The result needs to be interpreted. We will examine each of the two terms of $\tilde{B}_z(\xi,p)$ individually, since each represents an entirely different cause. If $B_0 = 0$, that is, there is no initial field ($h(\xi) = 0$), then

$$\tilde{B}_z(\xi,p) = \tilde{f}(p)\exp[-\sqrt{p}\,\xi] \,.$$

(6.102)

This equation describes the fraction of the field which diffuses from the surface into the half-space due to the boundary condition at $\xi = 0$. It shall be explained in Sect. 6.5.2. Conversely, if $\tilde{f}(p) = 0$ while $B_0 \neq 0$, then

$$\tilde{B}_z(\xi,p) = \frac{B_0}{2\sqrt{p}}\{-\exp[(-\sqrt{p})(\xi+\xi')] + \exp[-\sqrt{p}|\xi-\xi'|]\} \,.$$

(6.103)

This represents the fraction that stems from the initial condition and which satisfies the boundary condition $\tilde{B}_z(0,p) = 0$ at the surface $\xi = 0$. It will be discussed in Sect. 6.5.3.

6.5.2   Field Diffusion from the Surface into a Half-space (Impact of the Boundary Conditions)

If we remove the fraction of the field that is introduced by the initial condition, what remains is

$$\tilde{B}_z(\xi,p) = \tilde{f}(p)\exp[-\sqrt{p}\,\xi] \,.$$

(6.104)

Let us start with the special case of a δ-function

$$f(\tau) = B_1 \delta(\tau - \tau') \,. \tag{6.105}$$

Then

$$\tilde{f}(p) = B_1 \int_0^\infty \delta(\tau - \tau') \exp(-p\tau) d\tau = B_1 \exp(-p\tau') \tag{6.106}$$

and

$$\tilde{B}_z(\xi, p) = B_1 \exp[-p\tau' - \sqrt{p}\xi] \,. \tag{6.107}$$

This can be transformed back into the time domain (see table 6.1)

$$B_z(\xi, \tau) = \begin{cases} 0 & \text{for } \tau < \tau' \\ \dfrac{B_1 \xi \exp\!\left[-\dfrac{\xi^2}{4(\tau - \tau')}\right]}{2\sqrt{\pi(\tau - \tau')^3}} & \text{for } 0 \le \tau' < \tau \,. \end{cases} \tag{6.108}$$

This is the solution to the problem of a very strong field, acting very briefly at the surface at the time τ'. Now, we consider the general boundary condition $f(\tau)$, which can be assumed as the superposition of many δ-functions:

$$f(\tau) = \int_0^\infty f(\tau_0) \delta(\tau - \tau_0) d\tau \,. \tag{6.109}$$

Each one expands according to the result in (6.108), with the final result

$$\boxed{B_z(\xi, \tau) = \int_0^\tau f(\tau_0) \frac{\xi \exp\!\left[-\dfrac{\xi^2}{4(\tau - \tau_0)}\right]}{2\sqrt{\pi}\sqrt{(\tau - \tau_0)^3}} d\tau_0} \tag{6.110}$$

or

$$\boxed{B_z(x, t) = \int_0^t f(t_0) \frac{x\sqrt{\mu\kappa} \exp\!\left[-\dfrac{x^2\mu\kappa}{4(t - t_0)}\right]}{2\sqrt{\pi}\sqrt{(t - t_0)^3}} dt_0} \,. \tag{6.111}$$

With this, we have reduced the problem of a completely general boundary condition to an integral transform, which maps the field at the surface onto the field inside. It achieves this by means of Green's function given in eq. (6.108).

Notice that we have initially chosen a δ-function as the boundary condition and then obtained the solution (6.110) by superposing many δ-functions. One could have chosen a more formal approach, deriving (6.110) directly from (6.104) by means of the convolution theorem (6.53), (6.54). This is justified because of (see Tab. 6.1)

$$\mathcal{L}^{-1}\{\exp(-\sqrt{p}\xi)\} = \frac{\xi \exp\!\left[-\dfrac{\xi^2}{4\tau}\right]}{2\sqrt{\pi}\sqrt{\tau^3}} \,. \tag{6.112}$$

Before, we performed the superposition task in an intuitive way. Now, the convolution theorem does the job of supperpositioning the individual δ-impulses, which makes up the function $f(\tau)$. Eq. (6.110) describes the field's spatial and temporal progress in the half-space for an arbitrary field given at the surface, if the half-space is initially field free. Let us consider the simple **example** of a field that increases step-like, at time $\tau = 0$ and then remains constant:

$$f(\tau) = \begin{cases} 0 & \text{for } \tau < 0 \\ B_0 & \text{for } 0 \leq \tau . \end{cases} \tag{6.113}$$

The field is then

$$B_z(\xi, \tau) = B_0 \int_0^\tau \frac{\xi \exp\left[-\dfrac{\xi^2}{4(\tau - \tau_0)}\right]}{2\sqrt{\pi}\sqrt{(\tau - \tau_0)^3}} d\tau_0 . \tag{6.114}$$

Introducing the new variable u

$$u = \frac{\xi}{2\sqrt{(\tau - \tau_0)}} , \tag{6.115}$$

makes

$$\frac{du}{d\tau_0} = \frac{\xi}{4\sqrt{(\tau - \tau_0)^3}} , \tag{6.116}$$

and substituting gives

$$B_z(\xi, \tau) = B_0 \frac{2}{\sqrt{\pi}} \int_{\xi/2\sqrt{\tau}}^\infty \exp[-u^2] du$$

$$= B_0 \frac{2}{\sqrt{\pi}} \left[\int_0^\infty \exp[-u^2] du - \int_0^{\xi/2\sqrt{\tau}} \exp[-u^2] du \right]$$

$$B_z(\xi, \tau) = B_0 \left[1 - \frac{2}{\sqrt{\pi}} \int_0^{\xi/2\sqrt{\tau}} \exp[-u^2] du \right]$$

$$= B_0 \left[1 - \operatorname{erf}\left(\frac{\xi}{2\sqrt{\tau}}\right) \right]$$

$$= B_0 \operatorname{erfc}\left(\frac{\xi}{2\sqrt{\tau}}\right)$$

and with **natural dimensions**

$$B_z(x, t) = B_0 \operatorname{erfc}\left(\frac{x\sqrt{\mu\kappa}}{2\sqrt{t}}\right) . \tag{6.117}$$

Here, we have introduced the so-called error function (erf) (Fig. 6.19) and its complementary function the error function complement (erfc):

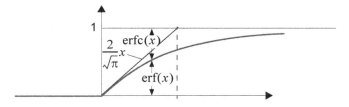

Fig. 6.19

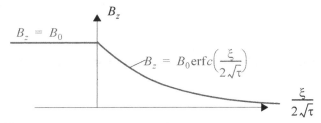

Fig. 6.20

$$\text{erf}(x) = \frac{2}{\sqrt{\pi}} \int_0^x \exp[-u^2]\,du \tag{6.118}$$

$$\text{erfc}(x) = 1 - \text{erf}(x) = \frac{2}{\sqrt{\pi}} \int_x^\infty \exp[-u^2]\,du \ . \tag{6.119}$$

These describe how the field, applied to the surface at time $t = 0$ and held constant thereafter, penetrates the half-space (Fig. 6.20). It is remarkable and an example for the above mentioned similarity theorems (Sect. 6.2.3), that the field only depends on $\xi/(2\sqrt{\tau})$ and not on the parameters ξ and τ individually. The field maintains its principal shape at all times (Fig. 6.20), Stretching more and more as time increases. When

$$\frac{\xi}{2\sqrt{\tau}} \approx 0.48$$

the value for erfc is

$$\text{erfc}\left(\frac{\xi}{2\sqrt{\tau}}\right) = \frac{1}{2} \ .$$

In a rough estimate, it can be stated that within the time τ, the field B_z has only reached as far as to the location

$$\xi \approx 2 \cdot 0.48\sqrt{\tau} \approx \sqrt{\tau} \ , \tag{6.120}$$

its half width. When returning to the variables with natural dimensions x and t, then with (6.66) we find how far the field (more precisely half of the field) has penetrated the half-space

$$x \approx \sqrt{\frac{t}{\mu \kappa}} \, , \tag{6.121}$$

and the time it takes to do so is

$$t \approx \mu \kappa x^2 \, . \tag{6.122}$$

This complies with our previous estimates (6.39) and (6.43). This simple example shall be sufficient for us here, but we will return in a later section (Sect. 6.5.4) to discuss the problem of the skin effect for the case of a field or current, periodic in time.

6.5.3 Diffusion of the Initial Field in the Half-Space (Impact of Initial Values)

The field according to (6.103) is

$$\tilde{B}_z(\xi, p) = \frac{-B_0 \exp[-\sqrt{p}(\xi + \xi')]}{2\sqrt{p}} + \frac{B_0 \exp[-\sqrt{p}|\xi - \xi'|]}{2\sqrt{p}} \, . \tag{6.123}$$

We examine the second term first. We know it very well from Sect. 6.4, eq. (6.86). The related time function according to (6.88) is

$$B_z(\xi, \tau) = B_0 \frac{\exp\left[-\dfrac{(\xi - \xi')^2}{4\tau}\right]}{\sqrt{4\pi\tau}} \, . \tag{6.124}$$

The first term in (6.123) is of the same kind, at least concerning its effect in the region $\xi > 0$, $\xi' > 0$. It describes a field in the positive half-space, which one can picture as a δ-function-like initial field at location $\xi = -\xi'$:

$$B_z(\xi, \tau) = -B_0 \frac{\exp\left[-\dfrac{(\xi + \xi')^2}{4\tau}\right]}{\sqrt{4\pi\tau}} \, . \tag{6.125}$$

The two fields mutually cancel at $\xi = 0$, which satisfies the boundary condition there. This is an example of an "image" field, which is necessary to satisfy the boundary conditions. Nevertheless, this image field is of a different kind than previous images. In the positive half-space at the time $\tau = 0$, one has the field

$$B_z(\xi, 0) = B_0 \delta(\xi - \xi') \tag{6.126}$$

and in the negative half-space ($\xi = -\xi'$), we have the field

$$B_z(\xi, 0) = -B_0 \delta(\xi + \xi') \, , \tag{6.127}$$

which is, of course, of fictitious nature.

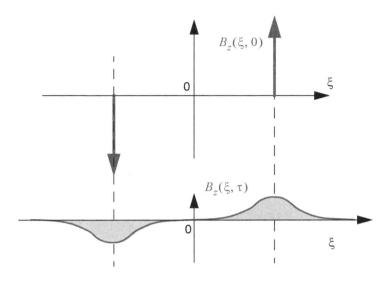

Fig. 6.21

Both fields widen with increasing time to Gaussian curves ((Fig. 6.21). The overall field at the time τ in the positive half-space is

$$B_z(\xi, \tau) = \frac{B_0}{\sqrt{4\pi\tau}}\left\{\exp\left[-\frac{(\xi-\xi')^2}{4\tau}\right] - \exp\left[-\frac{(\xi+\xi')^2}{4\tau}\right]\right\} . \tag{6.128}$$

In the negative half-space, this field is only fictitious. The real field there is actually zero $B_z = 0$. If there is an arbitrary initial field $h(\xi)$, then we need to add all fractions (in analogy to the discussion of Sect. 6.4).

$$B_z(\xi, \tau) = \int_0^\infty \frac{h(\xi_0)}{\sqrt{4\pi\tau}}\left\{\exp\left[-\frac{(\xi-\xi_0)^2}{4\tau}\right] - \exp\left[-\frac{(\xi+\xi_0)^2}{4\tau}\right]\right\}d\xi_0 . \tag{6.129}$$

This represents the general solution of the current problem.

Consider a simple, special case as an **example**:

$$h(\xi) = B_0 . \tag{6.130}$$

Substituting

$$u = \frac{\xi \pm \xi_0}{2\sqrt{\tau}} \tag{6.131}$$

and

$$\frac{du}{d\xi_0} = \pm\frac{1}{2\sqrt{\tau}} \tag{6.132}$$

gives

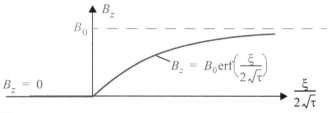

Fig. 6.22

$$B_z(\xi, \tau) = -\frac{B_0}{\sqrt{\pi}}\left\{\int_{\xi/2\sqrt{\tau}}^{-\infty} \exp[-u^2\,]du + \int_{\xi/2\sqrt{\tau}}^{+\infty} \exp[-u^2\,]du\right\}$$

$$= \frac{B_0}{\sqrt{\pi}}\left\{\int_{-\infty}^{\xi/2\sqrt{\tau}} \exp[-u^2\,]du - \int_{\xi/2\sqrt{\tau}}^{+\infty} \exp[-u^2\,]du\right\}$$

$$= \frac{2B_0}{\sqrt{\pi}}\int_0^{\xi/2\sqrt{\tau}} \exp[-u^2\,]du = B_0\mathrm{erf}\left(\frac{\xi}{2\sqrt{\tau}}\right)$$

i.e.,

$$B_z(x, t) = B_0\mathrm{erf}\left(\frac{x\sqrt{\mu\kappa}}{2\sqrt{t}}\right). \tag{6.133}$$

This must be so. It should be easy to see that the resulting field here has to complement the previously calculated field (6.117) in such a way, that the sum gives B_0. If there is an initial field B_0 inside and a field B_0 is applied on the surface also, then there is the same field everywhere and nothing will happen. If there is no field applied on the boundary, then eq. (6.133) describes how the field in the half-space decays. The progression always happens in a similar manner (Fig. 6.22).

6.5.4   Periodic Field and Skin Effect

Returning to the subject of Sect. 6.5.2, we apply a field, periodic in time to the surface of the half-space. Complex notation is particularly useful here to simplify periodic processes. We write

$$B_z(0, \tau) = f(\tau) = B_0[\cos(\Omega\tau) + i\sin(\Omega\tau)] = B_0\exp(i\Omega\tau). \tag{6.134}$$

The real part represents the physical field $B_0\cos(\Omega\tau)$ at the surface. Then

$$\tilde{f}(p) = \frac{B_0}{p - i\Omega} \tag{6.135}$$

and with (6.104)

$$\tilde{B}_z(\xi, p) = \frac{B_0}{p - i\Omega} \exp[-\sqrt{p}\xi] \ . \tag{6.136}$$

Ω is the dimensionless angular frequency, where

$$\Omega\tau = \omega t = \Omega \frac{t}{\mu\kappa l^2}$$

or

$$\Omega = \omega\mu\kappa l^2 \ . \tag{6.137}$$

The before mentioned table of Laplace transforms [5] contains the transform of (6.136) into the time domain.

$$B_z(\xi, \tau) = \frac{B_0}{2} \exp(i\Omega\tau)\left\{ \exp\left[-(1+i)\sqrt{\frac{\Omega\xi^2}{2}}\right]\text{erfc}\left[\frac{\xi}{2\sqrt{\tau}} - (1+i)\sqrt{\frac{\Omega\tau}{2}}\right]\right.$$

$$\left. + \exp\left[+(1+i)\sqrt{\frac{\Omega\xi^2}{2}}\right]\text{erfc}\left[\frac{\xi}{2\sqrt{\tau}} + (1+i)\sqrt{\frac{\Omega\tau}{2}}\right]\right\} \ . \tag{6.138}$$

For very large times ($\tau \to \infty$)

$$\text{erfc}\left[\frac{\xi}{2\sqrt{\tau}} - (1+i)\sqrt{\frac{\Omega\tau}{2}}\right] = 2$$

$$\text{erfc}\left[\frac{\xi}{2\sqrt{\tau}} + (1+i)\sqrt{\frac{\Omega\tau}{2}}\right] = 0$$

$$B_z(\xi, \tau) = B_0\exp\left[i\left(\Omega\tau - \sqrt{\frac{\Omega\xi^2}{2}}\right)\right]\exp\left[-\sqrt{\frac{\Omega\xi^2}{2}}\right]$$

and therefore the real physical field in the half-space is

$$\boxed{B_z(\xi, \tau) = B_0\cos\left[\left(\Omega\tau - \sqrt{\frac{\Omega\xi^2}{2}}\right)\right]\exp\left[-\sqrt{\frac{\Omega\xi^2}{2}}\right]} \ . \tag{6.139}$$

The field given by (6.139) remains, after the effects from the initial conditions have subsided (here, this is the field that vanishes at the beginning). The remaining field represents the so-called *steady state*. It alone would not satisfy the initial conditions at $\tau = 0$. A different initial condition would cause additional terms which would decay over time. The steady state will always be the one given by (6.139). The steady state shall occupy us for now. It shares the periodicity with the applied field, however, there is a spatial dependent phase shift and the amplitude decreases (damping) towards the inside of the half-space. Intuitively, both are to be expected.

If the interest is only in the steady state, it can be calculated rather easily. The starting point is

$$\frac{\partial^2}{\partial\xi^2}B_z(\xi, \tau) = \frac{\partial}{\partial\tau}B_z(\xi, \tau) \tag{6.140}$$

and the solution has to consist of exponential functions, therefore we try the Ansatz

$$B_z(\xi, \tau) = B_0 \exp[i(\Omega\tau - k\xi)]. \tag{6.141}$$

This requires

$$-k^2 = i\Omega$$

or

$$k = \sqrt{\Omega}\sqrt{-i} = \pm\left(\sqrt{\Omega}\frac{1-i}{\sqrt{2}}\right). \tag{6.142}$$

Thus, there are two solutions:

$$B_z(\xi, \tau) = B_0 \exp\left[i\left(\Omega\tau \mp \sqrt{\frac{\Omega}{2}}\xi^2\right) \mp \sqrt{\frac{\Omega}{2}}\xi^2\right]. \tag{6.143}$$

Only the sign on the top provides solutions which make physical sense. The bottom sign would cause a field which diverge for $\xi \to \infty$. Although this is a mathematically correct solution, one excludes it because Physics does not allow for such a field. Therefore, the field is given by:

$$B_z(\xi, \tau) = B_0 \exp\left[i\left(\Omega\tau - \sqrt{\frac{\Omega}{2}}\xi^2\right) - \sqrt{\frac{\Omega}{2}}\xi^2\right]. \tag{6.144}$$

Both, the real part as well as the imaginary part can be regarded as solutions. The real part just represents the previous solution of eq. (6.139). Returning to variables with dimensions, the field becomes

$$B_z(x, t) = B_0 \exp\left[-\sqrt{\frac{\mu\kappa\omega}{2}}x\right]\cos\left[\omega t - \sqrt{\frac{\mu\kappa\omega}{2}}x\right]. \tag{6.145}$$

The phase is constant when

$$\omega t - \sqrt{\frac{\mu\kappa\omega}{2}}x = \text{const.}$$

that is, for

$$\omega\,dt - \sqrt{\frac{\mu\kappa\omega}{2}}\,dx = 0$$

$$\frac{dx}{dt} = \sqrt{\frac{2\omega}{\mu\kappa}}. \tag{6.146}$$

This is the phase velocity, which represents the velocity with which the wave starting at the surface penetrates the half-space. The penetration depth, $i.e.$, the distance at which the amplitude has fallen to $1/e$ is given by

$$\sqrt{\frac{\mu\kappa\omega}{2}}d = 1$$

or

$$d = \sqrt{\frac{2}{\mu\kappa\omega}}. \tag{6.147}$$

With the exception of the factor $\sqrt{2}$, this is in entire harmony with our previous approximation (6.44). However, because of the very rough approximation in Sect. 6.2, which did not take into account the geometry of the set-up, we could not have expected an exact result.

From

$$\nabla \times \mathbf{H} = \mathbf{g}$$

follows for the corresponding current

$$g_y(x, t) = -\frac{\partial}{\partial t} H_z(x, t)$$

$$= H_0 \sqrt{\mu \kappa \omega} \exp\left[-\sqrt{\frac{\mu \kappa \omega}{2}} x\right] \cos\left[\omega t - \sqrt{\frac{\mu \kappa \omega}{2}} x + \frac{\pi}{4}\right] , \qquad (6.148)$$

where the trigonometric identity

$$\cos \alpha - \sin \alpha = \sqrt{2} \cos\left(\alpha + \frac{\pi}{4}\right) ,$$

was used. The time average of the squared current density is

$$\overline{g_y(x)^2} = \frac{H_0^2 \mu \kappa \omega}{2} \exp[-\sqrt{2\mu\kappa\omega} x] . \qquad (6.149)$$

This allows one to determine the average power that is transformed per unit of the surface of the half-space.

$$\int_0^\infty \frac{\overline{g_y(x)^2}}{\kappa} dx = H_0^2 \frac{\mu \omega}{2} \int_0^\infty \exp[-\sqrt{2\mu\kappa\omega} x] dx$$

$$= H_0^2 \frac{\mu \omega}{2\sqrt{2\mu\kappa\omega}} = H_0^2 \frac{\sqrt{\mu\omega}}{2\sqrt{2\kappa}} . \qquad (6.150)$$

According to (6.148) one writes

$$\int_a^b g_y(x) dx = -\int_a^b \frac{\partial H_z}{\partial x} dx = H_z(a) - H_z(b) \qquad (6.151)$$

and therefore, one gets for the overall current per unit length on the surface

$$\int_0^\infty g_y(x) dx = H_z(0) - H_z(\infty) = H_0 \cos \omega t . \qquad (6.152)$$

Squaring this and then taking the time average gives

$$\overline{\left(\int_0^\infty g_y(x) dx\right)^2} = \frac{H_0^2}{2} . \qquad (6.153)$$

Imagine that this current flows according to eq. (6.147) within the depth d, then the related power per unit area is

$$\frac{H_0^2}{2} \cdot R = \frac{H_0^2}{2} \cdot \frac{1}{\kappa d} = \frac{H_0^2}{2\kappa \sqrt{\dfrac{2}{\mu\kappa\omega}}} = \frac{H_0^2 \sqrt{\mu\omega}}{2\sqrt{2\kappa}} \; . \qquad\qquad (6.154)$$

This just describes the power calculated in eq. (6.150), which conjures up a scenario where the entire, effective current, flows with a uniform current density in a layer at the surface of thickness d. Fig. 6.23 compares *for example* for the time $t = 0$ the actual current distribution (Fig. 6.23a) with that of our model distribution (Fig. 6.23b).

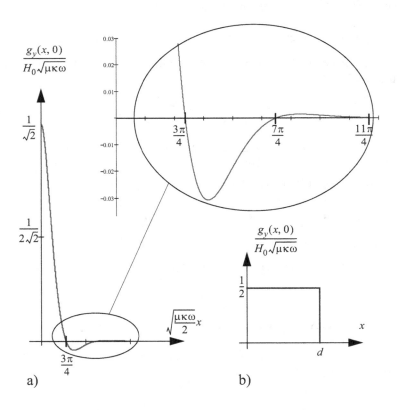

a)   b)

Fig. 6.23

6.6   Field Diffusion in a Plane Plate

6.6.1   General Solution

Now, we discuss the problem of a plane plate of thickness d, as illustrated in (Fig. 6.24). Then one has to solve the diffusion equation

$$\frac{\partial^2}{\partial \xi^2} B_z(\xi, \tau) = \frac{\partial}{\partial \tau} B_z(\xi, \tau) \tag{6.155}$$

with the boundary conditions

$$B_z(0, \tau) = f_1(\tau) \tag{6.156}$$

$$B_z(1, \tau) = f_2(\tau) \tag{6.157}$$

and the initial condition

$$B_z(\xi, 0) = h(\xi) \tag{6.158}$$

Here, one replaces the arbitrary length l by the thickness of the plate d, i.e.,

$$\tau = \frac{t}{\mu \kappa d^2} \tag{6.159}$$

and

$$\xi = \frac{x}{d} \qquad \text{for } 0 \le \xi \le 1 \ . \tag{6.160}$$

For $\tilde{B}_z(\xi, p)$, the formula (6.94) applies again

$$\frac{\partial^2}{\partial \xi^2} \tilde{B}_z(\xi, p) = p \tilde{B}_z(\xi, p) - h(\xi) \ . \tag{6.161}$$

Its general solution by eq. (6.95) is

$$\tilde{B}_z(\xi, p) = A_1 \exp[-\sqrt{p}\,\xi] + A_2 \exp[+\sqrt{p}\,\xi] - \int_0^\xi h(\xi_0) \frac{\sinh[\sqrt{p}(\xi - \xi_0)]}{\sqrt{p}} d\xi_0 \ . \tag{6.162}$$

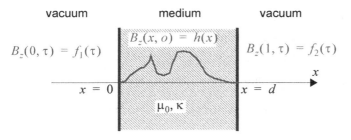

Fig. 6.24

In the p-domain, it has to satisfy the boundary conditions given in (6.156) and (6.157).

$$\tilde{B}_z(0, p) = \tilde{f}_1(p) \tag{6.163}$$

$$\tilde{B}_z(1, p) = \tilde{f}_2(p) . \tag{6.164}$$

This determines the constants A_1 and A_2, and the final result in the p-domain is

$$\left.\begin{aligned}
\tilde{B}_z(\xi, p) &= \tilde{f}_1(p)\frac{\sinh[\sqrt{p}(1-\xi)]}{\sinh[\sqrt{p}]} + \tilde{f}_2(p)\frac{\sinh[\sqrt{p}\xi]}{\sinh[\sqrt{p}]} \\[2mm]
&+ \frac{\sinh[\sqrt{p}\xi]}{\sinh[\sqrt{p}]}\int_0^1 h(\xi_0)\frac{\sinh[\sqrt{p}(1-\xi_0)]}{\sqrt{p}}d\xi_0 \\[2mm]
&- \int_0^\xi h(\xi_0)\frac{\sinh[\sqrt{p}(\xi-\xi_0)]}{\sqrt{p}}d\xi_0
\end{aligned}\right\} . \tag{6.165}$$

The field has three parts which we will discuss separately. The three parts are due to the two boundary conditions and the initial condition. It is possible to verify the correctness of this solution by substituting it into (6.161). That the boundary conditions (6.163) and (6.164) are satisfied can also be verified by inspection.

6.6.2   Diffusion of the Initial Field (Impact of Initial Condition)

We start our discussion with the last part, *i.e.*, the two terms of (6.165) which stem from $h(\xi)$. As we have found before, it is sufficient to study the special case

$$h(\xi) = B_0\delta(\xi - \xi') . \tag{6.166}$$

The reason is that the general case can be reduced to this special case. Without the terms with $\tilde{f}_1(p)$ and $\tilde{f}_2(p)$, the field becomes

$$\tilde{B}_z(\xi, p) = \begin{cases} B_0\dfrac{\sinh[\sqrt{p}\xi]\sinh[\sqrt{p}(1-\xi')]}{\sqrt{p}\sinh[\sqrt{p}]} & \text{for } \xi' > \xi \\[4mm] B_0\dfrac{\sinh[\sqrt{p}\xi']\sinh[\sqrt{p}(1-\xi)]}{\sqrt{p}\sinh[\sqrt{p}]} & \text{for } \xi' < \xi . \end{cases} \tag{6.167}$$

where we have used the relation

$$\sinh[\sqrt{p}\xi]\sinh[\sqrt{p}(1-\xi')] - \sinh[\sqrt{p}]\sinh[\sqrt{p}(\xi-\xi')]$$
$$= \sinh[\sqrt{p}\xi']\sinh[\sqrt{p}(1-\xi)] .$$

Transforming eq. (6.167) back into the time domain yields

$$B_z(\xi, \tau) = 2B_0 \sum_{n=1}^\infty \sin(n\pi\xi)\sin(n\pi\xi')\exp[-n^2\pi^2\tau] . \tag{6.168}$$

Its proof shall be postponed until the end of this section. Now, one can write the general solution for an arbitrary initial field

$$B_z(\xi, \tau) = 2 \sum_{n=1}^{\infty} \sin(n\pi\xi) \left\{ \int_0^1 h(\xi_0) \sin n\pi\xi_0 \, d\xi_0 \right\} \exp[-n^2\pi^2\tau] \ . \quad (6.169)$$

This result is in the form of a Fourier series. One could have derived this without the Laplace transform, had we chosen to start with an Ansatz of the form of a Fourier series. One can verify by inspection that eq. (6.169) solves our problem. First, every term satisfies the partial differential equation (6.155) and the boundary conditions $B_z(\xi, \tau) = 0$ for $\xi = 0$ and $\xi = 1$. Furthermore, for the time $\tau = 0$, the field is

$$B_z(\xi, 0) = \int_0^1 h(\xi_0) \left[\sum_{n=1}^{\infty} 2\sin n\pi\xi_0 \sin(n\pi\xi) \right] d\xi_0 = \int_0^1 h(\xi_0)\delta(\xi - \xi_0) d\xi_0$$

$$= h(\xi) \ , \quad (6.170)$$

that is, it satisfies the initial condition.

As specific **example**, consider

$$h(\xi) = B_0 \ , \quad (6.171)$$

for which one first finds

$$B_0 \int_0^1 \sin n\pi\xi_0 \, d\xi_0 = \frac{B_0}{n\pi}[1 - (-1)^n] \quad (6.172)$$

and therefore

$$B_z(\xi, \tau) = \sum_{n=1}^{\infty} \frac{2B_0}{n\pi}[1 - (-1)^n]\sin(n\pi\xi)\exp[-n^2\pi^2\tau] \ . \quad (6.173)$$

The result is in the form of an infinite series and converges extremely well if the time is not too small. For a sufficiently large times, the first term of the series already represents a rather useful approximation. For

$$\tau = \frac{t}{\mu\kappa d^2} \gg \frac{1}{\pi^2} \quad (6.174)$$

or

$$t \gg \frac{\mu\kappa d^2}{\pi^2} \ , \quad (6.175)$$

one has

$$B_z(x, t) \approx \frac{4B_0}{\pi}\sin\left(\frac{\pi x}{d}\right)\exp\left[-\frac{\pi^2 t}{\mu\kappa d^2}\right] \ , \quad (6.176)$$

that is, the field behaves as indicated in Fig. 6.25. Conversely, for small times, the series does not converge well. On a side note, it shall be mentioned that this series is closely related to the so-called θ-functions. There are relations between θ-functions that allow one to transform poorly converging series into well converging series.

Now, we return to the above announced proof of (6.167) and (6.168), where it was claimed that

$$
\mathcal{L}\left\{ 2 \sum_{n=1}^{\infty} \sin(n\pi\xi)\sin(n\pi\xi')\exp[-n^2\pi^2\tau] \right\}
$$

$$
= \begin{cases} \dfrac{\sinh[\sqrt{p}\,\xi]\sinh[\sqrt{p}(1-\xi')]}{\sqrt{p}\sinh[\sqrt{p}]} & \text{for } \xi' > \xi \\[3mm] \dfrac{\sinh[\sqrt{p}\,\xi']\sinh[\sqrt{p}(1-\xi)]}{\sqrt{p}\sinh[\sqrt{p}]} & \text{for } \xi' < \xi \end{cases} . \tag{6.177}
$$

It may be sufficient to prove one of the two cases. Let us pick the case $\xi' > \xi$. This equation meets the requirements to apply the residue theorem to find the inverse Laplace transform (see Sect. 6.3). According to (6.60), we need the residues for

$$
\tilde{f}(p)\exp(p\tau) = \frac{\sinh[\sqrt{p}\,\xi]\sinh[\sqrt{p}(1-\xi')]}{\sqrt{p}\sinh[\sqrt{p}]}\exp(p\tau) . \tag{6.178}
$$

Using

$$
\sinh(z) = -i\sin(iz) , \tag{6.179}
$$

one might as well write

$$
\tilde{f}(p)\exp(p\tau) = -i\frac{\sin[i\sqrt{p}\,\xi]\sin[i\sqrt{p}(1-\xi')]}{\sqrt{p}\sin[i\sqrt{p}]}\exp(p\tau) . \tag{6.180}
$$

The zeros of the denominator are at

$$
i\sqrt{p} = n\pi . \tag{6.181}
$$

However, notice that there is no pole for $n = 0$, $i.e.$ $p = 0$, as the nominator also vanishes with $p = 0$ and the limit $[\,\tilde{f}(p)\exp(p\tau)]_{p\to 0}$ has a finite value. The poles are therefore at

$$
i\sqrt{p} = n\pi, \qquad n \geq 1 . \tag{6.182}
$$

or in terms of p

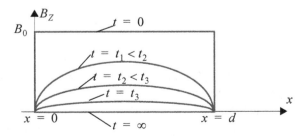

Fig. 6.25

$$p = -n^2\pi^2, \qquad n \geq 1 . \tag{6.183}$$

These are poles of order 1. There are no other poles. Its residues are

$$R_n = \lim_{p \to -n^2\pi^2} \frac{-i\sin(n\pi\xi)\sin[n\pi(1-\xi')]\exp[-n^2\pi^2\tau]}{\frac{n\pi}{i}\sin[i\sqrt{p}]}(p + n^2\pi^2) .$$

This requires l'Hospital's rule (6.63), by which we find

$$\lim_{p \to -n^2\pi^2} \frac{p + n^2\pi^2}{\sin[i\sqrt{p}]} = \lim_{p \to -n^2\pi^2} \frac{1}{\frac{i}{2\sqrt{p}}\cos[i\sqrt{p}]} = -\frac{2n\pi}{\cos(n\pi)} , \tag{6.184}$$

and its residue is

$$R_n = \frac{-2\sin(n\pi\xi)[\sin(n\pi)\cos(n\pi\xi') - \cos(n\pi)\sin(n\pi\xi')]\exp[-n^2\pi^2\tau]}{\cos(n\pi)}$$

$$= 2\sin(n\pi\xi)\sin(n\pi\xi')\exp[-n^2\pi^2\tau] . \tag{6.185}$$

Summing all these residues from $n = 1$ through $n = \infty$ totals to exactly what was claimed in (6.177).

6.6.3 Impact of Boundary Conditions

Now we study the impact of the boundary conditions at $\xi = 0$ and $\xi = 1$ (this relates to $x = 0$ and $x = d$, respectively). The first two parts from (6.165) responsible for the boundary conditions are

$$\tilde{B}_z(\xi, p) = \tilde{f}_1(p)\frac{\sinh[\sqrt{p}(1-\xi)]}{\sinh[\sqrt{p}]} + \tilde{f}_2(p)\frac{\sinh[\sqrt{p}\xi]}{\sinh[\sqrt{p}]} . \tag{6.186}$$

The convolution theorem shall be used to find the solution in the time domain, if we are able to find the inverse Laplace transform of the two functions

$$\frac{\sinh[\sqrt{p}(1-\xi)]}{\sinh[\sqrt{p}]} = \frac{\sin[i\sqrt{p}(1-\xi)]}{\sin[i\sqrt{p}]} .$$

and

$$\frac{\sinh[\sqrt{p}\xi]}{\sinh[\sqrt{p}]} = \frac{\sin[i\sqrt{p}\xi]}{\sin[i\sqrt{p}]} .$$

Both have poles of order 1 at

$$i\sqrt{p} = n\pi, \qquad n \geq 1 \tag{6.187}$$

or

$$p = -n^2\pi^2, \qquad n \geq 1 . \tag{6.188}$$

Again, there is no pole for $p = 0$ because both functions are finite there. Now one needs to find the residues for

$$\frac{\sin[i\sqrt{p}(1-\xi)]\exp(p\tau)}{\sin[i\sqrt{p}]} \qquad (6.189)$$

and

$$\frac{\sin[i\sqrt{p}\xi]\exp(p\tau)}{\sin[i\sqrt{p}]}. \qquad (6.190)$$

For the first case it is

$$R_n = \lim_{p \to -n^2\pi^2} \frac{\sin[n\pi(1-\xi)]\exp[-n^2\pi^2\tau]}{\sin[i\sqrt{p}]}(p+n^2\pi^2)$$

and by (6.184)

$$R_n = \frac{-2\pi n\sin[n\pi(1-\xi)]\exp[-n^2\pi^2\tau]}{\cos(n\pi)}$$

$$= \frac{-2\pi n[\sin(n\pi)\cos(n\pi\xi) - \cos(n\pi)\sin(n\pi\xi)]\exp[-n^2\pi^2\tau]}{\cos(n\pi)}$$

$$R_n = 2n\pi\sin(n\pi\xi)\exp[-n^2\pi^2\tau]. \qquad (6.191)$$

In like manner, one finds for the second case:

$$R_n = \lim_{p \to -n^2\pi^2} \frac{\sin(n\pi\xi)\exp[-n^2\pi^2\tau]}{\sin[i\sqrt{p}]}(p+n^2\pi^2)$$

$$= \frac{-2\pi n\sin(n\pi\xi)\exp[-n^2\pi^2\tau]}{\cos(n\pi)}$$

$$R_n = -2n\pi(-1)^n\sin(n\pi\xi)\exp[-n^2\pi^2\tau]. \qquad (6.192)$$

The residues of (6.191) belong to the function of (6.189) and the residues of (6.192) belong to the function (6.190). This enables one to write:

$$\boxed{\mathcal{L}\left\{\sum_{n=1}^{\infty} 2n\pi\sin(n\pi\xi)\exp[-n^2\pi^2\tau]\right\} = \frac{\sinh[\sqrt{p}(1-\xi)]}{\sinh[\sqrt{p}]}}, \qquad (6.193)$$

$$\boxed{\mathcal{L}\left\{\sum_{n=1}^{\infty} -2n\pi(-1)^n\sin(n\pi\xi)\exp[-n^2\pi^2\tau]\right\} = \frac{\sinh[\sqrt{p}\xi]}{\sinh[\sqrt{p}]}}. \qquad (6.194)$$

In principle, one of the formulas is sufficient because they are equivalent since

$$\sin[n\pi(1-\xi)] = \sin(n\pi)\cos(n\pi\xi) - \cos(n\pi)\sin(n\pi\xi) = -(-1)^n\sin(n\pi\xi)$$

.

Using this, the convolution theorem (6.54), and eq. (6.186), *i.e.* for $h(\xi) = 0$, yields

$$B_z(\xi, \tau) = \int_0^\tau f_1(\tau_0) \sum_{n=1}^\infty 2n\pi \sin(n\pi\xi) \exp[-n^2\pi^2(\tau - \tau_0)] d\tau_0$$

$$- \int_0^\tau f_2(\tau_0) \sum_{n=1}^\infty 2n\pi(-1)^n \sin(n\pi\xi) \exp[-n^2\pi^2(\tau - \tau_0)] d\tau_0$$

or slightly rewritten

$$B_z(\xi, \tau) = 2 \sum_{n=1}^\infty n\pi \sin(n\pi\xi) \exp[-n^2\pi^2\tau]$$

$$\cdot \int_0^\tau [f_1(\tau_0) - (-1)^n f_2(\tau_0)] \exp[n^2\pi^2\tau_0] d\tau_0 .$$

(6.195)

This solves the problem for arbitrary boundary conditions. The solution of the overall problem is the result of adding all contributions resulting from the initial condition (6.169) and all contributions due to the just calculated boundary conditions (6.195).

It is time for a simple **example**. The boundary conditions shall be

$$f_1(\tau) = 0$$

(6.196)

$$f_2(\tau) = \begin{cases} B_2 \\ 0 \end{cases} \quad \text{for} \quad \begin{aligned} \tau \geq 0 \\ \tau < 0 . \end{aligned}$$

(6.197)

Their Laplace transform is

$$\tilde{f}_1(p) = 0$$

(6.198)

$$\tilde{f}_2(p) = \frac{B_2}{p} ,$$

(6.199)

and therefore with (6.186)

$$\tilde{B}_z(\xi, p) = \frac{B_2}{p} \frac{\sinh[\sqrt{p}\xi]}{\sinh[\sqrt{p}]} .$$

(6.200)

This problem can be solved by the inverse transform of (6.200) as well as by applying (6.195). We shall do both. First, the inverse transform of (6.200) is carried out in a similar way as in the previous example. The difference is essentially that now, there is also a pole at $p = 0$ and all the other residues have an additional factor in the denominator of $p = -n^2\pi^2$ $(n \geq 1)$. The residue of

$$\frac{B_2}{p} \frac{\sinh[\sqrt{p}\xi]}{\sinh[\sqrt{p}]} \exp(p\tau)$$

at the pole $p = 0$ is

$$R_0 = B_2\xi$$

and the overall result is therefore

$$B_z(\xi, \tau) = B_2\xi + 2B_2 \sum_{n=1}^{\infty} \frac{(-1)^n \sin(n\pi\xi) \exp[-n^2\pi^2\tau]}{n\pi} . \qquad (6.201)$$

Now, taking the other approach, using eq. (6.195) as staring point, then

$$B_z(\xi, \tau) = 2 \sum_{n=1}^{\infty} n\pi \sin(n\pi\xi) \exp[-n^2\pi^2\tau]$$

$$\cdot B_2 \int_0^{\tau} \exp[n^2\pi^2\tau_0] d\tau_0 [-(-1)^n]$$

$$= 2B_2 \sum_{n=1}^{\infty} n\pi \sin(n\pi\xi) \exp[-n^2\pi^2\tau] \frac{\exp[n^2\pi^2\tau]-1}{n^2\pi^2} [-(-1)^n]$$

$$B_z(\xi, \tau) = 2B_2 \sum_{n=1}^{\infty} \left[-(-1)^n \frac{\sin(n\pi\xi)}{n\pi} \right]$$

$$+ 2B_2 \sum_{n=1}^{\infty} \left[(-1)^n \frac{\sin(n\pi\xi)\exp[n^2\pi^2\tau]}{n\pi} \right]$$

$$B_z(\xi, \tau) = B_2\xi + 2B_2 \sum_{n=1}^{\infty} \left[(-1)^n \frac{\sin(n\pi\xi)\exp[n^2\pi^2\tau]}{n\pi} \right] .$$

This is the same result as before, where it is to be noted that

$$\xi = \sum_{n=1}^{\infty} 2 \left[-(-1)^n \frac{\sin(n\pi\xi)}{n\pi} \right] \qquad \text{for } -1 < \xi < +1 . \qquad (6.202)$$

The proof can be provided by means of eqs. (3.118) and (3.120). We conclude that both methods produce the same result. There is another way to understand this. For the limit of large times, it is

$$\lim_{\tau \to \infty} B_z(\xi, \tau) = B_2\xi \qquad (6.203)$$

or

$$\lim_{t \to \infty} B_z(x, t) = B_2\frac{x}{d} . \qquad (6.204)$$

The consequence is that since the boundary conditions (6.196) and (6.197) are time independent, that for large times the field has to be time independent. This means that is must be

$$\frac{\partial^2}{\partial x^2} B_z(x) = 0 . \qquad (6.205)$$

Its general solution is

$$B_z(x) = ax + b . \qquad (6.206)$$

It follows from (6.196) that

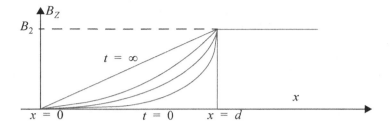

Fig. 6.26

$$b = 0 \; , \tag{6.207}$$

and from (6.196) follows that

$$a = \frac{B_2}{d} \; , \tag{6.208}$$

which leads to the solution (6.204). The complete solution (6.201) describes how the steady state (6.204) is gradually achieved. This is illustrated qualitatively in Fig. 6.26. If one considers the current, for which one can write:

$$g_y(x, t) = -\frac{\partial}{\partial x} H_z(x, t) = -\frac{1}{\mu_0} \frac{\partial}{\partial x} B_z(x, t) \; . \tag{6.209}$$

Initially, there is a surface current

$$g_y = -\frac{B_2}{\mu_0} \delta(x - d) \; , \tag{6.210}$$

which gradually spreads out in order to uniformly occupy the entire available space in its final state, the steady state:

$$g_y = -\frac{B_2}{\mu_0 d} \; . \tag{6.211}$$

The overall current per unit length remains unchanged

$$\frac{I}{l} = \int_0^d g_y(x)\,dx = -\frac{B_2}{\mu_0} \; . \tag{6.212}$$

It is fixed by the boundary conditions chosen in this example.

6.7   The Cylindrical Diffusion Problem

6.7.1   The Basic Formulas

We will now focus on problems that are important for practical applications, but nevertheless represent very simple specific cases, such as the skin effect in a cylindrical wire. The wire considered, is to be infinite in length and rotationally symmetric (*i.e.* a circular) cylinder. The magnetic field which surrounds it, or is present inside depends only on r. Consequently, all derivatives of φ and z have to vanish. The requirement that $\mathbf{B}$ has to be source free leads under these circumstances to the condition that the radial component B_r has to vanish. Otherwise, it would diverge at the axis. Eq. (3.32) gives

$$\nabla \bullet \mathbf{B} = \frac{1}{r}\frac{\partial}{\partial r}(rB_r) = 0 \ , \tag{6.213}$$

from which it follows that

$$B_r = \frac{\text{const.}}{r} \ . \tag{6.214}$$

Therefore,

$$\mathbf{B} = \langle 0, B_\varphi(r, t), B_x(r, t) \rangle \ . \tag{6.215}$$

The diffusion equation

$$\nabla^2 \mathbf{B} = \mu\kappa\frac{\partial \mathbf{B}}{\partial t} \tag{6.216}$$

for the assumptions made here and with (5.14) becomes

$$\left(\frac{1}{r}\frac{\partial}{\partial r}r\frac{\partial}{\partial r} - \frac{1}{r^2}\right)B_\varphi(r, t) = \mu\kappa\frac{\partial}{\partial t}B_\varphi(r, t) \tag{6.217}$$

$$\left(\frac{1}{r}\frac{\partial}{\partial r}r\frac{\partial}{\partial r}\right)B_z(r, t) = \mu\kappa\frac{\partial}{\partial t}B_z(r, t) \ . \tag{6.218}$$

Thus, we have to solve different equations for B_φ and B_z. Subsequently, we will discuss the field components separately. To simplify, we will limit our discussion to a solid, circular cylinder of radius r_0. We further simplify the problem by assuming a vanishing initial field,

$$[\mathbf{B}(r, t)]_{t = o} = \mathbf{B}(r, 0) = 0 \ . \tag{6.219}$$

It shall be noted that the general initial value problem with an arbitrary initial field can be solved without particular difficulties, *for example,* by expanding the initial field into a Fourier-Bessel series. The boundary conditions for our problem are

$$[\mathbf{B}(r, t)]_{r = r_0} = \mathbf{B}(r_0, t) = \mathbf{f}(t) \tag{6.220}$$

$$[\mathbf{B}(r, t)]_{r = 0} = \mathbf{B}(0, t) = \text{finite} \ . \tag{6.221}$$

We Introduce again the dimensionless variables

$$x = r/r_0 \tag{6.222}$$

and

$$\tau = \frac{t}{\mu\kappa r_0^2} \ . \tag{6.223}$$

The equations for B_φ and B_z read now:

$$\left(\frac{1}{x}\frac{\partial}{\partial x}x\frac{\partial}{\partial x}-\frac{1}{x^2}\right)B_\varphi(x,\tau)=\frac{\partial}{\partial\tau}B_\varphi(x,\tau) \tag{6.224}$$

$$\left(\frac{1}{x}\frac{\partial}{\partial x}x\frac{\partial}{\partial x}\right)B_z(x,\tau)=\frac{\partial}{\partial\tau}B_z(x,\tau)\ , \tag{6.225}$$

with the initial condition

$$\mathbf{B}(x,0)=0 \tag{6.226}$$

and the boundary conditions

$$\mathbf{B}(1,\tau)=\mathbf{f}(\tau) \tag{6.227}$$

$$\mathbf{B}(0,\tau)=\text{finite}\ . \tag{6.228}$$

This is the form for which eqs. (6.224) and (6.225) have to be solved – now, with the conditions (6.226) through (6.228). We will first solve for the longitudinal field B_z, and thereafter for the azimuthal field B_φ.

6.7.2   The longitudinal Field B$_z$

Consider an infinitely long cylinder in a space where a uniform magnetic field is created, that is, the field is oriented parallel to the cylinder axis:

$$B_z(\tau)=f_z(\tau) \tag{6.229}$$

If there is no field initially in the cylinder, then one has to solve the above specified problem. Instead of immediately solving for $B_z(x,\tau)$, we introduce its Laplace transformed field $\tilde{B}_z(x,p)$ and then, from (6.225) with the initial condition (6.226), we obtain the equation

$$\left(\frac{1}{x}\frac{\partial}{\partial x}x\frac{\partial}{\partial x}-p\right)\tilde{B}_z(x,p)=0\ , \tag{6.230}$$

and the boundary conditions in the p-domain

$$\tilde{B}_z(1,p)=\tilde{f}_z(p) \tag{6.231}$$

$$\tilde{B}_z(0,p)=\text{finite}\ . \tag{6.232}$$

This needs to be solved. Previously, in eq. (3.164), we had already met the Bessel differential equation:

$$\left[\frac{1}{\xi}\frac{\partial}{\partial\xi}\xi\frac{\partial}{\partial\xi}+\left(1-\frac{m^2}{\xi^2}\right)\right]Z_m(\xi)=0\ . \tag{6.233}$$

Substituting

$$\xi=xi\sqrt{p}\ . \tag{6.234}$$

in eq. (6.230) gives the new equation

$$\left[\frac{1}{\xi}\frac{\partial}{\partial\xi}\xi\frac{\partial}{\partial\xi}+1\right]\tilde{B}_z(\xi,p)=0\ . \tag{6.235}$$

This is Bessel's differential equation with index or order zero. Therefore

$$\tilde{B}_z(x, p) = AJ_0(xi\sqrt{p}) + BN_0(xi\sqrt{p}) ,$$ (6.236)

where A and B may depend on p but not on x. Because of the boundary conditions (6.232)

$$B = 0 ,$$ (6.237)

and because of the boundary condition (6.231), the other constant has to be

$$A = \frac{\tilde{f}_z(p)}{J_0(i\sqrt{p})} .$$ (6.238)

Therefore, the Laplace transformed field is

$$\tilde{B}_z(x, p) = \tilde{f}_z(p)\frac{J_0(xi\sqrt{p})}{J_0(i\sqrt{p})} .$$ (6.239)

We can obtain the general solution for arbitrary boundary conditions in the time domain from this formula, if we are able to find the inverse transform for $J_0(xi\sqrt{p})/J_0(i\sqrt{p})$. This is possible by means of the residue theorem. Notice the broad analogy to the problem of Sect. 6.6.3.

For the inverse transform, we need the residues of the function

$$\frac{J_0(xi\sqrt{p})}{J_0(i\sqrt{p})}\exp(p\tau) .$$ (6.240)

This has an infinite number of poles of order 1. If we use the notation introduced in Sect. 3.7.3.3, eq. (3.209) to address the zeros of J_0 by λ_{0n}, then it is for the poles

$$i\sqrt{p} = \lambda_{0n}$$ (6.241)

or

$$p = -\lambda_{0n}^2 .$$ (6.242)

The residues of the function in (6.240) are

$$R_n = \lim_{p \to -\lambda_{0n}^2} \frac{J_0(\lambda_{0n}x)\exp(-\lambda_{0n}^2\tau)}{J_0(i\sqrt{p})}(p + \lambda_{0n}^2)$$

$$= \lim_{p \to -\lambda_{0n}^2} \frac{J_0(\lambda_{0n}x)\exp(-\lambda_{0n}^2\tau)}{\frac{i}{2\sqrt{p}}J_0'(i\sqrt{p})} .$$

Using the identity

$$J_0' = -J_1$$

gives

$$R_n = \frac{2\lambda_{0n}J_0(\lambda_{0n}x)\exp(-\lambda_{0n}^2\tau)}{J_1(\lambda_{0n})} \tag{6.243}$$

and therefore, we have the transform

$$\boxed{\mathcal{L}\left[\sum_{n=1}^{\infty}\frac{2\lambda_{0n}J_0(\lambda_{0n}x)\exp(-\lambda_{0n}^2\tau)}{J_1(\lambda_{0n})}\right] = \frac{J_0(xi\sqrt{p})}{J_0(i\sqrt{p})}} \tag{6.244}$$

Applying the convolution theorem to eq. (6.239) yields

$$B_z(x,\tau) = \int_0^\tau f_z(\tau_0)\sum_{n=1}^{\infty}\frac{2\lambda_{0n}J_0(\lambda_{0n}x)\exp[-\lambda_{0n}^2(\tau-\tau_0)]}{J_1(\lambda_{0n})}d\tau_0$$

or

$$\boxed{B_z(x,\tau) = \sum_{n=1}^{\infty}\frac{2\lambda_{0n}J_0(\lambda_{0n}x)\exp[-\lambda_{0n}^2\tau]}{J_1(\lambda_{0n})}\int_0^\tau f_z(\tau_0)\exp(\lambda_{0n}^2\tau_0)d\tau_0}$$

$$\tag{6.245}$$

The result is in the form of a Fourier-Bessel series according to (3.213). Its coefficients are time dependent functions:

$$C_n = C_n(\tau) = \frac{2\lambda_{0n}\exp(-\lambda_{0n}^2\tau)}{J_1(\lambda_{0n})}\int_0^\tau f_z(\tau_0)\exp(\lambda_{0n}^2\tau_0)d\tau_0 \ . \tag{6.246}$$

The structure of eq. (6.245) is very similar to the corresponding structure of the plane equation (6.195).

The current in the cylinder is

$$g_\varphi(r) = -\frac{\partial H_z}{\partial r} = -\frac{1}{\mu_0}\frac{\partial B_z}{\partial r} = -\frac{1}{\mu_0 r_0}\frac{\partial B_z}{\partial x}$$

$$= \sum_{n=1}^{\infty}\frac{2\lambda_{0n}^2 J_1(\lambda_{0n}x)\exp(-\lambda_{0n}^2\tau)}{\mu_0 r_0 J_1(\lambda_{0n})}\int_0^\tau f_z(\tau_0)\exp(\lambda_{0n}^2\tau_0)d\tau_0 \ . \tag{6.247}$$

It is remarkable that this is not a Fourier-Bessel series, but rather a so-called *Dini series*, which we will not discuss further. Details for further study to the various types of series involving Bessel functions, including Dini series, can be found in [7].

As an **example**, let us consider the simple boundary condition

$$f_z(\tau) = B_0 \tag{6.248}$$

or

$$\tilde{f}_z(p) = B_0/p \ . \tag{6.249}$$

Then one writes the field with (6.239)

$$\tilde{B}_z(x, p) = \frac{B_0 J_0(xi\sqrt{p})}{p J_0(i\sqrt{p})} .$$ (6.250)

The inverse transform by means of the residue theorem occurs in similar manner as before for the functions $J_0(xi\sqrt{p})/J_0(i\sqrt{p})$. Recognizing the additional pole at $p = 0$ and the additional factor $p = -\lambda_{0n}^2$ in the denominator gives

$$B_z(x, \tau) = B_0 \left[1 - \sum_{n=1}^{\infty} \frac{2 J_0(\lambda_{0n} x) \exp(-\lambda_{0n}^2 \tau)}{\lambda_{0n} J_1(\lambda_{0n})} \right]$$ (6.251)

and

$$g_\varphi(x, \tau) = -\frac{1}{\mu_0 r_0} \frac{\partial B_z}{\partial x}$$

$$= -\frac{2 B_0}{\mu_0 r_0} \sum_{n=1}^{\infty} \frac{J_1(\lambda_{0n} x) \exp(-\lambda_{0n}^2 \tau)}{J_1(\lambda_{0n})} .$$ (6.252)

Another avenue is to start with the general solution (6.245) to obtain

$$B_z(x, \tau) = B_0 \sum_{n=1}^{\infty} \frac{2 \lambda_{0n} J_0(\lambda_{0n} x) \exp(-\lambda_{0n}^2 \tau)}{J_1(\lambda_{0n})} \left[\frac{\exp(\lambda_{0n}^2 \tau_0) - 1}{\lambda_{0n}^2} \right]$$

$$= B_0 \sum_{n=1}^{\infty} \frac{2 J_0(\lambda_{0n} x)}{\lambda_{0n} J_1(\lambda_{0n})} - B_0 \sum_{n=1}^{\infty} \frac{2 J_0(\lambda_{0n} x) \exp(-\lambda_{0n}^2 \tau)}{\lambda_{0n} J_1(\lambda_{0n})}$$

$$B_z(x, \tau) = B_0 \left[1 - \sum_{n=1}^{\infty} \frac{2 J_0(\lambda_{0n} x) \exp(-\lambda_{0n}^2 \tau)}{\lambda_{0n} J_1(\lambda_{0n})} \right] .$$ (6.253)

We have used the fact that

$$\sum_{n=1}^{\infty} \frac{2 J_0(\lambda_{0n} x)}{\lambda_{0n} J_1(\lambda_{0n})} = 1 .$$ (6.254)

This is the Fourier-Bessel series for the function 1 (the number one) in the range $0 \leq x < 1$. We shall prove this using the following Ansatz:

$$1 = \sum_{n=1}^{\infty} c_n J_0(\lambda_{0n} x) .$$

We multiply it with $x J_0(\lambda_{0n'} x)$ and integrate over x from 0 to 1:

$$\int_0^1 x J_0(\lambda_{0n'}) dx = \sum_{n=1}^{\infty} c_n \int_0^1 x J_0(\lambda_{0n} x) J_0(\lambda_{0n'}) dx$$

$$= \sum_{n=1}^{\infty} \frac{c_n}{2} [J_1(\lambda_{0n})]^2 \delta_{nn'} = \frac{c_{n'}}{2} [J_1(\lambda_{0n'})]^2 .$$

The last step is a result of the orthogonality relation (3.214). Furthermore, using our little set of formulas from Sect. 3.7.2 page 170 we find

$$\int_0^1 x J_0(\lambda_{0n'} x) dx = \frac{J_1(\lambda_{0n'})}{\lambda_{0n'}} .$$

The coefficients are therefore

$$c_{n'} = \frac{2}{\lambda_{0n'} J_1(\lambda_{0n'})} .$$

This proves our claim. The results of (6.251) and () agree. The expansion (6.254) makes the result (6.251) understandable. According to (6.251) for $\tau = 0$ one has

$$B_z(x, 0) = B_0 \left[1 - \sum_{n=1}^{\infty} \frac{2 J_0(\lambda_{0n} x)}{\lambda_{0n} J_1(\lambda_{0n})} \right] = B_0[1 - 1] = 0 ,$$

just as the initial condition requires.

The exponential terms may be neglected for very large times, which results in

$$[B_z(x, \tau)]_{\tau \to \infty} = B_0 .$$

Of course, this is necessary. For very large times, the uniform magnetic field occupies the entire space, including the inside of the cylinder. The currents which initially shielded the inside of the cylinder from the magnetic field decay over time. The time to decay is of the order of magnitude

$$\lambda_{01}^2 \tau = \frac{\lambda_{01}^2 t}{\mu_0 \kappa r_0^2} \approx 1 ,$$

$$t \approx \frac{\mu \kappa r_0^2}{\lambda_{01}^2} = \frac{\mu \kappa r_0^2}{(2.40)^2} . \tag{6.255}$$

Disregarding the factor (2.40^2) this agrees with our rough approximation (6.39).

6.7.3 The azimuthal Field B_φ

We will now discus the azimuthal field in a similar manner as the longitudinal field. Now, we create a magnetic field

$$[B_\varphi(x, \tau)]_{x=1} = B_\varphi(1, \tau) = f_\varphi(\tau) . \tag{6.256}$$

at the surface of a cylinder, while there is no initial field inside of the cylinder. Then eq. (6.224) applies to $B_\varphi(x, \tau)$. This results for $\tilde{B}_\varphi(x, p)$ in

$$\left(\frac{1}{x}\frac{\partial}{\partial x}x\frac{\partial}{\partial x} - \frac{1}{x^2} - p\right)\tilde{B}_\varphi(x,p) = 0 \ . \tag{6.257}$$

The boundary conditions are

$$\tilde{B}_\varphi(1,p) = \tilde{f}_\varphi(p) \tag{6.258}$$

$$\tilde{B}_\varphi(0,p) = \text{finite} \ . \tag{6.259}$$

The only difference to the longitudinal case is that now, J_1 replaces J_0. The solution is obtained in analogy to (6.239)

$$\tilde{B}_\varphi(x,p) = \tilde{f}_\varphi(p)\frac{J_1(xi\sqrt{p})}{J_1(i\sqrt{p})} \ . \tag{6.260}$$

For the inverse transform, we need the residues of the function

$$\frac{J_1(xi\sqrt{p})}{J_1(i\sqrt{p})}\exp(p\tau) \ . \tag{6.261}$$

Its poles are at

$$i\sqrt{p} = \lambda_{1n} \qquad (n \ge 1) \ . \tag{6.262}$$

or

$$p = -\lambda_{1n}^2 \qquad (n \ge 1) \ . \tag{6.263}$$

The zero $p = 0$ of $J_1(i\sqrt{p})$ is no pole but a removable singularity, since the numerator vanishes there as well. The ratio $J_1(xi\sqrt{p})/J_1(i\sqrt{p})$ remains finite ($= x$). The residues are then

$$R_n = \lim_{p \to -\lambda_{1n}^2} \frac{J_1(\lambda_{1n}x)\exp[-\lambda_{1n}^2\tau]}{J_1(i\sqrt{p})}(p + \lambda_{1n}^2)$$

$$= \lim_{p \to -\lambda_{1n}^2} \frac{J_1(\lambda_{1n}x)\exp[-\lambda_{1n}^2\tau]}{\frac{i}{2\sqrt{p}}J_1'(i\sqrt{p})} \ .$$

Using the identity

$$J_1'(z) = J_0(z) - \frac{1}{z}J_1(z)$$

and

$$J_1'(\lambda_{1n}) = J_0(\lambda_{1n})$$

gives

$$R_n = -\frac{2\lambda_{1n}J_1(\lambda_{1n}x)\exp[-\lambda_{1n}^2\tau]}{J_0(\lambda_{1n})} \ , \tag{6.264}$$

and similar to (6.244), the transformation is

$$\mathcal{L}\left[-\sum_{n=1}^{\infty}\frac{2\lambda_{1n}J_1(\lambda_{1n}x)\exp[-\lambda_{1n}^2\tau]]}{J_0(\lambda_{1n})}\right]=\frac{J_1(xi\sqrt{p})}{J_1(i\sqrt{p})}.\qquad(6.265)$$

Applying the convolution theorem to eq. (6.260) yields

$$B_\varphi(x,\tau)=-\sum_{n=1}^{\infty}\frac{2\lambda_{1n}J_1(\lambda_{1n}x)\exp[-\lambda_{1n}^2\tau]}{J_0(\lambda_{1n})}\int_0^\tau f_\varphi(\tau_0)\exp(\lambda_{1n}^2\tau_0)d\tau_0.$$

$$(6.266)$$

As an **example**, let us consider a similar case as before, now with the boundary condition

$$f_\varphi(\tau)=B_0 \qquad(6.267)$$

or

$$\tilde{f}_\varphi(p)=B_0/p . \qquad(6.268)$$

Then we write the field with (6.260)

$$\tilde{B}_\varphi(x,p)=\frac{B_0J_1(xi\sqrt{p})}{pJ_1(i\sqrt{p})}. \qquad(6.269)$$

The function

$$\frac{B_0J_1(xi\sqrt{p})}{pJ_1(i\sqrt{p})}\exp(p\tau) .$$

has poles at $p=-\lambda_{1n}^2$ and also at $p=0$. The residue there can be found most conveniently by finding the initial term of the expansion of J_1. According to (3.174) for small magnitudes of the argument J_1 it is approximately

$$J_1(xi\sqrt{p})\approx\frac{xi\sqrt{p}}{2} ,$$

and thus, for the residue there

$$R_0=\lim_{p\to0}\frac{B_0J_1(xi\sqrt{p})}{pJ_1(i\sqrt{p})}\exp(p\tau)p=B_0\frac{\frac{xi\sqrt{p}}{2}}{\frac{i\sqrt{p}}{2}}=B_0x . \qquad(6.270)$$

The remaining residues are found similarly to (6.264), now with an additional factor $p=-\lambda_{1n}^2$ in the denominator.

$$R_n=B_0\frac{2J_1(\lambda_{1n}x)\exp[-\lambda_{1n}^2\tau]}{\lambda_{1n}J_1(\lambda_{1n})}. \qquad(6.271)$$

Thus

$$B_\varphi(x, \tau) = B_0\left[x + \sum_{n=1}^{\infty} \frac{2J_1(\lambda_{1n}x)\exp[-\lambda_{1n}^2\tau]}{\lambda_{1n}J_0(\lambda_{1n})}\right] . \tag{6.272}$$

The same result can be derived using the general solution (6.266)

$$B_\varphi(x, \tau) = -\sum_{n=1}^{\infty} \frac{2B_0\lambda_{1n}J_1(\lambda_{1n}x)\exp(-\lambda_{1n}^2\tau)}{J_0(\lambda_{1n})}\left(\frac{\exp[\lambda_{1n}^2\tau]-1}{\lambda_{1n}^2}\right)$$

$$= -\sum_{n=1}^{\infty} \frac{2B_0J_1(\lambda_{1n}x)}{\lambda_{1n}J_0(\lambda_{1n})} + \sum_{n=1}^{\infty} \frac{2B_0J_1(\lambda_{1n}x)\exp[-\lambda_{1n}^2\tau]}{\lambda_{1n}J_0(\lambda_{1n})}$$

$$= B_0\left[x + \sum_{n=1}^{\infty} \frac{2J_1(\lambda_{1n}x)\exp[-\lambda_{1n}^2\tau]}{\lambda_{1n}J_0(\lambda_{1n})}\right] . \tag{6.273}$$

The Fourier-Bessel series for x in the range $0 \leq x < 1$ has to be used here:

$$x = -\sum_{n=1}^{\infty} \frac{2J_1(\lambda_{1n}x)}{\lambda_{1n}J_0(\lambda_{1n})} . \tag{6.274}$$

The proof is achieved using the formulas from Sect. 3.7.3.3 and is left for the reader as an exercise. Thus, both methods provide the same result. With (6.272) and (6.274) follows for the time $\tau = 0$

$$B_\varphi(x, 0) = B_0\left[x + \sum_{n=1}^{\infty} \frac{2J_1(\lambda_{1n}x)}{\lambda_{1n}J_0(\lambda_{1n})}\right] = B_0[x - x] = 0 ,$$

that is, the initial condition for a vanishing field inside the cylinder is satisfied, indeed. For very large times the field becomes as required:

$$[B_\varphi(x, 0)]_{\tau \to \infty} = B_0x = B_0\frac{r}{r_0} . \tag{6.275}$$

Prescribing B_φ on the boundary of the cylinder means that the total current I inside the cylinder is determined:

$$I = \int_0^{r_0} g_z(r)2\pi r\,dr = \frac{1}{\mu_0}2\pi r_0 B_0 . \tag{6.276}$$

The current I remains constant during the entire diffusion process, merely the current density $g_z(r, t)$ varies over time. Initially, the current flows entirely in the cylinder surface, and finally, the current has a uniform density inside the entire cylinder.

The field that linearly increases with the radius (6.275) corresponds to this final state. One can also derive the steady state directly from (6.217), from which follows for the steady state that

$$\left(\frac{1}{r}\frac{\partial}{\partial r}r\frac{\partial}{\partial r} - \frac{1}{r^2}\right)B_\varphi(r) = 0 , \tag{6.277}$$

with the general solution

$$B_\varphi(r) = Ar + \frac{B}{r} .$$ (6.278)

To avoid the divergence at $r = 0$ requires that

$$B = 0 ,$$

and the boundary condition at $r = r_0$ is satisfied if

$$A = B_0/r_0 .$$

Therefore, $B_\varphi = B_0 r/r_0$, as claimed.

6.7.4 Skin Effect in a Cylindrical Wire

As another special case of diffusion, we discuss the skin effect of a cylindrical wire in an azimuthal field $B_\varphi(x, \tau)$. There shall be no initial field inside the wire. The current

$$I = I_0 \cos(\Omega\tau)$$ (6.279)

shall start flowing through the wire at time $t = 0$ ($\tau = 0$). At the surface, this creates the field

$$B_\varphi = \frac{\mu_0 I_0}{2\pi r_0} \cos(\Omega\tau) = B_0 \cos(\Omega\tau)$$ (6.280)

$$B_0 = \frac{\mu_0 I_0}{2\pi r_0} .$$ (6.281)

Thus, the boundary conditions in the time and p-domain, respectively are

$$f_\varphi(\tau) = B_0 \cos(\Omega\tau)$$ (6.282)

$$\tilde{f}_\varphi(p) = B_0 \frac{p}{p^2 + \Omega^2} .$$ (6.283)

With (6.260), we obtain

$$\tilde{B}_\varphi(x, p) = \frac{B_0 p}{p^2 + \Omega^2} \frac{J_1(xi\sqrt{p})}{J_1(i\sqrt{p})} = \frac{B_0 p J_1(xi\sqrt{p})}{(p + i\Omega)(p - i\Omega)J_1(i\sqrt{p})} .$$ (6.284)

The function

$$\frac{B_0 p J_1(xi\sqrt{p})}{(p + i\Omega)(p - i\Omega)J_1(i\sqrt{p})} \exp(p\tau)$$ (6.285)

has poles of order 1 at

$$i\sqrt{p} = \lambda_{1n}$$ (6.286)

$$p = -\lambda_{1n}^2$$ (6.287)

and also at

$$p = \pm i\Omega \ . \tag{6.288}$$

Its residues at the locations $p = -\lambda_{1n}^2$ are

$$R_n = \frac{2B_0\lambda_{1n}^3 J_1(\lambda_{1n})}{(\lambda_{1n}^4 + \Omega^2)J_0(\lambda_{1n})} \exp(-\lambda_{1n}^2 \tau) \ . \tag{6.289}$$

This can be obtained from the residues (6.264) and the factor

$$\frac{B_0 p}{p^2 + \Omega^2} = -\frac{B_0\lambda_{1n}^2}{\lambda_{1n}^4 + \Omega^2} \ .$$

The residues of the poles at $p = \pm i\Omega$ are

$$R_\pm = \frac{B_0(\pm i\Omega)J_1(xi\sqrt{\pm i\Omega})}{2(\pm i\Omega)J_1(i\sqrt{\pm i\Omega})} \exp(\pm i\Omega\tau)$$

$$= \frac{B_0}{2} \frac{J_1(xi\sqrt{\pm i\Omega})}{J_1(i\sqrt{\pm i\Omega})} \exp(\pm i\Omega\tau) \ . \tag{6.290}$$

Therefore

$$B_\varphi(x, \tau) = \frac{B_0}{2}\left[\frac{J_1(x\sqrt{-i\Omega})}{J_1(\sqrt{-i\Omega})} \exp(i\Omega\tau) + \frac{J_1(x\sqrt{i\Omega})}{J_1(\sqrt{i\Omega})} \exp(-i\Omega\tau)\right]$$

$$+ B_0 \sum_{n=1}^{\infty} \frac{2\lambda_{1n}^3 J_1(\lambda_{1n}x)}{(\lambda_{1n}^4 + \Omega^2)J_0(\lambda_{1n})} \exp(-\lambda_{1n}^2 \tau) \ . \tag{6.291}$$

Again, the exponentially decaying terms are insignificant for large times. This gives the "steady state" for which one obtains

$$B_\varphi(x, \tau) = \frac{B_0}{2}\left[\frac{J_1(x\sqrt{-i\Omega})}{J_1(\sqrt{-i\Omega})} \exp(i\Omega\tau) + \frac{J_1(x\sqrt{i\Omega})}{J_1(\sqrt{i\Omega})} \exp(-i\Omega\tau)\right] \ . \tag{6.292}$$

Particularly for $x = 1$, $(r = r_0)$, in harmony with the boundary condition, the field becomes

$$B_\varphi(1, \tau) = \frac{B_0}{2}[\exp(i\Omega\tau) + \exp(-i\Omega\tau)] = B_0\cos(\Omega\tau) \ .$$

The functions $J_1(xi\sqrt{\mp i\Omega})$ are complex valued, however, one may split them into their real and imaginary parts. These functions have such great importance that new functions and names were introduced to address them, the *Kelvin functions* with "*ber*" (Bessel real part) and "*bei*" (Bessel imaginary part)

$$J_\nu(x\sqrt{\mp i\Omega}) = J_\nu(x\sqrt{\Omega}\sqrt{\mp i}) = \mathrm{ber}_\nu(x\sqrt{\Omega}) \pm i\,\mathrm{bei}_\nu(x\sqrt{\Omega}) \ . \tag{6.293}$$

Kelvin functions allow one to write $B_\varphi(x, \tau)$ as of (6.291) or (6.292) in a real form. We will leave the details for the reader's further study. Here, we will suffice with a short discussion of two limits (very low and very high frequencies).

1) The limit of very low Frequencies ($\Omega \ll 1$)

If $\Omega \ll 1$, then the magnitude of all arguments of the Bessel functions in (6.292) are much smaller than 1 because x is in the interval $0 \le x < 1$. Consequently

$$\frac{J_1(x\sqrt{\pm i\Omega})}{J_1(\sqrt{\pm i\Omega})} \approx x \ , \tag{6.294}$$

and therefore

$$B_\varphi(x, \tau) \approx \frac{B_0}{2}[x\exp(i\Omega\tau) + x\exp(-i\Omega\tau)] = B_0 x \cos(\Omega\tau) \ . \tag{6.295}$$

This should come as no surprise. The field is in phase everywhere because we assumed that the frequency of the field is small and its corresponding period is consequently large, that is, it is large against the time it takes to penetrate the cylinder. From

$$\Omega = \omega\mu\kappa r_0^2 \ll 1 \tag{6.296}$$

follows that

$$\frac{1}{\omega} \gg \mu\kappa r_0^2 \ . \tag{6.297}$$

Note eq. (6.137) with $l = r_0$. The amplitude increases linearly with x (*i.e.*, linear with the radius). Therefore, the current density inside the cylinder is spatially constant for all times. However, it oscillates over time with the period $\cos(\Omega\tau)$. In principle, this is the behavior of a direct current, which changes only slowly (slowly compared to the typical penetration depths, $\mu\kappa r_0^2$).

2) The limit of very high frequencies ($\Omega \gg 1$)

A zero order approximation for very large arguments is this:

$$\mathrm{ber}_1(x\sqrt{\Omega}) \approx \left[\frac{\exp\left(x\sqrt{\frac{\Omega}{2}}\right)\cos\left(x\sqrt{\frac{\Omega}{2}} + \frac{3\pi}{8}\right)}{\sqrt{2\pi x\sqrt{\Omega}}}\right] \tag{6.298}$$

$$\mathrm{bei}_1(x\sqrt{\Omega}) \approx \left[\frac{\exp\left(x\sqrt{\frac{\Omega}{2}}\right)\sin\left(x\sqrt{\frac{\Omega}{2}} + \frac{3\pi}{8}\right)}{\sqrt{2\pi x\sqrt{\Omega}}}\right] \ . \tag{6.299}$$

Using this in (6.292) and (6.293) gives

$$B_\varphi(x, \tau) \approx \frac{B_0 \exp\left[-(1-x)\sqrt{\frac{\Omega}{2}}\right] \cos\left(\Omega\tau-(1-x)\sqrt{\frac{\Omega}{2}}\right)}{\sqrt{x}}. \tag{6.300}$$

The reader is cautioned that this expression is valid only for $x\sqrt{\Omega} \gg 1$ (that is, even for large frequencies, its validity is restricted to x being not too small, *i.e.*, not too close to the axis). Take eq. (6.300), close to the surface where $r \approx r_0$ or $x \approx 1$, there one observes that the field behaves just like the planar case. For this purpose, compare the current result (6.300) with that of the half-space (6.139) and notice that there, the distance ξ from the surface of the half space corresponds here to the distance from the cylinder surface $r_0 - r$, or dimensionless: $(r_0 - r)/r_0 = (1 - r/r_0) = 1 - x$. From

$$\omega\mu\kappa r_0^2 = \Omega \gg 1 \tag{6.301}$$

follows that

$$r_0 \gg \frac{1}{\sqrt{\omega\mu\kappa}}, \tag{6.302}$$

that is, the penetration depth is very small versus the cylinder radius. Also plausible is, that under those circumstances the diffusion occurs as in the plane case. This is true, not only for cylinders, but for all kinds of shapes, as long as the frequency is large enough, or the interest is only in sufficiently thin penetration depths. From this perspective, the result (6.139) is of rather general significance.

Summarizing, we may conclude that the case of very low frequencies can be reduced to the case of direct currents, while the case of high frequencies can be reduced to a plane diffusion problem. For intermediate frequencies, there is no easy approximation. However, the behavior is qualitatively (not quantitatively) similar to the plane case that we have studied in Sect. 6.5.4: The wave is damped while penetrating the medium and it exhibits a phase difference.

6.8   Limits of the Quasi Stationary Theory

The quasi stationary theory is an approximation which is based on neglecting the displacement current in Maxwell's equations. We have mentioned already that all phenomena related to electromagnetic waves are neglected. It may sound paradox that we have encountered wave behavior in processes like the skin effect. Notice however, that these processes are enforced by the boundary conditions and are unrelated to electromagnetic waves which we will discuss later.

A typical behavior of propagation of waves is that this occurs with a certain finite velocity. We will discuss this in detail in the next chapter. Furthermore, the fundamental postulate of relativity, and thereby for the entire natural science altogether, is that there is no signal velocity higher than the speed of light in vacuum, $c_0 \approx 3 \cdot 10^8 ms^{-1}$. Consider *for example*, a field of the kind as discussed in Sect. 6.4, initially shaped like a δ-function, propagating in the infinite space.

Eq. (6.69) describes such a field. This equation indicates something very peculiar. At the time $\tau = 0$, the field exists only at one particular place, however, it is there infinitely large. Then the field already penetrates the entire space after an arbitrarily small period of time τ, even though it is very small at large distances. This seems to suggest that the signal velocity is infinitely large. Indeed, given sufficiently sensitive instruments, it should be possible to detect a field at very large distances after an extremely short time period, and use this to transmit signals, at least in principle, if these fields actually existed. Nonetheless, such is not the case. The infinite signal speed is typical for "diffusion processes", *i.e.*, for those processes described by the diffusion equation. However, physically they are not real. Formally – as we find – this results from neglecting the displacement current, which is equivalent to the assumption of infinite signal propagation velocity.

Another example we have dealt with shall be mentioned: The conductive half-space with no initial field. When a constant field B_0 is suddenly applied to the surface at the time $\tau = 0$, then according to eq. (6.117) or Fig. 6.20, this field penetrates the entire half-space after an arbitrarily short time period.

All this should not be interpreted in a manner, rendering all these fields as useless or totally wrong. Under appropriate preconditions, the quasi stationary theory provides an excellent approximation of the actual field behavior. The enormous velocity of the speed of light is thereby fundamental. The fields propagate with this speed, and the far away areas which have not been reached by the field are for many problems immaterial. Besides, even though these faraway fields described by the quasi stationary theory do not vanish, nevertheless, they are oftentimes so small that they are deemed to be insignificant. The qualitative conclusion of these remarks is that the quasi stationary theory is a useful approximation only for sufficiently large times or sufficiently slow or low frequency processes. To arrive at a quantitative statement, we return to the example of the field diffusion in the infinite space. In contrast to eq. (6.69), the field has to vanish if

$$(x'-x)^2 = l^2(\xi'-\xi)^2 > c^2 t^2 = \frac{t^2}{\mu\varepsilon}.$$

In practice, this is insignificant if

$$\exp\left[-\frac{(\xi'-\xi)^2}{4\tau}\right] = \exp\left[-\frac{(x'-x)^2}{4t}\mu\kappa\right] \ll 1 \qquad \text{for} \quad (x'-x)^2 = \frac{t^2}{\mu\varepsilon}$$

that is, if

$$(x'-x)^2\frac{\mu\kappa}{t} \gg 1 \qquad \text{for} \quad (x'-x)^2 = \frac{t^2}{\mu\varepsilon}.$$

Therefore, it must be:

$$\frac{\mu\kappa}{\mu\varepsilon}t \gg 1$$

$$t \gg \frac{\varepsilon}{\kappa} = t_r .$$

The time t_r is the relaxation time, discussed in Sect. 4.2. For the case of the skin effect, a necessary requirement is that the frequency is

$$\omega = \frac{2\pi}{t} \ll \frac{1}{t_r} = \frac{\kappa}{\varepsilon} .$$

These prerequisites and the related approximations can be dropped when considering the displacement current, as we will do in Chapter 7. Specifically in Sect. 7.12, we will return to the currently discussed problems and solve some of which exactly, from the perspective of the wave theory. The limits of the quasi stationary theory will become more apparent in this context. Then, we will also see, how the solutions of the quasi stationary approximation transform into the wave theory and vice versa.

7 Time Dependent Problems II (Electromagnetic Waves)

7.1 Wave Equations and their simplest Solutions

7.1.1 The Wave Equations

Now we consider Maxwell's equations in their complete form. Assume that the medium in which we will solve them is uniform, *i.e.*, ε, μ, κ are constant over all space.

$$\nabla \times \mathbf{H} = \mathbf{g} + \frac{\partial \mathbf{D}}{\partial t} \tag{7.1}$$

$$\nabla \times \mathbf{E} = -\frac{\partial \mathbf{B}}{\partial t} \tag{7.2}$$

$$\nabla \bullet \mathbf{B} = 0 \tag{7.3}$$

$$\nabla \bullet \mathbf{D} = \rho \tag{7.4}$$

$$\mathbf{D} = \varepsilon \mathbf{E} \tag{7.5}$$

$$\mathbf{g} = \kappa \mathbf{E} \tag{7.6}$$

$$\mathbf{B} = \mu \mathbf{H} \tag{7.7}$$

Taking the curl of (7.2) and substitute by use of (7.7), (7.1), (7.6), and (7.5) yields

$$\nabla \times (\nabla \times \mathbf{E}) = -\frac{\partial}{\partial t}(\nabla \times \mathbf{B}) = -\mu \frac{\partial}{\partial t}(\nabla \times \mathbf{H}) = -\mu \frac{\partial}{\partial t}(\mathbf{g} + \frac{\partial \mathbf{D}}{\partial t})$$

$$= -\mu \frac{\partial}{\partial t}(\kappa \mathbf{E} + \frac{\partial}{\partial t}\varepsilon \mathbf{E}) = -\mu \kappa \frac{\partial}{\partial t}\mathbf{E} - \mu \varepsilon \frac{\partial^2 \mathbf{E}}{\partial t^2}$$

Making use of the vector identity

$$\nabla \times (\nabla \times \mathbf{A}) = \nabla(\nabla \bullet \mathbf{A}) - \Delta \mathbf{A} = \nabla(\nabla \bullet \mathbf{A}) - \nabla^2 \mathbf{A}$$

gives

$$\boxed{\nabla^2 \mathbf{E} - \nabla(\nabla \bullet \mathbf{E}) = \mu \kappa \frac{\partial \mathbf{E}}{\partial t} + \mu \varepsilon \frac{\partial^2 \mathbf{E}}{\partial t^2}} \; . \tag{7.8}$$

On the other hand, taking the curl of (7.1) and substitute by use of (7.6), (7.5), (7.2), and (7.7) yields

G. Lehner, *Electromagnetic Field Theory for Engineers and Physicists*,
DOI 10.1007/978-3-540-76306-2_7, © Springer-Verlag Berlin Heidelberg 2010

$$\nabla\times(\nabla\times\mathbf{H}) = \nabla\times\mathbf{g} + \frac{\partial}{\partial t}(\nabla\times\mathbf{D}) = \kappa\nabla\times\mathbf{E} + \varepsilon\frac{\partial}{\partial t}(\nabla\times\mathbf{E})$$

$$= -\kappa\frac{\partial\mathbf{B}}{\partial t} - \varepsilon\frac{\partial^2\mathbf{B}}{\partial t^2} = \frac{1}{\mu}\nabla\times(\nabla\times\mathbf{B})$$

$$= \frac{1}{\mu}(\nabla(\nabla\bullet\mathbf{B}) - \nabla^2\mathbf{B})$$

and with (7.3) obtain

$$\boxed{\nabla^2\mathbf{B} = \mu\kappa\frac{\partial\mathbf{B}}{\partial t} + \mu\varepsilon\frac{\partial^2\mathbf{B}}{\partial t^2}} \;. \qquad (7.9)$$

The two equations (7.8) and (7.9) represent the so-called wave equations for **E** and **B** in their most general form for uniform media.

Inside a charge-free and non-conducting dielectric, vacuum (or free space) in particular, we have

$$\rho = 0 \;,$$

$$\mathbf{g} = 0 \;,$$

$$\kappa = 0 \;,$$

giving rise to the more specific wave equations

$$\boxed{\nabla^2\mathbf{E} = \mu\varepsilon\frac{\partial^2\mathbf{E}}{\partial t^2}} \qquad (7.10)$$

$$\boxed{\nabla^2\mathbf{B} = \mu\varepsilon\frac{\partial^2\mathbf{B}}{\partial t^2}} \;. \qquad (7.11)$$

7.1.2 The simplest Case: Plane Wave in an Insulator

Initially, only the simplest solutions of the wave equations (7.10) and (7.11) shall be examined. It shall be assumed that **E** and **B** depend on only one of the three Cartesian coordinates *for example*, z, but are time dependent.

$$\mathbf{E} = \mathbf{E}(z, t) = \langle E_x(z, t), E_y(z, t), E_z(z, t)\rangle \qquad (7.12)$$

$$\mathbf{B} = \mathbf{B}(z, t) = \langle B_x(z, t), B_y(z, t), B_z(z, t)\rangle \qquad (7.13)$$

Both fields have to be source free since $\rho = 0$, *i.e.*,

$$\nabla\bullet\mathbf{E} = \frac{\partial}{\partial z}E_z(z, t) = 0 \qquad (7.14)$$

$$\nabla\bullet\mathbf{B} = \frac{\partial}{\partial z}B_z(z, t) = 0 \;. \qquad (7.15)$$

It follows that

$$E_z = E_z(t) \qquad (7.16)$$

$$B_z = B_z(t) . \tag{7.17}$$

Later, we will find that neither E_z nor B_z may depend on t. It is possible that the space contains a field E_z or B_z, neither of which depend on time or space, and thus, do not interest us. We might as well assume that

$$E_z = 0 , \tag{7.18}$$

$$B_z = 0 . \tag{7.19}$$

Fields which, when a coordinate system is chosen properly, depend only on one Cartesian coordinate and time are called *plane waves*. Therefore, we note that plane waves have no field components in propagation direction (here the z-direction), that is, they are inevitably *transverse waves*. This results from above assumption, namely the absence of any volume charges. In the presence of volume charges, it is entirely possible to have plane waves with field components in propagation direction, the so-called *longitudinal waves*. The so-called "plasma waves", which are important in plasmas and solid state physics, are of this kind. We will limit our discussion to transverse waves. In this case, one only needs to consider the transverse field components E_x, E_y, B_x, and B_y:

$$\frac{\partial^2 E_x}{\partial z^2} = \varepsilon\mu\frac{\partial^2 E_x}{\partial t^2} \tag{7.20}$$

$$\frac{\partial^2 E_y}{\partial z^2} = \varepsilon\mu\frac{\partial^2 E_y}{\partial t^2} . \tag{7.21}$$

Almost obvious is that arbitrary functions f_x and g_x satisfy the associated wave equation

$$\boxed{E_x = f_x(z-ct) + g_x(z+ct)} . \tag{7.22}$$

This is *d'Alembert's solution* to the wave equation, where

$$c = \frac{1}{\sqrt{\varepsilon\mu}} \tag{7.23}$$

is the speed of light in the observed medium.

The proof is simple. Starting with the derivative

$$\frac{\partial E_x}{\partial z} = f'_x + g'_x ,$$

$$\frac{\partial^2 E_x}{\partial z^2} = f''_x + g''_x ,$$

and

$$\frac{\partial E_x}{\partial t} = -cf'_x + cg'_x$$

$$\frac{\partial^2 E_x}{\partial t^2} = c^2 f''_x + c^2 g''_x = c^2 \frac{\partial^2 E_x}{\partial z^2} = \frac{1}{\varepsilon\mu}\frac{\partial^2 E_x}{\partial z^2},$$

from which is obvious that this satisfies (7.20). This applies similarly to E_y and the components of $\mathbf{B}$ (B_x and B_y). On the other hand, the components of $\mathbf{B}$ and $\mathbf{E}$ are not independent of each other. From (7.2) follows that

$$\nabla\times\mathbf{E} = \begin{vmatrix} \mathbf{e}_x & \mathbf{e}_y & \mathbf{e}_z \\ 0 & 0 & \frac{\partial}{\partial z} \\ E_x & E_y & E_z \end{vmatrix} = \langle -\frac{\partial E_y}{\partial z}, \frac{\partial E_x}{\partial z}, 0\rangle = -\langle \frac{\partial B_x}{\partial t}, \frac{\partial B_y}{\partial t}, \frac{\partial B_z}{\partial t}\rangle.$$

A consequence thereof is that B_z has to be constant in time. Anticipating this, we have used this fact already above. Furthermore

$$\frac{\partial B_x}{\partial t} = \frac{\partial E_y}{\partial z} = f'_y(z-ct) + g'_y(z+ct)$$

$$\frac{\partial B_y}{\partial t} = -\frac{\partial E_x}{\partial z} = -f'_x(z-ct) - g'_x(z+ct)$$

or when integrating over the time

$$B_x = -\frac{1}{c}f_y(z-ct) + \frac{1}{c}g_y(z+ct) + F_x(z) \tag{7.24}$$

$$B_y = \frac{1}{c}f_x(z-ct) - \frac{1}{c}g_x(z+ct) + F_y(z). \tag{7.25}$$

On the other hand, using (7.1) for an insulator with $\mathbf{g} = 0$, we get

$$\nabla\times\mathbf{H} = \begin{vmatrix} \mathbf{e}_x & \mathbf{e}_y & \mathbf{e}_z \\ 0 & 0 & \frac{\partial}{\partial z} \\ H_x & H_y & H_z \end{vmatrix} = \langle -\frac{\partial H_y}{\partial z}, \frac{\partial H_x}{\partial z}, 0\rangle = \langle \frac{\partial D_x}{\partial t}, \frac{\partial D_y}{\partial t}, \frac{\partial D_z}{\partial t}\rangle.$$

This means, as claimed before, that D_z and thus E_z are also independent of t. Moreover

$$\frac{\partial B_x}{\partial z} = \varepsilon\mu\frac{\partial E_y}{\partial t} = \frac{1}{c^2}[-c f'_y(z-ct) + cg'_y(z+ct)]$$

$$= -\frac{1}{c}f'_y(z-ct) + \frac{1}{c}g'_y(z+ct)$$

$$\frac{\partial B_y}{\partial z} = -\varepsilon\mu\frac{\partial E_x}{\partial t} = -\frac{1}{c^2}[-c f'_x(z-ct) + cg'_x(z+ct)]$$

$$= \frac{1}{c}f'_x(z-ct) - \frac{1}{c}g'_x(z+ct).$$

Integrating for z gives

$$B_x = -\frac{1}{c} f_y(z - ct) + \frac{1}{c} g_y(z + ct) + G_x(t) \tag{7.26}$$

$$B_y = \frac{1}{c} f_x(z - ct) - \frac{1}{c} g_x(z + ct) + G_y(t) \,. \tag{7.27}$$

Comparing eqs. (7.24) through (7.27) with each other provides a condition for the integration constants, which is

$$\left.\begin{aligned} F_x(z) &= G_x(t) \\ F_y(z) &= G_y(t) \end{aligned}\right\} \,. \tag{7.28}$$

Consequently, these may not depend on either z or t, *i.e.*, they have to be constant in space and time. Disregarding such constant fields, the field of a plane wave has to be:

$$\left.\begin{aligned} E_x &= f_x(z - ct) + g_x(z + ct) \\ E_y &= f_y(z - ct) + g_y(z + ct) \\ E_z &= 0 \\ B_x &= -\frac{1}{c} f_y(z - ct) + \frac{1}{c} g_y(z + ct) \\ B_y &= \frac{1}{c} f_x(z - ct) - \frac{1}{c} g_x(z + ct) \\ B_z &= 0 \end{aligned}\right\} \,. \tag{7.29}$$

This plane wave has two parts, one which moves in positive z-direction without changing its shape:

$$\left.\begin{aligned} E_x &= f_x(z - ct) \\ E_y &= f_y(z - ct) \\ E_z &= 0 \\ B_x &= -\frac{1}{c} f_y(z - ct) \\ B_y &= \frac{1}{c} f_x(z - ct) \\ B_z &= 0 \end{aligned}\right\} \,. \tag{7.30}$$

Its propagation direction is

$$\mathbf{e}_a = \langle 0, 0, 1 \rangle$$

Expressed in a different form, one writes

$$\boxed{\mathbf{B} = \frac{\mathbf{e}_a \times \mathbf{E}}{c}} \tag{7.31}$$

or

$$H = \frac{e_a \times E}{\mu c} = \frac{e_a \times E}{\sqrt{\dfrac{\mu}{\varepsilon}}} = \frac{e_a \times E}{Z}, \tag{7.32}$$

where

$$Z = \sqrt{\frac{\mu}{\varepsilon}} \tag{7.33}$$

is the so-called *characteristic impedance* of the medium. The other part propagates also without changing its shape, but in the other direction, that of the negative z-axis:

$$
\left.
\begin{aligned}
E_x &= g_x(z + ct) \\
E_y &= g_y(z + ct) \\
E_z &= 0 \\
B_x &= \frac{1}{c}\, g_y(z + ct) \\
B_y &= -\frac{1}{c}\, g_x(z + ct) \\
B_z &= 0
\end{aligned}
\right\} .
$$

Writing

$$e_a = \langle 0, 0, -1 \rangle$$

allows to write this in the form of eqs. (7.31) through (7.33). These equations are generally applicable for any plane wave. Since they are in vector form, they are independent of a coordinate system. In a rotated coordinate system, the propagation direction would not be along the z-axis anymore, but eqs. (7.31) through (7.33) would still apply. Conversely, it is also true that

$$E = c(B \times e_a) = Z(H \times e_a) \tag{7.34}$$

i.e., the three vectors E, B (or H), and e_a (in this order) form a right handed system. To show this, one multiplies (7.31) in a vector product with e_a to obtain

$$B \times e_a = \frac{e_a \times E}{c} \times e_a = -e_a \frac{(e_a \cdot E)}{c} + \frac{E(e_a \cdot e_a)}{c} .$$

Applying

$$e_a \cdot E = 0$$

and

$$e_a \cdot e_a = 1 ,$$

completes the proof.

The characteristic impedance for vacuum is

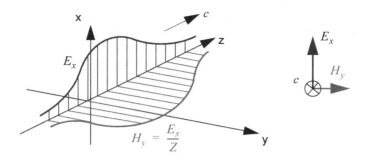

Fig. 7.1

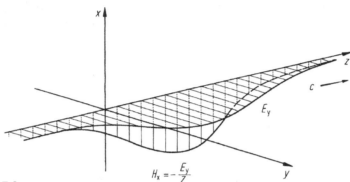

Fig. 7.2

$$Z = Z_0 = \sqrt{\frac{\mu_0}{\varepsilon_0}} = \sqrt{\frac{4\pi \cdot 10^{-7}\dfrac{Vs}{Am}}{8.855 \cdot 10^{-12}\dfrac{As}{Vm}}} \approx 377\Omega \approx 120\pi\Omega \ . \tag{7.35}$$

If $E_x = 0$ in eq. (7.29) (and therefore also $B_y = 0$) or if $E_y = 0$ (and therefore also $B_x = 0$), then the electromagnetic field oscillates only in a plane. These waves are called *linearly polarized*. A general plane wave can be constructed from two waves which are polarized perpendicular to each other. The result is then just the wave described in (7.29). Fig. 7.1 and Fig. 7.2 illustrate examples of both types of linearly polarized waves.

7.1.3 Harmonic Plane Waves

The waves of the kind depicted in Fig. 7.1 and Fig. 7.2 are also referred to as "wave trains" or "wave packets". They can be formed by superposition of appropriate sets of harmonic waves with specific wave lengths. The underlying formal concept is the possibility to represent any function as a Fourier integral, $i.e.$, by superposition of so-called harmonic waves. For instance

$$\left.\begin{array}{l} E_x(z, t) = E_{x0}\cos(\omega t - kz + \varphi) \\ H_y(z, t) = H_{y0}\cos(\omega t - kz + \varphi) \\ H_{y0} = \dfrac{E_{x0}}{Z} \end{array}\right\}.$$ (7.36)

This wave is sketched in Fig. 7.3.

Here, E_{x0} and H_{y0} are the amplitudes of the fields; φ is a phase angle which depends on where the origin of the chosen coordinate system is, and the choice of when the time is zero. ω is the angular frequency of the wave and k is the wave number. If f is its frequency, τ its period, and λ its wave length, then we may write

$$\omega = 2\pi f = \frac{2\pi}{\tau}$$ (7.37)

$$k = \frac{2\pi}{\lambda}.$$ (7.38)

The "phase velocity" of the wave is

$$c = \frac{1}{\sqrt{\varepsilon\mu}} = \lambda f = \frac{2\pi}{k} \cdot \frac{\omega}{2\pi} = \frac{\omega}{k}$$ (7.39)

This is a result of the wave equation (7.10) when using it in (7.36). We obtain

$$-k^2 E_x = -\frac{1}{c^2}\omega^2 E_x ,$$

which, again, gives (7.39). Moreover, the wave expressed in (7.36) can be regarded as a special case of (7.22). The reason is that

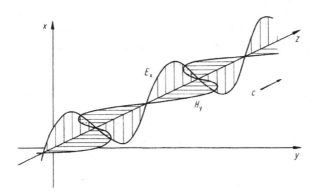

Fig. 7.3

$$\cos(\omega t - kz + \varphi) = \cos\left[\varphi - k\left(z - \frac{\omega}{k}t\right)\right] = \cos[\varphi - k(z - ct)] ,$$

i.e., a function of $(z - ct)$.

From (7.39) one obtains

$$\boxed{\omega = ck} . \tag{7.40}$$

The relation between ω and k is called the *dispersion relation*, here for the case of a plane electromagnetic wave in an ideal insulator. In other circumstances the relation between ω and k could be different, *i.e.*, the dispersion relation could be of the form:

$$\boxed{\omega = \omega(k)} . \tag{7.41}$$

In this general case, we have to associate different velocities with the wave. The velocity with which the phase propagates is still the phase velocity. The phase

$$\omega(k)t - kz + \varphi$$

remains constant for

$$z = z_0 + \frac{\omega(k)}{k}t .$$

The constant phase is then

$$\omega(k)t - kz_0 - \omega(k)t + \varphi = \varphi - kz_0 .$$

Therefore, the phase velocity in general is

$$\boxed{v_{ph} = \frac{\omega(k)}{k}} . \tag{7.42}$$

Dispersion accounts for the fact that the phase velocity according to (7.42) may be a function of the frequency (wave length). In the special case when the dispersion relation is of the form given in (7.40), then the phase velocity is the same for all frequencies (wave lengths), i.e., this is the dispersion free case.

In addition, the so-called *group velocity* v_G is also of great importance. It is defined by

$$\boxed{v_G = \frac{d}{dk}\omega(k)} . \tag{7.43}$$

For the specific case of the dispersion relation as of (7.40), both velocities coincide, both are equal to c:

$$v_{ph} = \frac{\omega}{k} = c = \frac{d\omega}{dk} = v_G . \tag{7.44}$$

In later sections, we will find dispersion relations for which this is not true. The significant quantity in these cases is not the phase velocity, but the group velocity, which describes the transmission of signals or the energy transfer. It relates to a group of waves (a wave packet), which is composed of individual waves having different wave lengths. In the case of (7.40), all individual waves travel with the same phase velocity $v_{ph} = c$. Under these circumstances, the wave packet

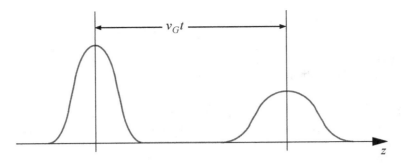

Fig. 7.4

maintains its shape despite of its propagation, which is also an immediate result of the wave equation, and which manifests itself in D'Alembert's solution. The subject becomes much more complicated when different parts of the wave propagate with different phase velocities. Then, general statements about the behavior of a wave packet are no longer possible. What is probable in this case is that its shape changes significantly over time. As a consequence, it may not even be possible to describe its motion with a *single* velocity. However, certain statements for wave packets of a narrow frequency band are possible. Narrow frequency band shall mean that the frequencies within the wave packet fall into a small frequency interval ($\omega, \omega + \Delta\omega$) where $\Delta\omega \ll \omega$. Then, the maximum of the wave packet travels with the velocity v_G (Fig. 7.4). The shape of the packet changes as it moves. To transmit a signal by means of waves requires a wave packet, and their signal velocity is, as already mentioned, the group velocity. A word of caution is here appropriate. $d\omega/dk$ may not always be interpreted as the signal velocity.

The harmonic plane waves are of fundamental theoretical importance because every possible wave can be composed of them by superposition. We will discuss a few examples below.

A plane harmonic wave may propagate in an arbitrary direction of the space. The propagation direction is typically described by the wave vector (propagation vector, wave number vector) $\mathbf{k}$. Its direction is the propagating direction and its magnitude is like before

$$|\mathbf{k}| = k = \frac{2\pi}{\lambda} \,. \tag{7.45}$$

This defines a plane wave in the following way:

$$\mathbf{E} = \mathbf{E}_0\cos(\omega t - \mathbf{k}\bullet\mathbf{r} + \varphi) \qquad \text{where} \quad \mathbf{E}_0\bullet\mathbf{k} = 0 \,. \tag{7.46}$$

The phase is thereby constant, if for a fixed point in time

$$\mathbf{k}\bullet\mathbf{r} = \text{const} \,. \tag{7.47}$$

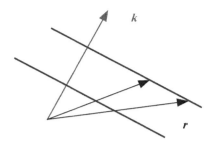

Fig. 7.5

These are the equations of planes to which **k** is perpendicular (Fig. 7.5). These waves are, indeed, plane waves which propagate in the direction defined by **k**.

7.1.4   Elliptic Polarization

Consider superimposing two linearly polarized waves of the same frequency, propagating in z-direction, but with different amplitude and a phase difference of φ.

$$E_x = E_{x0}\cos(\omega t - kz) \tag{7.48}$$

$$E_y = E_{y0}\cos(\omega t - kz + \varphi)$$

$$= E_{y0}[\cos(\omega t - kz)\cos\varphi - \sin(\omega t - kz)\sin\varphi] \ . \tag{7.49}$$

After eliminating $\cos(\omega t - kz)$ and $\sin(\omega t - kz)$ by means of these relations:

$$\cos(\omega t - kz) = \frac{E_x}{E_{x0}} \quad \text{and} \quad \sin(\omega t - kz) = \sqrt{1 - \left(\frac{E_x}{E_{x0}}\right)^2} \ ,$$

we obtain

$$\frac{E_x^2}{E_{x0}^2} + \frac{E_y^2}{E_{y0}^2} - \frac{2E_xE_y}{E_{x0}E_{y0}}\cos\varphi = \sin^2\varphi \ . \tag{7.50}$$

This is the equation for an ellipse, when regarding it as an equation for E_x, E_y. This means that in the planes $z = $ const., the tips of the **E** vector describe an elliptic path. These waves are therefore appropriately called *elliptically polarized* (or simply elliptic waves). Particularly for

$$\varphi = \frac{\pi}{2} \ , \tag{7.51}$$

or even more general, for

$$\varphi = \frac{2n + 1}{2}\pi \ , \tag{7.52}$$

we obtain

$$\frac{E_x^2}{E_{x0}^2} + \frac{E_y^2}{E_{y0}^2} = 1 \ . \tag{7.53}$$

This is the equation of an ellipse with its principal axis parallel the x- and y-axis. Further specializing, so that

$$E_{x0} = E_{y0} = E_0 \tag{7.54}$$

yields

$$E_x^2 + E_y^2 = E_0^2 \ . \tag{7.55}$$

that is the equation of a circle. This represents a *circularly polarized wave*. For $\varphi = n\pi$ one obtains a *linearly polarized wave*.

7.1.5   Standing Waves

Two waves with the same amplitude, wave length, and polarization travelling in opposite directions give

$$\left. \begin{aligned} E_x(z, t) &= E_{x0}\cos(\omega t - kz + \varphi_1) + E_{x0}\cos(\omega t + kz + \varphi_2) \\ H_y(z, t) &= \frac{E_{x0}}{Z}\cos(\omega t - kz + \varphi_1) - \frac{E_{x0}}{Z}\cos(\omega t + kz + \varphi_2) \end{aligned} \right\} \ . \tag{7.56}$$

Because of the trigonometric identities

$$\left. \begin{aligned} \cos\alpha + \cos\beta &= 2\cos\left(\frac{\alpha + \beta}{2}\right)\cos\left(\frac{\alpha - \beta}{2}\right) \\ \cos\alpha - \cos\beta &= -2\sin\left(\frac{\alpha + \beta}{2}\right)\sin\left(\frac{\alpha - \beta}{2}\right) \end{aligned} \right\} , \tag{7.57}$$

it follows that

$$\left. \begin{aligned} E_x(z, t) &= 2E_{x0}\cos\left(\omega t + \frac{\varphi_1 + \varphi_2}{2}\right)\cos\left(-kz + \frac{\varphi_1 - \varphi_2}{2}\right) \\ H_y(z, t) &= -\frac{2E_{x0}}{Z}\sin\left(\omega t + \frac{\varphi_1 + \varphi_2}{2}\right)\sin\left(-kz + \frac{\varphi_1 - \varphi_2}{2}\right) \end{aligned} \right\} \ . \tag{7.58}$$

This represents a *standing wave*, which, so to speak, oscillates in place (Fig. 7.6). With the appropriate choice of z and t, it is possible to let $\varphi_1 = \varphi_2 = 0$. Then

$$\left. \begin{aligned} E_x(z, t) &= 2E_{x0}\cos kz \cos \omega t \\ H_y(z, t) &= \frac{2E_{x0}}{Z}\sin kz \sin \omega t \end{aligned} \right\} \ . \tag{7.59}$$

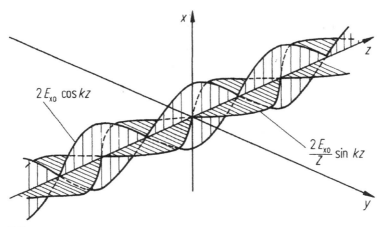

2 E_{x0} cos kz

$\dfrac{2 E_{x0}}{Z}$ sin kz

Fig. 7.6

Fig. 7.6 illustrates how the amplitudes of the oscillating field vary.
The zeros of the electric field E_x (which are called its *nodes* or *nodal points*) are located at

$$kz = \frac{2n+1}{2}\pi$$

$$z = \frac{2n+1}{2}\frac{\pi}{k} = \frac{2n+1}{4}\lambda$$

and thus

$$z = \pm\frac{1}{4}\lambda, \pm\frac{3}{4}\lambda, \pm\frac{5}{4}\lambda, \dots$$

The nodes of the magnetic field are at

$$kz = n\pi$$

$$z = n\frac{\pi}{k} = \frac{n}{2}\lambda$$

and thus

$$z = 0, \pm\frac{1}{2}\lambda, \pm\lambda, \pm\frac{3}{2}\lambda, \dots .$$

At the time $t = 0$, *for instance*, $H_y = 0$, while E_x assumes its maximum value. Conversely for $t = \tau/4$, it is $E_x = 0$, while H_y assumes its maximum value, etc.

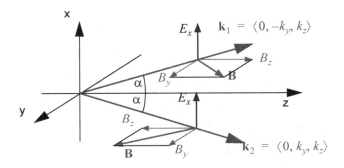

Fig. 7.7

7.1.6 TE- and TM Wave

As illustrated in Fig. 7.7, we now superimpose two plane waves with the same frequency, amplitude, and polarization, but with different propagation direction. It shall be for the first wave:

1)
$$\begin{cases} \mathbf{E}_1 = \mathbf{E}_0 \cos(\omega t - \mathbf{k}_1 \bullet \mathbf{r}) \\ \mathbf{B}_1 = \mathbf{B}_{01} \cos(\omega t - \mathbf{k}_1 \bullet \mathbf{r}) \end{cases}$$

where

$$\mathbf{k}_1 = \langle 0, -k_y, k_z \rangle$$

$$\mathbf{E}_0 = \langle E_{x0}, 0, 0 \rangle$$

$$\mathbf{B}_{01} = \frac{\mathbf{e}_{a1} \times \mathbf{E}_0}{c} = \frac{\mathbf{k}_1 \times \mathbf{E}_0}{k_1 c} = \frac{\mathbf{k}_1 \times \mathbf{E}_0}{\omega} = \frac{E_{x0}}{\omega} \langle 0, k_z, k_y \rangle$$

and for the second wave

2)
$$\begin{cases} \mathbf{E}_2 = \mathbf{E}_0 \cos(\omega t - \mathbf{k}_2 \bullet \mathbf{r}) \\ \mathbf{B}_2 = \mathbf{B}_{02} \cos(\omega t - \mathbf{k}_2 \bullet \mathbf{r}) \end{cases}$$

where

$$\mathbf{k}_2 = \langle 0, k_y, k_z \rangle$$

$$\mathbf{E}_0 = \langle E_{x0}, 0, 0 \rangle$$

$$\mathbf{B}_{02} = \frac{\mathbf{e}_{a2} \times \mathbf{E}_0}{c} = \frac{\mathbf{k}_2 \times \mathbf{E}_0}{\omega} = \frac{E_{x0}}{\omega} \langle 0, k_z, -k_y \rangle$$

When superposing these and using (7.57), we find

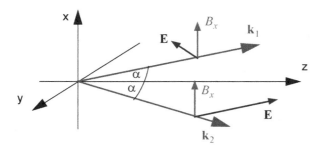

Fig. 7.8

$$\begin{aligned}
\mathbf{E} &= \mathbf{E}_0 \cos(\omega t + k_y y - k_z z) + \mathbf{E}_0 \cos(\omega t - k_y y - k_z z) \\
&= 2\mathbf{E}_0 \cos(\omega t - k_z z)\cos(k_y y)
\end{aligned}$$

and

$$\mathbf{B} = \mathbf{B}_{01}\cos(\omega t - \mathbf{k}_1 \bullet \mathbf{r}) + \mathbf{B}_{02}\cos(\omega t - \mathbf{k}_2 \bullet \mathbf{r})$$

or

$$\left.\begin{aligned}
E_x &= 2E_{x0}\cos(\omega t - k_z z)\cos(k_y y) \\
E_y &= 0 \\
E_z &= 0
\end{aligned}\right\} \tag{7.60}$$

and

$$\left.\begin{aligned}
B_x &= 0 \\
B_y &= 2E_{x0}\frac{k_z}{\omega}\cos(\omega t - k_z z)\cos(k_y y) \\
B_z &= -2E_{x0}\frac{k_y}{\omega}\sin(\omega t - k_z z)\sin(k_y y)
\end{aligned}\right\} \, . \tag{7.61}$$

The wave described by (7.60) and (7.61) is not a plane wave. It travels in z-direction. Its phase velocity is

$$v_{ph} = \frac{\omega}{k_z} \, . \tag{7.62}$$

Besides the transverse field components E_x and B_y, the magnetic field has also a longitudinal component B_z. Thus the field is transverse with respect to the electric field, but not with respect to the magnetic field. Such a wave is called *transverse electric wave* or abbreviated *TE wave* (also *H wave*). Its amplitudes are y-dependent.

One can treat the case illustrated in Fig. 7.8 in a similar manner. Now one superimposes the following two waves. First:

1) $\begin{cases} \mathbf{E}_1 = \mathbf{E}_{01}\cos(\omega t - \mathbf{k}_1 \bullet \mathbf{r}) \\ \mathbf{B}_1 = \mathbf{B}_0\cos(\omega t - \mathbf{k}_1 \bullet \mathbf{r}) \end{cases}$

where

$$\mathbf{k}_1 = \langle 0, -k_y, k_z \rangle$$
$$\mathbf{B}_0 = \langle B_{x0}, 0, 0 \rangle$$

$$\mathbf{E}_{01} = c\mathbf{B}_0 \times \mathbf{e}_a = c\mathbf{B}_0 \times \frac{\mathbf{k}_1}{k_1} = -\frac{cB_{x0}}{k_1}\langle 0, k_z, k_y \rangle$$

and for the second wave

2) $\begin{cases} \mathbf{E}_2 = \mathbf{E}_{02}\cos(\omega t - \mathbf{k}_2 \bullet \mathbf{r}) \\ \mathbf{B}_2 = \mathbf{B}_0\cos(\omega t - \mathbf{k}_2 \bullet \mathbf{r}) \end{cases}$

where

$$\mathbf{k}_2 = \langle 0, k_y, k_z \rangle$$
$$\mathbf{B}_0 = \langle B_{x0}, 0, 0 \rangle$$

$$\mathbf{E}_{02} = -\frac{cB_{x0}}{k_2}\langle 0, k_z, -k_y \rangle .$$

The superposition yields

$$\left. \begin{aligned} B_x &= 2B_{x0}\cos(\omega t - k_z z)\cos(k_y y) \\ B_y &= 0 \\ B_z &= 0 \end{aligned} \right\}$$ (7.63)

and

$$\left. \begin{aligned} E_x &= 0 \\ E_y &= -2cB_{x0}\frac{k_z}{k}\cos(\omega t - k_z z)\cos(k_y y) \\ E_z &= 2cB_{x0}\frac{k_y}{k}\sin(\omega t - k_z z)\sin(k_y y) \end{aligned} \right\} .$$ (7.64)

Where it was used that

$$k_1 = \sqrt{k_y^2 + k_z^2} = k_2 = k .$$

The wave described by (7.63) and (7.64) is, again, not a plane wave. It travels in the z-direction. It is transverse with respect to $\mathbf{B}$ but not with respect to $\mathbf{E}$. Such a wave is called a *transverse magnetic wave* or abbreviated *TM wave* (also *E wave*).

These waves are important in connection with wave guides to which we will return later. True for both kinds of waves (TE and TM) is that

$$v_{ph} = \frac{\omega}{k_z}$$

and

$$k^2 = k_y^2 + k_z^2 = \mu\varepsilon\omega^2 = \frac{\omega^2}{c^2} \ .$$

Therefore,

$$k_z = \sqrt{\frac{\omega^2}{c^2} - k_y^2} \tag{7.65}$$

and

$$v_{ph} = \frac{\omega}{\sqrt{\frac{\omega^2}{c^2} - k_y^2}} = \frac{c}{\sqrt{1 - \frac{k_y^2 c^2}{\omega^2}}} \geq c \ . \tag{7.66}$$

These waves are not dispersion free. Their phase velocity is higher than the speed of light for the respective medium, thus its phase velocity for vacuum is higher than the speed of light in vacuum. This is entirely possible and does not entail any contradiction to the theory of special relativity, according to which no signal velocity may exceed the speed of light in vacuum. The group velocity is what qualifies for the signal velocity, which is

$$v_G = \frac{d\omega}{dk_z} = \frac{1}{\frac{dk_z}{d\omega}} = c^2 \frac{\sqrt{\frac{\omega^2}{c^2} - k_y^2}}{\omega} = \frac{c^2}{v_{ph}} \ . \tag{7.67}$$

Consequently

$$v_G v_{ph} = c^2 \tag{7.68}$$

and

$$v_G \leq c \ , \tag{7.69}$$

as it must be. To obtain (7.68) directly from the dispersion relation by differentiating for k_z is also possible:

$$k_y^2 + k_z^2 = \mu\varepsilon\omega^2 = \frac{\omega^2}{c^2}$$

$$2k_z = \frac{1}{c^2} 2\omega \frac{d\omega}{dk_z}$$

$$\frac{\omega \, d\omega}{k_z \, dk_z} = v_G v_{ph} = c^2 \ .$$

In the limit $k_y = 0$, the waves (7.60), (7.61) as well as (7.63), (7.64) become plane waves. The longitudinal components of B_z and E_z vanish in the process.

The other limit, $k_z = 0$ results in standing waves of the kind discussed in Sect. 7.1.5.

7.1.7   Energy Density in and Energy Transfer by Waves

First, consider the plane wave

$$E_x = E_{x0}\cos(\omega t - kz) \ . \tag{7.70}$$

$$H_y = H_{y0}\cos(\omega t - kz) = \frac{E_{x0}}{Z}\cos(\omega t - kz) \ . \tag{7.71}$$

Then, the pointing vector discussed in Sect. 2.14 is according to (2.153)

$$\mathbf{S} = \mathbf{E} \times \mathbf{H} = \langle 0, 0, E_x H_y \rangle = \langle 0, 0, \frac{E_x^2}{Z} \rangle \ . \tag{7.72}$$

$\mathbf{S}$ is the energy flux density, *i.e.*, the electromagnetic energy transferred across an area per unit time and unit area. On the other hand, the energy density stored in the field of the wave is

$$\frac{\varepsilon E^2}{2} + \frac{\mu H^2}{2} = \frac{\varepsilon E_x^2}{2} + \frac{\mu E_x^2}{2Z^2} = \frac{\varepsilon E_x^2}{2} + \frac{\varepsilon E_x^2}{2} = \varepsilon E_x^2 \ . \tag{7.73}$$

Multiplying this with c, just yields S_z:

$$\varepsilon E_x^2 c = \varepsilon E_x^2 \frac{1}{\sqrt{\varepsilon\mu}} = \frac{E_x^2}{Z} \ . \tag{7.74}$$

We conclude that the energy stored in the field is transferred in z-direction with the speed of light.

The just discussed TE and TM waves are interesting examples. We limit our discussion to the TE wave described by the two equations (7.60) and (7.61). in this case we have

$$\mathbf{S} = \mathbf{E} \times \mathbf{H} = \langle 0, -E_x H_z, E_x H_y \rangle \ . \tag{7.75}$$

Observe that on average over time, there is no energy transfer in y-direction. However, there is an energy transfer in z-direction, as the time average of $E_x H_y$ does not vanish. The effective, time averaged energy transfer is

$$S_{z\,\text{eff}} = \frac{2E_{x0}^2 k_z \cos^2(k_y y)}{\mu\omega} \ . \tag{7.76}$$

Further averaging, now over the location dependency, we obtain

$$\overline{S_{z\,\text{eff}}} = \frac{E_{x0}^2 k_z}{\mu\omega} = \frac{\varepsilon E_{x0}^2 k_z}{\varepsilon\mu\omega} = \frac{\varepsilon E_{x0}^2 c^2 k_z}{\omega} = \varepsilon E_{x0}^2 v_G \ . \tag{7.77}$$

Furthermore, the spatial and temporal average of the energy density is

$$\frac{\varepsilon E_{x0}^2}{2} + \frac{E_{x0}^2(k_z^2 + k_y^2)}{2\mu\omega^2} = \frac{\varepsilon E_{x0}^2}{2} + \frac{E_{x0}^2}{2\mu c^2} = \varepsilon E_{x0}^2 \ . \tag{7.78}$$

Comparing the last two equations reveals that the available energy, averaged over time and over space, is transferred in the z-direction with the group velocity v_G (and not by the phase velocity v_{ph} which is in this case higher than the speed of light c).

Note: In case the fields are written in complex notation, then it is advisable to return to the real fields before calculating energy densities or the Poynting vector.

7.2 Plane Waves in a Conductive Medium

7.2.1 Wave Equations and Dispersion Relation

Now, we study a plane wave in a conductive medium which does not have any volume charges. Then with (7.8) and (7.9), we obtain

$$\nabla^2 \mathbf{E} = \mu\kappa \frac{\partial \mathbf{E}}{\partial t} + \mu\varepsilon \frac{\partial^2 \mathbf{E}}{\partial t^2} \tag{7.79}$$

$$\nabla^2 \mathbf{B} = \mu\kappa \frac{\partial \mathbf{B}}{\partial t} + \mu\varepsilon \frac{\partial^2 \mathbf{B}}{\partial t^2} \ . \tag{7.80}$$

For variation and to simplify, we take advantage of the complex notation which allows to describe the wave in the following form:

$$\mathbf{E} = \mathbf{E}_0 \exp[i(\omega t - \mathbf{k} \bullet \mathbf{r})] \tag{7.81}$$

$$\mathbf{B} = \mathbf{B}_0 \exp[i(\omega t - \mathbf{k} \bullet \mathbf{r})] \ . \tag{7.82}$$

The physically relevant fields are represented individually by either the real part or by the imaginary part thereof. Using this approach in either of the two eqs. (7.79) or (7.80) yields the same dispersion relation:

$$(-ik_x)^2 + (-ik_y)^2 + (-ik_z)^2 = \mu\kappa i\omega + \mu\varepsilon(i\omega)^2$$

or

$$\boxed{\mu\varepsilon\omega^2 - \mu\kappa i\omega - k^2 = 0} \ . \tag{7.83}$$

The operators of vector analysis can be expressed via multiplications when applying them to statements of the kind as expressed by (7.81) and (7.82). For instance, the divergence is

$$\nabla \bullet \mathbf{E} = -ik_x E_x - ik_y E_y - ik_z E_z = -i\mathbf{k} \bullet \mathbf{E} \ . \tag{7.84}$$

Furthermore

$$\nabla \times \mathbf{E} = \begin{vmatrix} \mathbf{e}_x & \mathbf{e}_y & \mathbf{e}_z \\ \dfrac{\partial}{\partial x} & \dfrac{\partial}{\partial y} & \dfrac{\partial}{\partial z} \\ E_x & E_y & E_z \end{vmatrix} = \begin{vmatrix} \mathbf{e}_x & \mathbf{e}_y & \mathbf{e}_z \\ -ik_x & -ik_y & -ik_z \\ E_x & E_y & E_z \end{vmatrix} = -i\mathbf{k} \times \mathbf{E} \ . \tag{7.85}$$

Both statements may be formally expressed by

$$\boxed{\nabla = -i\mathbf{k}} \ . \tag{7.86}$$

From it follows immediately

$$\operatorname{div}\mathbf{E} = \nabla\bullet\mathbf{E} = -i\mathbf{k}\bullet\mathbf{E}$$

and

$$\operatorname{curl}\mathbf{E} = \nabla\times\mathbf{E} = -i\mathbf{k}\times\mathbf{E} \ .$$

Furthermore

$$\Delta = \nabla\bullet\nabla = \nabla^2 = (-i\mathbf{k})\bullet(-i\mathbf{k}) = -k^2 \ . \tag{7.87}$$

With this, the two statements (7.81) and (7.82), together with the wave equations (7.79) and (7.80), again yield the dispersion relation (7.83). Next we will apply Maxwell's equations to the statements (7.81) and (7.82). Starting with (7.4), letting $\rho = 0$ gives

$$\nabla\bullet\mathbf{D} = \varepsilon\nabla\bullet\mathbf{E} = -i\varepsilon\mathbf{k}\bullet\mathbf{E} = 0 \ . \tag{7.88}$$

Consequently, $\mathbf{k}$ and $\mathbf{E}$ have to be perpendicular to each other, *i.e.*, the wave has to be transverse relative to $\mathbf{E}$. It also has to be transverse relative to $\mathbf{B}$, since it follows from (7.3) that

$$\nabla\bullet\mathbf{B} = -i\mathbf{k}\bullet\mathbf{B} = 0 \ . \tag{7.89}$$

Furthermore, from (7.2) it follows that

$$\nabla\times\mathbf{E} = -i\mathbf{k}\times\mathbf{E} = -\frac{\partial\mathbf{B}}{\partial t} = -i\omega\mathbf{B} \ ,$$

and thus

$$\mathbf{B} = \frac{\mathbf{k}\times\mathbf{E}}{\omega} \ . \tag{7.90}$$

As a special case, this formula contains relation (7.31), which was previously derived in a much more tedious manner. The generalization is grounded in the fact that according to (7.83), $\mathbf{k}$ is generally not a real valued vector. Finally, one needs to consider (7.1), which, when used in conjunction with (7.6) yields

$$-i\mathbf{k}\times\mathbf{B} = \mu\kappa\mathbf{E} + \mu\varepsilon i\omega\mathbf{E} \ . \tag{7.91}$$

Eq. (7.90) allows to eliminate $\mathbf{B}$ and obtain

$$-i\mathbf{k}\times\left(\frac{\mathbf{k}\times\mathbf{E}}{\omega}\right) = -\frac{i}{\omega}[\mathbf{k}(\mathbf{k}\bullet\mathbf{E}) - \mathbf{E}(\mathbf{k}\bullet\mathbf{k})] = \mu\kappa\mathbf{E} + \mu\varepsilon i\omega\mathbf{E}$$

Finally, with (7.88) one finds

$$\frac{ik^2}{\omega}\mathbf{E} \;=\; \mu\kappa\mathbf{E} + \mu\varepsilon i\omega\,\mathbf{E}\;,\tag{7.92}$$

i.e., again the dispersion relation (7.83).

With this, we have now again derived the properties for the special case of a plane wave in an insulator ($\kappa = 0$), however, this time in a formally much shorter and more elegant way than above. The current approach for the derivation highlights specifically the transverse nature of the wave with respect to $\mathbf{E}$ and is a consequence of $\mathbf{E}$ being source free. Remember that we had assumed that $\rho = 0$. The fact that $\mathbf{E}$ and $\mathbf{B}$ are perpendicular to each other as expressed in (7.90), is an immediate consequence of the law of induction.

For the case $\kappa = 0$, the dispersion relation (7.83) reduces to a result we already know:

$$k^2 \;=\; \mu\kappa\omega^2 \;=\; \frac{\omega^2}{c^2}$$

or

$$\omega \;=\; ck\;.$$

The relation is more complicated for a conductor. There are a number of different cases, of which we shall restrict our discussion to only two limiting cases.

7.2.2   The Process is Harmonic in Space

If a process is harmonic in space, then k is real valued and as a consequence, ω is complex valued. If we identify the real part of ω with η and its imaginary part with σ_1, then

$$\omega \;=\; \eta + i\sigma_1\;.\tag{7.93}$$

Rewriting eq. (7.83) using this new assignment gives

$$\mu\varepsilon(\eta^2 + 2\eta\sigma_1 i - \sigma_1^2) - \mu\kappa(i\eta - \sigma_1) - k^2 \;=\; 0\;.$$

For this to be true, both, the real and the imaginary part have to vanish independently

$$-k^2 + \kappa\mu\sigma_1 + \varepsilon\mu\eta^2 - \varepsilon\mu\sigma_1^2 \;=\; 0\tag{7.94}$$

$$-\eta\kappa\mu + 2\varepsilon\eta\mu\sigma_1 \;=\; 0\;.\tag{7.95}$$

Solving (7.95) for σ_1 gives

$$\sigma_1 \;=\; \frac{\eta\kappa\mu}{2\varepsilon\eta\mu} \;=\; \frac{\kappa}{2\varepsilon}\;.\tag{7.96}$$

Inserting this in (7.94) and solving for η gives

$$\eta \;=\; \pm\sqrt{\frac{k^2 + \varepsilon\mu\sigma_1^2 - \kappa\mu\sigma_1}{\varepsilon\mu}} \;=\; \pm\sqrt{\frac{k^2}{\varepsilon\mu} - \frac{\kappa^2}{4\varepsilon^2}}\;.\tag{7.97}$$

Definition (7.93) requires that η is real. This is only true if

$$k^2 \geq \frac{\kappa^2 \mu}{4\varepsilon} \ . \tag{7.98}$$

Then, the wave becomes

$$\mathbf{E} = \mathbf{E}_0 \exp[i(\eta + i\sigma_1)t - i\mathbf{k} \bullet \mathbf{r}]$$

$$= \mathbf{E}_0 \exp[i(\eta t - \mathbf{k} \bullet \mathbf{r})] \exp[-\sigma_1 t] \ . \tag{7.99}$$

This means that for the wave, η plays the role of the real valued angular frequency while σ_1 introduces exponential decay over time (damping).

Conversely, if

$$k^2 < \frac{\kappa^2 \mu}{4\varepsilon} \ , \tag{7.100}$$

then, ω becomes purely imaginary and we may write

$$\omega = i\sigma_2 \ . \tag{7.101}$$

Substituting this into (7.83) gives

$$-k^2 + \kappa \mu \sigma_2 - \varepsilon \mu \sigma_2^2 = 0 \ . \tag{7.102}$$

The two solutions when solving the quadratic equation (7.102) for σ_2 are

$$\sigma_2 = \frac{\kappa}{2\varepsilon} \pm \sqrt{\frac{\kappa^2}{4\varepsilon^2} - \frac{k^2}{\varepsilon \mu}} \ . \tag{7.103}$$

Finally, the wave becomes

$$\mathbf{E} = \mathbf{E}_0 \exp[-i(\mathbf{k} \bullet \mathbf{r})] \exp[-\sigma_2 t] \ . \tag{7.104}$$

To compare the two waves of (7.99) and (7.104) with each other is revealing. While the wave propagates for sufficiently large wave numbers (condition (7.98)), this is not the case for small wave numbers (condition (7.100)). The root cause for this peculiar behavior is that the diffusion process and the wave propagation process compete with each other. Take the wave equation in its form (7.79) or (7.80), then we find that in the statement of the form (7.81) and (7.82), the diffusion term basically behaves like

$$\mu \kappa \omega \ ,$$

while the wave propagation term behaves like

$$\mu \varepsilon \omega^2 \ .$$

If, *for example*, κ is very small, ε very large, then the diffusion term can be neglected. Then, according to (7.98), the wave propagates with the phase velocity η / k for almost all wave numbers k (except for extremely small ones). Conversely, if κ is very large and ε very small, then the diffusion term dominates. Now, for almost all wave numbers k (except for extremely large ones), the resulting expression has now the form of (7.104), which describes an exponentially decaying field and not a propagating wave.

7.2.3   The Process is Harmonic in Time

For a harmonic process, which in practise is the more important case, ω is real, while k is complex. One may write

$$k = \beta - i\alpha , \tag{7.105}$$

and with eq. (7.83) obtains

$$\varepsilon\mu\omega^2 - \kappa\mu\omega i - (\beta^2 - 2\alpha\beta i - \alpha^2) = 0 .$$

Separating the real and imaginary part gives

$$-\beta^2 + \alpha^2 + \varepsilon\mu\omega^2 = 0 \tag{7.106}$$

$$2\alpha\beta - \kappa\mu\omega = 0 . \tag{7.107}$$

Solving for α gives

$$\alpha = \frac{\kappa\mu\omega}{2\beta} ,$$

and the quadratic equation in β^2 is

$$\beta^4 - \varepsilon\mu\omega^2\beta^2 - \frac{\kappa^2\mu^2\omega^2}{4} = 0 . \tag{7.108}$$

Solving for β yields four solutions:

$$\beta = \pm\omega\sqrt{\frac{\varepsilon\mu}{2}\left(1 \pm \sqrt{1 + \frac{\kappa^2}{\omega^2\varepsilon^2}}\right)} .$$

In the definition, we have required that β is real. Consequently for the term under the radical, only the positive sign can be considered. Therefore

$$\boxed{\beta = \pm\omega\sqrt{\frac{\varepsilon\mu}{2}\left(\sqrt{1 + \frac{\kappa^2}{\omega^2\varepsilon^2}} + 1\right)}} . \tag{7.109}$$

With this and eqs. (7.106), (7.107) one obtains for α:

$$\boxed{\alpha = \pm\omega\sqrt{\frac{\varepsilon\mu}{2}\left(\sqrt{1 + \frac{\kappa^2}{\omega^2\varepsilon^2}} - 1\right)}} . \tag{7.110}$$

It shall be noted, that one could allow for imaginary values for β. However, this makes α imaginary as well and comparison of the results reveals that this does not provide anything new. Merely, α and β exchange their roles.

Using eqs. (7.105), (7.109), (7.110) in (7.81) gives

$$\mathbf{E} = \mathbf{E}_0\exp[i(\omega t - \beta z)]\exp[-\alpha z] , \tag{7.111}$$

having assumed that $\mathbf{k}$ is a vector in z-direction. This makes the real part of k, namely β, responsible for propagation of the wave, while its imaginary part α, controls the damping. Therefore α is called *damping constant*, and β the *phase constant*. Eq. (7.111) describes a damped plane wave. The plane $z =$ const. represents both: planes of constant phase, as well as planes of constant amplitude. A wave with these properties is also called a *homogeneous wave*.

This is by no means the most general case. This is obtained if

$$\mathbf{k} = \boldsymbol{\beta} - i\boldsymbol{\alpha} \;,\tag{7.112}$$

where $\boldsymbol{\alpha}$ and $\boldsymbol{\beta}$ are vectors for which the dispersion relation has to hold, *i.e.*,

$$\mu\varepsilon\omega^2 - \mu\kappa i\omega - (\beta^2 - i2\boldsymbol{\alpha}\bullet\boldsymbol{\beta} - \alpha^2) = 0 \;.\tag{7.113}$$

If the two vectors $\boldsymbol{\alpha}$ and $\boldsymbol{\beta}$ are parallel, then the appropriate choice of coordinate system allows to arrive at the same expression as stated in (7.111), and thus one obtains a homogeneous wave in the just defined sense. The planes of constant phase are perpendicular to $\boldsymbol{\beta}$. The planes of constant amplitude are perpendicular to $\boldsymbol{\alpha}$. If the two vectors point in different directions, then the result is an *inhomogeneous wave*. In order to be able to provide a simple **example**, let $\kappa = 0$ in (7.113). This forces $\boldsymbol{\alpha}$ and $\boldsymbol{\beta}$ to be perpendicular. One may assume, *for example,* that

$$\mathbf{k} = \langle b, 0, -ia\rangle \;,\tag{7.114}$$

which results in the wave

$$\mathbf{E} = \mathbf{E}_0 \exp[i(\omega t - bx)]\exp[-az] \;.\tag{7.115}$$

For the relation between b and a we have

$$-b^2 + a^2 + \mu\varepsilon\omega^2 = 0 \;.\tag{7.116}$$

Inhomogeneous waves are not transverse, as (7.90) clearly shows. Moreover, they are not plane waves in the sense of our definition because the amplitude is not constant in the planes where the phase is constant. Nevertheless, inhomogeneous waves are important. Oftentimes, they are necessary to satisfy boundary conditions, *e.g.*, for reflection problems. We will come back to such cases.

The magnetic field that belongs to the wave described by (7.111) results from (7.90). $\mathbf{B}$ is perpendicular to $\mathbf{E}$. However, there is a phase difference between $\mathbf{B}$ and $\mathbf{E}$ because $\mathbf{k}$ is a complex vector. There was no phase difference in the ideal insulator ($\kappa = 0$). If the electric field is

$$\mathbf{E}_0 = \langle E_{x0}, 0, 0\rangle \;,$$

then the magnetic field becomes

$$\mathbf{B} = \frac{\langle 0, 0, \beta - i\alpha\rangle}{\omega} \times \langle E_{x0}, 0, 0\rangle \exp[i(\omega t - \beta z)]\exp[-\alpha z]$$

$$= \langle 0, (\beta - i\alpha)E_{x0}, 0\rangle \frac{\exp[i(\omega t - \beta z)]\exp[-\alpha z]}{\omega} \;.$$

Notice that $\mathbf{B}$ has only a y-component

$$B_y = \frac{(\beta - i\alpha)E_{x0}}{\omega} \exp[i(\omega t - \beta z)]\exp[-\alpha z] \;.$$

If E_{x0} is real valued, then the electric field expressed in real value notation is

$$E_x = E_{x0}\cos(\omega t - \beta z)\exp[-\alpha z]\tag{7.117}$$

and the magnetic field is

$$B_y = \Re\left\{\frac{E_{x0}}{\omega}(\beta - i\alpha)[\cos(\omega t - \beta z) + i\sin(\omega t - \beta z)]\exp[-\alpha z]\right\},$$

i.e.,

$$B_y = \frac{E_{x0}}{\omega}[\beta\cos(\omega t - \beta z) + \alpha\sin(\omega t - \beta z)]\exp(-\alpha z)$$

$$= \frac{E_{x0}}{\omega}\sqrt{\alpha^2 + \beta^2}\left[\frac{\beta}{\sqrt{\alpha^2 + \beta^2}}\cos(\omega t - \beta z)\right.$$

$$\left. + \frac{\alpha}{\sqrt{\alpha^2 + \beta^2}}\sin(\omega t - \beta z)\right]\exp(-\alpha z).$$

Let

$$\frac{\alpha}{\sqrt{\alpha^2 + \beta^2}} = \sin\varphi,$$

$$\frac{\beta}{\sqrt{\alpha^2 + \beta^2}} = \cos\varphi,$$

and thus

$$\tan\varphi = \frac{\alpha}{\beta}, \tag{7.118}$$

which yields

$$B_y = B_{y0}\cos(\omega t - \beta z - \varphi)\exp[-\alpha z], \tag{7.119}$$

where

$$B_{y0} = \frac{E_{x0}}{\omega}\sqrt{\alpha^2 + \beta^2} = E_{x0}\sqrt{\mu\varepsilon}\sqrt[4]{1 + \frac{\kappa^2}{\omega^2\varepsilon^2}} \tag{7.120}$$

and

$$H_{y0} = \frac{E_{x0}}{Z}\sqrt[4]{1 + \frac{\kappa^2}{\omega^2\varepsilon^2}}. \tag{7.121}$$

Letting $\kappa = 0$ results in the previously discussed case of the ideal insulator. We will discuss two limits: If

$$\omega\varepsilon \ll \kappa \quad | \quad \kappa \ll \omega\varepsilon$$

or expressed differently by using the relaxation time t_r from Sect. 4.2, eq. (4.23)

$$\omega t_r \ll 1, \quad | \quad 1 \ll \omega t_r,$$

then the dominant term in the wave equation is

the diffusion term. | the wave propagation term.

430 Time Dependent Problems II (Electromagnetic Waves)

Using (7.109) and (7.110) yields

$$\alpha \approx \beta \approx \pm\sqrt{\frac{\mu\kappa\omega}{2}} \qquad \begin{aligned} &\alpha \approx \pm\frac{\kappa}{2}\sqrt{\frac{\mu}{\varepsilon}} \\ &\beta \approx \pm\omega\sqrt{\mu\varepsilon}\left(1 + \frac{\kappa^2}{8\omega^2\varepsilon^2}\right) \end{aligned}$$

and for the phase velocity

$$v_{ph} = \frac{\omega}{\beta} \approx \pm\sqrt{\frac{2\omega}{\mu\kappa}} \ . \qquad \begin{aligned} v_{ph} &\approx \frac{\pm 1}{\sqrt{\mu\varepsilon}\left(1 + \frac{\kappa^2}{8\omega^2\varepsilon^2}\right)} \\ &\approx \frac{\pm\left(1 - \frac{\kappa^2}{8\omega^2\varepsilon^2}\right)}{\sqrt{\mu\varepsilon}} \ . \end{aligned}$$

The dominating property in this case is

characteristics of a conductor . | characteristics of an insulator.

The material is therefore called a

conductor. | not an ideal insulator.

Whether a particular material should be considered more in terms of a conductor or an insulator with respect to a particular wave, does not only depend on the material constants κ and ε, but also on the frequency of the wave under consideration. The

conductor properties | insulator properties

become apparent for frequencies sufficiently

small. | large.

Apprehensibly, the reason lies in volume charge carrier mobility. When the frequency of the oscillating electric field is

small, | large,

the motion of the charges is

sufficiently fast | not fast enough

to be able to cancel the field created by the displaced volume charges.

It is impossible for an electrostatic field to exist in a conductor (Sect. 2.6), while slowly oscillating fields can penetrate the conductor only a very small distance. Even within the distance of one wave length, they are damped by a factor of $\exp(-2\pi) \approx 2 \cdot 10^{-3}$, which is the result of (7.111) for $\alpha = \beta$. Otherwise, the results we obtain here for the limit of small frequencies are identical to those obtained in Sect. 6.5.4. Take note of eq. (6.145).

The energy lost by damping is transformed into heat. The proof can be performed using the energy principle, which we will forgo here.

7.3   Reflection and Refraction

7.3.1   Reflection and Refraction for Insulators

Let a plane wave be incident on a plane boundary between two insulators ("media boundary"), then the various boundary conditions for $\mathbf{E}$, $\mathbf{D}$, $\mathbf{H}$, $\mathbf{B}$ have to be satisfied. Fig. 7.9 illustrates this. The boundary conditions can be satisfied, if one assumes that besides the incident wave ($\mathbf{k}_i$), there is also a wave reflected back into medium 1 ($\mathbf{k}_r$), and a transmitted wave ($\mathbf{k}_t$) into medium 2. Disregarding the case of total internal reflection, which we will discuss later, one has a wave of the form:

$$\mathbf{E}_i = \mathbf{E}_{i0} \exp[i(\omega_i t - \mathbf{k}_i \bullet \mathbf{r})] \tag{7.122}$$

$$\mathbf{E}_r = \mathbf{E}_{r0} \exp[i(\omega_r t - \mathbf{k}_r \bullet \mathbf{r})] \tag{7.123}$$

$$\mathbf{E}_t = \mathbf{E}_{t0} \exp[i(\omega_t t - \mathbf{k}_t \bullet \mathbf{r})]. \tag{7.124}$$

Certain field components have to be continuous at the media boundary. This results in certain relations between $\mathbf{E}_{i0}$, $\mathbf{E}_{r0}$, and $\mathbf{E}_{t0}$. Because the boundary conditions have to be satisfied at all times and for every point $\mathbf{r}_M$ on the boundary, all the phases of the exponential functions in (7.122) through (7.124) have be equal. In particular, it must be

$$\omega_i = \omega_r = \omega_t = \omega. \tag{7.125}$$

Consequently, there is only one frequency in both media. Without limiting the generality, it may be assumed that the origin of the coordinate system lies in the boundary between the media. Then the vectors $\mathbf{r}_M$ for all points on the boundary lay entirely on the boundary itself. Therefore, it has to be

$$\mathbf{k}_i \bullet \mathbf{r}_M = \mathbf{k}_r \bullet \mathbf{r}_M = \mathbf{k}_t \bullet \mathbf{r}_M. \tag{7.126}$$

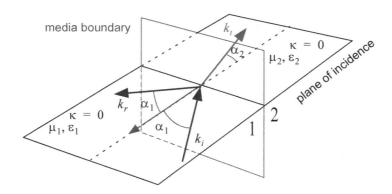

Fig. 7.9

From this we first obtain

$$(\mathbf{k}_i - \mathbf{k}_r) \bullet \mathbf{r}_M = 0 \; . \tag{7.127}$$

Consequently, the vector $\mathbf{k}_i - \mathbf{k}_r$ is perpendicular to the boundary and we may write

$$\mathbf{k}_i - \mathbf{k}_r = A\mathbf{n} \; , \tag{7.128}$$

where A is a constant. Therefore, the three vectors $\mathbf{k}_i$, $\mathbf{k}_r$, $\mathbf{n}$ lie in a plane, the so-called *plane of incidence*, they are *coplanar*. Also, the two vectors $\mathbf{k}_i$ and $\mathbf{k}_r$ have the same magnitude, *i.e.*, they share the same frequency $\omega = \omega_i = \omega_r$ and the same medium (medium 1). The components of $\mathbf{k}_i$ and $\mathbf{k}_r$ which are parallel to the boundary are obviously the same. Therefore, both angles are the same

$$\boxed{\alpha_i = \alpha_r = \alpha_1} \; . \tag{7.129}$$

This is the well-known law of refraction. Furthermore

$$(\mathbf{k}_i - \mathbf{k}_t) \bullet \mathbf{r}_M = 0 \; . \tag{7.130}$$

Therefore, besides the three vectors $\mathbf{k}_i$, $\mathbf{k}_r$, $\mathbf{n}$, also $\mathbf{k}_t$ lays in the plane of incidence. However, $\mathbf{k}_i$ and $\mathbf{k}_t$ have a different magnitude. From the dispersion relation (7.39) one finds that

$$\frac{\omega}{k_i} = \frac{1}{\sqrt{\varepsilon_1 \mu_1}} \tag{7.131}$$

$$\frac{\omega}{k_t} = \frac{1}{\sqrt{\varepsilon_2 \mu_2}} \; . \tag{7.132}$$

It follows from (7.130) that the tangential components of $\mathbf{k}_i$ and $\mathbf{k}_t$ must be equal, *i.e.*,

$$k_i \sin \alpha_i = k_t \sin \alpha_t$$

and applying (7.131) and (7.132) gives

$$\frac{k_i}{k_t} = \frac{\sin \alpha_t}{\sin \alpha_i} = \frac{\omega \sqrt{\varepsilon_1 \mu_1}}{\omega \sqrt{\varepsilon_2 \mu_2}} = \frac{\sqrt{\varepsilon_1 \mu_1}}{\sqrt{\varepsilon_2 \mu_2}} \; .$$

Renaming

$$\alpha_t = \alpha_2 \tag{7.133}$$

and

$$n = \frac{c_0}{c} = \frac{\sqrt{\varepsilon \mu}}{\sqrt{\varepsilon_0 \mu_0}} = \sqrt{\varepsilon_r \mu_r} \tag{7.134}$$

yields *Snell's law:*

$$\boxed{\frac{\sin \alpha_2}{\sin \alpha_1} = \frac{n_1}{n_2} = \frac{c_2}{c_1}} \; . \tag{7.135}$$

7.3.2 Fresnel's Equations for Insulators

Now we need to derive the relations between the amplitudes of the waves (7.122) through (7.124) from the boundary conditions for the fields. To do so, we need to distinguish two cases, whether the vector of the incident electric field lies in the plane of incidence, *i.e.*, is parallel to it, or whether it is perpendicular to the plane of incidence. Every wave can be decomposed into these two components. We call one case parallel polarization and the other perpendicular polarization. The discussion shall start with perpendicular polarization (Fig. 7.10). The components of **E** and **H** parallel to the boundary have to be continuous at the boundary. It must be

$$E_{i0} + E_{r0} = E_{t0} \tag{7.136}$$

$$H_{i0}\cos\alpha_1 - H_{r0}\cos\alpha_1 = H_{t0}\cos\alpha_2 . \tag{7.137}$$

With (7.32), (7.33) we get

$$H_0 = E_0/Z , \tag{7.138}$$

and using this with (7.137) gives

$$(E_{i0} - E_{r0})\frac{\cos\alpha_1}{Z_1} = E_{t0}\frac{\cos\alpha_2}{Z_2} . \tag{7.139}$$

With (7.136) and (7.139), we have two equations for two unknowns E_{r0} and E_{t0} (E_{i0} is considered as given). Solving for these yields

$$\left(\frac{E_{r0}}{E_{i0}}\right)_\perp = \frac{\dfrac{\cos\alpha_1}{Z_1} - \dfrac{\cos\alpha_2}{Z_2}}{\dfrac{\cos\alpha_1}{Z_1} + \dfrac{\cos\alpha_2}{Z_2}} = \frac{Z_2\cos\alpha_1 - Z_1\cos\alpha_2}{Z_2\cos\alpha_1 + Z_1\cos\alpha_2} \tag{7.140}$$

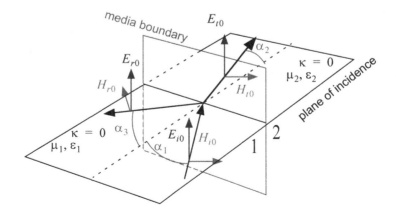

Fig. 7.10

$$\left(\frac{E_{t0}}{E_{i0}}\right)_{\perp} = \frac{2\dfrac{\cos\alpha_1}{Z_1}}{\dfrac{\cos\alpha_1}{Z_1} + \dfrac{\cos\alpha_2}{Z_2}} = \frac{2Z_2\cos\alpha_1}{Z_2\cos\alpha_1 + Z_1\cos\alpha_2} \quad [\,. \tag{7.141}$$

These are Fresnel's equations for the case of perpendicular polarization. They were derived only from the boundary conditions of the parallel components of **E** and **H**. The perpendicular components of **D** vanish, which makes them automatically continuous. The perpendicular components of **B** are continuous if

$$\mu_1(H_{i0} + H_{r0})\sin\alpha_1 = \mu_2 H_{t0}\sin\alpha_2 \tag{7.142}$$

or

$$\frac{\mu_1\sin\alpha_1}{Z_1}(E_{i0} + E_{r0}) = \frac{\mu_2\sin\alpha_2}{Z_2}E_{t0}\,.$$

Using (7.136) gives

$$\frac{\mu_1\sin\alpha_1}{Z_1} = \frac{\mu_2\sin\alpha_2}{Z_2}$$

or

$$\sqrt{\varepsilon_1\mu_1}\sin\alpha_1 = \sqrt{\varepsilon_2\mu_2}\sin\alpha_2\,,$$

that is, continuity of the perpendicular components of **B** is ensured by Snell's law and by the continuity of the parallel components of **E**.

Multiplying (7.136) by (7.139) yields

$$(E_{i0}^2 - E_{r0}^2)\frac{\cos\alpha_1}{Z_1} = E_{t0}^2\frac{\cos\alpha_2}{Z_2}\,. \tag{7.143}$$

In conjunction with (7.72), this gives

$$S_i\cos\alpha_1 - S_r\cos\alpha_1 = S_t\cos\alpha_2\,. \tag{7.144}$$

This equation expresses the conservation of energy principle. The incident field energy is passed in parts to the reflected wave and the remainder to the transmitted wave.

Now we approach the case of parallel polarization (Fig. 7.11). Here one has

$$H_{i0} - H_{r0} = H_{t0} \tag{7.145}$$

or

$$(E_{i0} - E_{r0})/Z_1 = E_{t0}/Z_2 \tag{7.146}$$

This ensures continuity of the parallel components of **H**. Those of **E** have to be continuous as well

$$(E_{i0} - E_{r0})\cos\alpha_1 = E_{t0}\cos\alpha_2\,. \tag{7.147}$$

The result are Fresnel's equations for the case of parallel polarization.

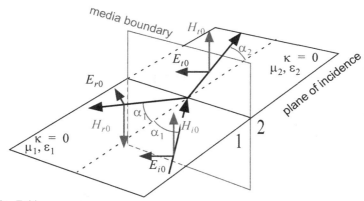

Fig. 7.11

$$\left(\frac{E_{r0}}{E_{i0}}\right)_{\parallel} = \frac{-\dfrac{\cos\alpha_1}{Z_2} + \dfrac{\cos\alpha_2}{Z_1}}{\dfrac{\cos\alpha_1}{Z_2} + \dfrac{\cos\alpha_2}{Z_1}} = \frac{-Z_1\cos\alpha_1 + Z_2\cos\alpha_2}{Z_1\cos\alpha_1 + Z_2\cos\alpha_2} , \qquad (7.148)$$

$$\left(\frac{E_{t0}}{E_{i0}}\right)_{\parallel} = \frac{2\dfrac{\cos\alpha_1}{Z_1}}{\dfrac{\cos\alpha_1}{Z_2} + \dfrac{\cos\alpha_2}{Z_1}} = \frac{2Z_2\cos\alpha_1}{Z_1\cos\alpha_1 + Z_2\cos\alpha_2} . \qquad (7.149)$$

The perpendicular components of **B** vanish and, thus, are continuous. Continuity of the perpendicular components of **D** is ensured by Snell's law and by the continuity of the parallel components of **H**:

$$\varepsilon_1(E_{i0} - E_{r0})\sin\alpha_1 = \varepsilon_2 E_{t0}\sin\alpha_2 . \qquad (7.150)$$

Using (7.146) gives

$$\varepsilon_1\sin\alpha_1 Z_1 = \varepsilon_2\sin\alpha_2 Z_2$$

or

$$\sqrt{\varepsilon_1\mu_1}\sin\alpha_1 = \sqrt{\varepsilon_2\mu_2}\sin\alpha_2 ,$$

in agreement with Snell's law.
Multiplying (7.146) by (7.147) and in conjunction with (7.143) and (7.144) yields the conservation of energy, as before.

7.3.3   Nonmagnetic Media

For the special case

$$\mu_1 = \mu_2 = \mu_0 \ ,$$

Fresnel's equations (7.140), (7.141), (7.148), and (7.149) simplify. Using

$$\frac{Z_1}{Z_2} = \frac{\sqrt{\dfrac{\mu_0}{\varepsilon_1}}}{\sqrt{\dfrac{\mu_0}{\varepsilon_2}}} = \frac{\sqrt{\varepsilon_2\mu_0}}{\sqrt{\varepsilon_1\mu_0}} = \frac{\sin\alpha_1}{\sin\alpha_2} \tag{7.151}$$

now yields

$$\boxed{\left(\frac{E_{r0}}{E_{i0}}\right)_{\perp} = \frac{\sin\alpha_2\cos\alpha_1 - \sin\alpha_1\cos\alpha_2}{\sin\alpha_2\cos\alpha_1 + \sin\alpha_1\cos\alpha_2} = \frac{\sin(\alpha_2 - \alpha_1)}{\sin(\alpha_2 + \alpha_1)}} \ , \tag{7.152}$$

$$\boxed{\left(\frac{E_{t0}}{E_{i0}}\right)_{\perp} = \frac{2\sin\alpha_2\cos\alpha_1}{\sin\alpha_2\cos\alpha_1 + \sin\alpha_1\cos\alpha_2} = \frac{2\sin\alpha_2\cos\alpha_1}{\sin(\alpha_2 + \alpha_1)}} \ , \tag{7.153}$$

and

$$\boxed{\begin{aligned}\left(\frac{E_{r0}}{E_{i0}}\right)_{\parallel} &= \frac{\sin\alpha_2\cos\alpha_2 - \sin\alpha_1\cos\alpha_1}{\sin\alpha_2\cos\alpha_2 + \sin\alpha_1\cos\alpha_1} = \frac{\sin 2\alpha_2 - \sin 2\alpha_1}{\sin 2\alpha_2 + \sin 2\alpha_1} \\ &= \frac{\sin(\alpha_2 - \alpha_1)\cos(\alpha_2 + \alpha_1)}{\sin(\alpha_2 + \alpha_1)\cos(\alpha_2 - \alpha_1)} = \frac{\tan(\alpha_2 - \alpha_1)}{\tan(\alpha_2 + \alpha_1)}\end{aligned}} \ , \tag{7.154}$$

$$\boxed{\left(\frac{E_{t0}}{E_{i0}}\right)_{\parallel} = \frac{2\sin\alpha_2\cos\alpha_1}{\sin\alpha_1\cos\alpha_1 + \sin\alpha_2\cos\alpha_2} = \frac{2\sin\alpha_2\cos\alpha_1}{\sin(\alpha_1 + \alpha_2)\cos(\alpha_1 - \alpha_2)}} \ . \tag{7.155}$$

If $\alpha_1 = \alpha_2$, which means that there is no media boundary and nothing noteworthy happens, i.e., the incident wave continues and there is no reflection. There is no reflection because of

$$\left(\frac{E_{r0}}{E_{i0}}\right)_{\perp} = \left(\frac{E_{r0}}{E_{i0}}\right)_{\parallel} = 0 \ ,$$

$$\left(\frac{E_{t0}}{E_{i0}}\right)_{\perp} = \left(\frac{E_{t0}}{E_{i0}}\right)_{\parallel} = 1 \ .$$

Less self-evident is that in the case of parallel polarization, there is no reflected wave when

$$\alpha_1 + \alpha_2 = \frac{\pi}{2} \ . \tag{7.156}$$

That this is true can be derived from (7.154):

$$\left(\frac{E_{r0}}{E_{i0}}\right)_{\parallel} = 0 \qquad \text{for } \alpha_1 + \alpha_2 = \frac{\pi}{2} . \tag{7.157}$$

Snell's law states that

$$\frac{\sin\alpha_1}{\sin\alpha_2} = \sqrt{\frac{\varepsilon_2}{\varepsilon_1}} = \frac{n_2}{n_1} = \frac{\sin\alpha_1}{\cos\left(\frac{\pi}{2} - \alpha_2\right)} = \frac{\sin\alpha_1}{\cos\alpha_1} = \tan\alpha_1 .$$

The thereby defined angle α_1

$$\tan\alpha_1 = \frac{n_2}{n_1} = \sqrt{\frac{\varepsilon_2}{\varepsilon_1}} , \tag{7.158}$$

is the so-called *Brewster angle,* also called the *polarization angle.* It is a remarkable angle because, if we shine unpolarized light or any unpolarized electromagnetic radiation, under this angle onto a boundary, then the polarization of the reflected radiation is completely perpendicularly polarized because the parallel polarized radiation passes through the boundary in its entirety without reflection. Modification of eq. (7.158) are necessary if $\mu_1 \neq \mu_2$. Furthermore, in such case there would also be a polarization angle for the perpendicularly polarized wave. We will discuss neither case any further.

Let us divide (7.143) by $E_{i0}^2 \cos\alpha_1 / Z_1$: This gives

$$1 - \left(\frac{E_{r0}}{E_{i0}}\right)^2 = \left(\frac{E_{t0}}{E_{i0}}\right)^2 \frac{Z_1 \cos\alpha_2}{Z_2 \cos\alpha_1} . \tag{7.159}$$

$(E_{r0}/E_{i0})^2$ is the fraction of incident energy which is reflected. Conversely, $(E_{t0}/E_{i0})^2 (Z_1 \cos\alpha_2 / Z_2 \cos\alpha_1)$ is the fraction of incident energy which passes through. Therefore, one defines the two quantities

$$R = \left(\frac{E_{r0}}{E_{i0}}\right)^2 \tag{7.160}$$

and

$$T = \left(\frac{E_{t0}}{E_{i0}}\right)^2 \frac{Z_1 \cos\alpha_2}{Z_2 \cos\alpha_1} . \tag{7.161}$$

R is called *reflectance* and T is called *transmittance*. Their sum is of course

$$T + R = 1 . \tag{7.162}$$

Take as an example the perpendicular incidence on a non magnetic material, *i.e.*,

$$\alpha_1 = \alpha_2 = 0 .$$

$$\cos\alpha_1 = \cos\alpha_2 = 1 \tag{7.163}$$

Then we use (7.140) and (7.148) together with (7.151) to obtain

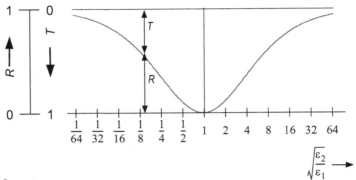

Fig. 7.12

$$\left(\frac{E_{r0}}{E_{i0}}\right)_{\perp} = \left(\frac{E_{r0}}{E_{i0}}\right)_{\parallel} = \frac{Z_2 - Z_1}{Z_2 + Z_1} = \frac{1 - \dfrac{Z_1}{Z_2}}{1 + \dfrac{Z_1}{Z_2}} = \frac{1 - \sqrt{\dfrac{\varepsilon_2}{\varepsilon_1}}}{1 + \sqrt{\dfrac{\varepsilon_2}{\varepsilon_1}}} \qquad (7.164)$$

and using (7.141) and (7.149) together with (7.151) to obtain

$$\left(\frac{E_{t0}}{E_{i0}}\right)_{\perp} = \left(\frac{E_{t0}}{E_{i0}}\right)_{\parallel} = \frac{2Z_2}{Z_1 + Z_2} = \frac{2}{1 + \dfrac{Z_1}{Z_2}} = \frac{2}{1 + \sqrt{\dfrac{\varepsilon_2}{\varepsilon_1}}} \ . \qquad (7.165)$$

Consequently for the reflectance we get

$$R_{\perp} = R_{\parallel} = R = \left(\frac{\sqrt{\dfrac{\varepsilon_2}{\varepsilon_1}} - 1}{\sqrt{\dfrac{\varepsilon_2}{\varepsilon_1}} + 1}\right)^2 \qquad (7.166)$$

and for the transmittance

$$T_{\perp} = T_{\parallel} = T = \frac{4 \cdot \sqrt{\dfrac{\varepsilon_2}{\varepsilon_1}}}{\left(\sqrt{\dfrac{\varepsilon_2}{\varepsilon_1}} + 1\right)^2} \ . \qquad (7.167)$$

Of course, in case of $\varepsilon_1 = \varepsilon_2$ it must be $R = 0$ and $T = 1$. The graph of R and T is illustrated in Fig. 7.12 for perpendicular incidence.

7.3.4   Total Reflection

Snell's law states

$$\frac{\sin\alpha_2}{\sin\alpha_1} = \frac{\sqrt{\varepsilon_1\mu_1}}{\sqrt{\varepsilon_2\mu_2}} = \frac{n_1}{n_2} = \frac{c_2}{c_1} .$$

(7.168)

The medium with the relatively lower speed of light is a higher-index medium, the one with the greater speed of light is a lower-index medium.

The lower-index medium belongs to the larger angle, while the higher-index medium corresponds to the smaller angle. When light travels into a higher-index medium, refraction is towards the normal direction, while when travelling into a lower-index medium, refraction is away from the normal (Fig. 7.13).For the refraction away from the normal we have

$$\sin\alpha_2 = \frac{c_2}{c_1}\sin\alpha_1 > \sin\alpha_1 .$$

(7.169)

Now, there are certain angles α_1 which require $\sin\alpha_2 > 1$. This is not possible for real angles. The result is, that in this case the entire incident energy is reflected. There is no refracted plane wave. The limit between ordinary reflection and this so-called *total reflection*, *i.e.*, the maximum possible angle α_1 for which regular reflection and refraction is possible is given by

$$\sin\alpha_2 = 1 = \frac{c_2}{c_1}\sin\alpha_c$$

i.e.,

$$\sin\alpha_c = \frac{c_1}{c_2} .$$

(7.170)

This angle α_c is called *critical angle* of total reflection.

It must be emphasized that this does not at all mean that there are no waves in medium 2. However this wave in medium two is not a plane (homogeneous) wave,

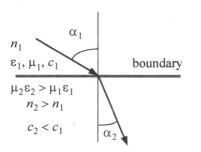

refraction towards the normal

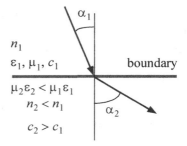

refraction away from the normal

Fig. 7.13

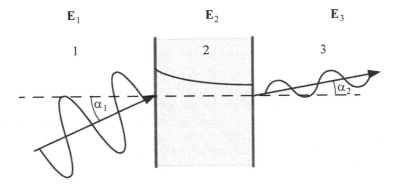

Fig. 7.14

but an inhomogeneous wave. This wave is necessary to satisfy the boundary conditions. It travels parallel to the boundary surface and decreases exponentially in the direction perpendicular to the boundary. The wave is of the kind discussed in Sect. 7.2.3, eq. (7.115). One can calculate amplitude, phase constant, and damping constant for the wave in medium 2, when starting from appropriate statements, including the boundary conditions. This is not difficult but tedious and therefore, will not be presented here. A formal solution would start from Snell's law. The angle α_2 becomes imaginary when $\sin\alpha_2 > 1$ and the refracted wave becomes inhomogeneous.

Of fundamental interest is a slightly modified case. This is illustrated in Fig. 7.14, which shows a wave in medium 1, incident on a thin layer (medium 2). Because of $(\alpha_1 > \alpha_{1c})$, this should be a case of total reflection. However, a certain fraction of energy is still passing through. What is happening here is that the incident wave causes in the medium 2, a wave which decreases exponentially in the direction perpendicular to the boundary. When the decaying wave has reached the other end of the thin layer, it has not decayed enough to be actually zero. Under certain conditions, this remainder of the wave can be the source of another propagating wave into medium 3. Depending on the thickness of the layer, the amplitude of this wave may be very small, however. This wave is necessary to satisfy the boundary conditions on the boundary between medium 2 and 3. Formally, this is analogous to Quantum Mechanics famous *tunnel effect*, which is highly important also for electrical engineering because of its significance for the properties of semiconductors. Fig. 7.14 sketches the behavior of the refracted wave for

$$\frac{\sqrt{\varepsilon_3\mu_3}}{\sqrt{\varepsilon_1\mu_1}} > \sin\alpha_1 > \frac{\sqrt{\varepsilon_2\mu_2}}{\sqrt{\varepsilon_1\mu_1}} \, ,$$

that is, α_1 is greater than the critical angle for medium 2, but is smaller than the critical angle for medium 3. Then, Snell's law applies for media 1 and 3, just as if there were no intermediate layer

$$\frac{\sin\alpha_3}{\sin\alpha_1} = \sqrt{\frac{\varepsilon_1\mu_1}{\varepsilon_3\mu_3}} \ .$$

7.3.5 Reflection at a Conducting Medium

The case of another inhomogeneous wave, will not be calculated, but only touched upon briefly. Consider a wave emerging from an insulator that is incident on a conductor. Here too, besides the reflected wave back into the insulator, we also need an inhomogeneous wave in the conductor to satisfy the boundary conditions between an insulator and the conductor. Fig. 7.15 illustrates this. The inhomogeneous wave is such that the propagation direction is given by Snell's law, but at the same time decays exponentially in the direction perpendicular to the media boundary. The locus of constant phase is perpendicular to the propagation direction.

With a coordinate system as defined by Fig. 7.15, the following Ansatz may serve to solve this problem:

$$\mathbf{E}_i = \mathbf{E}_{i0}\exp[i(\omega t - k_1 y\sin\alpha_1 - k_1 z\cos\alpha_1)] \tag{7.171}$$

$$\mathbf{E}_r = \mathbf{E}_{r0}\exp[i(\omega t - k_1 y\sin\alpha_1 + k_1 z\cos\alpha_1)] \tag{7.172}$$

$$\mathbf{E}_t = \mathbf{E}_{t0}\exp[i(\omega t - \beta_2 y\sin\alpha_2 - \beta_2 z\cos\alpha_2)]\exp[-\gamma_2 z] \ . \tag{7.173}$$

Besides the material constants $\mu_1, \varepsilon_1, \mu_2, \varepsilon_2$ and κ_2, the quantities ω, k_1 and α_1 have to be considered as given. The task is to find the appropriate $\alpha_2, \beta_2, \gamma_2$, which

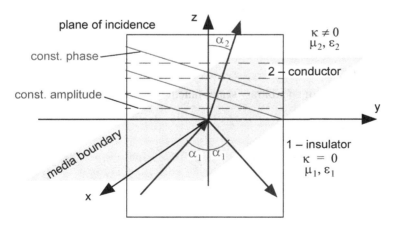

Fig. 7.15

is possible by means of the dispersion relation of the wave in eq. (7.173) and by means of the following phase relation which was obtained from the phases above

$$k_1 \sin\alpha_1 = \beta_2 \sin\alpha_2 \ . \tag{7.174}$$

This represents Snell's law in the here applicable from. Calculating the amplitudes E_{r0} and E_{t0} requires one to invoke the boundary conditions, applicable for the various field components, whereby the proceedings are analogous to Sect. 7.3.2. We shall pass on the tedious details of this calculation.

7.4   Potentials and their Wave Equations

7.4.1   The Inhomogeneous Wave Equation for A and φ

Starting point are again Maxwell's equations

$$\nabla \times \mathbf{E} = -\frac{\partial \mathbf{B}}{\partial t} \tag{7.175}$$

$$\nabla \times \mathbf{H} = \mathbf{g} + \frac{\partial \mathbf{D}}{\partial t} \tag{7.176}$$

$$\nabla \bullet \mathbf{B} = 0 \tag{7.177}$$

$$\nabla \bullet \mathbf{D} = \rho \ . \tag{7.178}$$

Because of (7.177), one may define as

$$\mathbf{B} = \nabla \times \mathbf{A} \ , \tag{7.179}$$

where $\mathbf{A}$ is a vector depending on space and time

$$\mathbf{A} = \mathbf{A}(\mathbf{r}, t) \ , \tag{7.180}$$

But $\mathbf{A}$ is not unique. For instance, another vector potential $\mathbf{A}_e$ may provide the same $\mathbf{B}$-field.

$$\mathbf{B} = \nabla \times \mathbf{A} = \nabla \times \mathbf{A}_e \ ,$$

therefore

$$\nabla \times (\mathbf{A} - \mathbf{A}_e) = 0 \ .$$

Consequently, by means of an arbitrary scalar function ψ we may write

$$\mathbf{A} - \mathbf{A}_e = -\nabla \psi \ . \tag{7.181}$$

Inserting (7.179) in (7.175) gives

$$\nabla \times \mathbf{E} + \frac{\partial}{\partial t} \nabla \times \mathbf{A} = \nabla \times \left(\mathbf{E} + \frac{\partial \mathbf{A}}{\partial t} \right) = 0$$

or

$$\mathbf{E} + \frac{\partial \mathbf{A}}{\partial t} = -\nabla \varphi(\mathbf{r}, t) \ . \tag{7.182}$$

Consequently, one may calculate **B** and **E** from **A** and φ, and considering (7.179) and (7.182) write:

$$\boxed{\mathbf{B} = \nabla \times \mathbf{A}}$$
(7.183)

$$\boxed{\mathbf{E} = -\nabla \varphi(\mathbf{r}, t) - \frac{\partial \mathbf{A}}{\partial t}}.$$
(7.184)

φ is not unique either. Instead of calculating **B** and **E** from **A** and φ, we might as well choose $\mathbf{A}_e$ and φ_e as starting point,

$$\mathbf{E} = -\nabla \varphi_e - \frac{\partial \mathbf{A}_e}{\partial t} = -\nabla \varphi - \frac{\partial \mathbf{A}}{\partial t}$$

$$= -\nabla \varphi - \frac{\partial \mathbf{A}_e}{\partial t} + \frac{\partial}{\partial t} \nabla \psi$$

inasmuch as

$$\varphi - \varphi_e = \frac{\partial \psi}{\partial t}.$$
(7.185)

Therefore, the relation between **A**, φ and $\mathbf{A}_e$, φ_e is given by (7.181) and (7.185). ψ is a function that may be chosen arbitrarily. Thus, there is substantial freedom when choosing the potentials **A** and φ. We use this freedom to require that the following condition be met:

$$\boxed{\nabla \bullet \mathbf{A} + \mu \varepsilon \frac{\partial \varphi}{\partial t} = 0}.$$
(7.186)

This condition is called the *Lorentz gauge*. For time independent problems, this reverts into the previously introduced Coulomb gauge (5.9). Should the potentials in a given problem not correspond to the Lorentz gauge, then by means of an appropriate function ψ, it is always possible to find potentials that do so.

The fields given by (7.183), (7.184) automatically satisfy two of Maxwell's equations, namely (7.175) and (7.177). To satisfy the other two requires some work. We start with (7.176):

$$\nabla \times \mathbf{H} = \nabla \times \frac{1}{\mu} (\nabla \times \mathbf{A}) = \mathbf{g} - \varepsilon \nabla \frac{\partial}{\partial t} \varphi - \varepsilon \frac{\partial^2 \mathbf{A}}{\partial t^2}$$

$$\nabla \times (\nabla \times \mathbf{A}) + \varepsilon \mu \nabla \frac{\partial}{\partial t} \varphi + \varepsilon \mu \frac{\partial^2 \mathbf{A}}{\partial t^2} = \mu \mathbf{g}$$

$$\nabla (\nabla \bullet \mathbf{A}) - \nabla^2 \mathbf{A} + \varepsilon \mu \nabla \frac{\partial}{\partial t} \varphi + \varepsilon \mu \frac{\partial^2 \mathbf{A}}{\partial t^2} = \mu \mathbf{g}.$$

With (7.186) we get

$$\boxed{\nabla^2 \mathbf{A} - \varepsilon \mu \frac{\partial^2 \mathbf{A}}{\partial t^2} = -\mu \mathbf{g}}.$$
(7.187)

On the other hand, using (7.178) yields

$$\nabla\bullet\varepsilon\left(-\nabla\varphi - \frac{\partial \mathbf{A}}{\partial t}\right) = \rho$$

$$-\varepsilon\nabla\bullet(\nabla\varphi) - \varepsilon\frac{\partial}{\partial t}\nabla\bullet\mathbf{A} = \rho \; .$$

Finally with (7.186), we get

$$\boxed{\nabla^2\varphi - \varepsilon\mu\frac{\partial^2\varphi}{\partial t^2} = -\frac{\rho}{\varepsilon}} \; . \qquad\qquad (7.188)$$

The two equations (7.187) and (7.188) are the wave equations for the potentials $\mathbf{A}$ and φ. They are inhomogeneous if there are currents or charges involved. The two *inhomogeneous wave equations* are equivalent to the four Maxwell equations. They serve to calculate the potentials $\mathbf{A}$ and φ when the current density $\mathbf{g}(\mathbf{r}, t)$ and charge density $\rho(\mathbf{r}, t)$ are given. With some limitations, these may be chosen largely arbitrarily. The continuity equation

$$\nabla\bullet\mathbf{g} + \frac{\partial}{\partial t}\rho = 0$$

is also a consequence of the inhomogeneous wave equations, just as it was before a consequence of Maxwell's equations. This can be immediately verified by inspection when calculating $\nabla\bullet\mathbf{g}$ in (7.187) and $\partial\rho/\partial t$ in (7.188) while also considering the assumed Lorentz gauge. If we try to solve eqs. (7.187), (7.188) with incompatible current densities $\mathbf{g}(\mathbf{r}, t)$ and charge densities $\rho(\mathbf{r}, t)$, then we end up in contradictions. For instance, the resulting potentials will not satisfy the Lorentz gauge as required.

As a particularly simple **example**, let us take a plane wave. The particular solution of the wave equations (7.187), (7.188) for $\mathbf{g} = 0$ and $\rho = 0$ are the following plane waves:

$$\mathbf{A} = \mathbf{A}_0\exp[i(\omega t - \mathbf{k}\bullet\mathbf{r})]$$

$$\varphi = \varphi_0\exp[i(\omega t - \mathbf{k}\bullet\mathbf{r})]$$

with the requirement that

$$\frac{\omega^2}{k^2} = c^2 = \frac{1}{\mu\varepsilon} \; .$$

The Lorentz gauge mandates

$$-i\mathbf{k}\bullet\mathbf{A}_0 + \frac{1}{c^2}i\omega\varphi_0 = 0 \; .$$

If we split $\mathbf{A}_0$ into a part parallel to $\mathbf{k}$ and one perpendicular to it:

$$\mathbf{A}_0 = \mathbf{A}_{0\|} + \mathbf{A}_{0\perp} \; ,$$

then we may write

$$-ikA_{0\|} + \frac{1}{c^2}i\omega\varphi_0 = 0 \; ,$$

i.e.,

$$\varphi_0 = \frac{kc^2}{\omega} A_{0\|} = cA_{0\|} .$$

Applying (7.184) gives

$$\mathbf{E} = (ik\varphi_0 - i\omega\,\mathbf{A}_0)\exp[i(\omega t - \mathbf{k} \bullet \mathbf{r})] ,$$

or

$$\mathbf{E} = (ik\varphi_0 - i\omega\,\mathbf{A}_{0\|} - i\omega\,\mathbf{A}_{0\perp})\exp[i(\omega t - \mathbf{k} \bullet \mathbf{r})]$$
$$= -i\omega\,\mathbf{A}_{0\perp}\exp[i(\omega t - \mathbf{k} \bullet \mathbf{r})] .$$

For the magnetic field, one obtains with (7.183) and in analogy to (7.85)

$$\mathbf{B} = \nabla\times\mathbf{A} = -i\mathbf{k} \times \mathbf{A} = -i\mathbf{k} \times \mathbf{A}_{0\perp}\exp[i(\omega t - \mathbf{k} \bullet \mathbf{r})] .$$

Thus, ultimately, $\mathbf{A}_{0\|}$ is irrelevant and may be dropped altogether. When $\mathbf{A}_{0\|} = 0$ then also $\varphi = 0$ and the fields result exclusively from $\mathbf{A}$:

$$\mathbf{E} = -i\omega\,\mathbf{A}$$

$$\mathbf{B} = -i\mathbf{k} \times \mathbf{A} = \frac{\mathbf{k} \times \mathbf{E}}{\omega} .$$

Thus, one can describe a plane wave by means of a vector potential $\mathbf{A}$ which is proportional to the electric wave:

$$\mathbf{A} = \frac{i}{\omega}\mathbf{E} .$$

7.4.2 Solution of the Inhomogeneous Wave Equations (Retardation)

In the presence of charges or currents, the wave equations for $\mathbf{A}$ and φ become inhomogeneous. This poses the question on how to solve these, which we will discuss for φ in conjunction with eq. (7.188). As long as we use Cartesian coordinates, this result can easily be ported to the three components of (7.187). First, however, we need to deal with the following equation:

$$\nabla^2\varphi(\mathbf{r}, t) - \varepsilon\mu\frac{\partial^2\varphi(\mathbf{r}, t)}{\partial t^2} = -\frac{\rho(\mathbf{r}, t)}{\varepsilon} . \tag{7.189}$$

We will start with a time dependent charge at the origin

$$\rho(\mathbf{r}, t) = Q(t)\delta(\mathbf{r}) , \tag{7.190}$$

and later generalize the result. This problem is spherically symmetric. Furthermore, there are no volume charges for $r > 0$. Therefore, using (3.43) we can write

$$\frac{1}{r^2}\frac{\partial}{\partial r}r^2\frac{\partial}{\partial r}\varphi(r, t) - \frac{1}{c^2}\frac{\partial^2}{\partial t^2}[\varphi(r, t)] = 0 .$$

Re-writing this a little gives

$$\frac{\partial^2}{\partial r^2}[r\varphi(r,t)] - \frac{1}{c^2}\frac{\partial^2}{\partial t^2}[r\varphi(r,t)] = 0 \ .$$

D'Alembert's general solution for this equation is

$$r\varphi(r,t) = f_1\left(t - \frac{r}{c}\right) + f_2\left(t + \frac{r}{c}\right) \ . \tag{7.191}$$

f_1 describes a process originating at the origin, while f_2 describes a process terminating at the origin. In other words, f_1 describes an effect traveling with the speed of light from the origin to a point a distance r away, where it arrives at time

$$t = \frac{r}{c} \ .$$

For this reason, f_1 is called the *retarded (delayed) solution.*
Conversely, f_2 may not be caused by processes at the origin, since they would have arrived even before their cause (namely at time $t = -r/c$). f_2 is therefore called the *advanced solution.* This contradicts the causality principle and, therefore, may not be considered. Only f_1 has physical meaning. This additional requirement is called the condition of outward radiation. Consequently, the potential has to be of the form

$$\varphi(r,t) = \frac{1}{r}f_1\left(t - \frac{r}{c}\right) \ . \tag{7.192}$$

Also, for $r \to 0$ the potential must be

$$\varphi(r,t) = \frac{Q(t)}{4\pi\varepsilon r} \ . \tag{7.193}$$

Overall, the potential for the time dependent charge at the origin becomes:

$$\varphi(r,t) = \frac{Q\left(t - \frac{r}{c}\right)}{4\pi\varepsilon r} \ . \tag{7.194}$$

The potential of an arbitrary charge distribution can be calculated by superposition. Consider the volume charges

$$\rho(\mathbf{r},t) \ .$$

Their share in the volume element $d\tau'$ at location $\mathbf{r}'$ is

$$dQ(\mathbf{r}',t) = \rho(\mathbf{r}',t)d\tau' \ .$$

Its contribution to the potential is given by

$$d\varphi(r,t) = \frac{\rho\left(\mathbf{r}', t - \frac{|\mathbf{r} - \mathbf{r}'|}{c}\right)d\tau'}{4\pi\varepsilon|\mathbf{r} - \mathbf{r}'|} \ .$$

Overall, the potential becomes then

$$\varphi = \int \frac{\rho\left(\mathbf{r}', t - \frac{|\mathbf{r} - \mathbf{r}'|}{c}\right)}{4\pi\varepsilon|\mathbf{r} - \mathbf{r}'|} \, d\tau' \,. \tag{7.195}$$

In similar manner we find the solution to the components of the vector potential

$$A_x = \int \frac{\mu g_x\left(\mathbf{r}', t - \frac{|\mathbf{r} - \mathbf{r}'|}{c}\right)}{4\pi|\mathbf{r} - \mathbf{r}'|} \, d\tau'$$

$$A_y = \int \frac{\mu g_y\left(\mathbf{r}', t - \frac{|\mathbf{r} - \mathbf{r}'|}{c}\right)}{4\pi|\mathbf{r} - \mathbf{r}'|} \, d\tau' \tag{7.196}$$

$$A_z = \int \frac{\mu g_z\left(\mathbf{r}', t - \frac{|\mathbf{r} - \mathbf{r}'|}{c}\right)}{4\pi|\mathbf{r} - \mathbf{r}'|} \, d\tau'$$

It shall be emphasized again, that the quantities $\mathbf{g}(\mathbf{r}, t)$ and $\rho(\mathbf{r}, t)$ in the solutions (7.195) and (7.196) have to satisfy the continuity equation. If, and only if this is the case, then the potentials described in (7.195) and (7.196) satisfy the Lorentz gauge. Similarly to the case of magnetostatics (see Sect. 5.1), the Lorentz gauge is not a consequence of continuity (as sometimes claimed). For every gauge choice, there will be contradictions if the continuity equation is violated.

Let us specifically look at a charged particle moving along a given trajectory $\mathbf{r}_0(t)$ then:

$$\rho(t) = Q(t)\delta[\mathbf{r} - \mathbf{r}_0(t)]$$

and

$$\mathbf{g}(t) = Q(t)\mathbf{v}_0(t)\delta[\mathbf{r} - \mathbf{r}_0(t)] \,.$$

With this, eqs. (7.195) and (7.196) yield the so-called Liénard-Wiechert potentials, which we will discuss in appendix A4.

7.4.3 The Electric Hertz Vector

The Lorentz gauge (7.186) makes it seemingly dispensable to work with two potentials $\mathbf{A}$ and φ, since φ can be eliminated by means of this formula. We will pursue this and define a new vector $\mathbf{\Pi}_e$, the *electric Hertz vector* in the following way:

$$\mu\varepsilon\frac{\partial\mathbf{\Pi}_e}{\partial t} = \mathbf{A} \,. \tag{7.197}$$

It follows form (7.186) that

$$\mu\varepsilon\frac{\partial}{\partial t}(\nabla\bullet\mathbf{\Pi}_e) + \mu\varepsilon\frac{\partial\varphi}{\partial t} = 0 \,.$$

and we may substitute

$$\boxed{\varphi = -\nabla\bullet\mathbf{\Pi}_e}\ .\tag{7.198}$$

Thus, we can express φ and $\mathbf{A}$ by means of $\mathbf{\Pi}_e$. The fields result from (7.183) and (7.184):

$$\boxed{\mathbf{B} = \mu\varepsilon\nabla\times\frac{\partial\mathbf{\Pi}_e}{\partial t}}\ ,\tag{7.199}$$

$$\boxed{\mathbf{E} = \nabla(\nabla\bullet\mathbf{\Pi}_e) - \mu\varepsilon\frac{\partial^2\mathbf{\Pi}_e}{\partial t^2}}\ .\tag{7.200}$$

7.4.4 Vector Potential for *D* and Magnetic Hertz Vector

For a region, free of volume charges, one finds according to (7.178)

$$\nabla\bullet\mathbf{D} = 0$$

and therefore, $\mathbf{D}$ may be expressed as

$$\boxed{\mathbf{D} = -\nabla\times\mathbf{A}^*}\ .\tag{7.201}$$

(Note that we use $\mathbf{A}^*$ to represent the electric vector potential. The asterisk shall differentiate it from the magnetic vector potential $\mathbf{A}$ and here, is not meant to indicate the complex conjugate. This applies also to the quantity φ^*, which we will introduce shortly.) If the region is current free, then according to (7.176)

$$\nabla\times\mathbf{H} = \frac{\partial\mathbf{D}}{\partial t} = \frac{\partial}{\partial t}(-\nabla\times\mathbf{A}^*)\ ,$$

$$\nabla\times\left(\mathbf{H} + \frac{\partial}{\partial t}\mathbf{A}^*\right) = 0\ ,$$

i.e.,

$$\boxed{\mathbf{H} = -\nabla\varphi^* - \frac{\partial}{\partial t}\mathbf{A}^*}\ .\tag{7.202}$$

The two eqs. (7.201) and (7.202) are analogous to eqs. (7.183), (7.184). $\mathbf{A}^*$ is a new kind of vector potential and φ^* a scalar potential for $\mathbf{H}$, which we have met before under a different notation (see Sect. 5.1, eq. (5.22) in particular). With the fields of (7.201) and (7.202), eqs. (7.176) and (7.178) are automatically satisfied. One also needs to consider the other two of Maxwell's equations. Using (7.175) one gets:

$$\nabla\times\left(-\frac{1}{\varepsilon}\nabla\times\mathbf{A}^*\right) = \mu\frac{\partial}{\partial t}\nabla\varphi^* + \mu\frac{\partial^2\mathbf{A}^*}{\partial t^2}$$

or

$$-\nabla(\nabla\bullet\mathbf{A}^*) + \nabla^2\mathbf{A}^* = \mu\varepsilon\nabla\frac{\partial}{\partial t}\varphi^* + \mu\varepsilon\frac{\partial^2\mathbf{A}^*}{\partial t^2}\ .$$

Gradient and time derivative commute. Regarding $\mathbf{A}^*$ and φ^*, one has similar freedom as for $\mathbf{A}$ and φ. Choosing again the Lorentz gauge,

$$\boxed{\nabla \bullet \mathbf{A}^* + \mu\varepsilon\frac{\partial}{\partial t}\varphi^* = 0}, \tag{7.203}$$

from which one obtains

$$\boxed{\nabla^2\mathbf{A}^* - \mu\varepsilon\frac{\partial^2\mathbf{A}^*}{\partial t^2} = 0}. \tag{7.204}$$

Finally, with (7.177) one finds

$$\nabla \bullet \left(-\mu\nabla\varphi^* - \mu\frac{\partial\mathbf{A}^*}{\partial t}\right) = -\mu\nabla^2\varphi^* + \mu\mu\varepsilon\frac{\partial^2\varphi^*}{\partial t^2} = 0$$

i.e.,

$$\boxed{\nabla^2\varphi^* - \mu\varepsilon\frac{\partial^2\varphi^*}{\partial t^2} = 0}. \tag{7.205}$$

We conclude that the Lorentz gauge (7.203) is responsible that the homogeneous wave equations (7.204) and (7.205) for $\mathbf{A}^*$ and φ^* emerge from the Maxwell's equations (7.175) and (7.177).

Now defining

$$\boxed{\mathbf{A}^* = \mu\varepsilon\frac{\partial\mathbf{\Pi}_m}{\partial t}} \tag{7.206}$$

and

$$\boxed{\varphi^* = -\nabla \bullet \mathbf{\Pi}_m}. \tag{7.207}$$

This satisfies the Lorentz gauge (7.203) and one can calculate the fields of (7.201) and (7.202) in the following way

$$\boxed{\mathbf{H} = \nabla(\nabla \bullet \mathbf{\Pi}_m) - \mu\varepsilon\frac{\partial^2\mathbf{\Pi}_m}{\partial t^2}} \tag{7.208}$$

$$\boxed{\mathbf{D} = -\mu\varepsilon\nabla \times \frac{\partial\mathbf{\Pi}_m}{\partial t}}. \tag{7.209}$$

These two expressions should be compared to (7.199) and (7.200). The vector introduced here is called the *magnetic Hertz vector* or also the *Fitzgerald vector*.

Because all this is largely in analogy to the previous section, we will be brief here. It shall be emphasized once more, that every formula in this section has its "dual" formula in a previous section.

According to (7.197) and (7.198), as well as (7.206) and (7.207), the Hertz vectors allow one to calculate the potentials $\mathbf{A}$ and φ as well as $\mathbf{A}^*$ and φ^*. Thus one might say, the Hertz vectors are potentials to calculate potentials, which sometimes is accounted for by calling them *super potentials*.

7.4.5 Hertz Vectors and Dipole Moments

The two Hertz vectors are a useful tool for many field calculations, and we will take advantage of them in this context.

In the general case, it is possible to calculate the fields from $\mathbf{\Pi}_m$ and $\mathbf{\Pi}_e$, and then use superposition, so that according to (7.199), (7.200) and (7.208), (7.209), one obtains

$$\mathbf{H} = \varepsilon \nabla \times \frac{\partial \mathbf{\Pi}_e}{\partial t} + \nabla(\nabla \bullet \mathbf{\Pi}_m) - \mu\varepsilon \frac{\partial^2 \mathbf{\Pi}_m}{\partial t^2} \qquad (7.210)$$

$$\mathbf{E} = -\mu \nabla \times \frac{\partial \mathbf{\Pi}_m}{\partial t} + \nabla(\nabla \bullet \mathbf{\Pi}_e) - \mu\varepsilon \frac{\partial^2 \mathbf{\Pi}_e}{\partial t^2} \ . \qquad (7.211)$$

Now, based on these fields, we want to discuss the case when there are "permanent" electric or magnetic dipoles present, besides the just described polarization or magnetization effects caused by ε and μ. Allow these "permanent" dipole moments to be time dependent. The word "permanent" shall mean that these thereby addressed dipoles are not caused or "induced" by an applied electric or magnetic field. Maxwell's equations (7.175) through (7.178) apply,

$$\mathbf{B} = \mu\mathbf{H} + \mathbf{M} \qquad (7.212)$$

$$\mathbf{D} = \varepsilon\mathbf{E} + \mathbf{P} \ . \qquad (7.213)$$

The induced polarization and magnetization effects are contained in the constants ε and μ. $\mathbf{M}$ and $\mathbf{P}$ represent the permanent fraction of the fields. Using this in Maxwell's equations gives:

$$\nabla \times \mathbf{E} = -\mu \frac{\partial}{\partial t}\mathbf{H} - \frac{\partial}{\partial t}\mathbf{M} \qquad (7.214)$$

$$\nabla \times \mathbf{H} = \mathbf{g} + \varepsilon\frac{\partial}{\partial t}\mathbf{E} + \frac{\partial}{\partial t}\mathbf{P} \qquad (7.215)$$

$$\mu\nabla \bullet \mathbf{H} + \nabla \bullet \mathbf{M} = 0 \qquad (7.216)$$

$$\varepsilon\nabla \bullet \mathbf{E} + \nabla \bullet \mathbf{P} = \rho \ . \qquad (7.217)$$

This is a very interesting form of Maxwell's equations. It contains the polarization current $\partial\mathbf{P}/\partial t$ which results from the displacement current $\partial\mathbf{D}/\partial t$, if $\mathbf{D}$ is chosen according to (7.213). However, this is not necessary, as we have seen in Sect. 2.13. The polarization current might contain an additional source free term. This would mean that during polarization, the transport of charges does not take the shortest route. Also interesting are the terms $\partial\mathbf{M}/\partial t$ and $\nabla \bullet \mathbf{M}$. If one takes advantage of the concept of magnetic charges, then we might think of these in terms of $\mathbf{g}_m$ and ρ_m, as of eqs. 1.82, i.e., magnetic current densities and magnetic volume charges. Although these charges are entirely fictitious, they may still be useful.

Now, we will find out if these equations can be satisfied by means of fields given by (7.210) and (7.211). We substitute these and after some algebra, which we skip, find in this order:

$$\frac{\partial}{\partial t}\left(\nabla^2\Pi_m - \mu\varepsilon\frac{\partial^2\Pi_m}{\partial t^2} + \frac{\mathbf{M}}{\mu}\right) = 0 \ , \tag{7.218}$$

$$\frac{\partial}{\partial t}\left(\nabla^2\Pi_e - \mu\varepsilon\frac{\partial^2\Pi_e}{\partial t^2} + \frac{\mathbf{P}}{\varepsilon}\right) = 0 \ , \tag{7.219}$$

$$\nabla\bullet\left(\nabla^2\Pi_m - \mu\varepsilon\frac{\partial^2\Pi_m}{\partial t^2} + \frac{\mathbf{M}}{\mu}\right) = 0 \ , \tag{7.220}$$

$$\nabla\bullet\left(\nabla^2\Pi_e - \mu\varepsilon\frac{\partial^2\Pi_e}{\partial t^2} + \frac{\mathbf{P}}{\varepsilon}\right) = 0 \ , \tag{7.221}$$

where we assumed that

$\rho = 0$

$\mathbf{g} = 0$.

Thus, there are no free charges and no free currents (nevertheless, there are bound charges and bound currents, both magnetization currents and polarization currents. From (7.218) and (7.220) we would at first conclude that

$$\nabla^2\Pi_m(\mathbf{r}, t) - \mu\varepsilon\frac{\partial^2\Pi_m(\mathbf{r}, t)}{\partial t^2} + \frac{\mathbf{M}(\mathbf{r}, t)}{\mu} = \nabla\times\mathbf{C}(\mathbf{r}) \ .$$

In this case, $\mathbf{C}$ is an arbitrary time independent vector. Considering a time dependent magnetization $\mathbf{M}$ as the only cause of potentially existing fields (described by Π_m), then it turns out that $\mathbf{C}(\mathbf{r})$ must vanish. A similar approach for the two equations (7.219) and (7.221) eventually yields

$$\boxed{\nabla^2\Pi_m - \mu\varepsilon\frac{\partial^2\Pi_m}{\partial t^2} = -\frac{\mathbf{M}}{\mu}} \tag{7.222}$$

$$\boxed{\nabla^2\Pi_e - \mu\varepsilon\frac{\partial^2\Pi_e}{\partial t^2} = -\frac{\mathbf{P}}{\varepsilon}} \ . \tag{7.223}$$

These make it obvious that the wave equation also applies to the Hertz vectors. The electric polarization and "magnetic polarization" (= Magnetization) are the inhomogeneities. Therefore, the Hertz vectors are also called *polarization potentials*. Solving eqs (7.222) and (7.223) for given $\mathbf{M}$ or $\mathbf{P}$ is carried out in the same manner as described in Sect. 7.4.2, *i.e.*, in analogy to the results (7.195) and (7.196).

The homogeneous wave equations apply to regions where $\mathbf{M}$ or $\mathbf{P}$ vanish:

$$\nabla^2\Pi_m - \mu\varepsilon\frac{\partial^2\Pi_m}{\partial t^2} = \nabla(\nabla\bullet\Pi_m) - \nabla\times(\nabla\times\Pi_m) - \mu\varepsilon\frac{\partial^2\Pi_m}{\partial t^2} = 0$$

$$\nabla^2 \Pi_e - \mu\varepsilon\frac{\partial^2 \Pi_e}{\partial t^2} = \nabla(\nabla\bullet\Pi_e) - \nabla\times(\nabla\times\Pi_e) - \mu\varepsilon\frac{\partial^2 \Pi_e}{\partial t^2} = 0 ,$$

i.e.,

$$\nabla(\nabla\bullet\Pi_m) - \mu\varepsilon\frac{\partial^2 \Pi_m}{\partial t^2} = \nabla\times(\nabla\times\Pi_m)$$

$$\nabla(\nabla\bullet\Pi_e) - \mu\varepsilon\frac{\partial^2 \Pi_e}{\partial t^2} = \nabla\times(\nabla\times\Pi_e) .$$

Now, instead of (7.210) and (7.211), and under the condition that $\mathbf{M} = 0$ as well as $\mathbf{P} = 0$, we write

$$\mathbf{H} = \nabla\times(\nabla\times\Pi_m) + \varepsilon\nabla\times\frac{\partial \Pi_e}{\partial t} \qquad (7.224)$$

$$\mathbf{E} = \nabla\times(\nabla\times\Pi_e) - \mu\nabla\times\frac{\partial \Pi_m}{\partial t} . \qquad (7.225)$$

The next few sections shall be used to discuss radiation of an oscillating electric dipole (dipole antenna) and the radiation of an oscillating magnetic dipole (frame antenna). Furthermore, we shall study the wave propagation in cylindrical wave guides. For all this, we will find that the just developed methods and terminology will be extremely useful.

7.4.6   Potentials for Uniformly Conductive Media without Volume Charges

In the above sections, we have discussed Maxwell's equations for given current densities $\mathbf{g}$ and volume charges ρ. The current densities can not be prescribed inside a uniformly conductive medium, rather – applying Ohm's law – one has

$$\mathbf{g} = \kappa\mathbf{E} .$$

Volume charges, one the other hand, will vanish very quickly and therefore, shall be neglected. Thus, taking Maxwell's equations in the following form:

$$\nabla\times\mathbf{E} = -\frac{\partial\mathbf{B}}{\partial t} \qquad (7.226)$$

$$\nabla\times\mathbf{H} = \kappa\mathbf{E} + \frac{\partial\mathbf{D}}{\partial t} \qquad (7.227)$$

$$\nabla\bullet\mathbf{B} = 0 \qquad (7.228)$$

$$\nabla\bullet\mathbf{D} = 0 . \qquad (7.229)$$

These are solved with the Ansatz

$$\mathbf{B} = \nabla\times\mathbf{A} \qquad (7.230)$$

$$\mathbf{E} = -\nabla\varphi - \frac{\partial\mathbf{A}}{\partial t} ,\tag{7.231}$$

which automatically satisfy eqs. (7.226) and (7.228). The other two, eqs. (7.227) and (7.229), in conjunction with the gauge choice

$$\nabla\bullet\mathbf{A} + \mu\kappa\varphi + \mu\varepsilon\frac{\partial\varphi}{\partial t} = 0 ,\tag{7.232}$$

yield the homogeneous equations

$$\nabla^2\mathbf{A} - \mu\kappa\frac{\partial\mathbf{A}}{\partial t} - \mu\varepsilon\frac{\partial^2\mathbf{A}}{\partial t^2} = 0\tag{7.233}$$

$$\nabla^2\varphi - \mu\kappa\frac{\partial\varphi}{\partial t} - \mu\varepsilon\frac{\partial^2\varphi}{\partial t^2} = 0 .\tag{7.234}$$

Letting

$$\mathbf{A} = \left(\mu\kappa + \mu\varepsilon\frac{\partial}{\partial t}\right)\mathbf{\Pi}_e\tag{7.235}$$

$$\varphi = -\nabla\bullet\mathbf{\Pi}_e ,\tag{7.236}$$

satisfies the gauge condition (7.232). Then, the wave equations (7.233), (7.234) yields

$$\left(\mu\kappa + \mu\varepsilon\frac{\partial}{\partial t}\right)\left[\nabla^2\mathbf{\Pi}_e - \mu\kappa\frac{\partial\mathbf{\Pi}_e}{\partial t} - \mu\varepsilon\frac{\partial^2\mathbf{\Pi}_e}{\partial t^2}\right] = 0$$

$$\nabla\bullet\left[\nabla^2\mathbf{\Pi}_e - \mu\kappa\frac{\partial\mathbf{\Pi}_e}{\partial t} - \mu\varepsilon\frac{\partial^2\mathbf{\Pi}_e}{\partial t^2}\right] = 0$$

which is satisfied if

$$\boxed{\nabla^2\mathbf{\Pi}_e - \mu\kappa\frac{\partial\mathbf{\Pi}_e}{\partial t} - \mu\varepsilon\frac{\partial^2\mathbf{\Pi}_e}{\partial t^2} = 0} .\tag{7.237}$$

With this, one can calculate $\mathbf{B}$ and $\mathbf{E}$ from $\mathbf{\Pi}_e$ in the following manner:

$$\boxed{\mathbf{B} = \left(\mu\kappa + \mu\varepsilon\frac{\partial}{\partial t}\right)\nabla\times\mathbf{\Pi}_e}\tag{7.238}$$

$$\boxed{\mathbf{E} = \nabla(\nabla\bullet\mathbf{\Pi}_e) - \left(\mu\kappa\frac{\partial}{\partial t} + \mu\varepsilon\frac{\partial^2}{\partial t^2}\right)\mathbf{\Pi}_e = \nabla\times(\nabla\times\mathbf{\Pi}_e)} .\tag{7.239}$$

A different approach is also possible. Now, starting with the Ansatz

$$\mathbf{D} = -\nabla\times\mathbf{A}^*\tag{7.240}$$

$$\mathbf{H} = -\nabla\varphi^* - \left(\frac{\kappa}{\varepsilon} + \frac{\partial}{\partial t}\right)\mathbf{A}^* ,\tag{7.241}$$

which satisfy the two eqs. (7.227) and (7.229), while the other two eqs. (7.226) and (7.228), together with the gauge choice

$$\nabla \bullet \mathbf{A}^* + \mu \varepsilon \frac{\partial}{\partial t} \varphi^* = 0 , \tag{7.242}$$

yields the homogeneous equations

$$\nabla^2 \mathbf{A}^* - \mu \kappa \frac{\partial \mathbf{A}^*}{\partial t} - \mu \varepsilon \frac{\partial^2 \mathbf{A}^*}{\partial t^2} = 0 \tag{7.243}$$

$$\nabla^2 \varphi^* - \mu \kappa \frac{\partial \varphi^*}{\partial t} - \mu \varepsilon \frac{\partial^2 \varphi^*}{\partial t^2} = 0 . \tag{7.244}$$

Letting

$$\mathbf{A}^* = \mu \varepsilon \frac{\partial}{\partial t} \mathbf{\Pi}_m \tag{7.245}$$

$$\varphi^* = -\nabla \bullet \mathbf{\Pi}_m , \tag{7.246}$$

satisfies the gauge condition (7.242). Now, the wave equations (7.243) and (7.244) yield

$$\frac{\partial}{\partial t}\left[\nabla^2 \mathbf{\Pi}_m - \mu \kappa \frac{\partial \mathbf{\Pi}_m}{\partial t} - \mu \varepsilon \frac{\partial^2 \mathbf{\Pi}_m}{\partial t^2} \right] = 0$$

$$\nabla \bullet \left[\nabla^2 \mathbf{\Pi}_m - \mu \kappa \frac{\partial \mathbf{\Pi}_m}{\partial t} - \mu \varepsilon \frac{\partial^2 \mathbf{\Pi}_m}{\partial t^2} \right] = 0$$

$$\boxed{\nabla^2 \mathbf{\Pi}_m - \mu \kappa \frac{\partial \mathbf{\Pi}_m}{\partial t} - \mu \varepsilon \frac{\partial^2 \mathbf{\Pi}_m}{\partial t^2} = 0} . \tag{7.247}$$

And finally, one obtains

$$\boxed{\mathbf{D} = -\mu \varepsilon \frac{\partial}{\partial t} \nabla \times \mathbf{\Pi}_m} . \tag{7.248}$$

$$\boxed{\mathbf{H} = \nabla(\nabla \bullet \mathbf{\Pi}_m) - \left(\mu \kappa \frac{\partial}{\partial t} + \mu \varepsilon \frac{\partial^2}{\partial t^2}\right) \mathbf{\Pi}_m = \nabla \times (\nabla \times \mathbf{\Pi}_m)} . \tag{7.249}$$

7.5   Hertz's Dipole

7.5.1   Fields of Oscillating Dipoles

Consider a dipole located at the origin, oriented along the z-axis, and oscillating in time:

$$\mathbf{p} = \mathbf{e}_z p_0 \sin \omega t . \tag{7.250}$$

Its polarization, defined as the spatial density of the dipole moment is

$$\mathbf{P} = p_0 \sin(\omega t)\delta(\mathbf{r})\mathbf{e}_z \ . \tag{7.251}$$

The temporal change of the dipole moment is linked to currents. According to Fig. 7.16, we get

$$p = lQ \tag{7.252}$$

or

$$\frac{dp}{dt} = l\frac{dQ}{dt} = lI = \omega p_0 \cos\omega t \tag{7.253}$$

and therefore

$$I = \frac{\omega p_0}{l}\cos\omega t = I_0 \cos\omega t \ , \tag{7.254}$$

where

$$I_0 = \frac{\omega p_0}{l} \ . \tag{7.255}$$

To calculate the field of the oscillating dipole, one uses eq. (7.223), which may be solved similarly to (7.188) and whose solution was (7.195). Since $\mathbf{P}$ has only a z-component, the electric Hertz vector reduces to

$$\Pi_{ex} = 0 \ . \tag{7.256}$$

$$\Pi_{ey} = 0 \ . \tag{7.257}$$

$$\Pi_{ez} = \frac{1}{4\pi\varepsilon} \int \frac{P_z\left(\mathbf{r}', t - \dfrac{|\mathbf{r} - \mathbf{r}'|}{c}\right)}{|\mathbf{r} - \mathbf{r}'|} d\tau'$$

$$= \frac{1}{4\pi\varepsilon} \int \frac{p_0 \sin\left[\omega\left(t - \dfrac{|\mathbf{r} - \mathbf{r}'|}{c}\right)\right]\delta(\mathbf{r}')}{|\mathbf{r} - \mathbf{r}'|} d\tau'$$

$$\Pi_{ez} = \frac{p_0}{4\pi\varepsilon r}\sin\left[\omega\left(t - \frac{r}{c}\right)\right] \ . \tag{7.258}$$

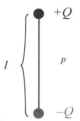

Fig. 7.16

r is the distance from the observation point to the origin and, hence, to the oscillating dipole. To use spherical coordinates is advisable. The transformed Hertz vector is

$$\Pi_{er} = \Pi_{ez}\cos\theta = \frac{p_0\cos\theta}{4\pi\varepsilon r}\sin\left[\omega\left(t-\frac{r}{c}\right)\right] \tag{7.259}$$

$$\Pi_{e\theta} = -\Pi_{ez}\sin\theta = -\frac{p_0\sin\theta}{4\pi\varepsilon r}\sin\left[\omega\left(t-\frac{r}{c}\right)\right] \tag{7.260}$$

$$\boxed{\Pi_{e\varphi} = 0} . \tag{7.261}$$

The fields are calculated according to (7.210), (7.211) and (7.224), (7.225), respectively, and yield

$$\mathbf{H} = \varepsilon\nabla\times\frac{\partial\mathbf{\Pi}_e}{\partial t} \tag{7.262}$$

$$\mathbf{E} = \nabla(\nabla\bullet\mathbf{\Pi}_e) - \mu\varepsilon\frac{\partial^2\mathbf{\Pi}_e}{\partial t^2} = \nabla\times(\nabla\times\mathbf{\Pi}_e) , \tag{7.263}$$

where it has to be pointed out that the expression to the far right of (7.263) is only valid for locations where $\mathbf{P} = 0$ (*i.e.*, in our case, outside the origin). By means of the formulas from Sect. 3.3.3 and some algebra, which we skip here, one obtains:

$$\mathbf{E} = \begin{bmatrix} E_r \\ E_\theta \\ E_\varphi \end{bmatrix} = \begin{bmatrix} \frac{2p_0\cos\theta}{4\pi\varepsilon}\left\{\frac{1}{r^3}\sin\left[\omega\left(t-\frac{r}{c}\right)\right] + \frac{\omega}{cr^2}\cos\left[\omega\left(t-\frac{r}{c}\right)\right]\right\} \\ \frac{p_0\sin\theta}{4\pi\varepsilon}\left\{\left(\frac{1}{r^3} - \frac{\omega^2}{rc^2}\right)\sin\left[\omega\left(t-\frac{r}{c}\right)\right] + \frac{\omega}{cr^2}\cos\left[\omega\left(t-\frac{r}{c}\right)\right]\right\} \\ 0 \end{bmatrix} \tag{7.264}$$

$$\mathbf{H} = \begin{bmatrix} H_r \\ H_\theta \\ H_\varphi \end{bmatrix} = \begin{bmatrix} 0 \\ 0 \\ \frac{\omega p_0\sin\theta}{4\pi}\left\{\left(-\frac{\omega}{rc}\right)\sin\left[\omega\left(t-\frac{r}{c}\right)\right] + \frac{1}{r^2}\cos\left[\omega\left(t-\frac{r}{c}\right)\right]\right\} \end{bmatrix} \tag{7.265}$$

Of course, we might as well use (7.197) and (7.198), to calculate $\mathbf{A}$ and φ:

$$\mathbf{A} = \mu\varepsilon\frac{\partial\mathbf{\Pi}_e}{\partial t} = \begin{bmatrix} A_r \\ A_\theta \\ A_\varphi \end{bmatrix} = \begin{bmatrix} \dfrac{\mu p_0 \omega \cos\theta}{4\pi r}\cos\left[\omega\left(t - \dfrac{r}{c}\right)\right] \\ -\dfrac{\mu p_0 \omega \sin\theta}{4\pi r}\cos\left[\omega\left(t - \dfrac{r}{c}\right)\right] \\ 0 \end{bmatrix} \qquad (7.266)$$

$$\varphi = -\nabla\bullet\mathbf{\Pi}_e = \frac{p_0\cos\theta}{4\pi\varepsilon}\left\{\frac{1}{r^2}\sin\left[\omega\left(t - \frac{r}{c}\right)\right] + \frac{\omega}{cr}\cos\left[\omega\left(t - \frac{r}{c}\right)\right]\right\}. \qquad (7.267)$$

The effects of retardation occur in the ubiquitous argument $t - r/c$. If the speed of light were infinite, there would be no retardation. An interesting task is to determine what kind of fields would result in this limit ($c \to \infty$). One finds

$$\mathbf{E} = \begin{bmatrix} E_r \\ E_\theta \\ E_\varphi \end{bmatrix} = \begin{bmatrix} \dfrac{2p_0\cos\theta\sin\omega t}{4\pi\varepsilon r^3} \\ \dfrac{p_0\sin\theta\sin\omega t}{4\pi\varepsilon r^3} \\ 0 \end{bmatrix} = \begin{bmatrix} \dfrac{2p\cos\theta}{4\pi\varepsilon r^3} \\ \dfrac{p\sin\theta}{4\pi\varepsilon r^3} \\ 0 \end{bmatrix} \qquad (7.268)$$

and

$$\mathbf{H} = \begin{bmatrix} H_r \\ H_\theta \\ H_\varphi \end{bmatrix} = \begin{bmatrix} 0 \\ 0 \\ \dfrac{\omega p_0\sin\theta\cos\omega t}{4\pi r^2} \end{bmatrix}. \qquad (7.269)$$

Comparison with eqs. (2.63) reveals that this limit results in the "static" dipole field. Its time dependency follows exactly all changes of the dipole at the origin, *i.e.*, all changes are instantaneously apparent everywhere in the entire space, as expected for an infinite speed of light. For better appreciation of (7.269), we express H_φ in a slightly different form. Applying eqs. (7.254) and (7.255) gives

$$H_\varphi = \frac{I_0 l\sin\theta\cos\omega t}{4\pi r^2} = \frac{Il\sin\theta}{4\pi r^2}. \qquad (7.270)$$

According to (5.20), this can be understood as the field of the current I in the conductor element $l\mathbf{e}_z$, *i.e.*, it represents the field corresponding to Biot-Savart's law and is noticeable instantaneously in the entire space. In contrast to magnetostatics, here it is permissible to regard current carrying line elements (*i.e.*, currents with sources), because we also consider the correlated time dependent charges (here, represented by the time dependent dipole).

The magnetic field has according to (7.265) only an azimuthal component. Thus, the magnetic field lines are concentric circles around the z-axis. The electric field lines are situated in the meridian plane $\varphi = \text{const.}$. Their equations can be stated, whereby it is beneficial to apply some analogy from the discussion of Sect. 5.11. With (7.263) we may write

$$\mathbf{E} = \nabla \times \mathbf{C} \ . \tag{7.271}$$

if

$$\mathbf{C} = \nabla \times \mathbf{\Pi}_e = \begin{bmatrix} C_r \\ C_\theta \\ C_\varphi \end{bmatrix} = \begin{bmatrix} 0 \\ 0 \\ \dfrac{p_0 \sin\theta}{4\pi\varepsilon r}\left(\dfrac{1}{r}\sin\left[\omega\left(t-\dfrac{r}{c}\right)\right] + \dfrac{\omega}{c}\cos\left[\omega\left(t-\dfrac{r}{c}\right)\right]\right) \end{bmatrix}, \tag{7.272}$$

$\mathbf{C}$ has only an azimuthal component and therefore, the electric field becomes

$$\mathbf{E} = \begin{bmatrix} E_r \\ E_\theta \\ E_\varphi \end{bmatrix} = \begin{bmatrix} \dfrac{1}{r\sin\theta}\dfrac{\partial}{\partial\theta}(\mathbf{C}_\varphi \sin\theta) \\ -\dfrac{1}{r}\dfrac{\partial}{\partial r}(r\mathbf{C}_\varphi) \\ 0 \end{bmatrix}. \tag{7.273}$$

Now, take the function $r\sin\theta\,\mathbf{C}_\varphi$ (it corresponds to the function $r\mathbf{A}_\varphi$ of Sect. 5.11, but notice that there, we had used cylindrical coordinates, while here we use spherical coordinates, which is the reason why $r\sin\theta$ replaces r). We find its gradient with (3.41)

$$\nabla(r\sin\theta\mathbf{C}_\varphi) = \begin{bmatrix} \dfrac{\partial}{\partial r}(r\sin\theta\mathbf{C}_\varphi) \\ \dfrac{1}{r}\dfrac{\partial}{\partial\theta}(r\sin\theta\mathbf{C}_\varphi) \\ 0 \end{bmatrix} = \begin{bmatrix} \sin\theta\dfrac{\partial}{\partial r}(r\mathbf{C}_\varphi) \\ \dfrac{\partial}{\partial\theta}(\sin\theta\mathbf{C}_\varphi) \\ 0 \end{bmatrix}. \tag{7.274}$$

This shows that

$$\mathbf{E} \bullet \nabla(r\sin\theta\mathbf{C}_\varphi) = 0 \ .$$

Thus $\mathbf{E}$ is perpendicular to the gradient of $r\sin\theta\mathbf{C}_\varphi$, *i.e.*, $\mathbf{E}$ lies within the lines of the meridian planes, along which $r\sin\theta\mathbf{C}_\varphi$ is constant. In other words, the function $r\sin\theta\mathbf{C}_\varphi$ can be regarded as flux function. With

$$\frac{\omega}{c} = k \ , \tag{7.275}$$

one can rewrite this:

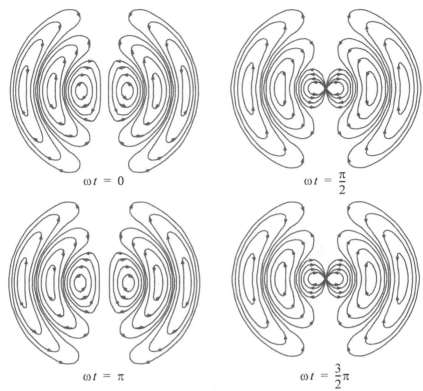

$\omega t = 0$

$\omega t = \dfrac{\pi}{2}$

$\omega t = \pi$

$\omega t = \dfrac{3}{2}\pi$

Fig. 7.17

$$r\sin\theta\,\mathbf{C}_\varphi = \frac{p_0 k\sin^2\theta}{4\pi\varepsilon}\left\{\frac{1}{kr}\sin[\omega t - kr] + \cos[\omega t - kr]\right\}$$

$$= \frac{p_0 k\sin^2\theta}{4\pi\varepsilon}\sqrt{\frac{1}{(kr)^2} + 1}\,\sin[\omega t - kr + \arctan(kr)] .$$

Omitting the insignificant factor $p_0 k/4\pi\varepsilon$, one can write the equations for the field lines in the form

$$\sin^2\theta\sqrt{\frac{1}{(kr)^2} + 1}\,\sin[\omega t - kr + \arctan(kr)] = \text{constant} . \tag{7.276}$$

Fig. 7.17. illustrates several field lines for the dipole.

There are many ways to express the dipole fields of eqs. (7.264) and (7.265). The following one is sometimes particularly useful:

$$E_r = \frac{2p_0\cos\theta\sqrt{1 + (kr)^2}}{4\pi\varepsilon r^3}\sin[\omega t - kr + \chi_r] \tag{7.277}$$

$$E_\theta = \frac{p_0 \sin\theta \sqrt{1-(kr)^2+(kr)^4}}{4\pi\varepsilon r^3} \sin[\omega t - kr + \chi_\theta] \qquad (7.278)$$

$$H_\varphi = \frac{\omega p_0 \sin\theta \sqrt{1+(kr)^2}}{4\pi r^2} \cos[\omega t - kr + \chi_\varphi] \; , \qquad (7.279)$$

where

$$\chi_r = \chi_\varphi = \arctan(kr) \qquad (7.280)$$

$$\chi_\theta = \text{atan}\left[\frac{kr}{1-(kr)^2}\right] . \qquad (7.281)$$

Notice that the phase angles χ_r, χ_θ, and χ_φ are function of r.

7.5.2   Far Field and Radiation Power

The representation of the dipole fields in eqs. (7.264) and (7.265) is a little difficult to grasp. However, we will discover that only few of these terms are of sufficient interest, at least with respect to radiation of electromagnetic waves by the oscillating dipole. The components of **E** contain terms that are proportional to r^{-1}, r^{-2}, and r^{-3}, while H_φ contain terms that are proportional to r^{-1} and r^{-2}. This result is particularly peculiar but also important and results from retardation, $i.e.$, is a consequence of the finite nature of c. In the "static" case, $i.e.$ if the speed of light were infinite, then **E** would be proportional to r^{-3} and H_φ to r^{-2}, $i.e.$, it would become very small for large distances.

When analyzing the energy flux through the surface of a sphere where a dipole oscillates in its center reveals, that at large distances only those terms with r^{-1} make non vanishing contributions to the field of **E** and **H**. The reason is that the corresponding part of the Poynting vector is proportional to r^{-2}, while the surface of a sphere is proportional to r^2. All other terms of the Poynting vector decay faster (namely by r^{-3}, r^{-4}, r^{-5}). Therefore, in the following we will take interest only in the *far field* of the oscillating dipole.

$$\mathbf{E} = \begin{bmatrix} E_r \\ E_\theta \\ E_\varphi \end{bmatrix} = \begin{bmatrix} 0 \\ \dfrac{-p_0\omega^2\sin\theta}{4\pi\varepsilon c^2 r}\sin[\omega t - kr] \\ 0 \end{bmatrix} , \qquad (7.282)$$

$$\mathbf{H} = \begin{bmatrix} H_r \\ H_\theta \\ H_\varphi \end{bmatrix} = \begin{bmatrix} 0 \\ 0 \\ \dfrac{-p_0\omega^2\sin\theta}{4\pi c r}\sin[\omega t - kr] \end{bmatrix} = \begin{bmatrix} 0 \\ 0 \\ \dfrac{E_\theta}{Z} \end{bmatrix} . \qquad (7.283)$$

This is now a much simpler representation of the field, which behaves basically like a plane wave (Fig. 7.18). It is purely transverse, **E** and **H** are perpendicular, etc. However, its r and θ dependency distinguishes it from a plane wave.

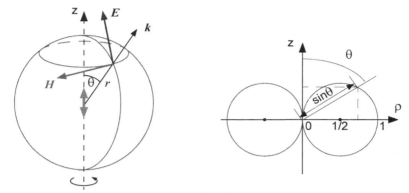

Fig. 7.18   Fig. 7.19

The angular distribution of the dipole radiation, given by $\sin\theta$ is illustrated in Fig. 7.19 by a polar plot. The amplitude of the field vanishes in z-direction, *i.e.*, the direction of the dipole orientation. It is an easy task to verify that the endpoints corresponding to $\sin\theta$ are circles

$$\left(\rho-\frac{1}{2}\right)^2 + z^2 = \frac{1}{4}$$

(7.284)

The Poynting vector of the far field is

$$\mathbf{S} = \mathbf{E} \times \mathbf{H} = \begin{vmatrix} \mathbf{e}_r & \mathbf{e}_\theta & \mathbf{e}_\varphi \\ 0 & E_\theta & 0 \\ 0 & 0 & H_\varphi \end{vmatrix} = \langle E_\theta H_\varphi, 0, 0 \rangle .$$

(7.285)

It only has an r-component, which is

$$S_r = E_\theta H_\varphi = \frac{E_\theta^2}{Z} = \frac{p_0^2 \omega^4 \sin^2\theta \sin^2(\omega t - kr)}{Z 16\pi^2 \varepsilon^2 c^4 r^2}$$

$$= \frac{p_0^2 \omega^4 \sin^2\theta \sin^2(\omega t - kr)}{16\pi^2 \varepsilon c^3 r^2} .$$

(7.286)

S_r is the energy radiated per unit time and unit area, *i.e.*, the radiation power per unit area. It depends on the direction and is proportional to $\sin^2\theta$. To find the total radiation power requires to integrate over the surface of a sphere.

$$P = \int_{\text{surface of sphere}} \mathbf{S} \bullet d\mathbf{A} = \int_{\text{surface of sphere}} S_r dA .$$

(7.287)

Substituting

$$dA = r^2 \sin^2\theta\, d\theta\, d\varphi \tag{7.288}$$

gives initially

$$P = \frac{p_0^2 \omega^4 \sin^2(\omega t - kr)}{16\pi^2 \varepsilon c^3} \int_0^{2\pi} d\varphi \int_0^{\pi} \sin^3\theta\, d\theta \ .$$

Obviously

$$\int_0^{2\pi} d\varphi = 2\pi$$

and

$$\int_0^{\pi} \sin^3\theta\, d\theta = \int_{-1}^{+1} \sin^2\theta\, d(\cos\theta) = \int_{-1}^{+1}(1-x^2)\,dx = \frac{4}{3} \ .$$

Therefore

$$\boxed{P = \frac{p_0^2 \omega^4 \sin^2(\omega t - kr)}{6\pi\varepsilon c^3}} \ . \tag{7.289}$$

P is always positive. Its time average is the effective value or root mean square (rms):

$$\boxed{P_{rms} = \frac{p_0^2 \omega^4}{12\pi\varepsilon c^3}} \ . \tag{7.290}$$

Eq. (7.255) allows one to eliminate p_0 and replace it with I_0:

$$P_{rms} = \frac{I_0^2 l^2 \omega^2}{12\pi\varepsilon c^3} = \frac{I_0^2 l^2 k^2}{12\pi}\sqrt{\frac{\mu}{\varepsilon}} = \frac{I_0^2 l^2 \pi}{3\lambda^2} Z \ . \tag{7.291}$$

Introducing the RMS current

$$I_0^2 = 2I_{rms}^2$$

allows to write the RMS power as

$$P_{rms} = \frac{2\pi}{3} Z\left(\frac{l}{\lambda}\right)^2 I_{rms}^2 \ . \tag{7.292}$$

Finally, if one defines the so-called *radiation resistance* R_s by

$$\boxed{P_{rms} = R_s I_{rms}^2} \ , \tag{7.293}$$

then because of (7.292) one obtains

$$\boxed{R_s = \frac{2\pi}{3} Z\left(\frac{l}{\lambda}\right)^2} \ . \tag{7.294}$$

More specifically, R_s for vacuum is

$$R_s = \frac{2\pi}{3}\left(\frac{l}{\lambda}\right)^2 377\Omega = \left(\frac{l}{\lambda}\right)^2 790\Omega \ . \tag{7.295}$$

This describes the radiation of an oscillating dipole, the so-called dipole antenna.

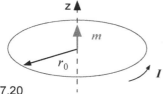

Fig. 7.20

The antenna gain G is defined as the ratio of the maximum radiation power $S_{r\max}$ (direction of maximum power) over the average radiation power $\langle S_r \rangle$ (averaging all directions). In the current case, with (7.286) and (7.289), this ratio is

$$G = \frac{S_{r\max}}{\langle S_r \rangle} = \frac{\dfrac{p_0^2 \omega^4 \sin^2(\omega t - kr)}{16\pi^2 \varepsilon c^3 r^2}}{\dfrac{1}{4\pi r^2} \dfrac{p_0^2 \omega^4 \sin^2(\omega t - kr)}{6\pi \varepsilon c^3}} = \frac{3}{2} . \tag{7.296}$$

Now, we proceed from the dipole antenna to the dual frame antenna.

7.6 Frame Antenna

An oscillating magnetic dipole radiates electromagnetic waves. Consider a magnetic dipole

$$\mathbf{m} = \mathbf{e}_z m_0 \sin \omega t \tag{7.297}$$

at the origin (Fig. 7.20). Its magnetization

$$\mathbf{M} = \mathbf{e}_z m_0 \sin \omega t \delta(\mathbf{r}) \tag{7.298}$$

is caused by a circular current

$$I = I_0 \sin \omega t , \tag{7.299}$$

where

$$m_0 = \mu I_0 r_0^2 \pi . \tag{7.300}$$

This arrangement is also called *frame antenna*. The discussion here parallels almost completely the discussion of the dipole antenna in the previous section. Therefore, one may be rather brief. First, one needs to solve the wave equation (7.222) with the specific magnetization given in (7.298), which in analogy to eqs (7.256) through (7.258) yields

$$\boxed{\Pi_{mx} = 0} , \tag{7.301}$$

$$\boxed{\Pi_{my} = 0} , \tag{7.302}$$

$$\boxed{\Pi_{mz} = \frac{m_0 \sin\left[\omega\left(t - \dfrac{r}{c}\right)\right]}{4\pi \mu r}} . \tag{7.303}$$

The fields can now be calculated from this and eqs. (7.210) and (7.211), or if outside of the origin also from (7.224) and (7.225). Thus:

$$\mathbf{H} = \nabla(\nabla\bullet\boldsymbol{\Pi}_m) - \mu\varepsilon\frac{\partial^2\boldsymbol{\Pi}_m}{\partial t^2} = \nabla\times(\nabla\times\boldsymbol{\Pi}_m) \,, \tag{7.304}$$

$$\mathbf{E} = -\mu\nabla\times\frac{\partial\boldsymbol{\Pi}_m}{\partial t} \,. \tag{7.305}$$

Some more algebra, which we skip, finally yields

$$\mathbf{H} = \begin{bmatrix} H_r \\ H_\theta \\ H_\varphi \end{bmatrix} = \begin{bmatrix} \dfrac{2m_0\cos\theta}{4\pi\mu}\left\{\dfrac{1}{r^3}\sin\left[\omega\left(t-\dfrac{r}{c}\right)\right] + \dfrac{\omega}{cr^2}\cos\left[\omega\left(t-\dfrac{r}{c}\right)\right]\right\} \\ \dfrac{m_0\sin\theta}{4\pi\mu}\left\{\left(\dfrac{1}{r^3}-\dfrac{\omega^2}{rc^2}\right)\sin\left[\omega\left(t-\dfrac{r}{c}\right)\right] + \dfrac{\omega}{cr^2}\cos\left[\omega\left(t-\dfrac{r}{c}\right)\right]\right\} \\ 0 \end{bmatrix} \tag{7.306}$$

$$\mathbf{E} = \begin{bmatrix} E_r \\ E_\theta \\ E_\varphi \end{bmatrix} = \begin{bmatrix} 0 \\ 0 \\ -\dfrac{\omega m_0\sin\theta}{4\pi}\left\{-\dfrac{\omega}{cr}\sin\left[\omega\left(t-\dfrac{r}{c}\right)\right] + \dfrac{1}{r^2}\cos\left[\omega\left(t-\dfrac{r}{c}\right)\right]\right\} \end{bmatrix} . \tag{7.307}$$

Here, as in Sect. 7.5, it is most advantageous to carry out the calculations in spherical coordinates. The result can easily be verified by comparison of eqs. (7.304) through (7.307) with the similar eqs. (7.262) through (7.265). Obviously, one only needs to exchange a few quantities: $\mathbf{E}$ and $\mathbf{H}$, replace p_0 by m_0, ε by μ, and observe the sign change during the transition between E_φ and H_φ.

Again, one is mostly interested in the far field region:

$$\mathbf{H} = \begin{bmatrix} H_r \\ H_\theta \\ H_\varphi \end{bmatrix} = \begin{bmatrix} 0 \\ -\dfrac{m_0\sin\theta\omega^2}{4\mu\pi c^2 r}\sin\left[\omega\left(t-\dfrac{r}{c}\right)\right] \\ 0 \end{bmatrix} , \tag{7.308}$$

$$\mathbf{E} = \begin{bmatrix} E_r \\ E_\theta \\ E_\varphi \end{bmatrix} = \begin{bmatrix} 0 \\ 0 \\ +\dfrac{m_0\omega^2\sin\theta}{4\pi cr}\sin\left[\omega\left(t-\dfrac{r}{c}\right)\right] \end{bmatrix} = \begin{bmatrix} 0 \\ 0 \\ -ZH_\theta \end{bmatrix} . \tag{7.309}$$

The Poynting vector of the far field is

$$S = E \times H = \begin{vmatrix} \mathbf{e}_r & \mathbf{e}_\theta & \mathbf{e}_\varphi \\ 0 & 0 & E_\varphi \\ 0 & H_\theta & 0 \end{vmatrix} = \begin{bmatrix} -E_\varphi H_\theta \\ 0 \\ 0 \end{bmatrix}. \tag{7.310}$$

It only has a radial component

$$S_r = -E_\varphi H_\theta = \frac{E_\varphi^2}{Z} = \frac{m_0^2 \omega^4 \sin^2\theta \sin^2\omega\left(t - \frac{r}{c}\right)}{Z \cdot 16\pi^2 c^2 r^2}$$

$$= \frac{m_0^2 \omega^4 \sin^2\theta \sin^2\omega\left(t - \frac{r}{c}\right)}{16\pi^2 \mu c^3 r^2}. \tag{7.311}$$

The radiated power is obtained thereof by integrating over the surface of a sphere.

$$P = \frac{m_0^2 \omega^4 \sin^2\omega\left(t - \frac{r}{c}\right)}{6\pi\mu c^3}, \tag{7.312}$$

and its time average is

$$P_{rms} = \frac{m_0^2 \omega^4}{12\pi\mu c^3}. \tag{7.313}$$

Under consideration of (7.300), the RMS power results in

$$P_{rms} = \frac{\mu I_0^2 r_0^4 \omega^4 \pi}{12 c^3} = Z \frac{I_0^2 r_0^4 k^4 \pi}{12}$$

$$= \frac{\pi}{6}(r_0 k)^4 Z \cdot I_{rms}^2 = R_s \cdot I_{rms}^2, \tag{7.314}$$

having defined the radiation resistance to

$$R_s = \frac{\pi}{6}(r_0 k)^4 Z = \frac{\pi}{6}\left(\frac{2\pi r_0}{\lambda}\right)^4 Z. \tag{7.315}$$

The radiation resistance for vacuum is

$$R_s = \left(\frac{2\pi r_0}{\lambda}\right)^4 \frac{\pi}{6} \cdot 377\Omega = \left(\frac{2\pi r_0}{\lambda}\right)^4 \cdot 198\Omega \tag{7.316}$$

The antenna gain is again

$$G = \frac{3}{2}. \tag{7.317}$$

7.7   Waves in Cylindrical Wave Guides

7.7.1   Basic Equations

In this section, we will discuss propagation of waves in cylindrical wave guides of arbitrary cross sections (Fig. 7.21). Its interior consists of a uniform, but not necessarily ideal dielectric (*i.e.*, it may be $\kappa \neq 0$). The outside space shall be of infinite conductivity. We look for waves of the form

$$\mathbf{E} = \mathbf{E}_0(x, y)\exp[i(\omega t - k_z z)] \tag{7.318}$$

$$\mathbf{H} = \mathbf{H}_0(x, y)\exp[i(\omega t - k_z z)] \ . \tag{7.319}$$

There shall be no volume charges. Then, Maxwell's equations apply in the following form:

$$\nabla \times \mathbf{E} = -\mu\frac{\partial \mathbf{H}}{\partial t} \tag{7.320}$$

$$\nabla \times \mathbf{H} = \kappa \mathbf{E} + \varepsilon\frac{\partial \mathbf{E}}{\partial t} \tag{7.321}$$

$$\nabla \bullet \mathbf{E} = 0 \tag{7.322}$$

$$\nabla \bullet \mathbf{H} = 0 \ . \tag{7.323}$$

In light of the Ansatz in (7.318) and (7.319), one makes use the following relations

$$\frac{\partial}{\partial z} = -ik_z \tag{7.324}$$

$$\frac{\partial}{\partial t} = i\omega \ . \tag{7.325}$$

Applying this to (7.320) through (7.323) yields

$$\frac{\partial E_z}{\partial y} + ik_z E_y = -i\omega\mu H_x \tag{7.326}$$

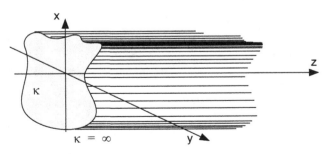

Fig. 7.21 κ

$$-\frac{\partial E_z}{\partial x} - ik_z E_x = -i\omega\mu H_y \tag{7.327}$$

$$\frac{\partial E_y}{\partial x} - \frac{\partial E_x}{\partial y} = -i\omega\mu H_z \tag{7.328}$$

$$\frac{\partial H_z}{\partial y} + ik_z H_y = (\kappa + i\omega\varepsilon)E_x \tag{7.329}$$

$$-\frac{\partial H_z}{\partial x} - ik_z H_x = (\kappa + i\omega\varepsilon)E_y \tag{7.330}$$

$$\frac{\partial H_y}{\partial x} - \frac{\partial H_x}{\partial y} = (\kappa + i\omega\varepsilon)E_z \tag{7.331}$$

$$\frac{\partial E_x}{\partial x} + \frac{\partial E_y}{\partial y} - ik_z E_z = 0 \tag{7.332}$$

$$\frac{\partial H_x}{\partial x} + \frac{\partial H_y}{\partial y} - ik_z H_z = 0 \ . \tag{7.333}$$

These equations require an additional remark. If an Ansatz is proportional to $\exp(i\omega t)$, then

$$\nabla\times\mathbf{E} = -i\omega\mu\mathbf{H}, \qquad \nabla\times\mathbf{H} = (\kappa + i\omega\varepsilon)\mathbf{E} \ ,$$

and when taking the divergence, it immediately follows that

$$\nabla\bullet\mathbf{H} = 0, \qquad \nabla\bullet\mathbf{E} = 0 \ .$$

Consequently, eqs. (7.322) and (7.323) or (7.332) and (7.333) are dependent on the other equations and thus are really redundant. One might as well omit them.
Eqs. (7.326) and (7.330) allow to calculate E_y and H_x as functions of H_z and E_z. In similar manner, we may use eqs. (7.327) and (7.329) to calculate E_x and H_y as functions of H_z and E_z. The result is

$$\boxed{E_x = \frac{-ik_z\dfrac{\partial E_z}{\partial x} - i\omega\mu\dfrac{\partial H_z}{\partial y}}{N}} \tag{7.334}$$

$$\boxed{E_y = \frac{-ik_z\dfrac{\partial E_z}{\partial y} + i\omega\mu\dfrac{\partial H_z}{\partial x}}{N}} \tag{7.335}$$

$$\boxed{H_x = \frac{(\kappa + i\omega\varepsilon)\dfrac{\partial E_z}{\partial y} - ik_z\dfrac{\partial H_z}{\partial x}}{N}} \tag{7.336}$$

$$\boxed{H_y = \frac{-(\kappa + i\omega\varepsilon)\dfrac{\partial E_z}{\partial x} - ik_z\dfrac{\partial H_z}{\partial y}}{N}} \ . \tag{7.337}$$

Each of these equations share the same denominator for which we introduce N as an abbreviation. Thus:

$$\boxed{N = \omega^2 \varepsilon \mu - k_z^2 - i\omega\kappa\mu}$$
(7.338)

One obtains the so-called Helmholtz equations when substituting these results into (7.331)

$$\frac{\partial^2 E_z}{\partial x^2} + \frac{\partial^2 E_z}{\partial y^2} + NE_z = 0$$
(7.339)

and (7.328)

$$\frac{\partial^2 H_z}{\partial x^2} + \frac{\partial^2 H_z}{\partial y^2} + NH_z = 0 .$$
(7.340)

We could have obtained the result (7.339), (7.340) directly from (7.8) and (7.9) by means of the Ansatz of (7.318) and (7.319). We introduce the two dimensional Laplacian in the x-y plane ∇_2^2 or Δ_2 to write:

$$\boxed{\Delta_2 E_z + NE_z = 0}$$
(7.341)

$$\boxed{\Delta_2 H_z + NH_z = 0} .$$
(7.342)

This approach reduces the problem to solving these two, two-dimensional Helmholtz equations in the x-y-plane. After having found H_z and E_z from these equations and under consideration of the boundary conditions, all other field components can be obtained from (7.334) through (7.337).

Some restrictions apply. Initially we have to require that the denominator does not vanish. Note however, that $H_z = 0$ or $E_z = 0$ does not necessarily cause the other components to vanish. They will not vanish if the denominator N vanishes simultaneously. In other words: For pure transverse waves, waves which are transverse with respect to both **E** and **H**, the so-called *TEM waves*, the following dispersion relation applies.

$$\boxed{N = \omega^2 \varepsilon \mu - k_z^2 - i\omega\kappa\mu = 0} .$$
(7.343)

We have met this relation before when we studied plane waves, which are a special case of TEM waves (see Section 7.2, in particular eq. 7.83). We will initially exclude TEM waves from our discussion, but will return to them later.

If $N \neq 0$, then at least one of the two quantities H_z or E_z has to be non-zero. It is possible to compose an arbitrary wave from a combination of waves were $H_z = 0$ but $E_z \neq 0$, as well as if $E_z = 0$ but $H_z \neq 0$. One set is transverse with respect to **H** and called *TM waves*, while the other set is transverse with respect to **H** and called *TE waves*.

Overall, there are three different types of waves, TM waves, TE waves, and TEM waves. We shall discuss each one separately in above order.

We have already encountered simple cases of TM and TE waves in Sect. 7.1.6.

7.7.2   TM Waves

For $H_z = 0$, we have to solve the Helmholtz equation for E_z (7.341) and the remaining field components result from (7.334) through (7.337). The same result can be obtained when starting from an electric Hertz vector which only has a z-component (Π_{ez}). It has to satisfy the wave equation (7.237), which in turn, leads to an equation of type (7.341):

$$\boxed{\Delta_2 \Pi_{ez} + N \Pi_{ez} = 0} .$$
(7.344)

For the fields, eqs. (7.238) and (7.239) apply. And specifically for the current case one obtains

$$\mathbf{E} = \nabla(-ik_z \Pi_{ez}) + (\varepsilon \mu \omega^2 - i\omega \kappa \mu) \Pi_{ez} \mathbf{e}_z$$
(7.345)

$$\mathbf{H} = (\nabla \times [i\varepsilon\omega + \kappa]) \Pi_{ez} \mathbf{e}_z .$$
(7.346)

For the individual components of $\mathbf{E}$ and $\mathbf{H}$ we obtain:

$$\boxed{\begin{array}{ll} E_x = -ik_z \dfrac{\partial \Pi_{ez}}{\partial x} & H_x = (\kappa + i\varepsilon\omega)\dfrac{\partial \Pi_{ez}}{\partial y} \\[2mm] E_y = -ik_z \dfrac{\partial \Pi_{ez}}{\partial y} & H_y = -(\kappa + i\varepsilon\omega)\dfrac{\partial \Pi_{ez}}{\partial x} \\[2mm] E_z = N \Pi_{ez} & H_z = 0 . \end{array}}$$
(7.347)

These are exactly those fields, which one would obtain from (7.334) through (7.337) if we let $H_z = 0$.

Obviously,

$$\mathbf{E} \bullet \mathbf{H} = 0 ,$$
(7.348)

revealing that $\mathbf{E}$ and $\mathbf{H}$ are perpendicular.

The tangential component of $\mathbf{E}$ has to vanish on the boundary towards the infinitely conducting medium. The perpendicular component of $\mathbf{H}$ has to vanish there as well. This is a consequence of the mandate for continuity of the respective components and thereby, all field components have to vanish in the ideal conductor. Therefore on the boundary, it has to be $E_z = 0$, *i.e.* is must be

$$\boxed{\Pi_{ez} = 0 \qquad (\text{boundary})} .$$
(7.349)

This condition automatically satisfies all other conditions. A consequence of (7.347) is that $\mathbf{E}$ is perpendicular to the boundary, *i.e.*, does not have a parallel component. $\mathbf{H}$, on the other hand, is perpendicular to $\mathbf{E}$ everywhere, thus $\mathbf{H}$ has no component perpendicular to the boundary. Therefore, we need to solve eq. (7.344) with the boundary conditions (7.349). Thus, the problem of a TM wave is a two-dimensional Dirichlet boundary value problem.

Our conclusions are independent from our previous choice of Cartesian coordinates. It is permissible to transform to any other coordinate system.

Another conclusion of eqs. (7.347) is that Π_{ez} is constant on the magnetic field lines, and thus can be regarded as a flux function.

7.7.3   TE Waves

TE waves can be dealt with in similar manner, if we start from a magnetic Hertz vector which has only a z-component (Π_{mz}). It satisfies the wave equation (7.247), which for the current case takes the from

$$\boxed{\Delta_2\Pi_{mz} + N\Pi_{mz} = 0} \ . \tag{7.350}$$

The field equations are based on (7.248) and (7.249) and yield for our specific case:

$$\mathbf{E} = -i\omega\mu\nabla\times(\Pi_{mz}\mathbf{e}_z) \tag{7.351}$$

$$\mathbf{H} = \nabla(-ik_z\Pi_{mz}) + (\varepsilon\mu\omega^2 - i\omega\kappa\mu)\Pi_{mz}\mathbf{e}_z \ . \tag{7.352}$$

For the individual components we obtain:

$$\boxed{\begin{aligned}
E_x &= -i\omega\mu\frac{\partial\Pi_{mz}}{\partial y} & H_x &= -ik_z\frac{\partial\Pi_{mz}}{\partial x} \\
E_y &= +i\omega\mu\frac{\partial\Pi_{mz}}{\partial x} & H_y &= -ik_z\frac{\partial\Pi_{mz}}{\partial y} \\
E_z &= 0 & H_z &= N\Pi_{mz} \ .
\end{aligned}} \tag{7.353}$$

Here again:

$$\mathbf{E} \bullet \mathbf{H} = 0 \ . \tag{7.354}$$

On the boundary, $\mathbf{H}$ may not have a perpendicular component. From (7.353) follows that on the boundary it has to be:

$$\boxed{(\nabla\Pi_{mz})_n = \frac{\partial\Pi_{mz}}{\partial n} = 0 \qquad \text{(boundary)}} \ , \tag{7.355}$$

where the subscript n shall describe the normal component. This is now a Neumann boundary value problem. This also ensures that the parallel components of $\mathbf{E}$ vanish on the boundary, because $\mathbf{E}$ and $\mathbf{H}$ are perpendicular to each other. Π_{mz} is constant along the electric field lines and thus represents their flux function.

7.7.4   TEM Waves

Now, we shall assume that all z-components vanish:

$$H_z = 0 \tag{7.356}$$

$$E_z = 0 \ . \tag{7.357}$$

With this, eqs. (7.326) through (7.333) yield:

$$k_z E_y = -\omega\mu H_x$$

$$k_z E_x = +\omega\mu H_y$$

$$\frac{\partial E_y}{\partial x} - \frac{\partial E_x}{\partial y} = 0$$

$$k_z H_y = +(\omega\varepsilon - i\kappa)E_x$$

$$k_z H_x = -(\omega\varepsilon - i\kappa)E_y$$

$$\frac{\partial H_y}{\partial x} - \frac{\partial H_x}{\partial y} = 0$$

$$\frac{\partial E_x}{\partial x} + \frac{\partial E_y}{\partial y} = 0$$

$$\frac{\partial H_x}{\partial x} + \frac{\partial H_y}{\partial y} = 0$$

(7.358)

When eliminating, *for example*, E_x and E_y by means of the first two equations, one realizes that all equations are satisfied if

$$N = \omega^2\varepsilon\mu - k_z^2 - i\omega\kappa\mu = 0 \ , \tag{7.359}$$

as claimed earlier (7.343) and if it is also

$$\frac{\partial H_x}{\partial x} + \frac{\partial H_y}{\partial y} = 0 \tag{7.360}$$

and

$$\frac{\partial H_y}{\partial x} - \frac{\partial H_x}{\partial y} = 0 \ . \tag{7.361}$$

The last two equations signify that the **H** field is both, source-free and irrotational. It is irrotational because there are no currents in z-direction which could create a curl. Notice that eq. (7.361) is derived from (7.331) which contains κE_z, the free current in the conductor and $i\omega\varepsilon E_z$, the displacement current in z-direction.

Another consequence from (7.358) is that for TEM waves we also have:

$$\mathbf{E} \bullet \mathbf{H} = 0 \ . \tag{7.362}$$

One can show that the fields given in (7.347) and (7.353) satisfy eqs. (7.358), if we take $N = 0$ while Π_{ez} and Π_{mz} satisfy the corresponding eqs. (7.344) and (7.350), respectively. This context also requires one to check the boundary conditions given in eqs. (7.349) and (7.355). While (7.355) remains unchanged, now Π_{ez} is no longer required to vanish on the boundary. It suffices if Π_{ez} is constant:

$$\Pi_{ez} = \text{const.} \quad \text{(on boundary)} \ . \tag{7.363}$$

TEM waves can not exist in every wave guide. To appreciate this, let us look at a wave guide with a "simply connected" cross section (Fig. 7.22). The **H** lines on the

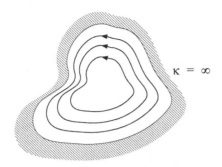

Fig. 7.22

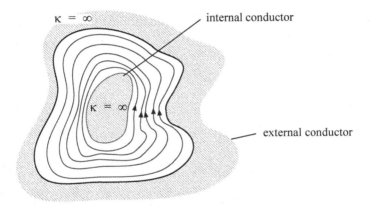

Fig. 7.23

boundary have to be parallel to it and their path has to qualitatively look like
sketched in Fig. 7.22. However, this is impossible because the integral $\oint \mathbf{H} \bullet d\mathbf{s}$
would be nonzero, even though there are no currents inside which could account
for this. The situation changes if we introduce a multiply connected cross section as
shown in Fig. 7.23. In this case, the cause for a non-vanishing integral $\oint \mathbf{H} \bullet d\mathbf{s}$
could be attributed to one or more internal conductors. In practice, wave guides of
this kind occur frequently, *for example*, in the form of a coaxial cable. The so-
called *transmission theory* solves these kind of problems by means of the
telegrapher's equations. However, transmission theory is an approximation and
does not allow to describe all types of waves, possible in such a wave guide. Only
field theory allows for this. We will return later to the relation between field theory
and transmission theory.

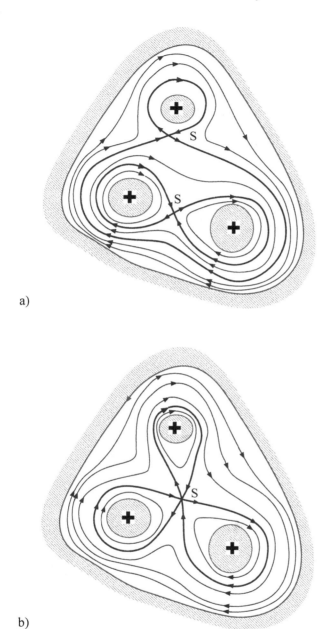

a)

b)

Fig. 7.24

Fig. 7.24 illustrates qualitatively the structure of the magnetic field in wave guides with several internal conductors – the figure shows three conductors. They

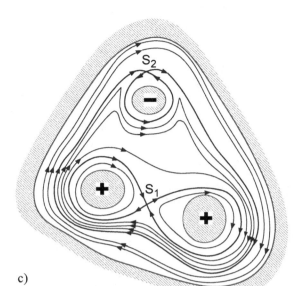

c)

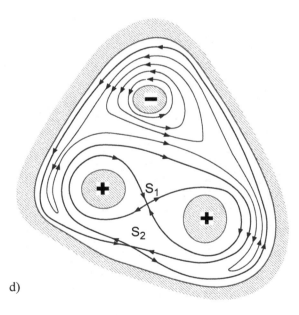

d)

Fig. 7.24

may carry currents, all in the same direction (Fig. 7.24a,b) or different directions (Fig. 7.24c,d). In the general case, there will be two stagnation lines where the field vanishes (Fig. 7.24a,c,d). For special values of currents, the two lines may degenerate to a single one (Fig. 7.24b). The force lines (separatrices) which pass through the stagnation lines separate the regions of the cross section into sections of different field structure, as we have encountered before in electrostatics (Sect. 2.4) and in magnetostatics (Sect. 5.2.1). The difference between Fig. 7.24c and d is merely that the overall current is positive in Fig. 7.24c and negative in Fig. 7.24d.

7.8   Rectangular Wave Guide

7.8.1   Separation of Variables

Our first example in applying the theory of the previous section shall be for a wave in a wave guide of the shape as shown in Fig. 7.25, with $a \geq b$. We will limit the discussion to the case of an insulator ($\kappa = 0$). Π_z (or more specifically Π_{ez} in case of TM waves and Π_{mz} for TE waves) has to satisfy the Helmholtz equation (7.344) or (7.350), respectively.

$$\left(\frac{\partial^2}{\partial x^2} + \frac{\partial^2}{\partial y^2} + \varepsilon\mu\omega^2 - k_z^2\right)\Pi_z = 0 \tag{7.364}$$

If one now separates according to the model given in Sect. 3.5, then we obtain Π_z in the form:

$$\Pi_z = X(x) \cdot Y(y) \cdot Z(z)$$

$$\Pi_z = (C_1 \sin k_x x + C_2 \cos k_x x)(C_3 \sin k_y y + C_4 \cos k_y y)\exp[i(\omega t - k_z z)], \tag{7.365}$$

where initially C_1, C_2, C_3, C_4 are arbitrary constants. Furthermore, to satisfy (7.364), the dispersion relation has to be fulfilled

$$k_x^2 + k_y^2 + k_z^2 = \varepsilon\mu\omega^2 . \tag{7.366}$$

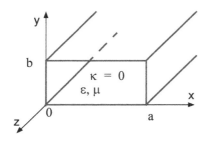

Fig. 7.25

7.8.2   TM Waves in a Rectangular Wave Guide

Because of the boundary conditions (7.349), $\Pi_{ez} = 0$ for $x = 0$, $x = a$, $y = 0$, and $y = b$. This requires that $C_2 = C_4 = 0$. Furthermore, only certain values for k_x and k_y are possible, namely, as before in Sect. 3.5.1, eqs. (3.70) and (3.71):

$$k_x = \frac{n\pi}{a} \qquad (n \text{ whole numbers}) \tag{7.367}$$

$$k_y = \frac{m\pi}{b} \qquad (m \text{ whole numbers}) . \tag{7.368}$$

This leaves

$$\Pi_{ez} = C_e \sin k_x x \sin k_y y \exp[i(\omega t - k_z z)] , \tag{7.369}$$

from which the fields can be calculated according to (7.347):

$$\left.\begin{aligned}
E_x &= -ik_x k_z C_e \cos k_x x \sin k_y y \\
E_y &= -ik_y k_z C_e \sin k_x x \cos k_y y \\
E_z &= (k_x^2 + k_y^2) C_e \sin k_x x \sin k_y y \\
H_x &= +i\omega\varepsilon k_y C_e \sin k_x x \cos k_y y \\
H_y &= -i\omega\varepsilon k_x C_e \cos k_x x \sin k_y y \\
H_z &= 0
\end{aligned}\right\} \cdot \exp[i(\omega t - k_z z)] . \tag{7.370}$$

All possible TM waves are described by this set of equations. Two whole numbers, n, m correspond to every possible type of wave. This wave is called TM_{nm} wave. Obviously, in order for not all fields to vanish, it has to have $n \geq 1$ and $m \geq 1$. Ergo, there are no TM_{00}, TM_{01}, or TM_{10} waves.
The dispersion relation (7.366), together with (7.367) and (7.368) yields

$$k_z^2 = \varepsilon\mu\omega^2 - k_x^2 - k_y^2 = \varepsilon\mu\omega^2 - \frac{\pi^2 n^2}{a^2} - \frac{\pi^2 m^2}{b^2} . \tag{7.371}$$

Form this one finds the phase velocity of the wave

$$v_{ph} = \frac{\omega}{k_z} = \frac{\omega}{\sqrt{\dfrac{\omega^2}{c^2} - \dfrac{\pi^2 n^2}{a^2} - \dfrac{\pi^2 m^2}{b^2}}} , \tag{7.372}$$

and for its group velocity

$$v_G = \frac{d\omega}{dk_z} = \frac{c^2}{v_{ph}} , \tag{7.373}$$

so that

$$v_G v_{ph} = c^2 . \tag{7.374}$$

We have found this result before in a special case, eq. (7.68). If one calculates the average energy per unit length for the wave guide and multiply this by v_G, then one obtains the same energy transfer as it results from the z-component of the Poynting vector, averaged over time and space (cross section). We conclude, here as well, the group velocity can be regarded as the velocity of the energy transfer.

If λ_z is the wave length inside the wave guide in the z-direction, and λ its corresponding wave length in the free space, then

$$\varepsilon\mu\omega^2 = \frac{\omega^2}{c^2} = k^2 = \left(\frac{2\pi}{\lambda}\right)^2 \tag{7.375}$$

and

$$\lambda_z = \frac{2\pi}{k_z} = \frac{2\pi}{\sqrt{\left(\frac{2\pi}{\lambda}\right)^2 - \frac{\pi^2 n^2}{a^2} - \frac{\pi^2 m^2}{b^2}}}$$

$$\lambda_z = \frac{\lambda}{\sqrt{1 - \left(\frac{\lambda}{2}\right)^2 \left(\frac{n^2}{a^2} + \frac{m^2}{b^2}\right)}} . \tag{7.376}$$

This means that λ_z is always greater than λ. λ_z even becomes infinite for

$$\lambda = \lambda_c = \frac{2}{\sqrt{\frac{n^2}{a^2} + \frac{m^2}{b^2}}} = \frac{2ab}{\sqrt{n^2 b^2 + m^2 a^2}} , \tag{7.377}$$

λ_z (or k_z) becomes imaginary for $\lambda > \lambda_c$. The related fields can not propagate inside the wave guide. The related wave length λ_c is called the critical wave length of the TM_{nm} wave. It is also the largest free space wave length for which this type of wave can exist. The related angular frequency, the *cutoff* frequency ω_c, is the lowest frequency for which propagation inside the wave guide is possible:

$$\omega_c = \frac{2\pi c}{\lambda_c} = \pi c \sqrt{\frac{n^2}{a^2} + \frac{m^2}{b^2}} . \tag{7.378}$$

The largest among all possible critical wave lengths belongs to the TM_{11} wave, which is

$$(\lambda_c)_{TM_{11}} = \frac{2ab}{\sqrt{a^2 + b^2}} . \tag{7.379}$$

Fig. 7.26 provides a qualitative picture of the fields for a few wave types, namely their projection onto the cross section. Take into account when interpreting these pictures, that the electric field has a z-component also. When the electric fields in those pictures appear to have sources or sinks, then this deception is simply caused by the projection, while in reality the field is diverted at those points into the z-direction. These z-fields are the displacement currents, which create the magnetic fields of the wave.

At the surface, the electric field has perpendicular components and the magnetic field has tangential components, which cause surface charges and surface

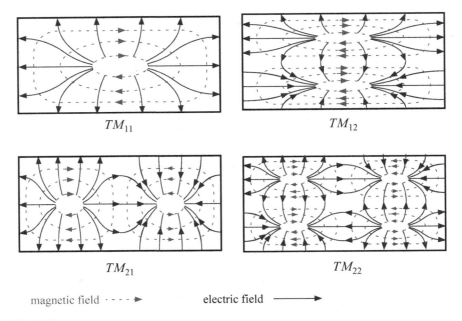

TM_{11} TM_{12}

TM_{21} TM_{22}

magnetic field $\cdots\cdots\blacktriangleright$ electric field $\longrightarrow$

Fig. 7.26

currents there. Their fields can be calculated from the boundary conditions, which are time dependent. Of course, currents and charges have to meet the continuity equation, which in the current case takes the form:

$$\nabla_2 \bullet \mathbf{k} + \frac{\partial \sigma}{\partial t} = 0 \ ,$$

where $\nabla_2 \bullet$ shall signify the two dimensional divergence operator in a plane. Of course, $\mathbf{k}$ replaces $\mathbf{g}$ and σ replaces ρ.

7.8.3   TE Waves in a Rectangular Wave Guide

In this case, the boundary condition (7.355) requires that $C_1 = C_3 = 0$. Furthermore, the wave numbers k_x and k_y have to satisfy (7.367) and (7.368). Therefore:

$$\Pi_{mz} = C_m \cos k_x x \cos k_y y \exp[i(\omega t - k_z z)] \ . \tag{7.380}$$

With (7.353) the TE_{nm} wave becomes:

$$\left.\begin{aligned}
E_x &= +i\omega\mu k_y C_m \cos k_x x \sin k_y y \\
E_y &= -i\omega\mu k_x C_m \sin k_x x \cos k_y y \\
E_z &= 0 \\
H_x &= +ik_x k_z C_m \sin k_x x \cos k_y y \\
H_y &= +ik_y k_z C_m \cos k_x x \sin k_y y \\
H_z &= +(k_x^2 + k_y^2) C_m \cos k_x x \cos k_y y
\end{aligned}\right\} \cdot \exp[i(\omega t - k_z z)] . \tag{7.381}$$

Formulas (7.371) through (7.378) apply to TE waves unchanged.

In contrast to the TM wave, the TE wave has non-vanishing fields, if at least one of the two integers n, m is nonzero. That is to say, there is no TE_{00} wave, however, TE_{01} and TE_{10} waves exist. The largest critical wave length belongs to the TE_{10} wave and is determined by (7.377)

$$(\lambda_c)_{TE_{10}} = 2a , \tag{7.382}$$

while the relation for the TE_{10} wave is

$$(\lambda_c)_{TE_{01}} = 2b \le 2a = (\lambda_c)_{TE_{10}} . \tag{7.383}$$

Fig. 7.27 provides a few pictures of various TE wave modes. What was said in conjunction with Fig. 7.26 applies here in like manner. Of course, the magnetic field lines do not have sources or sinks. Whenever their projection onto the cross-sectional area suggests otherwise, then this is because the projection hides the z-component of the field.

7.8.4 TEM waves

Recall from our general discussion in Sect. 7.7.4 that no TEM wave can exist in simply connected rectangular wave guides (as is the case in every simply connected wave guide of any shape). This is also obvious from (7.370) and (7.381). For the z-components of **E** and **H** to vanish requires that

$$k_x^2 + k_y^2 = 0 .$$

This, in turn requires that $k_x = k_y = 0$, whereby all other field components vanish as well. We know already that neither TM_{00} nor TE_{00} waves exist.

On the other hand, TEM waves do exist in multiply connected rectangular wave guides, *for example*, of the kind depicted in Fig. 7.28. The related theory is rather complicated, however, and shall not be discussed in this text. A remark to avoid misconception: TEM waves are possible between infinite parallel plates (Fig. 7.29). The difference between these and an ordinary rectangular wave guide is that the infinite plates do not impose a limit on H_x. A possible wave is therefore

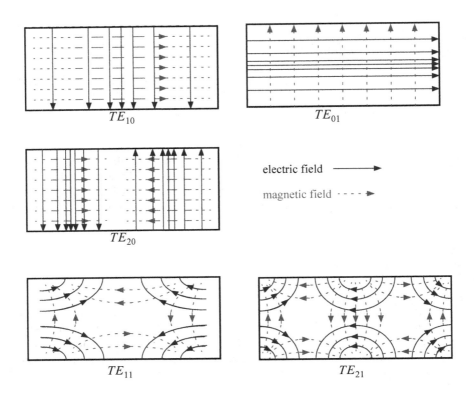

Fig. 7.27

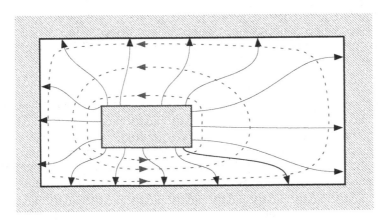

Fig. 7.28

$$H_x = \frac{E_0}{Z} \exp[i(\omega t - k_z z)]$$

$$E_y = -E_0 \exp[i(\omega t - k_z z)] \ .$$

This is nothing other than a plane wave, which however, does not extend to infinity in the y-direction., *i.e.*, in the direction of the electric field vector. This is possible because a perpendicular component of **E** is permissible on the boundary. There, it causes the necessary surface charges. A different polarization is not possible because H_y and E_x have to vanish on the boundary. Notice, that the case of infinite parallel plates is not a result of the limit of a rectangular wave guide with $a \to \infty$, at least not with respect to the TEM waves. The rectangular wave guide does not allow TEM waves for any value of a, not even for infinitely large a. Of course, all kinds of TM and TE waves are possible between the parallel plates. Those can definitely be derived from simply connected rectangular wave guides as the limit $a \to \infty$.

7.9 Rectangular Cavities

A cavity bounded by conductors is a *resonant cavity*. It can hold an electromagnetic oscillation. The calculation of the various oscillations that are possible in such a resonant cavity is in general mathematically difficult. However, for a cuboid cavity, the problem can be reduced to the just discussed rectangular wave guide.

  If we let two like waves propagate in opposite directions, then their superposition gives a standing wave, just as we had discussed for a plane wave in Sect. 7.1.5.

A standing TM wave can be described by

$$\Pi_{ez} = C_e \sin k_x x \sin k_y y \cos k_z z \exp[i\omega t] \ . \tag{7.384}$$

  A standing TE wave can be described by

$$\Pi_{mz} = C_m \cos k_x x \cos k_y y \sin k_z z \exp[i\omega t] \ . \tag{7.385}$$

Π_{ez} and Π_{mz} satisfy their corresponding wave equations for $\kappa = 0$ if

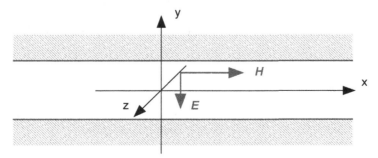

Fig. 7.29

$$k_x^2 + k_y^2 + k_z^2 = k^2 = \varepsilon\mu\omega^2 = \frac{\omega^2}{c^2} \ . \tag{7.386}$$

The corresponding fields can be found with (7.238), (7.239) and (7.248), (7.249). For TM waves

$$\left.\begin{aligned}
E_x &= -k_x k_z C_e \cos k_x x \sin k_y y \sin k_z z \\
E_y &= -k_y k_z C_e \sin k_x x \cos k_y y \sin k_z z \\
E_z &= (k_x^2 + k_y^2) C_e \sin k_x x \sin k_y y \cos k_z z \\
H_x &= +i\omega\varepsilon k_y C_e \sin k_x x \cos k_y y \cos k_z z \\
H_y &= -i\omega\varepsilon k_x C_e \cos k_x x \sin k_y y \cos k_z z \\
H_z &= 0
\end{aligned}\right\} \cdot \exp[i\omega t] \tag{7.387}$$

and for TE waves

$$\left.\begin{aligned}
E_x &= +i\omega\mu k_y C_m \cos k_x x \sin k_y y \sin k_z z \\
E_y &= -i\omega\mu k_x C_m \sin k_x x \cos k_y y \sin k_z z \\
E_z &= 0 \\
H_x &= -k_x k_z C_m \sin k_x x \cos k_y y \cos k_z z \\
H_y &= -k_y k_z C_m \cos k_x x \sin k_y y \cos k_z z \\
H_z &= (k_x^2 + k_y^2) C_m \cos k_x x \cos k_y y \sin k_z z
\end{aligned}\right\} \cdot \exp[i\omega t] \ . \tag{7.388}$$

All these fields satisfy the required boundary conditions of the previously discussed rectangular wave guide, if

$$k_x = \frac{n\pi}{a} \tag{7.389}$$

and

$$k_y = \frac{m\pi}{b} \ . \tag{7.390}$$

We can create a cuboidal resonant cavity if we cut a piece of length d from this wave guide and insert a reflecting wall (i.e., an infinitely conducting wall) at $z = 0$ and $z = d$ (Fig. 7.30). The additional walls impose additional boundary conditions:

$$\left.\begin{aligned}
E_x &= E_y = 0 \\
H_z &= 0
\end{aligned}\right\} \quad \begin{aligned} &\text{for} \quad z = 0 \\ &\text{and} \quad z = d \ . \end{aligned}$$

Our Ansatz already satisfies the boundary condition for $z = 0$, which was chosen with this in mind. For $z = d$, the boundary conditions require

$$k_z = \frac{p\pi}{d} \qquad p \text{ whole number} \ . \tag{7.391}$$

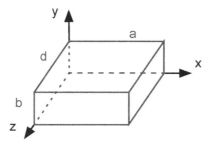

Fig. 7.30

Overall, using (7.386) and (7.389) through (7.391), yield the following resonant angular frequencies

$$\omega_{nmp} = c\pi\sqrt{\left(\frac{n}{a}\right)^2 + \left(\frac{m}{b}\right)^2 + \left(\frac{p}{d}\right)^2} \qquad n, m, p \text{ whole numbers} \qquad (7.392)$$

as result for the TM_{nmp} and TE_{nmp} waves of the resonant cavity.

The totality of all resonant frequencies (eigenfrequencies, *i.e.* frequencies to the respective eigenvalues) is obtained by permutation of all permissible combinations of n, m, p. The requirement for these combinations is that at least two of the numbers have to be non-zero. Furthermore, there may be no TE_{nm0}, no TM_{0mp}, and no TM_{n0p}, while, TE_{0mp}, TE_{n0p}, and TM_{nm0} waves are permissible, as is obvious from (7.387) and (7.388), respectively.

A resonant cavity has certain common properties with an LC oscillator (neglecting the fact that the LC oscillator looses energy by radiation, while the resonant cavity is unable to radiate energy, at least in the approximation of the ideal conducting walls). Both possess a constant overall energy, which is composed of electric energy ($1/2 \cdot CV^2$, for the LC oscillator) and of magnetic energy ($1/2 \cdot LI^2$, for the LC oscillator) whereby the two forms of energy mutually transform continuously. The energy for the resonant cavity in case of the TM wave is derived from (7.387)

$$\left.\begin{aligned} W_{mag.} &= W_t\sin^2\omega t \\ W_{el.} &= W_t\cos^2\omega t \end{aligned}\right\} \qquad (7.393)$$

and for the TE wave, derived from (7.388) (just reversed)

$$\left.\begin{aligned} W_{mag.} &= W_t\cos^2\omega t \\ W_{el.} &= W_t\sin^2\omega t \end{aligned}\right\}. \qquad (7.394)$$

In both cases, W_t is the total energy. This behavior is illustrated in Fig. 7.31.

Apart from the fact that a resonant cavity possesses an infinite number of resonant frequencies, while an LC oscillator has only one, the major other difference is that the magnetic and electric fields of the resonant cavity are spatially

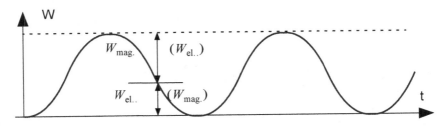

Fig. 7.31

united, while they are spatially separated for the LC oscillator, namely in the two distinct components. Nevertheless, the transition is gradual. It is possible to start from an LC oscillator and make a smooth transition into a resonant cavity (Fig. 7.32).

Notice that for the standing waves of this section as described by eqs. (7.384), (7.385), (7.387), (7.388) (contrary to the previous sections) it is not permissible to replace the differential operator by $-ik_z$. The reason is that the standing wave is created by superposition of the two functions $\exp[-ik_zz]$ and $\exp[+ik_zz]$, which, in the said equation., resulted in the sine and cosine functions, respectively. This had prevented us to obtain the fields of eq. (7.387) and (7.388) from eqs. (7.347) and (7.353), rather we had to go back to the more generally applicable eqs. (7.238), (7.239), (7.248), and (7.249).

It shall also be pointed out that eqs. (7.389) (7.390), (7.391) – just as analogous equations for mechanical oscillations and waves – have a plausible

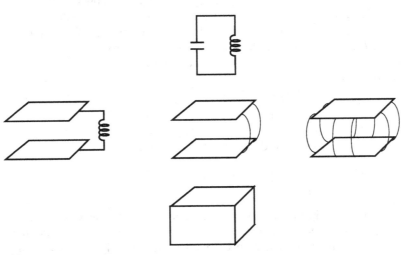

Fig. 7.32

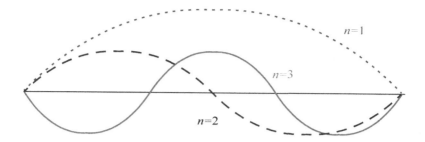

Fig. 7.33

explanation. Consider, *for example*, a string, mounted at both ends. The relation for this vibrating string is

$$n\frac{\lambda}{2} = a,$$

where a is the length of the string and n is a whole number (Fig. 7.33). For a wave guide, this applies for waves in x and y direction, while it applies to all three directions in case of the resonant cavity. The waves in the x and y direction of a wave guide are standing waves, while the wave in z direction is travelling. In case of the resonant cavity, all three directions carry standing waves. Intuitively, one may picture a standing wave by superposition of an incident wave and waves, reflected at the boundary (the reader shall thereby be reminded of sects. 7.1.5 and 7.1.6).

7.10 Circular Wave Guide

7.10.1 Separation of Variables

For circular cylindrical wave guides, whether simply or multiply connected (Fig. 7.34), the best approach is to use cylindrical coordinates. According to eq. (3.33), the two-dimensional Laplacian Δ_2 in the x-y-plane takes the form

$$\Delta_2 = \frac{1}{r}\frac{\partial}{\partial r}r\frac{\partial}{\partial r} + \frac{1}{r^2}\frac{\partial^2}{\partial \varphi^2} \ . \tag{7.395}$$

The formulas applicable to Π_{ez} and Π_{mz} are (7.344) and (7.350), respectively, where N is given by (7.338). Therefore we obtain

$$\left(\frac{1}{r}\frac{\partial}{\partial r}r\frac{\partial}{\partial r} + \frac{1}{r^2}\frac{\partial^2}{\partial \varphi^2} + N\right)\Pi_z = 0 \ . \tag{7.396}$$

Separating like in Sect. 3.7 gives

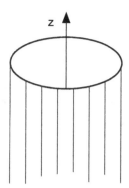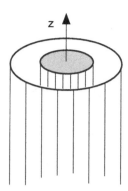

Fig. 7.34

$$\Pi_z = Z_m(r\sqrt{N})\cos m\varphi\exp[i(\omega t - k_z z)] \ . \tag{7.397}$$

We could have picked a sine instead of the cosine, but this does not constitute a significant difference. Z_m is a general cylinder function:

$$Z_m = C_1 J_m + C_2 N_m \ . \tag{7.398}$$

The specific form of Π_z given in (7.397) allows to solve an abundance of problems. Notice, that in general, it has to be $C_2 = 0$, if the area under consideration includes the z-axis, since N_m diverges for $r \to 0$. However, if the area does not include the z-axis, then the complete solution (7.398) has to be used.

Fig. 7.35 shows a number of arrangements, which can be solved using (7.397). First, there is the "normal" cylindrical wave guide (Fig. 7.35a), characterized by an ideal insulator surrounded by an ideal conductor. From this, the coaxial cable (Fig. 7.35b) emerges when another ideal conductor is inserted inside the insulator. Fig. 7.35c shows the so-called *Sommerfeld conductor*, where a medium of finite conductivity is surrounded by an ideal insulator. The *Harms-Goubau conductor* of Fig. 7.35d has a conductor of finite conductivity at its center and is surrounded by two layers of dielectrics with different permittivity. Another interesting case is the so-called *light conductor* or *light guide*, shown in Fig. 7.35e, an arrangement important in optics and has gained significance in telecommunications as well. Subsequently, we will limit our discussion to ordinary wave guides and the coaxial cable. The limit of ideal insulators and media with infinite conductivity can – with the exception of super conductors – not be realized at all or only approximately. Using these limits, however, simplifies the theoretical analysis significantly, without misrepresenting the material points of the results.

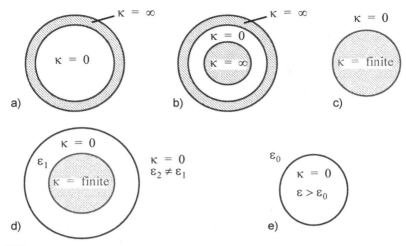

Fig. 7.35

7.10.2 TM Waves in a Circular Cylindrical Wave Guide

For the ordinary wave guide, we have to let $C_2 = 0$ in eq. (7.398). Here, we are interested in TM waves and thus with (7.397) find

$$\Pi_{ez} = C_e J_m(r\sqrt{N})\cos m\varphi \exp[i(\omega t - k_z z)] \ . \tag{7.399}$$

According to (7.349) we also have to ensure that

$$(\Pi_{ez})_{\text{boundary}} = 0 \ . \tag{7.400}$$

It follows that $\sqrt{N}$ may not assume every value. If, for instance, r_0 is the radius of the wave guide, then it must be

$$J_m(r_0\sqrt{N}) = 0 \ . \tag{7.401}$$

We have already found the zeros λ_{mn} of J_m in Sect. 3.7.3.3.:

$$\lambda_{mn} = r_0\sqrt{N} \ . \tag{7.402}$$

If we only consider ideal insulators ($\kappa = 0$), then with (7.338) we obtain

$$\varepsilon\mu\omega^2 - k_z^2 = \frac{\lambda_{mn}^2}{r_0^2} \ . \tag{7.403}$$

The associated wave is called the TM_{mn} wave of the wave guide. Its fields are derived from eqs. (7.238) and (7.239):

$$
\left.
\begin{aligned}
E_r &= -ik_z\sqrt{N}C_e J'_m(r\sqrt{N})\cos m\varphi \\[4pt]
E_\varphi &= +\frac{ik_z m}{r}C_e J_m(r\sqrt{N})\sin m\varphi \\[4pt]
E_z &= +NC_e J_m(r\sqrt{N})\cos m\varphi \\[4pt]
H_r &= -\frac{i\omega\varepsilon m}{r}C_e J_m(r\sqrt{N})\sin m\varphi \\[4pt]
H_\varphi &= -i\omega\varepsilon\sqrt{N}C_e J'_m(r\sqrt{N})\cos m\varphi \\[4pt]
H_z &= 0
\end{aligned}
\right\} \cdot \exp[i(\omega t - k_z z)] \ .
\tag{7.404}
$$

Notice that $N = \varepsilon\mu\omega^2 - k_z^2$ has to assume one of the values allowed by eq. (7.402). It becomes clear from inspection that with this restriction, E_φ, E_z, and H_r vanish when $r = r_0$, as it must be. J'_m is the derivative of the Bessel function, where the subject of derivative is its entire argument. TM_{mn} waves exist for $m \geq 0$, $n > 0$ (i.e., $N > 0$). $n = 0$, $N = 0$ would result in a TEM wave, which is not possible in the current case, and even more so, according to our general conclusions regarding TEM waves.

Similar to the rectangular wave guide, we find the phase velocity from the dispersion relation, here eq. (7.403)

$$
v_{ph} = \frac{\omega}{k_z} = \frac{\omega}{\sqrt{\varepsilon\mu\omega^2 - \dfrac{\lambda_{mn}^2}{r_0^2}}}
\tag{7.405}
$$

and the group velocity is

$$
v_G = \frac{d\omega}{dk_z} = \frac{c^2}{v_{ph}} \ .
\tag{7.406}
$$

The wave length in z-direction, the propagation direction in the wave guide is

$$
\lambda_z = \frac{2\pi}{k_z} = \frac{2\pi}{\sqrt{\dfrac{\omega^2}{c^2} - \left(\dfrac{\lambda_{mn}}{r_0}\right)^2}} - \frac{1}{\sqrt{\dfrac{1}{\lambda^2} - \left(\dfrac{\lambda_{mn}}{2\pi r_0}\right)^2}} \ .
\tag{7.407}
$$

λ is the related free space wave length. The largest possible free space wave length of the TM_{mn} wave is the critical wave length

$$
\lambda_c = \frac{2\pi r_0}{\lambda_{mn}} \ .
\tag{7.408}
$$

A few values of λ_{mn} are listed in Tab. 7.1. Notice that the quantities with the double index λ_{mn} represent the zeros of the Bessel functions. They are dimensionless and should not be confused with the wave length λ, λ_z, λ_c, etc.

Table 7.1   **The values** λ_{mn} **of** $J_m(\lambda_{mn}) = 0$

n m	1	2	3	4	5
0	2.40483	5.52008	8.65373	11.79153	14.93092
1	3.83171	7.01559	10.17347	13.32369	16.47063
2	5.13562	8.41724	11.61984	14.79595	17.95982
3	6.38016	9.76102	13.01520	16.22347	19.40942
4	7.58834	11.06471	14.37254	17.61597	20.82693
5	8.77148	12.33860	15.70017	18.98013	22.21780
6	9.93611	13.58929	17.00382	20.32079	23.58608

7.10.3   TE Waves in a Circular Cylindrical Wave Guide

For TE waves we have

$$\Pi_{mz} = C_m J_m(r\sqrt{N})\cos m\varphi \exp[i(\omega t - k_z z)] \ , \tag{7.409}$$

where

$$\left(\frac{\partial \Pi_{mz}}{\partial r}\right)_{r=r_0} = 0 \ , \tag{7.410}$$

which requires that

$$J_m'(r_0\sqrt{N}) = 0 \ . \tag{7.411}$$

If one addresses the zeros of J_m' in their order by μ_{mn}, then one obtains the dispersion relation

$$r_0\sqrt{N} = \mu_{mn} \ , \tag{7.412}$$

and for $\kappa = 0$ it becomes

$$\varepsilon\mu\omega^2 - k_z^2 = \frac{\mu_{mn}^2}{r_0^2} \ . \tag{7.413}$$

In analogy to (7.405) and (7.406), one finds here again

$$v_{ph} \cdot v_G = c^2 \ . \tag{7.414}$$

The related fields of the TE_{mn} wave are determined by (7.248) and (7.249)

$$
\left.
\begin{aligned}
E_r &= \frac{i\omega\mu m}{r} C_m J_m(r\sqrt{N})\sin m\varphi \\
E_\varphi &= i\omega\mu\sqrt{N} C_m J_m'(r\sqrt{N})\cos m\varphi \\
E_z &= 0 \\
H_r &= -ik_z\sqrt{N} C_m J_m'(r\sqrt{N})\cos m\varphi \\
H_\varphi &= \frac{ik_z m}{r} C_m J_m(r\sqrt{N})\sin m\varphi \\
H_z &= N C_m J_m(r\sqrt{N})\cos m\varphi
\end{aligned}
\right\} \cdot \exp[i(\omega t - k_z z)] \ .
\tag{7.415}
$$

N has to assume a value permitted by eq. (7.412). Eq. (7.407) applies to λ_z, if λ_{mn} is replaced by μ_{mn}. The critical wave length is found in analogy to (7.408)

$$
\lambda_c = \frac{2\pi r_0}{\mu_{mn}} \ .
\tag{7.416}
$$

A number of values for μ_{mn} is given in Tab. 7.2. This table and the one for λ_{mn}

Table 7.2   The values μ_{mn} of $J_m'(\mu_{mn}) = 0$

n m	1	2	3	4	5
0	3.8317	7.0156	10.1735	13.3237	16.4706
1	1.8412	5.3314	8.5363	11.7060	14.8636
2	3.0542	6.7061	9.9695	13.1704	16.3475
3	4.2012	8.0152	11.3459	14.5859	17.7888
4	5.3175	9.2824	12.6819	15.9641	19.1960
5	6.4156	10.5199	13.9872	17.3128	20.5755
6	7.5013	11.7349	15.2682	18.6374	21.9318

make it particularly evident that the largest critical wave length of all TM and TE waves belongs to the TE_{11} wave, with a value of

$$
(\lambda_c)_{TE_{11}} = \frac{2\pi r_0}{\mu_{11}} = \frac{2\pi r_0}{1.84} = 3.41 r_0 \ .
\tag{7.417}
$$

Finally, we will present a few graphs, illustrating the field of some TM and TE waves, possible in a circular cylindric wave guide (Fig. 7.36, Fig. 7.37).

7.10.4   The Coaxial Cable

An in depth discussion of the coaxial cable requires one to start with the solutions (7.397), (7.398). However, we will not present the coaxial cable in its full

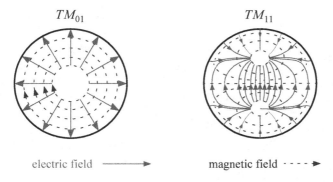

electric field ⟶   magnetic field · · · · ▶

Fig. 7.36

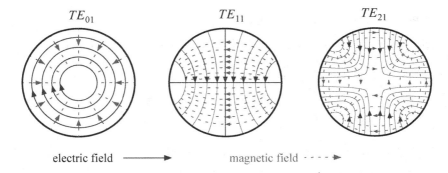

electric field ⟶   magnetic field · · · · ▶

Fig. 7.37

generality, which would provide all of its TE and TM waves. We will limit ourselves to the case $N = 0$, *i.e.*, its TEM wave, which requires to solve eq. (7.396) in its form

$$\left(\frac{1}{r}\frac{\partial}{\partial r}r\frac{\partial}{\partial r} + \frac{1}{r^2}\frac{\partial^2}{\partial \varphi^2}\right)\Pi_z = 0 \ . \tag{7.418}$$

The Ansatz

$$\Pi_z = R(r)\cos m\varphi \exp[i(\omega t - k_z z)] \tag{7.419}$$

yields an equation for $R(r)$:

$$\left(\frac{1}{r}\frac{\partial}{\partial r}r\frac{\partial}{\partial r} - \frac{m^2}{r^2}\right)R(r) = 0 \ . \tag{7.420}$$

Its general solution is

$$R(r) = C_1 + C_2 \ln r/r_0 \qquad \text{for} \quad m = 0 \tag{7.421}$$

or

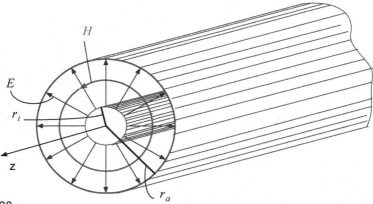

Fig. 7.38

$$R(r) = C_1 r^m + C_2 r^{-m} \qquad \text{for} \quad m > 0 \; , \tag{7.422}$$

which can also be obtained from the cylindrical functions when taking a limit. We have met eq. (7.420) and its solutions (7.421) and (7.422) before in Sect. 3.7.3.5, eqs. (3.262) through (3.264). The result for the Hertz vector is

$$\Pi_z = \left(C_1 + C_2 \ln\frac{r}{r_0}\right) \exp[i(\omega t - k_z z)] \qquad \text{for} \quad m = 0 \tag{7.423}$$

and

$$\Pi_z = (C_1 r^m + C_2 r^{-m}) \cos m\varphi \, \exp[i(\omega t - k_z z)] \qquad \text{for} \quad m > 0 \; . \tag{7.424}$$

Now one has the freedom to take Π_z as either the z-component of the electric or the magnetic Hertz vector and calculate the fields from (7.345), (7.346) or (7.351), (7.352), respectively. Either way, after considering the boundary conditions (7.363) and (7.364) at the outer conductor ($r = r_a$) and at the inner conductor ($r = r_i$) of the coaxial cable, it turns out that there is only one case where not all fields are identically zero (Fig. 7.38). This case requires that $m = 0$ and if one regards Π_z as Π_{ez}, (7.345), (7.346) yield the fields (for $\kappa = 0$):

$$\left.\begin{aligned}
E_r &= -\frac{ik_z}{r}C_2 \exp[i(\omega t - k_z z)] \\
E_\varphi &= 0 \\
E_z &= 0 \\
H_r &= 0 \\
H_\varphi &= -\frac{i\omega\varepsilon}{r}C_2 \exp[i(\omega t - k_z z)] \\
H_z &= 0
\end{aligned}\right\} . \tag{7.425}$$

This describes an electric field which is purely radial and a magnetic field that is purely azimuthal. Fig. 7.38 illustrates this. Since

$$N = \varepsilon\mu\omega^2 - k_z^2 = 0 \tag{7.426}$$

one finds

$$H_\varphi = \frac{E_r}{Z} , \tag{7.427}$$

i.e., the wave behaves in this respect like a plane wave. There is no difficulty in satisfying the boundary conditions, because there are neither components of **E** parallel to the boundary, nor perpendicular components of **H**. E_r must decrease as $1/r$ so that **E** remains source-free, while H_φ needs to decrease as $1/r$ so that **H** remains irrotational.

For the sake of completeness, it shall be noted that the just derived TEM wave can also be obtained from

$$\Pi_{mz} = (D_1 + D_2\varphi)\exp[i(\omega t - k_z z)] .$$

For $m = 0$

$$\Pi_z = \left(C_1 + C_2\ln\frac{r}{r_0}\right)(D_1 + D_2\varphi)\exp[i(\omega t - k_z z)] ,$$

is the most general solution of (7.418). If we take Π_z for Π_{ez}, then it has to be $D_2 = 0$ because of (7.363), which in conjunction with $D_1 = 1$ provides the fields given in (7.425). If we take Π_z for Π_{mz}, then it has to be $C_2 = 0$ because of (7.355), which in conjunction with $C_1 = 1$ yields Π_{mz} as given above. With the appropriate choice of D_2, this provides the fields (7.425).

Wave types are also called modes. The just found TEM mode of the coaxial cable is very similar to the TEM wave between two parallel plates. The two converge for $r_a - r_i \ll r_a$. We could think of Fig. 7.29 as being a piece of an annular region with a very large radius.

According to (7.426), the TEM wave propagates with the speed of light for that particular dielectric:

$$v_{ph} = \frac{\omega}{k_z} = \frac{1}{\sqrt{\varepsilon\mu}} = c . \tag{7.428}$$

The just found wave is also used in *transmission theory*, where the *telegrapher's equation* is used instead of the wave equation. To better illustrate the relation, we shall now derive the telegrapher's equation and briefly discuss it.

7.10.5 Telegrapher's Equation

To derive the telegrapher's equation, one imagines the transmission line as being a continuous analog of a network circuit, where a transmission line is nothing more than a multiply connected wave guide of arbitrary cross section. The coaxial cable of Fig. 7.38 is a special case thereof. The two-gate shown in Fig. 7.39 corresponds to a line element of length dz.

The quantities G', C', L', R' are in the same order: conductance, capacitance, inductance, resistance, all per unit length of the transmission line. With $dz \to 0$ it follows from Fig. 7.39:

$$G'V + C'\frac{\partial V}{\partial t} + \frac{\partial I}{\partial z} = 0 \tag{7.429}$$

$$L'\frac{\partial I}{\partial t} + R'I + \frac{\partial V}{\partial z} = 0 \ . \tag{7.430}$$

Differentiating the first equation for z yields:

$$G'\frac{\partial V}{\partial z} + C'\frac{\partial^2 V}{\partial t \partial z} + \frac{\partial^2 I}{\partial z^2} = 0 \ .$$

Substituting $\partial V/\partial z$ from the second equation yields

$$G'\left(-L'\frac{\partial I}{\partial t} - R'I\right) + C'\frac{\partial}{\partial t}\left(-L'\frac{\partial I}{\partial t} - R'I\right) + \frac{\partial^2 I}{\partial z^2} = 0 \ ,$$

or after rearranging

$$\boxed{\frac{\partial^2 I}{\partial z^2} = L'C'\frac{\partial^2 I}{\partial t^2} + (R'C' + L'G')\frac{\partial I}{\partial t} + G'R'I} \ . \tag{7.431}$$

Similarly for the second equation

$$L'\frac{\partial}{\partial t}\frac{\partial I}{\partial z} + R'\frac{\partial I}{\partial z} + \frac{\partial^2 V}{\partial z^2} = 0 \ ,$$

and substituting $\partial I/\partial z$ from the first equation yields

$$\boxed{\frac{\partial^2 V}{\partial z^2} = L'C'\frac{\partial^2 V}{\partial t^2} + (R'C' + L'G')\frac{\partial V}{\partial t} + G'R'V} \ . \tag{7.432}$$

Notice that $I(z, t)$ and $V(z, t)$ satisfy the same telegrapher's equation. In the simplest case, the line is free of losses, this means that

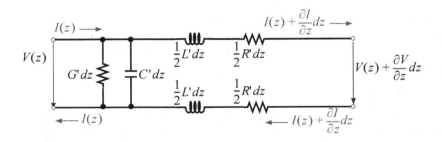

Fig. 7.39

$$G' = 0 \tag{7.433}$$

$$R' = 0 \tag{7.434}$$

and therefore

$$\frac{\partial^2 V}{\partial z^2} = L'C'\frac{\partial^2 V}{\partial t^2} \tag{7.435}$$

$$\frac{\partial^2 I}{\partial z^2} = L'C'\frac{\partial^2 I}{\partial t^2} \ . \tag{7.436}$$

One can prove that it must always be

$$L'C' = \varepsilon\mu \ . \tag{7.437}$$

We will skip the general proof. For the case of the coaxial cable, (7.437) can be verified by means of eqs. (2.98) and (5.210), when replacing ε_0 by ε and μ_0 by μ .

For V we try the following Ansatz

$$V(z, t) = V_0 \exp[i(\omega t - k_z z)] \tag{7.438}$$

and obtain from (7.435)

$$- k_z^2 + \varepsilon\mu\omega^2 = 0 \ ,$$

i.e., the usual dispersion relation for TEM waves in multiply connected wave guides (*for example*, in a coaxial cable, although above relation is by no means limited to the coaxial cable).

The current I creates the magnetic field (in case of the coaxial cable **H** is purely azimuthal), the EMF V creates the electric field between inside and outside conductor (for the coaxial cable, **E** is purely radial).

We will not advance any deeper into the matter of transmission theory, as our purpose was only to highlight the relation between transmission theory and field theory.

7.11 The Wave Guide as a Variational Problem

From a mathematical perspective, solving wave guide problems means to solve the Helmholtz equation in its form (7.344) or (7.350)

$$\Delta_2 \Pi_z(x, y) + N\Pi_z(x, y) = 0 \ , \tag{7.439}$$

where Δ_2 is the two-dimensional Laplacian, here in Cartesian coordinates:

$$\Delta_2 = \frac{\partial^2}{\partial x^2} + \frac{\partial^2}{\partial y^2} \ . \tag{7.440}$$

We might as well use a different coordinate system. In any case, according to (7.349), the boundary mandates for TM waves

$$\Pi_z = 0 , \tag{7.441}$$

and according to (7.355) for TE waves

$$\frac{\partial \Pi_z}{\partial n} = 0 \; . \tag{7.442}$$

From solving a number of example problems above, we found that there are only certain functions Π_z that satisfy the boundary conditions. Related are certain values of N. These functions are the *eigenfunctions* and the related N-values are the *eigenvalues* of the problem. The eigenfunctions form a complete system of orthogonal functions (Sect. 3.6). These are trigonometric functions in case of the rectangular wave guides and in case of circular cylindrical wave guides, these are Bessel functions. These functions are the basis to expand other functions in form of Fourier or Fourier-Bessel series, as we have done multiple times above.

Now address the eigenvalues by

$$N_1, N_2, N_3, \ldots$$

and their related eigenfunctions by

$$\Pi_1, \Pi_2, \Pi_3, \ldots \; ,$$

where we still refer to the z-component, but have dropped the index z to simplify. The order of the eigenvalues shall be such that

$$N_1 < N_2 < N_3 < N_4 < \ldots \; . \tag{7.443}$$

It is possible to show that the various eigenfunctions belonging to different eigenvalues are, indeed, orthogonal to each other. Also possible is to have several eigenfunctions to the same eigenvalue where the eigenfunctions are still orthogonal. Then the eigenfunctions and eigenvalues are said to be "degenerate". For simplicity reasons, we will exclude this case from our discussion, which does not limit the possible conclusions we will draw. Now, we will consider two eigenfunctions Π_i and Π_k with their eigenvalues N_i and N_k. With (7.439), one finds:

$$\Delta_2 \Pi_i + N_i \Pi_i = 0 \tag{7.444}$$

$$\Delta_2 \Pi_k + N_k \Pi_k = 0 \; . \tag{7.445}$$

Multiplying the first equation by Π_i and the second by Π_k yields

$$\Pi_k \Delta_2 \Pi_i + N_i \Pi_i \Pi_k = 0 \tag{7.446}$$

$$\Pi_i \Delta_2 \Pi_k + N_k \Pi_i \Pi_k = 0 \; . \tag{7.447}$$

Then we subtract them and integrate

$$\int (\Pi_k \Delta_2 \Pi_i - \Pi_i \Delta_2 \Pi_k) dA = -(N_i - N_k) \int \Pi_i \Pi_k dA \; , \tag{7.448}$$

where the area of integration is the cross section. Analogous to Green's theorem in the three-dimensional space (3.47), there is a Green's theorem for the plane (which can be derived from the one in three-dimensional space). It reads as follows (see Sect. 3.4.2):

$$\int (\Pi_k \Delta_2 \Pi_i - \Pi_i \Delta_2 \Pi_k) dA \;=\; \oint \left(\Pi_k \frac{\partial \Pi_i}{\partial n} - \Pi_i \frac{\partial \Pi_k}{\partial n} \right) ds \; . \tag{7.449}$$

Therefore, with the boundary conditions (7.441) or (7.442), we have

$$-(N_i - N_k) \int \Pi_i \Pi_k dA \;=\; \oint \left(\Pi_k \frac{\partial \Pi_i}{\partial n} - \Pi_i \frac{\partial \Pi_k}{\partial n} \right) ds \;=\; 0 \; ,$$

i.e., $N_i \neq N_k$ requires that

$$\int \Pi_i \Pi_k dA \;=\; 0 \; , \tag{7.450}$$

which proves our claim that the two functions Π_i and Π_k are orthogonal.

Now, consider an arbitrary, well behaved function ϕ and its series representation:

$$\phi \;=\; \sum_{i=1}^{\infty} a_i \Pi_i \; . \tag{7.451}$$

For the case of a rectangular wave guide, this would be a two-dimensional Fourier series, for the case of a TM wave in a circular cylinder, this would be a double series, where the φ-dependency is represented by a Fourier series while the r-dependency expands in a Fourier-Bessel series.

Let us investigate the following expression:

$$F \;=\; -\, \frac{\int \phi \Delta_2 \phi \, dA}{\int \phi^2 \, dA} \; . \tag{7.452}$$

Using (7.451) we obtain

$$F \;=\; -\, \frac{\int \displaystyle\sum_{i=1}^{\infty} a_i \Pi_i \Delta_2 \sum_{k=1}^{\infty} a_k \Pi_k \, dA}{\int \displaystyle\sum_{i,k=1}^{\infty} a_i a_k \Pi_i \Pi_k \, dA}$$

$$=\; \frac{\displaystyle\sum_{i,k=1}^{\infty} a_i a_k \int \Pi_i N_k \Pi_k dA}{\displaystyle\sum_{i,k=1}^{\infty} a_i a_k \int \Pi_i \Pi_k dA} \;=\; \frac{\displaystyle\sum_{i,k=1}^{\infty} a_i a_k N_k \, \delta_{ik} C_k}{\displaystyle\sum_{i,k=1}^{\infty} a_i a_k \, \delta_{ik} C_k}$$

$$F \;=\; \frac{\displaystyle\sum_{i=1}^{\infty} a_i^2 N_i C_i}{\displaystyle\sum_{i=1}^{\infty} a_i^2 C_i} \;\geq\; \frac{\displaystyle\sum_{i=1}^{\infty} a_i^2 N_1 C_i}{\displaystyle\sum_{i=1}^{\infty} a_i^2 C_i} \;=\; N_1 \; . \tag{7.453}$$

We have used the fact that (7.450) permits us to write

$$\int \Pi_i \Pi_k dA = \delta_{ik} C_k ,$$ (7.454)

where C_k is an arbitrary normalizing factor. We have determined that

$$F \geq N_1 .$$ (7.455)

The equal sign applies if, and only if

$$\phi = \Pi_1 .$$

In other words: N_1 is nothing else than the lowest value which the expression F may take and the function ϕ for which it assumes this value is the eigenfunction Π_1. Now we have transformed the wave guide problem into a variational problem. The task is now to find the function ϕ that fulfills the given boundary conditions, while minimizing the expression F. Then, after finding Π_1 we continue to determine Π_2. Now, we need to find the function ϕ which makes the expression F as small as possible, while ϕ still needs to satisfy the given boundary conditions and, in addition, has to be perpendicular to Π_1. This requires $a_1 = 0$ and thus obtain

$$F = \frac{\sum_{i=2}^{\infty} a_i^2 N_i C_i}{\sum_{i=2}^{\infty} a_i^2 C_i} \geq \frac{\sum_{i=2}^{\infty} a_i^2 N_2 C_i}{\sum_{i=2}^{\infty} a_i^2 C_i} = N_2 .$$ (7.456)

The equal sign now applies if $\phi = \Pi_2$.

This, from a formal perspective interesting remark, shall conclude the problem of waves in waveguides. Many more problems in physics, and in electromagnetic field theory in particular, can be regarded as variational problems. This is very useful because variation problems are a very good basis for approximation methods and numerical calculations. We will revisit this subject in Chapter 8.

7.12 Boundary and Initial Value Problems

In Chapter 6, we discussed the quasi-stationary approximation, while Chapter 7 was dedicated to the complete Maxwellian equations. From a formal perspective, the quasi-stationary case requires to solve the diffusion equation, in our case the wave equation, in one form or another. In its general form, the wave equation also contains a diffusion term. For instance, consider the magnetic field $\mathbf{B}$, then with (7.9) we write

$$\nabla^2 \mathbf{B} - \mu \kappa \frac{\partial \mathbf{B}}{\partial t} - \mu \varepsilon \frac{\partial^2 \mathbf{B}}{\partial t^2} = 0 ,$$ (7.457)

from which emerges the diffusion equation when neglecting the propagation terms

$$\nabla^2 \mathbf{B} = \mu \kappa \frac{\partial \mathbf{B}}{\partial t} = 0 .$$ (7.458)

At various times, we have discussed the limits of the quasi stationary theory and the competing effects of diffusion and wave propagation (see Sect. 6.8. and also Sect. 6.2). We shall revisit this discussion once more. For instance, one can solve the general wave equation (7.457) by the same methods which we have used in Chapter 6 to solve the diffusion equation. Then retroactively, we may let $\kappa \to 0$, which results in an undamped wave propagation in an ideal insulator, or conversely, let $\varepsilon \to 0$ to obtain the limit of diffusion.

To illustrate this, we will revisit two examples which we have already solved in Chapter 6, but now we include the wave propagation term. First, the problem of an initial field in the infinite, uniform space (Sect. 6.4) and second, the problem of the half-space (Sect. 6.5).

At the same time, the solution of these two problems shall demonstrate the general usefulness of methods previously used to solve initial value and boundary value problems.

7.12.1   The Initial Value Problem of the Infinite, Uniform Space

The task is to solve the problem given in Sect. 6.4 starting from the wave equation (7.457). Now, the order of the differential equation with respect to the time is two and, compared to before, this requires an additional initial value condition. We want to find $B_z(x, t)$ for which the wave equation has the form

$$\frac{\partial^2 B_z(x, t)}{\partial x^2} - \mu\kappa\frac{\partial B_z(x, t)}{\partial t} - \mu\varepsilon\frac{\partial^2 B_z(x, t)}{\partial t^2} = 0 . \tag{7.459}$$

Furthermore, the initial value and the boundary conditions shall be:

$$[B_z(x, t)]_{x \to +\infty} = B_z(\infty, t) = \text{finite} \tag{7.460}$$

$$[B_z(x, t)]_{x \to -\infty} = B_z(-\infty, t) = \text{finite} \tag{7.461}$$

$$B_z(x, 0) = h_1(x) \tag{7.462}$$

$$\left[\frac{\partial B_z(x, t)}{\partial t}\right]_{t \to 0} = h_2(x) . \tag{7.463}$$

The Laplace transform of eq. (7.459) gives

$$\frac{\partial^2 \tilde{B}_z(x, p)}{\partial x^2} - \mu\kappa[p\tilde{B}_z(x, p) - h_1(x)] - \mu\varepsilon[p^2\tilde{B}_z(x, p) - ph_1(x) - h_2(x)] = 0 \tag{7.464}$$

where $\tilde{B}_z(x, p)$ has to satisfy the boundary conditions

$$\tilde{B}_z(+\infty, p) = \text{finite} \tag{7.465}$$

$$\tilde{B}_z(-\infty, p) = \text{finite} . \tag{7.466}$$

The transition from (7.459) to (7.464) is based on (6.51), together with the initial values given by (7.462) and (7.463). The general solution of (7.464) is obtained analogous to the solution of (6.79) by (6.80). Here one obtains:

$$\tilde{B}_z(x,p) = A_1 \exp[-\sqrt{\mu\kappa p + \mu\varepsilon p^2}x] + A_2 \exp[+\sqrt{\mu\kappa p + \mu\varepsilon p^2}x]$$

$$-\int_{-\infty}^{x} [(\mu\kappa + \mu\varepsilon p)h_1(x_0) + \mu\varepsilon h_2(x_0)]$$

$$\cdot \frac{\sinh[\sqrt{\mu\kappa p + \mu\varepsilon p^2}(x-x_0)]}{\sqrt{\mu\kappa p + \mu\varepsilon p^2}} dx_0 . \qquad (7.467)$$

We limit ourselves to the special case where

$$h_1(x) = F\delta(x) \qquad (7.468)$$

$$h_2(x) = 0 . \qquad (7.469)$$

It shall be noted that the dimension of h_1 is that of B, while, because of the δ-function, F has the dimensions of B multiplied by a length. Using (7.468) and (7.469) in (7.467) gives

$$\tilde{B}_z(x,p) = \begin{cases} A_1 \exp[-\sqrt{\mu\kappa p + \mu\varepsilon p^2}x] + A_2 \exp[+\sqrt{\mu\kappa p + \mu\varepsilon p^2}x] \\ \qquad\qquad\qquad\qquad\qquad\qquad \text{for } x < 0 \\ A_1 \exp[-\sqrt{\mu\kappa p + \mu\varepsilon p^2}x] + A_2 \exp[+\sqrt{\mu\kappa p + \mu\varepsilon p^2}x] \\ \qquad -F\sqrt{\dfrac{\mu\kappa + \mu\varepsilon p}{p}}\sinh\sqrt{\mu\kappa p + \mu\varepsilon p^2}x \\ \qquad\qquad\qquad\qquad\qquad\qquad \text{for } x > 0 . \end{cases}$$

$$(7.470)$$

In order to satisfy the boundary conditions (7.465) and (7.466) one chooses

$$A_1 = 0 \qquad (7.471)$$

$$A_2 = \frac{F}{2}\sqrt{\frac{\mu\kappa + \mu\varepsilon p}{p}} \qquad (7.472)$$

which yields

$$\tilde{B}_z(x,p) = \frac{F}{2}\sqrt{\frac{\mu\kappa + \mu\varepsilon p}{p}}\exp[-\sqrt{\mu\kappa p + \mu\varepsilon p^2}|x|] . \qquad (7.473)$$

The inverse Laplace transform gives the function

$$B_z(x,t) = F\exp\left[-\frac{\kappa t}{2\varepsilon}\right]\left\{\frac{1}{2}\delta(x-ct) + \frac{1}{2}\delta(x+ct)\right.$$

$$+ H(ct - |x|)\left[\frac{\kappa}{4\varepsilon c}I_0\left(\frac{\kappa\sqrt{c^2t^2 - x^2}}{2\varepsilon c}\right) + \frac{\kappa t}{4\varepsilon\sqrt{c^2t^2 - x^2}}I_1\left(\frac{\kappa\sqrt{c^2t^2 - x^2}}{2\varepsilon c}\right)\right]\right\} .$$

$$(7.474)$$

The symbol H represents Heaviside's step function, defined in (3.55). In this case one has

$$H(ct - |x|) = \begin{cases} 1 & \text{for} \quad -ct < x < ct \\ 0 & \text{for} \quad \begin{cases} x < -ct \\ x > +ct \ . \end{cases} \end{cases}$$
(7.475)

The easiest proof is by means of the Laplace transform of (7.474) and then to find expression (7.473) thereof. This could be done *for example*, by use of [5], vol. I p. 200, eq. (5) and (9), as well as p. 129 eq. (5). The latter formula is also called "damping theorem".

Next, we will examine various limits of the problem's solution (7.474).

The limit $\kappa = 0$ leaves only

$$B_z(x, t) = \frac{F}{2}[\delta(x - ct) + \delta(x + ct)] \ .$$
(7.476)

The field, initially located at the origin, travels in equal halves in the positive and the negative z-direction. The reason for this even split is found in the initial conditions (7.468) and (7.469). In case of different initial conditions, the field would split in a different manner into the left and right travelling field (wave). In any case, both fractions will move without changing their shape. The reason is that these are ordinary plane waves travelling inside an ideal insulator (dielectric), as outlined in Sect. 7.1.2 and they behave exactly as expected. Of course, the separation into left and right travelling parts is possible in an arbitrary manner and is specified by the initial conditions. Fig. 7.40 illustrates this motion. It is possible to study the limit of pure diffusion. To achieve this, one lets $\varepsilon \to 0$ in eq. (7.474), or what amounts to the same, let $c \to \infty$. By means of the asymptotic formula (3.181) for I_0 and I_1, one obtains

$$B_z(x, t) = F\sqrt{\frac{\mu\kappa}{4\pi t}}\exp\left[-\frac{\mu\kappa x^2}{4t}\right],$$
(7.477)

which would also result from (6.78) when letting $h(x_0) = F\delta(x_0)$. In this case, the field spreads to an increasingly widening Gaussian curve (Fig. 7.41). By this, one has found both limiting cases (Fig. 7.40 and Fig. 7.41) from the general solution. The general case is given by damped δ-functions travelling left and right with the speed of light. There is no field in front of them. However, they drag an

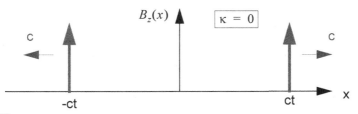

Fig. 7.40

umbrella of a diffusion field behind them. Fig. 7.42 illustrates this combination of diffusion and wave propagation. One could say that it describes a compromise of the trend shown in Figs. (7.40) and (7.41).

Although it would have been useful for practical reasons, to carry out these calculations with dimensionless quantities, we have refrained from doing so here (in contrast to the derivations of Sect. 6.4). This allowed to keep the dependency on quantities with dimensions like, *for example*, ε and κ immediately visible, which simplifies the limiting process. If one wants to save writing by use of dimensionless quantities, the best way to proceed is this:

$$\tau = \frac{t}{t_r} , \tag{7.478}$$

$$\xi = \frac{x}{x_0} , \tag{7.479}$$

with

$$t_r = \frac{\varepsilon}{\kappa} , \tag{7.480}$$

$$x_0 = ct_r = \frac{\varepsilon}{\kappa\sqrt{\varepsilon\mu}} = \frac{1}{\kappa}\sqrt{\frac{\varepsilon}{\mu}} , \tag{7.481}$$

from which we can derive the wave equation as

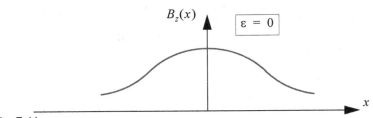

Fig. 7.41

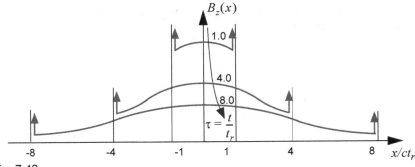

Fig. 7.42

$$\frac{\partial^2 B_z(\xi, \tau)}{\partial \xi^2} - \frac{\partial B_z(\xi, \tau)}{\partial \tau} - \frac{\partial^2 B_z(\xi, \tau)}{\partial \tau^2} = 0 \ . \tag{7.482}$$

As a, so-to-speak, natural unit of time, we see the relaxation time t_r, and as a natural unit of the length, we take the distance through which light travels during this time.

The Fourier transformation is another very useful tool to solve problems of the kind discussed here. This solution decomposes the initial field (wave packet) into its Fourier components. The previously discussed dispersion relation (Sect. 7.2) defines the behavior of each of these components, which allows the superposition of these packets at a later time. We will not discuss this method here, but it is nicely presented in detail in [8].

7.12.2 The Boundary Value Problem of the Half-Space

We shall revisit the problem of Sect. 6.5. Again, the task is to solve the wave equation

$$\frac{\partial^2 B_z}{\partial x^2} - \mu\kappa\frac{\partial B_z}{\partial t} - \mu\varepsilon\frac{\partial^2 B_z}{\partial t^2} = 0 \tag{7.483}$$

now, however, in the half space $x > 0$ with the boundary conditions:

$$B_z(\infty, t) = \text{finite} \tag{7.484}$$

$$B_z(0, t) = f(t) \tag{7.485}$$

and the initial conditions:

$$B_z(x, 0) = 0 \tag{7.486}$$

$$\left.\frac{\partial B_z(x, t)}{\partial t}\right|_{t=0} = 0 \ . \tag{7.487}$$

The problem we have solved in Sect 6.5 was more general with respect to the initial conditions. Conversely, there, we had one initial condition less. Using the Laplace transform on eq. (7.483) gives:

$$\frac{\partial^2 \tilde{B}_z(x, p)}{\partial x^2} = (\mu\kappa p + \mu\varepsilon p^2)\tilde{B}_z(x, p) \ , \tag{7.488}$$

while the boundary conditions in the p-domain become

$$\tilde{B}_z(\infty, p) = \text{finite} \tag{7.489}$$

$$\tilde{B}_z(0, p) = \tilde{f}(p) \ . \tag{7.490}$$

For the solution we obtain

$$\tilde{B}_z(x, p) = \tilde{f}(p)\exp[-\sqrt{\mu\kappa p + \mu\varepsilon p^2}x] \ . \tag{7.491}$$

Now, we choose

$$f(t) = G\delta(t) \tag{7.492}$$

which results in

$$\tilde{f}(p) = G \tag{7.493}$$

and therefore

$$\tilde{B}_z(x, p) = G \exp[-\sqrt{\mu\kappa p + \mu\varepsilon p^2}\,x]. \tag{7.494}$$

The inverse Laplace transformation yields:

$$B_z(x, t) = G \exp\left[-\frac{\kappa x}{2\varepsilon c}\right]\delta\left(t - \frac{x}{c}\right)$$

$$+ G\frac{\kappa x \exp\left[-\frac{\kappa t}{2\varepsilon}\right]I_1\left(\frac{\sqrt{c^2 t^2 - x^2}}{2\varepsilon c}\right)H\left(t - \frac{x}{c}\right)}{2\varepsilon\sqrt{c^2 t^2 - x^2}} . \tag{7.495}$$

Here again, the best approach to prove this is by the transform of $B_z(x, t)$ according to eq. (7.495). We achieve this by using *for example*, [5], Vol. I, p. 200, equation (8), and p. 129, eq. (5), which is the previously mentioned "damping theorem".

Specifically $\kappa = 0$ gives

$$B_z(x, t) = G\delta\left(t - \frac{x}{c}\right), \tag{7.496}$$

that is, the field of the very narrow impulse, created on the surface at $t = 0$ penetrates the medium with the speed of light. (Fig. 7.43). Conversely, if $\varepsilon = 0$, then taking the limit and by means of eq. (3.181) one obtains

$$B_z(x, t) = G\frac{x\sqrt{\mu\kappa}\exp\left[-\frac{\mu\kappa x^2}{4t}\right]}{\sqrt{4\pi t^3}}, \tag{7.497}$$

that is, just the result obtained when using eq. (6.111) and eq. (7.492). This corresponds to a pure diffusion process (Fig. 7.44).

The general case involves a damped δ-function and a diffusing field, which the δ-function drags behind itself. There is no field in front of the δ-function.

Fig. 7.43

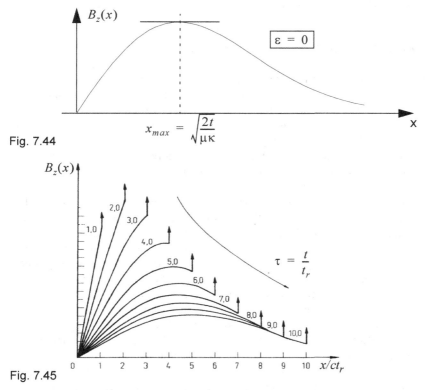

Fig. 7.44

Fig. 7.45

Various states of this process are illustrated in Fig. 7.45. For times $t \gg 4t_r$, the field looks more and more like the one shown in Fig. 7.44

8 Numerical Methods

8.1 Introduction

This text has introduced the terminology of electromagnetic field theory, which governs the relations between the various field quantities – Maxwell's equations in particular – and a few of the methods that are suitable to analytically solve field theoretical problems. It is obvious, however, that for many problems, analytical methods may only allow for an approximate solution or may not be solvable this way at all. If we need to solve such problems, a different methodology has to be employed. Sometimes, if the problem under consideration differs only slightly from one that can be solved exactly, then perturbation theory may be applied. We shall not discuss perturbation in this text. A discussion of that topic can be found at Morse Feshbach [9]. In contrast, the various types of numerical methods are generally applicable, at least in principle. In light of the ever increasing potential of computer power, these methods become more and more attractive and constitute a fruitful area for field theory as well. The subject is so vast, that we can only touch on the basic ideas. On the other hand, numerical methods are so important that it should not go unmentioned and therefore, the following shall be dedicated to describe the most important numerical methods, including some simple examples. Although we will focus on electrostatic problems, these methods are applicable to all areas of field theory, in particular to magnetostatics and to time dependent problems. It shall particularly be illustrated that, and how the various methods relate to the analytical methods formally, and in to the field theoretical problems from a content perspective. It is advisable to always work numerical and analytical problems in parallel. This is an important basis for successful work in this area of study. Such approach fosters a deeper and clearer understanding that is sufficiently critical of the problem and its results. Areas for possible errors can be recognized more easily and testing of the created programs can focus on the critical aspects of the problem and thus be more efficient.

The following sections (8.2 though 8.5) prepare for the later sections (8.6 though 8.10), where the various numerical methods shall be explained.

8.2 Basics of Potential Theory

8.2.1 Boundary Value Problems and Integral Equations

The potential theory only briefly described in Sect. 3.4, is a basic building block for both, the analytic as well as the numerical methods. This is particularly true for Kirchhoff's theorem (see 3.4.7, eq. (3.57)), which because of its far reaching significance could easily be called the principal theorem of potential theory.

G. Lehner, *Electromagnetic Field Theory for Engineers and Physicists*,
DOI 10.1007/978-3-540-76306-2_8, © Springer-Verlag Berlin Heidelberg 2010

$$4\pi\varphi(\mathbf{r}) = \int_V \frac{\rho(\mathbf{r}')d\tau'}{\varepsilon_0|\mathbf{r} - \mathbf{r}'|} + \oint \frac{1}{|\mathbf{r} - \mathbf{r}'|}\frac{\partial}{\partial n'}\varphi(\mathbf{r}')dA' - \oint\varphi(\mathbf{r}')\frac{\partial}{\partial n'}\frac{1}{|\mathbf{r} - \mathbf{r}'|}d\mathbf{A}'$$ (8.1)

This equation is valid for the three-dimensional space, where it addresses a point inside a given region. We have already emphasized that the equation in this form is not suitable to solve boundary values problems by arbitrarily specifying φ and $\partial\varphi/\partial n$ on the surface so that φ inside could be found. Only one or the other boundary condition is allowed on any one surface element (in case of a mixed boundary value problem, it has to be clear for which part of the area φ and for which part $\partial\varphi/\partial n$ applies. Nevertheless, we may use this equation to solve boundary value problems (analytically, as well as numerically). Instead of using eq. (8.1), for points on the surface one has

$$C(\mathbf{r})\varphi(\mathbf{r}) = \int_V \frac{\rho(\mathbf{r}')d\tau'}{\varepsilon_0|\mathbf{r} - \mathbf{r}'|} + \oint \frac{1}{|\mathbf{r} - \mathbf{r}'|}\frac{\partial\varphi(\mathbf{r}')}{\partial n'}dA' - \oint\varphi(\mathbf{r}')\frac{\partial}{\partial n'}\frac{1}{|\mathbf{r} - \mathbf{r}'|}d\mathbf{A}'$$ (8.2)

$C(\mathbf{r})$ is a factor that for objects with smooth surfaces everywhere has a constant value of 2π. These are characterized by tangential planes which are uniquely defined everywhere. For instance, the surface of a sphere is smooth. However, this is not true for the surface of a cube (which has edges where it is obviously not smooth), or that of a cone (which has a tip). Consider the three-dimensional δ-function as the limit of a series of functions, which, within a given volume V, takes on the constant value $1/V$ and vanishes outside of the volume. This makes it intuitively clear that the integral necessary for the derivation of eq. (8.1), and of eq. (3.57) in Sect. 3.4.7 becomes

$$\int_V \varphi(\mathbf{r})4\pi\delta(\mathbf{r} - \mathbf{r}')d\tau = C(\mathbf{r}')\varphi(\mathbf{r}') = \Omega\varphi(\mathbf{r}') \;,$$

where Ω is the solid angle that arises at point $\mathbf{r}'$ of the surface inside the region. For smooth surfaces, we have $C = 2\pi$. On the edges of a cube the relation is $C = \Omega = \pi$ and at its corners $C = \Omega = \pi/2$. This allows one to include the singular points of not-smooth surfaces as well.

A similar approach is possible for two-dimensional (plane) problems. In this case, the so-called fundamental solution of the three-dimensional space

$$\psi = \frac{1}{|\mathbf{r} - \mathbf{r}'|} \;,$$ (8.3)

that is, the Coulomb potential of a point charge, has to be replaced by its two-dimensional analogue, which is the potential of an infinitely long, straight, and uniform line charge. Disregarding constant factors this is given by

$$\psi = -\ln|\mathbf{r} - \mathbf{r}'| \;,$$ (8.4)

where

$$\nabla^2\ln|\mathbf{r} - \mathbf{r}'| = 2\pi\delta(\mathbf{r} - \mathbf{r}')$$ (8.5)

With this, instead of eq. (8.1), we obtain for points inside a two-dimensional region

$$2\pi\varphi(\mathbf{r})=-\int_A \frac{\rho(\mathbf{r}')}{\varepsilon_0}\ln|\mathbf{r}-\mathbf{r}'|\,dA' -\oint \ln|\mathbf{r}-\mathbf{r}'|\frac{\partial\varphi(\mathbf{r}')}{\partial n'}\,ds' +\oint\varphi(\mathbf{r}')\frac{\partial}{\partial n'}\ln|\mathbf{r}-\mathbf{r}'|\,ds'$$

(8.6)

The last two integrals are line integrals along the path element ds' which constitutes the boundary of the area. On the boundary itself one has

$$C(\mathbf{r})\varphi(\mathbf{r})=-\int_A \frac{\rho(\mathbf{r}')}{\varepsilon_0}\ln|\mathbf{r}-\mathbf{r}'|\,dA' -\oint \ln|\mathbf{r}-\mathbf{r}'|\frac{\partial\varphi(\mathbf{r}')}{\partial n'}\,ds'$$
$$+ \oint\varphi(\mathbf{r}')\frac{\partial}{\partial n'}\ln|\mathbf{r}-\mathbf{r}'|\,ds'$$

(8.7)

The quantities $\mathbf{r}$ and $\mathbf{r}'$ in equations (8.4) through (8.7) are two-dimensional vectors in a plane.

$C(\mathbf{r})$ represents the plane angle seen from inside the area. In case of a smooth boundary, $C = \pi$, for the corners of a rectangle we have $C = \pi/2$, etc.

In case of one-dimensional problems, the fundamental solution can be expressed in the form

$$\psi(x,x') = \frac{1}{2}|x-x'| = \begin{cases} \frac{1}{2}(x'-x) & \text{for} \quad x \le x' \\ \frac{1}{2}(x-x') & \text{for} \quad x \ge x'. \end{cases}$$

(8.8)

This case yields

$$\psi' = \frac{\partial}{\partial x}\psi = \left.\begin{cases} -\dfrac{1}{2} & \text{for} \quad x < x' \\ 0 & \text{for} \quad x = x' \\ +\dfrac{1}{2} & \text{for} \quad x > x' \end{cases}\right\} = -\frac{1}{2}+ H(x-x') ,$$

(8.9)

and by Sect. 3.4.5

$$\psi'' = \nabla^2\psi = \delta(x-x') .$$

(8.10)

The discontinuity of ψ' at $x = x'$ is typical and of immense significance. This is similar to the discontinuities of $(\partial/\partial n')(1/|\mathbf{r}-\mathbf{r}'|)$ or $(\partial/\partial n')(\ln|\mathbf{r}-\mathbf{r}'|)$ in case of three- or two dimensions, respectively. The value of the function ψ' at location $x = x'$ is the average of the left and right sided limit at this point.

If one wishes to solve a Dirichlet or Neumann boundary value problem, to start with eq. (8.2) or eq. (8.7) is advisable in order to find the boundary values of $\partial\varphi/\partial n$ from those of φ or vice versa. This method provides compatible values for $\partial\varphi/\partial n$ and φ over the entire boundary. With these, eq. (8.1) or eq. (8.6) finally allow one to determine the potential φ in the entire region. The boundary value problem is thereby reduced to these integral equations.

If, in case of a Dirichlet problem, φ is given, then eq. (8.2) or eq. (8.7) represent the so-called *Fredholm Integral equation of the 1st kind* for $\partial\varphi/\partial n$. In case of a Neumann problem, eq. (8.2) or eq. (8.7) represent the *Fredholm Integral equation of the 2nd kind* for φ.

There is another way how boundary value problems can be reduced to integral equations. Consider a surface bounding a region carrying a surface charge density $\sigma(\mathbf{r})$. Then in the three-dimensional case (the one- and two-dimensional cases can be solved analogously), the potential inside the volume and on the boundary is

$$\boxed{\varphi(\mathbf{r}) = \oint \frac{\sigma(\mathbf{r}')}{4\pi\varepsilon_0|\mathbf{r} - \mathbf{r}'|} dA'} \ . \tag{8.11}$$

The perpendicular component of the electric field is discontinuous at the surface, as we have already discovered before (Sect. 2.5.3 and 2.10). On the inside of the surface, it is therefore

$$\boxed{\frac{\partial \varphi(\mathbf{r})}{\partial n} = \oint \frac{\sigma(\mathbf{r}')}{4\pi\varepsilon_0} \frac{\partial}{\partial n} \frac{1}{|\mathbf{r} - \mathbf{r}'|} dA' + \frac{\sigma(\mathbf{r})}{2\varepsilon_0}} \ . \tag{8.12}$$

If there is a double layer or dipole layer on the surface with the surface density of the dipole moment τ, then the potential is

$$\boxed{\varphi(\mathbf{r}) = \oint \frac{\tau(\mathbf{r})}{4\pi\varepsilon_0} \frac{\partial}{\partial n'} \frac{1}{|\mathbf{r} - \mathbf{r}'|} dA'} \ . \tag{8.13}$$

In this case, even φ is discontinuous at the surface (see Sect. 2.5.3, eq. (2.73), eq. (2.82)). The potential on the inside of the double layer is

$$\boxed{\varphi(\mathbf{r}) = \oint \frac{\tau(\mathbf{r}')}{4\pi\varepsilon_0} \frac{\partial}{\partial n'} \frac{1}{|\mathbf{r} - \mathbf{r}'|} dA' - \frac{\tau(\mathbf{r})}{2\varepsilon_0}} \ . \tag{8.14}$$

From a formal perspective, both discontinuities are a consequence of the fact that $(\partial/\partial n')(1/|\mathbf{r} - \mathbf{r}'|)$ is discontinuous on the surface. Eqs (8.12) and (8.14) are valid for an approach of the surface from the inside, in order to calculate the surface integrals. The additional terms of Eqs (8.12) and (8.14) carry a factor 1/2, which is a result of the fact that the value on the surface is obtained as an average of the both sided limit, which can be regarded as dividing the discontinuities σ/ε_0 and τ/ε_0 in half.

Eqs. (8.11) through (8.14) may be used to calculate the solutions of Dirichlet, Neumann, or mixed boundary value problems. We start by just considering points on the boundary. For Dirichlet problems, one finds either $\sigma(\mathbf{r}')$ from (8.11) or $\tau(\mathbf{r}')$ from (8.14). From there, either use (8.11) or (8.13) to find the potential in the entire region. In case of Neumann problems, one calculates $\sigma(\mathbf{r}')$ by means of (8.12) and then finds the potential in the entire region by (8.11). Again, depending on the type of problem, one needs to solve either the Fredholm Integral equations of the 1st or 2nd kind.

These integral equations are of fundamental significance for field theory. For the Fredholm Integral equations, there exists a well established mathematical theory, which is also plausible because it is entirely analogous to the theory of linear algebraic equations. The Fredholm Integral equations are nothing else than the continuum analogues of linear algebraic equations. For instance, there are theorems about the existence or non-existence of solutions, as well as their uniqueness or multiplicity, which are an almost verbatim copy of their analogue in linear systems of equations. They form the basis for many fundamental theorems of potential theory, *for example* about the existence of solutions of boundary value problems (see *for example* [10 through 16]). An analytical solution of the equations is possible in some cases, which is what shall be demonstrated in the following by an example. Finally – and this is important for our case – they are also suitable for numerical evaluation, which results in the boundary element method and is presented in Sect 8.8.

8.2.2 Examples

8.2.2.1 The One-dimensional Problem

This basic problem illustrates how to use eqs (8.1) through (8.7) in the not so trivial 2-D and 3D cases. We use Green's integral theorem in one-dimension

$$\int_a^b (\psi\varphi'' - \varphi\psi'')dx = [\psi\varphi' - \varphi\psi']_a^b \tag{8.15}$$

and let ψ'' be according to eq. (8.10). Then we find for points inside

$$\varphi(x') = \int_a^b \psi\nabla^2\varphi\,dx + [\varphi\psi' - \psi\varphi']_a^b \qquad a < x' < b \tag{8.16}$$

and on the surface (*i.e.* for $x' = a$ or $x' = b$)

$$\frac{1}{2}\varphi(x') = \int_a^b \psi\nabla^2\varphi\,dx + [\varphi\psi' - \psi\varphi']_a^b . \tag{8.17}$$

The factor $1/2$ corresponds to the factors $C = 2\pi$ or $C = \pi$ in eqs (8.2) and (8.7), respectively.
We chose a simple **example** where

$$\nabla^2\varphi = Ax , \quad 0 \le x \le 1 , \tag{8.18}$$

and the Dirichlet boundary conditions

$$\varphi(0) = \varphi(1) = 0 . \tag{8.19}$$

The exact solution to this problem is

$$\varphi = \frac{Ax(x^2 - 1)}{6} . \tag{8.20}$$

This solution shall be obtained by means of the one-dimensional Kirchhoff theorem, eqs. (8.16) and (8.17). We start from eq. (8.18) with eq. (8.8)

$$\int_0^1 \psi\nabla^2\varphi\,dx = \frac{A}{12}(2x'^3 - 3x' + 2) . \tag{8.21}$$

From eq. (8.17), and the fact that the surface consists of only two points, it follows that

$$\frac{1}{2}\varphi(0) = 0 = \frac{A}{6} - [\psi(1,0)\varphi'(1) - \psi(0,0)\varphi'(0)]$$

$$\frac{1}{2}\varphi(1) = 0 = \frac{A}{12} - [\psi(1,1)\varphi'(1) - \psi(0,1)\varphi'(0)]$$

or

$$\frac{A}{6} = \left[\frac{1}{2}\cdot\varphi'(1) - 0\cdot\varphi'(0)\right]$$

$$\frac{A}{12} = \left[0\cdot\varphi'(1) - \frac{1}{2}\cdot\varphi'(0)\right],$$

with the solutions

$$\varphi'(0) = -\frac{A}{6}, \quad \varphi'(1) = \frac{A}{3}. \tag{8.22}$$

Thereby, the boundary values of φ determine the boundary values of $\partial\varphi/\partial n$ (the perpendicular derivatives). Together with eq. (8.16), they provide the solution

$$\varphi(x') = \frac{A}{12}(2x'^3 - 3x' + 2) - \frac{A}{3}\cdot\frac{1-x'}{2} - \frac{A}{6}\cdot\frac{x'}{2} = \frac{Ax'(x'^2-1)}{6},$$

which is the same as previously given in eq. (8.20).
As our second **example** we choose

$$\nabla^2\varphi = 0, \quad 0 \le x \le 1, \tag{8.23}$$

where

$$\varphi(0) = A, \quad \varphi(1) = B. \tag{8.24}$$

Again, using eq. (8.17) gives

$$\frac{1}{2}\varphi(0) = \frac{A}{2} = [B\psi'(1,0) - A\psi'(0,0) - \varphi'(1)\psi(1,0) + \varphi'(0)\psi(0,0)]$$

$$\frac{1}{2}\varphi(1) = \frac{B}{2} = [B\psi'(1,1) - A\psi'(0,1) - \varphi'(1)\psi(1,1) + \varphi'(0)\psi(0,1)].$$

Note that here, $\psi'(x,x')$ is discontinuous for $x = x'$ and by eq. (8.9)

$$\psi'(0,0) = 0, \quad \psi'(1,1) = 0.$$

Furthermore:

$$\frac{A}{2} = \frac{B}{2} - \frac{\varphi'(1)}{2},$$

$$\frac{B}{2} = \frac{A}{2} + \frac{\varphi'(0)}{2},$$

and

$$\varphi'(0) = \varphi'(1) = B - A.$$

This means that by eq. (8.16)

$$\varphi(x') = \frac{B}{2} - (B-A)\frac{1-x'}{2} + \frac{A}{2} + (B-A)\frac{x'}{2}$$

$$= A + (B-A)x' \ . \tag{8.25}$$

Of course, we obtain the linear potential, necessary for this case.

It shall be noted that the fundamental solution can be chosen differently. Important is only that it satisfies eq. (8.10). Therefore, the function

$$\psi(x, x') = \begin{cases} x'(1+x) & \text{for} & x \le x' \\ x(1+x') & \text{for} & x \ge x' \end{cases}$$

with

$$\psi'(x, x') = \frac{\partial \psi}{\partial x} = \begin{cases} x' & \text{for} & x < x' \\ \frac{1}{2} + x' & \text{for} & x = x' \\ 1 + x' & \text{for} & x > x' \end{cases} = x' + H(x - x')$$

can also be chosen as the fundamental solution. Naturally, both of our examples will give the same result, which to verify is left for the reader as an exercise.

8.2.2.2 Dirichlet's Boundary Value Problem of a Sphere

In this example, we will use the above presented integral equations to solve the inner Dirichlet boundary value problem of a sphere, for which Laplace's equation holds (that is no charges) in three different ways. This requires an expansion of the fundamental solution (*i.e.* the inverse distance) by means of spherical harmonics. By eq. (3.324) we write

$$\frac{1}{|\mathbf{r} - \mathbf{r}_0|} = \sum_{n=0}^{\infty} \sum_{m=0}^{n} (2 - \delta_{0m}) \begin{cases} \dfrac{r^n}{r_0^{n+1}} \\ \dfrac{r_0^n}{r^{n+1}} \end{cases} \frac{(n-m)!}{(n+m)!}$$

$$\cdot P_n^m(\cos\theta)P_n^m(\cos\theta_0)\cos[m(\varphi - \varphi_0)] \ . \tag{8.26}$$

The upper term is valid for $r \le r_0$, while the lower one holds for $r \ge r_0$. The perpendicular gradient is

$$\frac{\partial}{\partial n_0}\frac{1}{|\mathbf{r}-\mathbf{r}_0|} = \sum_{n=0}^{\infty}\sum_{m=0}^{n}(2-\delta_{0m})\begin{cases}-(n+1)\dfrac{r^n}{r_0^{n+2}}\\[2mm]-\dfrac{1}{2r_0^2}\\[2mm]n\dfrac{r_0^{n-1}}{r^{n+1}}\end{cases}\frac{(n-m)!}{(n+m)!}$$

$$\cdot P_n^m(\cos\theta)P_n^m(\cos\theta_0)\cos[m(\varphi-\varphi_0)] \ . \tag{8.27}$$

Significant is that the expression is discontinuous for $r = r_0$. Now, the upper term is valid for $r < r_0$, the middle term is for $r = r_0$, and the lower one holds for $r > r_0$. This discontinuity is the main reason for presenting this example here. From a formal perspective, this is the reason for the discontinuity of the electric field at charged surfaces and that of the potential at dipole layers which was demonstrated in Sect. 2.5.3 and has culminated in the integrals eq. (8.12) and eq. (8.14). This shall be demonstrated again by means of the present example. Note the frequently used relation:

$$\frac{\partial}{\partial n}\frac{1}{|\mathbf{r}-\mathbf{r}'|} = -\frac{\partial}{\partial n'}\frac{1}{|\mathbf{r}-\mathbf{r}'|} \ .$$

We also need the completeness relation for the spherical harmonics. It results form expanding $\delta(\theta-\theta')\delta(\varphi-\varphi')$ in terms of spherical harmonics by means of eq. (3.300) and the corresponding expression, now using sine instead of cosine:

$$\sum_{n=0}^{\infty}\sum_{m=0}^{n}\frac{(2n+1)(n-m)!}{4\pi(n+m)!}(2-\delta_{0m})P_n^m(\cos\theta)P_n^m(\cos\theta')\cos[m(\varphi-\varphi')]\sin\theta'$$

$$= \delta(\theta-\theta')\delta(\varphi-\varphi') \ .$$

We start by analyzing the integral in eq. (8.12). With the exception of the sign, the perpendicular gradient contained therein is given by eq. (8.27). For $r = r_0$ we find that at the inside and the outside boundary, respectively:

$$\frac{\partial}{\partial n'}\frac{1}{|\mathbf{r}-\mathbf{r}'|} = -\sum_{n=0}^{\infty}\sum_{m=0}^{n}\frac{(n-m)!}{(n+m)!}(2-\delta_{0m})P_n^m(\cos\theta)P_n^m(\cos\theta')$$

$$\cdot \cos[m(\varphi-\varphi')]\begin{cases}-\dfrac{(n+1)}{r_0^2}\\[3mm]+\dfrac{n}{r_0^2}\end{cases} \ .$$

Next, one calculates the integral at the outside as well as on the inside boundary, or the difference between the two integrals. One then finds the value of discontinuity to be

$$-\oint \frac{\sigma(\mathbf{r}')}{4\pi\varepsilon_0} \sum_{n=0}^{\infty} \sum_{m=0}^{n} \frac{(n-m)!}{(n+m)!}(2-\delta_{0m})P_n^m(\cos\theta)P_n^m(\cos\theta')\cos[m(\varphi-\varphi')]$$

$$\cdot \frac{2n+1}{r_0^2}r_0^2\sin\theta'd\theta'd\varphi' = -\oint \frac{\sigma(\mathbf{r}')}{\varepsilon_0}\delta(\theta-\theta')\delta(\varphi-\varphi')d\theta'd\varphi' = -\frac{\sigma(\mathbf{r}')}{\varepsilon_0}.$$

From eq. (8.13), in complete analogy, while recognizing the changed sign, one sees that the discontinuities of the integral is

$$\oint \frac{\tau(\mathbf{r}')}{\varepsilon_0}\delta(\theta-\theta')\delta(\varphi-\varphi')d\theta'd\varphi' = \frac{\tau(\mathbf{r}')}{\varepsilon_0}.$$

Obviously, these discontinuities are local properties of each individual surface element. For instance, suppose that $\sigma \neq 0$ is true only for one particular surface area element. Therefore, the results apply in general to an arbitrary surface.

Suppose $\phi(\mathbf{r})$ is given on the entire surface and if the coordinate system is chosen appropriately, then the boundary values can be expanded in terms of spherical harmonics:

$$\phi(\mathbf{r}) = \sum_{n=0}^{\infty} \sum_{m=0}^{n} B_{nm}P_n^m(\cos\theta)\cos(m\varphi) . \tag{8.28}$$

For $\sigma(r_0)$ we use the Ansatz

$$\sigma(r_0) = \sum_{n=0}^{\infty} \sum_{m=0}^{n} C_{nm}P_n^m(\cos\theta_0)\cos(m\varphi_0) , \tag{8.29}$$

which allows to determine $\sigma(r_0)$ by means of eq. (8.11) and eq. (8.26), as well as the orthogonality relation eq. (3.300). One finds

$$C_{nm} = \frac{(2n+1)\varepsilon_0 B_{nm}}{r_0} . \tag{8.30}$$

This solves the problem. One can now use eq. (8.11), together with eq. (8.29) and eq. (8.30) to write the solution for the interior of the sphere

$$\phi(\mathbf{r}) = \sum_{n=0}^{\infty} \sum_{m=0}^{n} B_{nm}\left(\frac{r}{r_0}\right)^n P_n^m(\cos\theta)\cos(m\varphi) , \tag{8.31}$$

which was expected based on eq. (8.28). The purpose of this example was to demonstrate that the integral relation eq. (8.11) is suitable to calculate the correct result.

In similar manner, one may use eq. (8.14) and the Ansatz

$$\tau(r_0) = \sum_{n=0}^{\infty} \sum_{m=0}^{n} D_{nm}P_n^m(\cos\theta_0)\cos(m\varphi_0) \tag{8.32}$$

to determine D_{nm}. Then

$$D_{nm} = -\frac{(2n+1)\varepsilon_0 B_{nm}}{n+1} \,,$$ (8.33)

which solves the problem as well. Again, one finds the potential (8.31).

Finally, with the Ansatz

$$\left(\frac{\partial\phi}{\partial n}\right)_{r=r_0} = \sum_{n=0}^{\infty}\sum_{m=0}^{n} E_{nm}P_n^m(\cos\theta_0)\cos(m\varphi_0) \,,$$ (8.34)

we may base the analysis on eq. (8.2), where we let $C = 2\pi$. One obtains

$$E_{nm} = \frac{n}{r_0}B_{nm}$$ (8.35)

and using eq. (8.1) gives again the potential (8.31).

Of course, this approach is applicable to arbitrarily shaped surfaces. If analytical solutions do not exist, then the problem can be solved numerically. To do this we partition the surface (the "boundary") into small surface elements ("boundary elements"). This leads to the boundary value method, which we will discuss in Sect. 8.8.

8.2.3 Mean Value Theorems of Potential Theory

The so-called mean value theorems of potential theory are a consequence of Kirchhoff's theorems. They illustrate important relations in a clear and intuitive manner. Consider a spherically shaped area, centered at the origin with radius R that does not contain charges ($\nabla^2\varphi = 0$). Then by eq. (8.1)

$$\varphi(0) = \frac{1}{4\pi}\oint\frac{1}{R}\frac{\partial}{\partial n'}\varphi(\mathbf{r}')dA' - \frac{1}{4\pi}\oint\varphi(\mathbf{r}')\frac{\partial}{\partial n'}\frac{1}{|\mathbf{r}'|}dA' \,.$$

Because there are no charges, one has

$$\oint\frac{\partial}{\partial n'}\varphi(\mathbf{r}')dA' = 0 \,.$$

Furthermore

$$\frac{\partial}{\partial n'}\frac{1}{|\mathbf{r}'|} = -\left(\frac{\mathbf{r}'}{|\mathbf{r}'|^3}\right)_{n'} = -\frac{1}{R^2} \,.$$

This gives

$$\boxed{\varphi(0) = \frac{1}{4\pi R^2}\oint\varphi(\mathbf{r}')dA' = \langle\varphi\rangle_A} \,,$$ (8.36)

i.e., the potential at the center of a charge-free sphere is equal to the mean value of the potential on the surface of the sphere. We might as well consider the entire volume of the sphere and apply the mean value theorem onto all spherical shells that are contained within the sphere and obtain

$$\varphi(0) = \frac{1}{4\pi R^3}\int_V \varphi(\mathbf{r}')d\tau' = \langle\varphi\rangle_V \ . \tag{8.37}$$

This applies in much the same way to two-dimensional spherical regions. Eq. (8.6) for the center of a circle reads:

$$\varphi(0) = -\frac{\ln R}{2\pi}\oint\frac{\partial\varphi(\mathbf{r}')}{\partial n'}ds' + \frac{1}{2\pi}\oint\varphi(\mathbf{r}')\frac{\partial}{\partial n'}\ln|\mathbf{r}'|ds' \ .$$

Using

$$\oint\frac{\partial\varphi(\mathbf{r}')}{\partial n'}ds' = 0$$

and

$$\frac{\partial}{\partial n'}\ln|\mathbf{r}'| = \left(\frac{1}{|\mathbf{r}'|}\frac{\mathbf{r}'}{|\mathbf{r}'|}\right)_{n'} = \frac{1}{R}$$

gives

$$\varphi(0) = \frac{1}{2\pi R}\oint\varphi(\mathbf{r}')ds' = \langle\varphi\rangle_s \ , \tag{8.38}$$

and for the entire area of the circle

$$\varphi(0) = \frac{1}{\pi R^2}\int\varphi(\mathbf{r}')dA' = \langle\varphi\rangle_A \ . \tag{8.39}$$

There are analogous statements for the one-dimensional case

$$\nabla^2\varphi(x) = 0 \ ,$$

which has only a linear solution that is

$$\varphi = A + Bx \ .$$

This case yields

$$\varphi\left(\frac{a+b}{2}\right) = \frac{\varphi(a) + \varphi(b)}{2} \tag{8.40}$$

and

$$\varphi\left(\frac{a+b}{2}\right) = \frac{1}{b-a}\int_a^b\varphi(x)dx \ . \tag{8.41}$$

The conclusion is, that the solution of the Laplace equation in one-, two-, and three dimensions has the property that its values at the center of a line, circle, sphere are equal to the mean value of the corresponding values on its boundary, which is the average of the line, the entire circle, or the entire sphere. We will encounter this fact again when we will discuss the method of finite differences.

8.3   Boundary Value Problems as Variational Problems

8.3.1   Variational Integrals and Euler's Equations

Many problems that are formulated with differential equations can be reduced to variational problems. This is also true for some problems of the electromagnetic field theory. We already encountered one such example when studying electromagnetic waves inside a cylindrical wave guide (Sect. 7.11). The general task in Variational Calculus is to find the function (or functions) which constitutes an extremum of its dependent integral, which is called *functional*, meaning a function of functions. Consider the following integral that spans a region in space:

$$I(u) = \int_V F(x, y, z, u, u_x, u_y, u_z) d\tau = \text{extremum} \tag{8.42}$$

Initially, we require that $u = u(x, y, z))$ vanishes on the boundary. Later, we will consider different boundary conditions. u_x, u_y, u_z represent the partial derivatives of u with respect to x, y, z. Let u be the function that gives an extremum to the integral. We define a differing, so-called *reference function*

$$\bar{u} = u + \varepsilon f, \tag{8.43}$$

where f represents a mostly arbitrary function, which vanishes on the boundary and its derivative is continuous. The reference function $\bar{u}$ satisfies the same boundary conditions as u. If we replace u in $I(u)$ by $\bar{u}$, then we obtain a function that depends on ε and whose derivative with respect to ε vanishes for $\varepsilon = 0$, *i.e.*, for $\bar{u} = u$. This gives

$$I(\varepsilon) = \int_V F(x, y, z, u + \varepsilon f, u_x + \varepsilon f_x, u_y + \varepsilon f_y, u_z + \varepsilon f_z) d\tau \tag{8.44}$$

and

$$\left[\frac{\partial I(\varepsilon)}{\partial \varepsilon} \right]_{\varepsilon = 0} = \int_V (F_u f + F_{u_x} f_x + F_{u_y} f_y + F_{u_z} f_z) d\tau = 0,$$

whereby

$$F_{u_x} f_x = \frac{\partial F}{\partial u_x} \frac{\partial f}{\partial x} = \frac{\partial}{\partial x} \left(f \frac{\partial F}{\partial u_x} \right) - f \left(\frac{\partial^2 F}{\partial u_x \partial x} \right).$$

Using the same procedure for $F_{u_y} f_y$ and $F_{u_z} f_z$ one obtains

$$\int_V \left(F_u - \frac{\partial^2 F}{\partial u_x \partial x} - \frac{\partial^2 F}{\partial u_y \partial y} - \frac{\partial^2 F}{\partial u_z \partial z} \right) f d\tau$$

$$+ \int_V \left(\frac{\partial}{\partial x} \left(f \frac{\partial F}{\partial u_x} \right) + \frac{\partial}{\partial y} \left(f \frac{\partial F}{\partial u_y} \right) + \frac{\partial}{\partial z} \left(f \frac{\partial F}{\partial u_z} \right) \right) f d\tau = 0.$$

The second integral vanishes because of Gauss' integral theorem and because of the boundary condition for f (*i.e.* $f = 0$ on the boundary). Since f is an arbitrary function, the integrand of the first integral has to vanish as its integral must do so and therefore:

$$\boxed{\frac{\partial F}{\partial u} - \frac{\partial}{\partial x}\frac{\partial F}{\partial u_x} - \frac{\partial}{\partial y}\frac{\partial F}{\partial u_y} - \frac{\partial}{\partial z}\frac{\partial F}{\partial u_z} = 0}. \qquad (8.45)$$

This is the so-called *Euler differential equation* (also known as *Euler-Lagrange equation*). Its solution is the solution of the variational problem given by eq. (8.42) and vice versa.

As an **example**, consider the following integral

$$\boxed{I = \int_V F \; d\tau = \int_V \left[\left(\frac{\partial u}{\partial x}\right)^2 + \left(\frac{\partial u}{\partial y}\right)^2 + \left(\frac{\partial u}{\partial z}\right)^2 - 2u(\mathbf{r})g(\mathbf{r}) - u^2(\mathbf{r})h(\mathbf{r}) \right] \; d\tau}.$$

$$(8.46)$$

The corresponding Euler-Lagrange equation is

$$\boxed{\nabla^2 u + g + hu = 0}, \qquad (8.47)$$

which as a special case contains Laplace's equation ($g = 0, h = 0$), Poisson's equation ($h = 0$), and the Helmholtz equation ($g = 0, h = $ const.). In case of the Laplace equation, the integral

$$\int_V (\nabla u)^2 \; d\tau = \int_V E^2 \; d\tau$$

possesses the very interesting property to assume an extremum (a minimum, specifically). This also minimizes the integral

$$\frac{\varepsilon_0}{2}\int_V (\nabla u)^2 \; d\tau = \int_V \frac{\varepsilon_0 E^2}{2} \; d\tau. \qquad (8.48)$$

This expresses the remarkable theorem that the electric field in a charge-free region arranges itself such that the electrostatic energy it contains is minimized (compatible with the boundary conditions).

The reader is reminded, that we have started with a Dirichlet problem and then introduced competing reference functions which had to satisfy the boundary conditions, who's task was to facilitate minimization of the functional. Therefore, not every arbitrary solution of the corresponding Euler-Lagrange equation is a suitable solution, but only the one which satisfies the Dirichlet boundary conditions. The next question is how to proceed when the boundary condition is given in a different form. This question leads to the need to distinguish between *essential boundary conditions* and *natural boundary conditions* . Suppose the boundary is A, partitioned into the parts A_1 and A_2. Let the essential Dirichlet boundary condition

$$u = b(\mathbf{r}) \qquad (8.49)$$

apply to part A_1 and the natural boundary condition

$$\frac{\partial u}{\partial n} + d(\mathbf{r})u = e(\mathbf{r}) \qquad (8.50)$$

shall apply on the remaining part of the boundary A_2. For $d = 0$, this condition reduces to the Neumann boundary condition (in case of $b = 0$ or $e = 0$, these boundary conditions are called homogeneous, otherwise they are inhomogeneous).

The distinction between essential and natural boundary conditions is based on the fact that the reference functions used in the variation must satisfy the essential boundary conditions, but do not need to satisfy the natural boundary conditions. In order to satisfy the natural boundary conditions, the variational integral has to be supplemented by a boundary integral, *i.e.*, eq. (8.46) is replaced by

$$I = \int_V [(\nabla u)^2 - 2u(\mathbf{r})g(\mathbf{r}) - u^2(\mathbf{r})h(\mathbf{r})] \, d\tau + \int_{A_2} [d(\mathbf{r})u^2(\mathbf{r}) - 2u(\mathbf{r})e(\mathbf{r})] dA.$$

(8.51)

Notice that the boundary integral applies to the surface A_2 only, that is, the part where the natural boundary conditions (8.50) apply. We will skip the proof of this, which can be found *for example,* at Davies [17]. If $d = 0$ and $e = 0$, then the boundary integral is obsolete, which means that in case of Neumann boundary conditions, only the volume integral remains. Its variation automatically results in the solution which satisfies the homogeneous Neumann boundary conditions. Unlike the Dirichlet boundary conditions, these do not require the choice of suitable reference functions to ensure a solution. We will demonstrate this difference by means of simple examples.

The reader shall be reminded that the Neumann boundary condition may not be prescribed entirely arbitrarily. For the inner (but not the outer) Neumann problem it has to be

$$\oint \mathbf{D} \bullet d A = -\varepsilon \oint \frac{\partial \varphi}{\partial n} dA = Q,$$

(8.52)

that is, the electric flux has to be compatible with the total charge inside the region (this is irrelevant for the outer problem because there, the electric flux is assumed to extend to infinity and the flux through the imaginary surface at infinity may take any value).

One can avoid the need to choose the reference functions in case of the Dirichlet boundary condition and start with arbitrary functions when supplementing the variational integral by appropriate supplemental conditions and consider these by means of *Lagrange parameters (Lagrange multipliers)*. We will illustrate this with an example.

The variational integrals are very well suited to obtain both an exact, as well as an approximative solution of the corresponding problem. Oftentimes, astonishingly accurate results can be obtained even with relatively simple means. They also provide starting points for the method of Finite Elements (Sect. 8.7).

A practically important method to obtain approximate solutions of the variation problems is the so-called *Ritz method* also known as *Rayleigh-Ritz method*. It is based on writing the solution as a series of linear independent functions φ_i in the following form:

$$u = \sum_{i=1}^{n} c_i \varphi_i.$$

(8.53)

Substitute this series into the variational integral and let the derivatives of the resulting function $I(c_1, ..., c_n)$ with respect to c_i be zero:

$$\frac{\partial I(c_1, ..., c_n)}{\partial c_i} = 0, \qquad i = 1, ..., n \ . \tag{8.54}$$

If the functions φ_i form a complete basis within the region where the functions u are defined, then this allows to obtain the exact solution.

8.3.2   Examples

8.3.2.1 Poisson's Equation

As our first example to demonstrate the variational calculation, consider the one-dimensional Poisson equation

$$\nabla^2 u = -g(x) = a + bx, \qquad 0 \leq x \leq 1 \tag{8.55}$$

with its general solution

$$u = A + Bx + \frac{a}{2}x^2 + \frac{b}{6}x^3 \ . \tag{8.56}$$

Initially, the boundary conditions shall be left undefined. Later we will solve the problem under the following boundary conditions

a)   First, the problem shall be solved by means of the variational integral for the Dirichlet boundary condition

$$u(0) = \gamma, \qquad u(1) = \delta \ . \tag{8.57}$$

The series solution eq. (8.53) shall have the form

$$u(x) = A + Bx + Cx^2 + Dx^3 \ . \tag{8.58}$$

It must satisfy the essential boundary conditions. This requires that

$$u(x) = \gamma + (\delta - \gamma - C - D)x + Cx^2 + Dx^3 \ . \tag{8.59}$$

Next, the coefficients C and D have to be chosen such that the variational integral is minimized:

$$I = \int_0^1 \{[u'(x)]^2 + 2u(x) \cdot (a + bx)\} \, dx = I(C, D) \ . \tag{8.60}$$

Calculating this integral and letting its derivative with respect to C and D be zero gives

$$C = \frac{a}{2}, \qquad D = \frac{b}{6} \tag{8.61}$$

and

$$u(x) = \gamma + \left(\delta - \gamma - \frac{a}{2} - \frac{b}{6}\right)x + \frac{a}{2}x^2 + \frac{b}{6}x^3 \ . \tag{8.62}$$

This represents the exact solution because the statement (8.58) is flexible enough to contain this as a special case.

b) The same problem can be solved in a different manner. The boundary conditions are introduced here as supplementary conditions when varying the integral and considered by means of the Lagrange multiplier λ and μ. Then with (8.58), the integral gives

$$I(A, B, C, D) = B^2 + \frac{4}{3}C^2 + \frac{9}{5}D^2 + 2BC + 2BD + 3CD$$

$$+ 2\left[aA + \frac{1}{2}aB + \frac{1}{3}aC + \frac{1}{4}aD + \frac{1}{2}bA + \frac{1}{3}bB + \frac{1}{4}bC + \frac{1}{5}bD\right] . \tag{8.63}$$

The supplemental conditions that correspond to the boundary conditions are

$$A - \gamma = 0, \qquad A + B + C + D - \delta = 0 . \tag{8.64}$$

This makes the functional which we have to vary

$$F(A, B, C, D) = I(A, B, C, D) + \lambda(A - \gamma) + \mu(A + B + C + D - \delta) . \tag{8.65}$$

The next step is to take the derivative with respect to A, B, C, D, λ, and μ:

$$\left. \begin{aligned} \partial F/\partial A &= 2a + b + \lambda + \mu = 0 \\ \partial F/\partial B &= 2B + 2C + 2D + a + \frac{2}{3}b + \mu = 0 \\ \partial F/\partial C &= 2B + \frac{8}{3}C + 3D + \frac{2}{3}a + \frac{1}{3}b + \mu = 0 \\ \partial F/\partial D &= 2B + 3C + \frac{18}{5}D + \frac{1}{2}a + \frac{2}{5}b + \mu = 0 \\ \partial F/\partial \lambda &= A - \gamma = 0 \\ \partial F/\partial \mu &= A + B + C + D - \delta = 0 \end{aligned} \right\} . \tag{8.66}$$

Eliminating λ and μ gives

$$A = \gamma, \qquad B = \delta - \gamma - \frac{a}{2} - \frac{b}{6}, \qquad C = \frac{a}{2}, \qquad D = \frac{b}{6} \tag{8.67}$$

which, again, is the solution eq. (8.62).

c) Now, we will cover Neumann's boundary value problem. Because of eq. (8.52), the Neumann boundary conditions have to be chosen such that

$$u'(1) - u'(0) = \frac{2a + b}{2} . \tag{8.68}$$

This means, we require that

$$u'(1) = \beta, \qquad u'(0) = \beta - \frac{2a + b}{2}, \tag{8.69}$$

Now, by eq. (8.51), we have to add the expression

$$-2 \oint u \frac{\partial u}{\partial n} dA = -2[uu']_0^1$$

$$= 2A\left(\beta - \frac{2a+b}{2}\right) - 2(A+B+C+D)\beta \tag{8.70}$$

to the variational integral. Then from the functional

$$G(A, B, C, D) = I(A, B, C, D) + 2A\left(\beta - \frac{2a+b}{2}\right) - 2(A+B+C+D)\beta \tag{8.71}$$

we obtain the equations

$$\left. \begin{aligned} \partial G/\partial A &= 2a+b+2\beta - (2a+b) - 2\beta = 0 \\ \partial G/\partial B &= 2B+2C+2D+a+\frac{2}{3}b-2\beta = 0 \\ \partial G/\partial C &= 2B+\frac{8}{3}C+3D+\frac{2}{3}a+\frac{1}{2}b-2\beta = 0 \\ \partial G/\partial D &= 2B+3C+\frac{18}{5}D+\frac{1}{2}a+\frac{2}{5}b-2\beta = 0 \end{aligned} \right\}. \tag{8.72}$$

The first of these four equations is identically satisfied. The reason is that the Neumann boundary conditions were chosen in the allowed manner. Had we not considered eq. (8.68), that condition would have to be imposed posteriorly. The remaining three equations give

$$B = \beta - \frac{2a+b}{2}, \qquad C = \frac{a}{2}, \qquad D = \frac{b}{6}, \tag{8.73}$$

while A may be chosen arbitrarily (since in case of a Neumann problem u is only specified up to a constant). Therefore:

$$u(x) = A + \left(\beta - \frac{2a+b}{2}\right)x + \frac{a}{2}x^2 + \frac{b}{6}x^3. \tag{8.74}$$

This represents the exact solution of the problem. In contrast to the Dirichlet boundary conditions, the Neumann boundary conditions had to be supplemented by the boundary integral (where in this one-dimensional case, this consists of only two terms). Omitting the boundary integral in an attempt to satisfy the Neumann boundary conditions by means of a reference function leads to an incorrect result. The reason is that the wrong quantity is minimized. The reader is encouraged to try this and thereby verify the claim.

Thus far, in above examples, we have found the exact solution. Now, consider a simplified Ansatz, which only allows to solve by approximation:

$$u(x) = A + Bx + Cx^2. \tag{8.75}$$

d)   When considering the Dirichlet boundary value problem with the boundary conditions eq. (8.57) we have to require

$$u(x) = \gamma + (\delta - \gamma - C)x + Cx^2 .$$ (8.76)

The variational integral obtained yields

$$C = \frac{2a+b}{4}$$

and therefore

$$u(x) = \gamma + \left(\delta - \gamma - \frac{2a+b}{4}\right)x + \frac{2a+b}{4}x^2 .$$ (8.77)

e) Considering the Neumann boundary value problem with the boundary conditions eq. (8.69) gives the first three of the eqs. (8.72) when $D = 0$ and thus

$$A = \text{arbitrary} , \quad B = \beta - \frac{12a+7b}{12} , \quad C = \frac{2a+b}{4} ,$$ (8.78)

and

$$u(x) = A + \left(\beta - \frac{12a+7b}{12}\right)x + \frac{2a+b}{4}x^2 .$$ (8.79)

This solution neither satisfies the Poisson equation nor the boundary conditions exactly, but only approximately.

f) To satisfy the boundary conditions of the Neumann problem exactly, requires one to find suitable reference functions. This requires a consideration of the additional boundary integral for the variation, because otherwise, the solution would be incorrect. For the present case, this leads to the trial function

$$u(x) = A + \left(\beta - \frac{2a+b}{2}\right)x + \frac{2a+b}{4}x^2 .$$ (8.80)

There are no more variational parameters left. The parameter A is arbitrary and the variation does not depend on it. When comparing the two approximations (8.79) and (8.80) with the exact solution (8.74), then we observe that the approximation (8.79) more closely resembles the exact solution and therefore can be regarded as the better of the two approximations. This can be generalized. When starting from the same Ansatz, then the case where the Neumann boundary conditions are considered exactly, is generally the less accurate approximation. This is still the case, even if variational parameters remain after the Neumann boundary conditions were applied. A better approach to determine the variational parameters is, to leave this to the variation integral, which defines the approximation and yields an optimized compromise for the solution.

8.3.2.2 Helmholtz Equation

As our second **example**, we shall employ the Helmholtz equation, which describes *for example*, the waves in wave guides and was discussed in sects. 7.7 through 7.11. The Helmholtz equation is

$$\nabla^2 u + Nu = 0 \tag{8.81}$$

and the associated variational integral for homogeneous boundary conditions is

$$I = \int_V [(\nabla u)^2 - Nu^2] d\tau \ . \tag{8.82}$$

This integral is homogeneous in u, *i.e.* the solution to the variational problem is specified up to a constant. When dividing eq. (8.82) by $\int u^2 d\tau$ (N is constant), one obtains

$$\bar{I} = \frac{I}{\int u^2 d\tau} = \frac{\int (\nabla u)^2 d\tau}{\int u^2 d\tau} - N = -\frac{\int u \nabla^2 u d\tau}{\int u^2 d\tau} - N \ . \tag{8.83}$$

Here we have the functional F:

$$F = -\frac{\int u \nabla^2 u d\tau}{\int u^2 d\tau} \ , \tag{8.84}$$

which was introduced for two-dimensional problems in Sect. 7.11, eq. (7.452). One can show that minimizing the integral (8.82) leads to the same result as that of the functional (8.84).

Next, we will analyze the two-dimensional problem of a wave in a rectilinear wave guide, where, because of its separability, we limit the discussion to one spatial coordinate.

a)   We start with the Dirichlet boundary value problem using the Ansatz

$$u(x) = c_1 \varphi_1 + c_2 \varphi_2 = c_1 x(1-x) + c_2 x \left(\frac{1}{2} - x\right)(1-x) \tag{8.85}$$

which satisfies the homogeneous boundary conditions

$$u(0) = 0 \ , \quad u(1) = 0 \ . \tag{8.86}$$

This represents a TM wave. We could use the same Ansatz for the y-dependency. Both functions φ_1 and φ_2 exhibit the qualitative characteristics that can be expected from the eigenfunctions which belong to the two lowest eigenvalues: no zero or one zero, respectively within the range $0 < x < 1$. Using the integrals

$$\int_0^1 \varphi_1^2(x) dx = \frac{1}{30}, \qquad \int_0^1 \varphi_2^2(x) dx = \frac{1}{840}$$

$$\int_0^1 \varphi_1'^2(x) dx = \frac{1}{3}, \qquad \int_0^1 \varphi_2'^2(x) dx = \frac{1}{20} \qquad \Bigg\}, \tag{8.87}$$

$$\int_0^1 \varphi_1(x)\varphi_2(x) dx = 0, \qquad \int_0^1 \varphi'_1(x)\varphi'_2(x) dx = 0$$

the integral (8.82) becomes

$$I = c_1^2 \left(\frac{1}{3} - \frac{N}{30}\right) + c_2^2 \left(\frac{1}{20} - \frac{N}{840}\right) \ . \tag{8.88}$$

One obtains

$$\frac{\partial I}{\partial c_1} = 2c_1\left(\frac{1}{3} - \frac{N}{30}\right) = 0$$

$$\frac{\partial I}{\partial c_2} = 2c_2\left(\frac{1}{20} - \frac{N}{840}\right) = 0$$

(8.89)

These simple equations rest on the fact that φ_1 and φ_2 are orthogonal to each other, as eqs. (8.87) reveal. This allows the eigenvalue equation to be diagonalized immediately and, thus, becomes trivial to solve. One obtains two eigenvalues. Either

or
$$\begin{matrix} N = N_1 = 10 & c_1 \text{ arbitrary} & c_2 = 0 \\ N = N_2 = 42 & c_2 \text{ arbitrary} & c_1 = 0 \end{matrix}$$

(8.90)

The exact eigenfunctions and eigenvalues are known (see Sect. 7.8.1)

$$\sin(\pi x) \text{ with } N_1 = \pi^2 = 9.8696\ldots$$

(8.91)

and

$$\sin(2\pi x) \text{ with } N_2 = 4\pi^2 = 39.4784\ldots$$

(8.92)

The values calculated by approximation are always too large (N_1 by about 1.3% and N_2 by about 6.4%). When choosing c_1 and c_2 such that

$$\varphi_1\left(\frac{1}{2}\right) = \varphi_2\left(\frac{1}{4}\right) = 1 \text{ then } \sin(2\pi x) \text{ is approximated by } \frac{64}{3}x\left(\frac{1}{2}-x\right)(1-x)$$

and $\sin(\pi x)$ by $4x(1-x)$. Better approximations yield more accurate (*i.e.* smaller) eigenvalues.

b)   Now we attempt to improve the lowest eigenvalue. For this purpose, we choose a more flexible Ansatz

$$u(x) = c_1\varphi_1 + c_2\varphi_2 = c_1 x(1-x) + c_2 x^2(1-x)^2$$

(8.93)

and with

$$\int_0^1 \varphi_1^2(x)dx = \frac{1}{30}, \int_0^1 \varphi_2^2(x)dx = \frac{1}{630}, \int_0^1 \varphi_1(x)\varphi_2(x)dx = \frac{1}{140}$$

$$\int_0^1 \varphi'^2_1(x)dx = \frac{1}{3}, \int_0^1 \varphi'^2_2(x)dx = \frac{2}{105}, \int_0^1 \varphi'_1(x)\varphi'_2(x)dx = \frac{1}{15}$$

(8.94)

one obtains

$$I = c_1^2\left(\frac{1}{3} - \frac{N}{30}\right) + 2c_1 c_2\left(\frac{1}{15} - \frac{N}{140}\right) + c_2^2\left(\frac{2}{105} - \frac{N}{630}\right),$$

(8.95)

which gives

$$\frac{\partial I}{\partial c_1} = 2c_1\left(\frac{1}{3} - \frac{N}{30}\right) + 2c_2\left(\frac{1}{15} - \frac{N}{140}\right) = 0$$

$$\frac{\partial I}{\partial c_2} = 2c_1\left(\frac{1}{15} - \frac{N}{140}\right) + 2c_2\left(\frac{2}{105} - \frac{N}{630}\right) = 0$$

(8.96)

Non-trivial solutions exist only if the determinant of the coefficients vanishes. This results in a quadratic equation for the eigenvalues N, with the two solutions:

$$N = 56 \pm \sqrt{2128} = 56 \pm 46.13025038 .$$

We are only interested in the smaller eigenvalue

$$N = 9.869749622 ,$$

(8.97)

which is a rather close approximation to the exact value $\pi^2 = 9.86960\ldots$. The ratio of the coefficients becomes

$$\frac{c_2}{c_1} = \frac{\dfrac{1}{3} - \dfrac{N}{30}}{\dfrac{1}{15} - \dfrac{N}{140}} = 1.133140 ,$$

(8.98)

which is, when normalized for $x = 1/2$, the eigenfunction to the lowest eigenvalue

$$u = 3.117000\ldots \cdot [x(1-x) + 1.133140 \cdot x^2(1-x)^2] .$$

(8.99)

c)   Next for the TE wave we, choose a different trial function:

$$u(x) = A + Bx + Cx^2 + Dx^3 .$$

(8.100)

This gives

$$I = \left(B^2 + \frac{4}{3}C^2 + \frac{9}{5}D^2 + 2BC + 2BD + 3CD\right)$$

$$- N\left(A^2 + \frac{1}{3}B^2 + \frac{1}{5}C^2 + \frac{1}{7}D^2\right.$$

(8.101)

$$\left. + AB + \frac{2}{3}AC + \frac{1}{2}AD + \frac{1}{2}BC + \frac{2}{5}BD + \frac{1}{3}CD\right)$$

and

$$\frac{\partial I}{\partial A} = -N\left(2A + B + \frac{2}{3}C + \frac{1}{2}D\right) = 0$$

$$\frac{\partial I}{\partial B} = 2B + 2C + 2D - N\left(A + \frac{2}{3}B + \frac{1}{2}C + \frac{2}{5}D\right) = 0$$

$$\frac{\partial I}{\partial C} = 2B + \frac{8}{3}C + 3D - N\left(\frac{2}{3}A + \frac{1}{2}B + \frac{2}{5}C + \frac{1}{3}D\right) = 0 \tag{8.102}$$

$$\frac{\partial I}{\partial D} = 2B + 3C + \frac{18}{5}D - N\left(\frac{1}{2}A + \frac{2}{5}B + \frac{1}{3}C + \frac{2}{7}D\right) = 0 .$$

The eigenvalue equation is of Order 4. However, by inspection, one can see immediately that $N_0 = 0$ is an eigenvalue. Furthermore, one can find that another eigenvalue is $N = 60$. This leaves a quadratic equation with the remaining two eigenvalues

$$N = 90 \pm \sqrt{6420} .$$

We are interested in the lower one

$$N_1 = 9.87509750\ldots . \tag{8.103}$$

The eigenfunction corresponding to the eigenvalue $N_0 = 0$ is

$$u_0 = A , \tag{8.104}$$

where A is arbitrary. This represents the trivial solution of the Neumann boundary value problem, an important result for TE_{01} or TE_{10} waves. The eigenfunction corresponding to the eigenvalue $N_1 = 9.87509750$, when normalized for $x = 0$ is

$$u_1 = 6.45533624 \cdot \left(0.15491059 + 0.02351213 \cdot x - x^2 + \frac{2}{3}x^3\right) . \tag{8.105}$$

Of course, the eigenfunctions u_0 and u_1 are orthogonal to each other.

In our example, we have not imposed any boundary conditions. Thus, we obtain solutions which approximately satisfy the homogeneous Neumann boundary conditions. ($u_1'(0) = u_1'(1) = 0.157$). This represents a TE wave. The exact solution for u_1 is $\cos \pi x$.

Fig. 8.1 compares the functions

$\varphi_1 = 4x(1-x)$ and $\varphi_2 = \frac{64}{3}x\left(\frac{1}{2} - x\right)(1 - x)$ according to eq. (8.85), as well as

$u(x)$ given in eq. (8.99) with $\sin(\pi x)$ and $\sin(2\pi x)$. Fig. 8.2 compares the function $u_1(x)$ as of eq. (8.105) with $\cos(\pi x)$. The difference between $\sin(\pi x)$ and $u(x)$ as of eq. (8.99) is so small that it is hardly visible in Fig. 8.1. Similarly, the difference between $\cos(\pi x)$ and $u_1(x)$ by eq. (8.105) is also so small that it is just recognizable in Fig. 8.2.

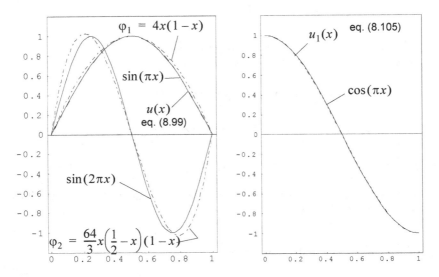

Fig. 8.1   Fig. 8.2

8.4   Method of Weighted Residuals

The method of weighted residuals is a very general method for exact or approximate solutions to all kinds of problems. This method is sometimes also called the momentum method (even though this is also the name for a more specific method).

Many problems can be stated in the form

$$Lu = f .$$

(8.106)

L is initially an arbitrary linear operator, *for example*, a differential operator. If u is an exact solution to eq. (8.106), then the *residual* is

$$R = Lu - f = 0 .$$

(8.107)

If u is only an approximate solution, then the *residual* is

$$R = Lu - f \neq 0 .$$

(8.108)

One might try to approximate u by a set of linear independent functions φ_i

$$u = \sum_{i=1}^{n} c_i \varphi_i .$$

(8.109)

This gives

$$R = \sum_{i=1}^{n} c_i (L\varphi_i) - f .$$

(8.110)

The approximation is better, the smaller R is. In general, R is a function of space. The strategy is to analyze appropriately chosen averages of R in order to determine criteria that allow one to find the best coefficients c_i, such that the functions φ_i in eq. (8.109) – the so-called *basis functions* – yield the desired approximation. The method of weighted residuals consists of defining at least n so-called *weight functions* w_k, whereby the mean values of the so formed integrals must vanish when integrating over the entire volume:

$$\int_V R w_k \, d\tau = \sum_{i=1}^{n} c_i \int_V w_k L \varphi_i d\tau - \int_V w_k f \, d\tau = 0 \; . \tag{8.111}$$

If the number of basis functions is equal to the number of weight functions, then one obtains n linearly independent equations which determine the n coefficients c_i. If the number of weight functions is larger, then the solution is overdetermined, and it may be inconsistent and is initially not solvable (if, indeed, the equations are linearly independent). In this case, as in curve fitting, the method of least squares is employed to find the best fitting coefficients.

Various variants of this method exist. The basis and weight functions may be identical, which leads to the *Galerkin method*. If the weight functions are the eigenfunctions of the given operator and they form a complete set of basis functions, then one obtains the usual representation of the solution by expanding them with respect to these functions (as was presented in detail in Chapter 3). With the exception of the Monte-Carlo method, all the numerical methods which we will discuss subsequently can be regarded as special cases of the method of weighted residuals.

A few specific methods shall now be described briefly.

8.4.1 Collocation Method

The δ-functions can also serve as weight functions. This has the advantage that it makes the integration trivial, which otherwise can be difficult at times. One obtains a system of equations, which lets the residual exactly vanish at the so-called collocation points, however, not at all other points. Usually, the number of δ-functions and thereby the number of collocation points is the same as the number of basis functions. If this number is larger, it is called an *overdetermined collocation*. An approximation method frequently used in field theory, the so-called image method or method of virtual charges, is (including various generalizations and modifications) a typical collocation method.

The procedure shall be demonstrated by means of a simple **example**. Consider the one-dimensional Poisson equation

$$\nabla^2 u(x) = x^2 \qquad 0 \le x \le 1 \tag{8.112}$$

with Dirichlet boundary conditions

$$u(0) = u(1) = 0 \; . \tag{8.113}$$

Its exact solution is easy to find

$$u(x) = -\frac{x(1-x^3)}{12}.$$ (8.114)

Now one uses an approximation consisting of two basis functions

$$u(x) = c_1 x(1-x) + c_2 x^2(1-x).$$ (8.115)

Both satisfy the boundary conditions. The residual is

$$R(x) = \nabla^2 u - x^2 = 2(c_2 - c_1) - 6c_2 x - x^2.$$ (8.116)

The parameters c_1 and c_2 are determined by collocation at the two points $x_1 = 1/3$ and $x_2 = 2/3$:

$$\int_0^1 R(x)\delta\left(x - \frac{1}{3}\right) dx = R\left(\frac{1}{3}\right) = 2(c_2 - c_1) - 2c_2 - \frac{1}{9} = 0$$

$$\int_0^1 R(x)\delta\left(x - \frac{2}{3}\right) dx = R\left(\frac{2}{3}\right) = 2(c_2 - c_1) - 4c_2 - \frac{4}{9} = 0$$

that is

$$-2c_1 + 0c_2 = \frac{1}{9}$$

$$-2c_1 - 2c_2 = \frac{4}{9}$$

therefore

$$c_1 = -\frac{1}{18}, \qquad c_2 = -\frac{1}{6}$$

and

$$u(x) = -\frac{x(1-x)(1+3x)}{18}.$$ (8.117)

If one chooses the three collocation points $x_1 = 1/4, x_2 = 1/2, x_3 = 3/4$, then one gets three equations for c_1 and c_2:

$$-2c_1 + \frac{1}{2}c_2 = \frac{1}{16}$$

$$-2c_1 - 1c_2 = \frac{1}{4}$$ (8.118)

$$-2c_1 - \frac{5}{2}c_2 = \frac{9}{16}.$$

We minimize the sum over the error squares (least error squares),

$$\left(-2c_1 + \frac{1}{2}c_2 - \frac{1}{16}\right)^2 + \left(-2c_1 - c_2 - \frac{1}{4}\right)^2 + \left(-2c_1 - \frac{5}{2}c_2 - \frac{9}{16}\right)^2 = \min.$$

Differentiating this with respect to c_1 and c_2 gives two equations

$$12c_1 + 6c_2 = -\frac{14}{8}$$

$$6c_1 + \frac{15}{2}c_2 = -\frac{13}{8} .$$

(8.119)

Eqs. (8.119) can be obtained directly from (8.118) when employing matrix notation

$$A\begin{bmatrix} c_1 \\ c_2 \end{bmatrix} = \begin{bmatrix} -2 & 1/2 \\ -2 & -1 \\ -2 & -5/2 \end{bmatrix}\begin{bmatrix} c_1 \\ c_2 \end{bmatrix} = \begin{bmatrix} 1/16 \\ 1/4 \\ 9/16 \end{bmatrix} ,$$

(8.120)

and multiply this by the transposed coefficient matrix $\bar{A}$, i.e.,

$$\bar{A} = \begin{bmatrix} -2 & -2 & -2 \\ 1/2 & -1 & -5/2 \end{bmatrix} .$$

(8.121)

Equation

$$\bar{A}A\begin{bmatrix} c_1 \\ c_2 \end{bmatrix} = \bar{A}\begin{bmatrix} 1/16 \\ 1/4 \\ 9/16 \end{bmatrix}$$

is another representation of (8.119), which leads to

$$\begin{bmatrix} c_1 \\ c_2 \end{bmatrix} = (\bar{A}A)^{-1}\bar{A}\begin{bmatrix} 1/16 \\ 1/4 \\ 9/16 \end{bmatrix} = \begin{bmatrix} 5/36 & -1/9 \\ -1/9 & 2/9 \end{bmatrix}\begin{bmatrix} -14/8 \\ -13/8 \end{bmatrix}$$

$$c_1 = -1/16 , \quad c_2 = -1/6$$

and

$$u(x) = -\frac{x(1-x)(3+8x)}{48} .$$

(8.122)

The matrix $(\bar{A}A)^{-1}\bar{A}$, encountered here is frequently used in curve fitting and is sometimes called the half-inverse. More on this can be found in [18].

8.4.2   Method of Fractional Regions

With this method, the region is separated in several subregions V_i, with the mandate that the integrals over all subregions vanish

$$\int_{V_i} R d\tau = 0 , \quad i = 1, ...n .$$

(8.123)

The weight functions are constant, but each one is non-zero in only one of the regions. In our example, one chooses two regions, $0 \leq x \leq 1/2$ and $1/2 \leq x \leq 1$ which gives

$$\int_0^{1/2} R\,dx = -c_1 + \frac{1}{4}c_2 - \frac{1}{24} = 0$$

$$\int_{1/2}^1 R\,dx = -c_1 - \frac{5}{4}c_2 - \frac{7}{24} = 0$$

that is

$$c_1 = -1/12 , \quad c_2 = -1/6$$

and

$$u(x) = -\frac{x(1-x)(1+2x)}{12} . \tag{8.124}$$

8.4.3 Momentum Method

In case of the momentum method (in the strict sense of the term), the factors $x^0, x^1, \ldots$ serve as weight functions, *i.e.*, one defines

$$\int R(x) \cdot x^i dx = 0, \quad i = 0, \ldots, n-1 . \tag{8.125}$$

For our current example, with $n = 2$ one obtains:

$$\int_0^1 R(x)\,dx = -2c_1 - c_2 - \frac{1}{3} = 0 ,$$

$$\int_0^1 R(x)x\,dx = -c_1 - c_2 - \frac{1}{4} = 0 ,$$

which results again in

$$c_1 = -\frac{1}{12} , \; c_2 = -\frac{1}{6} ,$$

and $u(x)$ according to eq. (8.124).

8.4.4 Method of Least Squares

Another possibility is to apply the method of the least squares directly onto R. This differs from the overdetermined collocation method. Then, the integral

$$I = \int R^2 d\tau \tag{8.126}$$

needs to be minimized. For our example, this gives

$$I = 4c_1^2 + 4c_1c_2 + 4c_2^2 + \frac{4}{3}c_1 + \frac{5}{3}c_2 + \frac{1}{5}$$

and

$$\frac{\partial I}{\partial c_1} = 8c_1 + 4c_2 + \frac{4}{3} = 0$$

$$\frac{\partial I}{\partial c_2} = 4c_1 + 8c_2 + \frac{5}{3} = 0$$

where

$$c_1 = -\frac{1}{12} \quad c_2 = -\frac{1}{6},$$

which again, now for the third time, yields the approximation solution eq. (8.124).

8.4.5 Galerkin Method

Of particular importance is the Galerkin method. Here, the basis and weight functions are identical. The Galerkin method is equivalent to the Rayleigh-Ritz method for the case of problems that can also be treated as variational problems; *i.e.,* see 8.3.1. It leads to the very important method of Finite Elements when choosing special basis functions (see Sect. 8.7) .

For illustration purposes, we will focus on an already used example. Let

$$\int_0^1 R(x)x(1-x)dx = -\frac{1}{3}c_1 - \frac{1}{6}c_2 - \frac{1}{20} = 0 \;,$$

$$\int_0^1 R(x)x^2(1-x)dx = -\frac{1}{6}c_1 - \frac{2}{15}c_2 - \frac{1}{30} = 0 \;,$$

to obtain

$$c_1 = -\frac{1}{15} \quad c_2 = -\frac{1}{6},$$

that is

$$u(x) = -\frac{x(1-x)(2+5x)}{30} \;. \tag{8.127}$$

If one tackles the problem with the Rayleigh-Ritz method, then the integral to be minimized is

$$I = \int_0^1 \left[\left(\frac{du}{dx}\right)^2 + 2ux^2\right]dx = \frac{1}{3}c_1^2 + \frac{2}{15}c_2^2 + 2\left(\frac{1}{6}c_1c_2 + \frac{1}{20}c_1 + \frac{1}{30}c_2\right)$$

and one obtains the same system of equations as above:

$$\frac{\partial I}{\partial c_1} = 2\left(\frac{1}{3}c_1 + \frac{1}{6}c_2 + \frac{1}{20}\right) = 0 \;,$$

$$\frac{\partial I}{\partial c_2} = 2\left(\frac{1}{6}c_1 + \frac{2}{15}c_2 + \frac{1}{30}\right) = 0 \;.$$

It should be easy to convince yourself that this is no coincidence, but that the results always coincide.

As a further **example**, we will solve the above problem with the generalized boundary conditions:

$$u(0) = u_0, \quad u(1) = u_3$$

and we will also use a different set of basis functions, namely

$$u = \begin{cases} (1-3x)u_0 + 3xu_1 & \text{for} & 0 \le x \le 1/3 \\ (2-3x)u_1 + (3x-1)u_2 & \text{for} & 1/3 \le x \le 2/3 \\ (3-3x)u_2 + (3x-2)u_3 & \text{for} & 2/3 \le x \le 1 \end{cases} \tag{8.128}$$

and

$$u' = \frac{du}{dx} = \begin{cases} 3(u_1-u_0) & \text{for} & 0 \le x \le 1/3 \\ 3(u_2-u_1) & \text{for} & 1/3 \le x \le 2/3 \\ 3(u_3-u_2) & \text{for} & 2/3 \le x \le 1. \end{cases} \tag{8.129}$$

This Ansatz represents a piecewise linear approximation of the function values u_0, u_1, u_2, u_3 between the locations $x = 0, 1/3, 2/3, 1$. The values of the functions occur as coefficients of the particular basis functions. The basis functions are each different from zero only in one of the subregions. This is a simple example of the method of finite elements, which we will discuss later in detail. Here, we chose special and very simple (namely linear) *form functions*.

It is now possible to use the Galerkin method or the Rayleigh-Ritz method to calculate u_1 and u_2 (u_0 and u_3 are given by the boundary conditions), which has to lead to the same result as before. In case of the Galerkin method, the residue contains the second derivative of u, $\nabla^2 u = u''$. According to eq. (8.129), the first derivatives u' are discontinuous at the sampling points. This means that the second derivatives are δ-functions. The necessary integrals can be calculated by means of these δ-functions. These integrals can be avoided when integrating by parts

$$\int_0^1 u(x)u''(x)dx = -\int_0^1 [u'(x)]^2 dx + [u(x)u'(x)]_0^1. \tag{8.130}$$

This situation is referred to as the *"weak" formulation* of the problem. We will use the variational method, which only needs $u'(x)$ from start, that is, it represents the weak formulation from the beginning. The calculation is tedious but without any problems and yields

$$I = \int_0^1 \{[u'(x)]^2 + 2x^2 u(x)\}dx$$

$$= 3(u_0^2 + 2u_1^2 + 2u_2^2 + u_3^2 - 2u_0u_1 - 2u_1u_2 - 2u_2u_3)$$

$$+ \left(\frac{1}{3}\right)^4 \left(\frac{1}{2}u_0 + 7u_1 + 25u_2 + \frac{43}{2}u_3\right) \tag{8.131}$$

and

$$\frac{\partial I}{\partial u_1} = -6u_0 + 12u_1 - 6u_2 + 7\left(\frac{1}{3}\right)^4 = 0$$

$$\frac{\partial I}{\partial u_2} = -6u_1 + 12u_2 - 6u_3 + 25\left(\frac{1}{3}\right)^4 = 0$$

(8.132)

i.e.

$$u_1 = \frac{2}{3}u_0 + \frac{1}{3}u_3 - \frac{13}{6}\left(\frac{1}{3}\right)^4$$

$$u_2 = \frac{1}{3}u_0 + \frac{2}{3}u_3 - \frac{19}{6}\left(\frac{1}{3}\right)^4 \ .$$

(8.133)

The exact solution of the problem is

$$u = u_0 + \left(u_3 - u_0 - \frac{1}{12}\right)x + \frac{1}{12}x^4 \ .$$

(8.134)

This means that at the two sampling points, we obtain the exact value for u_1 and u_2. We get the intermediate values by linear interpolation. For the special case

$$u_0 = u_3 = 0$$

we find

$$u_1 = -\frac{13}{486}, \qquad u_2 = -\frac{19}{486} \ .$$

(8.135)

Written in a slightly different form, the two eqs. (8.132) read

$$u_0 - 2u_1 + u_2 = \frac{7}{6}\left(\frac{1}{3}\right)^4$$

$$u_1 - 2u_2 + u_3 = \frac{25}{6}\left(\frac{1}{3}\right)^4$$

(8.136)

This exhibits great similarity with the result obtained by means of the finite differences (as we shall see later). For $h = 1/3$, one obtains from eq. (8.162) of the upcoming Section 8.6.1 the following result:

$$u_0 - 2u_1 + u_2 = h^2\left(\frac{1}{3}\right)^2 = \left(\frac{1}{3}\right)^4$$

$$u_1 - 2u_2 + u_3 = h^2\left(\frac{2}{3}\right)^2 = 4\left(\frac{1}{3}\right)^4$$

(8.137)

The difference between them manifests itself in the right sides of the equations, which stems from the inhomogeneity x^2. Eqs (8.137) give

———— Exact Solution
· · · · · Collocation Method
— — — Overdetermined Collocation
— · — Fractional Regions, Momentum Method, Least Error Squares
— · · — Galerkin, Variation
—☐— Finite Elements
—✕— Finite Differences

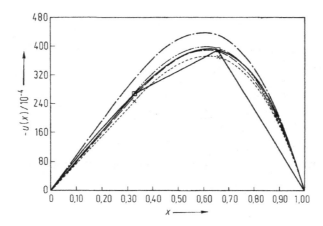

Fig. 8.3

$$u_1 = \frac{2}{3}u_0 + \frac{1}{3}u_3 - 2\left(\frac{1}{3}\right)^4$$

$$u_2 = \frac{1}{3}u_0 + \frac{2}{3}u_3 - 3\left(\frac{1}{3}\right)^4$$

(8.138)

and for $u_0 = u_3 = 0$

$$u_1 = -\frac{2}{81} \ , \qquad u_2 = -\frac{1}{27} \ .$$

(8.139)

Now, we have approximated solutions of eq. (8.112) with the boundary condition (8.113) by many different ways. Fig. 8.3 compares the exact solution (8.114) with the various approximations, namely:

- Eq. (8.117)   Collocation
- Eq. (8.122)   Overdetermined Collocation
- Eq. (8.124)   Method of Fractional Regions, Momentum Method, Method of Least Error Squares
- Eq. (8.127)   Galerkin Method, Rayleigh-Ritz Method
- Eq. (8.135)   Rayleigh-Ritz Method with different trial functions
- Eq. (8.139)   Method of finite differences.

Fig. 8.3 clearly illustrates that the various solutions are of different quality. On the other hand, it also becomes clear that by means of only a few simple trial functions, one can obtain very useful approximations. Nevertheless, one should not jump to conclusions with respect to the quality of the various methods based on generalizations of the results depicted in Fig. 8.3.

The method of the weighted residuals may be used in various variants. With the exception of the Monte-Carlo method, it can be regarded as the starting point of all the numerical methods which we will discuss in the following, namely the method of finite differences, the method of finite elements, boundary element method, and the method of image charges. Only the Monte-Carlo method is based on an entirely different approach. Nevertheless, for the field theoretical problems which we consider here, this method is ultimately equivalent to the method of finite differences.

8.5   Random-Walk Processes

In anticipation of the Monte-Carlo method, we consider simple random-walk processes. These are special, stochastic processes, which represent a very useful and intuitive model for many theoretical, and in practice interesting problems within the Theory of Probability and in Physics.

We start from a one-dimensional discrete random-walk process. For this purpose, consider an infinitely long straight line with equidistant grid points, which we identify by whole numbers from $-\infty$ through $+\infty$ (Fig. 8.4). At time $t = 0$, a particle (or a person) shall be located at point 0. Within defined time intervals, the particle moves in steps towards the right or left, each direction with the probability p or q, respectively. Of course,

$$p + q = 1 \ . \tag{8.140}$$

We want to know what is the probability to find the particle at any particular location after n steps. The probability to be at location $2m - n$ after n steps is, we claim

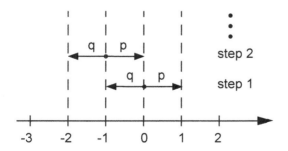

Fig. 8.4

$$w_{2m-n,\,n} = p^m q^{n-m} \binom{n}{m} .$$ (8.141)

In order to arrive at point $2m-n$ requires to move m of the total n steps to the right and $n-m$ steps to the left. The order of the steps is immaterial. The m and the $n-m$ steps, respectively, may be selected from the n steps in $\binom{n}{m}$ different ways.

Remarkable is that because of eq. (8.140):

$$(p+q)^n = \sum_{m=0}^{n} p^m q^{n-m} \binom{n}{m} = 1^n = 1 .$$ (8.142)

Therefore, the function $(p+q)^n$ is called the generating function of the probabilities w. Eq. (8.142) can be explained plausibly by the fact that when expanding $(p+q)^n$, the occurrence of the product $p^m q^{n-m}$ is the same as picking m or n - m elements of the total number of n elements, when disregarding the order. This explains the relation to the probabilities. At the same time, eq. (8.142) reveals that the sum of the probabilities equals 1, as it must be.

To know the generating function is very helpful. Consider, *for example*, the average of the passed locations after n steps and the average of its square:

$$\langle x \rangle = \sum_{m=0}^{n} (2m-n) \binom{n}{m} p^m q^{n-m}$$

$$\langle x^2 \rangle = \sum_{m=0}^{n} (2m-n)^2 \binom{n}{m} p^m q^{n-m}$$ (8.143)

From eq. (8.142) follows that

$$\frac{d}{dp}(p+q)^n = n(p+q)^{n-1} = \sum_{m=0}^{n} m p^{m-1} q^{n-m} \binom{n}{m} = n$$

and

$$\frac{d^2}{dp^2}(p+q)^n = n(n-1)(p+q)^{n-2} = \sum_{m=0}^{n} m(m-1) p^{m-2} q^{n-m} \binom{n}{m}$$

$$= n(n-1) .$$

Therefore

$$\sum_{m=0}^{n} m p^m q^{n-m} \binom{n}{m} = np$$ (8.144)

$$\sum_{m=0}^{n} m(m-1) p^m q^{n-m} \binom{n}{m} = n(n-1)p^2 .$$ (8.145)

Adding the two equations gives

$$\sum_{m=0}^{n} m^2 p^m q^{n-m} \binom{n}{m} = n(n-1)p^2 + np .$$ (8.146)

Using eqs. (8.144) and (8.146) one finds

$$\langle x \rangle = n(2p-1)$$ (8.147)

$$\langle x^2 \rangle = \quad = n^2(2p-1)^2 + 4np(1-p)$$. (8.148)

For the symmetric random-walk process one has $p = q = 1/2$ and

$$\langle x \rangle = 0$$ (8.149)

$$\langle x^2 \rangle = n, \qquad \sqrt{\langle x^2 \rangle} = \sqrt{n}$$. (8.150)

Symmetry explains the vanishing of $\langle x \rangle$. Eq. (8.150) is very interesting. If the steps occur at equidistant points in time, then $\sqrt{n} \sim \sqrt{t}$. The square root of the squared average of the surpassed distance is thus proportional to $\sqrt{t}$, and not to t. This is a typical property of diffusion processes, which can be regarded as generalized random-walk processes. We already discovered this fact, while covering the diffusion equation in Sect. 6.2.3, eq. (6.39) in particular. Of course, for the limit $p = 1$ (or $p = 0$), because in this case all the steps proceed in only one direction – either left or right – we obtain:

$$\langle x \rangle = \pm n, \quad \langle x^2 \rangle = n^2 \ .$$ (8.151)

This represents a directed motion and not a random-walk.

From a purely formal perspective, the random-walk problem may be treated as the solution to a difference equation for probabilities. Let $w_{i,k}$ be the probability that after k steps, a particle is found at location i. Then

$$w_{i,k} = pw_{i-1,k-1} + qw_{i+1,k-1} \ .$$ (8.152)

The methods to solve such difference equations are entirely analogous to the solution of differential equations. Both require initial and boundary conditions in order to make the solutions unique. The probabilities given above (8.141) satisfy the difference equations. They represent their only solution for the infinite one-dimensional space when the probabilities vanish at infinity and the initial condition is

$$w_{i,0} = \delta_{i,0} \ .$$ (8.153)

Conversely, one might want to analyze finite regions which limit the possible motion of the particle and note *for example*, that a particle arriving at $i = a$ or $i = b$ is absorbed or reflected, respectively (or absorbed or reflected with a certain probability). Based on difference equations, boundary conditions, and initial conditions one can calculate the various probabilities on a one-, two-, or three-dimensional lattice. This constitutes a wide field with many interesting results, which we shall not discuss here. However, when solving field theoretical problems by means of the Monte-Carlo method, then one-, two-, or three-dimensional discrete and symmetric random-walk processes with absorbing walls will be applied.

Here, we consider a few simple probabilities that will be required later. Fig. 8.5 shows a finite, one-dimensional lattice with absorbing walls at 0 and 4, and with three internal points. Let a particle carry out a random-walk process starting at

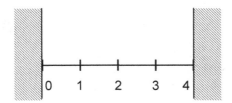

Fig. 8.5

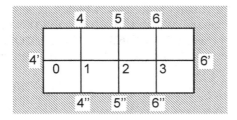

Fig. 8.6

one of the inner points. We ask, what is the probability W_{ik} for a particle starting at point i to be absorbed at point k, if k is a boundary point or, how often it passes the point k, if it is an inner point. The reader may convince himself that the probabilities are as follows:

$$W_{10} = \frac{3}{4}, \quad W_{11} = \frac{1}{2}, \quad W_{12} = 1, \quad W_{13} = \frac{1}{2}, \quad W_{14} = \frac{1}{4}$$
$$W_{20} = \frac{1}{2}, \quad W_{21} = 1, \quad W_{22} = 1, \quad W_{23} = 1, \quad W_{24} = \frac{1}{2}$$
$$W_{30} = \frac{1}{4}, \quad W_{31} = \frac{1}{2}, \quad W_{32} = 1, \quad W_{33} = \frac{1}{2}, \quad W_{34} = \frac{3}{4}$$
(8.154)

When we include the "passing" of the initial point, this would make $W_{11} = 3/2$, $W_{22} = 2$, $W_{33} = 3/2$.

In case of the two (three) dimensional symmetric random-walk, the particles move with the equal probability of 1/4 (1/6), respectively, towards one of the four (six) neighboring points. Fig. 8.6 shows a simple two-dimensional lattice with absorbing walls. What is the probability W_{ik} for a particle starting at point i, to be absorbed at point k? Note that this question, unlike above, does not ask for the frequency of passing inner points. Initially, to simplify, we note that there are only three significant boundary points. Because of symmetry, the probabilities of being absorbed at 4, 4', or 4" are the same. The same is true for 5 and 5", as well as for 6, 6', and 6". Furthermore, and this is left for the reader to verify, we have

$$W_{14} = W_{36} = \frac{15}{56}, \quad W_{15} = W_{35} = \frac{4}{56}, \quad W_{16} = W_{34} = \frac{1}{56}$$

$$W_{24} = W_{26} = \frac{4}{56}, \quad W_{25} = \frac{16}{56}, \qquad \left.\right\} \quad . \qquad (8.155)$$

Of course it must be

$$6W_{24} + 2W_{25} = 1$$

$$3W_{14} + 2W_{15} + 3W_{16} = 1$$

and so on.

8.6 Method of Finite Differences

8.6.1 Fundamental Relations

The method of finite differences is one of the oldest numerical methods. The procedure shall be illustrated by using the Poisson equation as an example. To simplify, we start from a rectangular region and use Cartesian coordinates x, y (*i.e.*, we consider a two-dimensional problem). As shown in Fig. 8.7 , the area is "discretized", *i.e.* only the potentials $\varphi_{i,j}$ that occur at the corners of small squares at the gridpoints x_i, x_j will be considered. The side of these squares have the length h. Expanding into a Taylor series gives

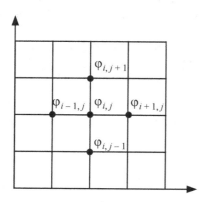

Fig. 8.7

$$\varphi_{i+1,j} = \varphi_{i,j} + \frac{\partial \varphi_{i,j}}{\partial x}h + \frac{1}{2}\frac{\partial^2 \varphi_{i,j}}{\partial x^2}h^2 + 0(h^3)$$

$$\varphi_{i-1,j} = \varphi_{i,j} - \frac{\partial \varphi_{i,j}}{\partial x}h + \frac{1}{2}\frac{\partial^2 \varphi_{i,j}}{\partial x^2}h^2 + 0(h^3)$$

$$\varphi_{i,j+1} = \varphi_{i,j} + \frac{\partial \varphi_{i,j}}{\partial y}h + \frac{1}{2}\frac{\partial^2 \varphi_{i,j}}{\partial y^2}h^2 + 0(h^3)$$

$$\varphi_{i,j-1} = \varphi_{i,j} - \frac{\partial \varphi_{i,j}}{\partial y}h + \frac{1}{2}\frac{\partial^2 \varphi_{i,j}}{\partial y^2}h^2 + 0(h^3).$$

Adding these four equations while disregarding the terms of higher order gives

$$4\varphi_{i,j} + h^2\nabla^2\varphi_{i,j} \cong \varphi_{i+1,j} + \varphi_{i-1,j} + \varphi_{i,j+1} + \varphi_{i,j-1} \qquad (8.156)$$

and with the Poisson equation

$$\nabla^2\varphi_{i,j} = -g_{i,j} \qquad (8.157)$$

$$\boxed{\varphi_{i,j} \cong \frac{1}{4}(\varphi_{i+1,j} + \varphi_{i-1,j} + \varphi_{i,j+1} + \varphi_{i,j-1}) + \frac{h^2 g_{i,j}}{4}} . \qquad (8.158)$$

For $g = 0$, this results in the so-called *five-point formula*

$$\boxed{\varphi_{i,j} \cong \frac{1}{4}(\varphi_{i+1,j} + \varphi_{i-1,j} + \varphi_{i,j+1} + \varphi_{i,j-1})} . \qquad (8.159)$$

It interlinks the potentials at the five points shown in Fig. 8.7. It expresses the potential at each gridpoint (not at the boundary points) in terms of the mean value of the potentials at its four neighbor points. Because of the corresponding mean value theorem (8.38), this is not surprising. If one regards the gridpoint with the potential $\varphi_{i,j}$ as the center of a circle with radius h, then the current approximation of the potentials on the circumference are represented by the potentials at the four mentioned points, whose averaging gives the potential at the center. The five-point formula is nothing less than the corresponding two-dimensional mean value theorem in a discretized form.

This can easily be generalized to three-dimensional problems by analogously defined gridpoints x_i, y_i, z_i. Each gridpoint now has six neighboring points. Expanding this in a Taylor series gives now six, instead of the previous four addends. Their summation gives

$$\boxed{\begin{array}{l}\varphi_{i,j,k} \cong \frac{1}{6}(\varphi_{i+1,j,k} + \varphi_{i-1,j,k} + \varphi_{i,j+1,k} + \varphi_{i,j-1,k} + \varphi_{i,j,k+1} + \varphi_{i,j,k-1}) \\[2mm] \quad + \frac{h^2 g_{i,j,k}}{6} . \end{array}}$$

$$(8.160)$$

For $g = 0$, *i.e.*, for the Laplace equation, this leads to the *seven point formula*

$$\varphi_{i,j,k} \cong \frac{1}{6}(\varphi_{i+1,j,k} + \varphi_{i-1,j,k} + \varphi_{i,j+1,k} + \varphi_{i,j-1,k} + \varphi_{i,j,k+1} + \varphi_{i,j,k-1})$$

(8.161)

This represents the three-dimensional mean value theorem, eq (8.36), in discrete form.

For the one-dimensional case we have of course

$$\varphi_i \cong \frac{1}{2}(\varphi_{i+1} + \varphi_{i-1}) + \frac{h^2 g_i}{2}$$

(8.162)

and for $g = 0$

$$\varphi_i \cong \frac{1}{2}(\varphi_{i+1} + \varphi_{i-1}) \,.$$

(8.163)

For comparison purposes, we have already used eq (8.162) in a previous section – eq. (8.137), $h = 1/3$, $g = -x^2$.

Writing the corresponding equation, *i.e.*, one of eqs. (8.158) through (8.163), depending on the kind of problem, for every internal gridpoint of a one- or two-dimensional lattice, then one obtains a linear system of algebraic equations. For the Laplace equation, it is homogeneous. When prescribing the potential at the boundary, this results in a solvable inhomogeneous system of equations with a unique solution. The number of equations is equal to the number of grid points inside and thereby equal to the number of unknowns (that is, the potentials at the inner gridpoints). One can prove that the coefficient matrix exhibits the necessary properties (that is, its determinant does not vanish). This reveals that the uniqueness theorem, proven in the potential theory for the solution of the Dirichlet boundary value problem (Sect. 3.4.3), because of the discretization, is related to the theorems of linear algebra.

The solution to Neumann or mixed boundary value problems is similar, but more tedious, and will not be discussed here.

The given relations need to be adopted in case of arbitrarily shaped boundaries, which requires to adjust the varying distances of the gridpoints from the boundary ("*boundary region formulas*"), which adds to complicating matters. This too, will be skipped here.

In any case, for all such problems, one obtains uniquely solvable systems of linear equations. To solve a pure Neumann boundary value problem requires one to fix a constant, *for example,* by prescribing the potential at a gridpoint. The results become increasingly accurate (within certain limits) as the granularity decreases, *i.e.*, the smaller the lattice constant h is chosen. Of course, this increases the computing effort as the number of unknowns increases. An advantage is thereby that the obtained coefficient matrix is only *lightly populated*, that is, it consists of mostly vanishing elements. The voluminous systems of equations can either be solved directly (*for example*, by Gauss elimination, by left-right decomposition, etc.) or by means of iteration methods (*for example*, Jacobi method, Gauss-Seidel method, or the relaxation method, etc.)

The discretization described above is not the only one possible. It is also possible to obtain more accurate relations when one includes more points during the discretization. *For example*, consider the Laplace operator. We start with a one-dimensional function $\varphi(x)$, where *for example*,

$$\varphi_i' \approx \frac{\varphi_{i+1} - \varphi_i}{h}$$

or as well

$$\varphi_i' \approx \frac{\varphi_i - \varphi_{i-1}}{h} .$$

Adding gives

$$\varphi_i' \approx \frac{\varphi_{i+1} - \varphi_{i-1}}{2h} .$$

Analogously we find

$$\varphi_i' \approx \frac{\varphi_{i+2} - \varphi_{i-2}}{4h} .$$

Mixing the last two relations by using the weights a and b gives

$$\varphi_i' \approx \frac{2a(\varphi_{i+1} - \varphi_{i-1}) + b(\varphi_{i+2} - \varphi_{i-2})}{4h(a+b)} . \tag{8.164}$$

Choosing $a = 4$ and $b = -1$ gives

$$\varphi_i' \approx \frac{8(\varphi_{i+1} - \varphi_{i-1}) - (\varphi_{i+2} - \varphi_{i-2})}{12h} . \tag{8.165}$$

Proceeding in an analogous way for the second derivatives gives

$$\varphi_i'' \approx \frac{\varphi_{i+1} - 2\varphi_i + \varphi_{i-1}}{h^2}$$

$$\varphi_i'' \approx \frac{\varphi_{i+2} - 2\varphi_i + \varphi_{i-2}}{4h^2}$$

$$\varphi_i'' \approx \frac{4a(\varphi_{i+1} + \varphi_{i-1}) + b(\varphi_{i+2} + \varphi_{i-2}) - 2(4a+b)\varphi_i}{4h^2(a+b)} , \tag{8.166}$$

and with $a = 4$ and $b = -1$

$$\varphi_i'' \approx \frac{16(\varphi_{i+1} + \varphi_{i-1}) - (\varphi_{i+2} + \varphi_{i-2}) - 30\varphi_i}{12h^2} . \tag{8.167}$$

For the one-dimensional case, eq. (8.167) is already a five-point formula, *i.e.*, in order to express φ'', it uses the values of φ at five points. Analogously, for the two- and three-dimensional case, this results in 9 or 13 point formulas. The best value for a and b in each case is determined by the remainder term of the Taylor expansion, which should be of the highest possible order (*i.e.* as small as possible). More details are provided *for example*, by Marsal [19]. Another interesting

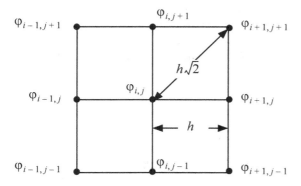

Fig. 8.8

representation of the two-dimensional Laplace operator is the following.
According to eq. (8.156) for the nine gridpoints presented in Fig. 8.8 , we have

$$\nabla^2\varphi_{i,j} \approx \frac{\varphi_{i+1,j} + \varphi_{i-1,j} + \varphi_{i,j+1} + \varphi_{i,j-1} - 4\varphi_{i,j}}{h^2}$$

and also

$$\nabla^2\varphi_{i,j} \approx \frac{\varphi_{i+1,j+1} + \varphi_{i+1,j-1} + \varphi_{i-1,j+1} + \varphi_{i-1,j-1} - 4\varphi_{i,j}}{2h^2} .$$

Therefore

$$\nabla^2\varphi_{i,j} \approx \frac{2a(\varphi_{i+1,j} + \varphi_{i-1,j} + \varphi_{i,j+1} + \varphi_{i,j-1}) + b(\varphi_{i+1,j+1} + \varphi_{i+1,j-1} + \varphi_{i-1,j+1} + \varphi_{i-1,j-1}) - 4(2a+b)\varphi_{i,j}}{2h^2(a+b)}$$

and with $a = 2$, $b = 1$

$$\nabla^2\varphi_{i,j} \approx \frac{4(\varphi_{i+1,j} + \varphi_{i-1,j} + \varphi_{i,j+1} + \varphi_{i,j-1}) + (\varphi_{i+1,j+1} + \varphi_{i+1,j-1} + \varphi_{i-1,j+1} + \varphi_{i-1,j-1}) - 20\varphi_{i,j}}{6h^2} .$$

(8.168)

In particular for $\nabla^2\varphi = 0$

$$\varphi_{i,j} \approx \frac{4(\varphi_{i+1,j} + \varphi_{i-1,j} + \varphi_{i,j+1} + \varphi_{i,j-1}) + (\varphi_{i+1,j+1} + \varphi_{i+1,j-1} + \varphi_{i-1,j+1} + \varphi_{i-1,j-1})}{20} .$$

(8.169)

Further details and similar relations can also be found in Marsal [19].

8.6.2   An Example

As an example, we shall apply the method to the Dirichlet boundary value problem depicted in Fig. 8.9. There are nine gridpoints inside the square area. The Laplace equation shall be solved for the boundary conditions of $\varphi = 100$ at the top boundary and $\varphi = 0$ at the remaining boundaries (all are dimensionless quantities)

Its solution is

$$\varphi = \frac{400}{\pi} \sum_{n = 1, 3, 5, \dots} \frac{1}{n} \frac{\sinh\left(\frac{n\pi y}{d}\right) \sin\left(\frac{n\pi x}{d}\right)}{\sinh(n\pi)} , \tag{8.170}$$

where d represents the sides of the square. Evaluating this result at the gridpoints yields the following potentials:

$$\varphi_1 = 43.20833, \qquad \varphi_2 = 54.052922, \qquad \varphi_3 = 18.202833,$$

$$\varphi_4 = 25, \qquad \varphi_5 = 6.797166, \qquad \varphi_6 = 9.541422 \tag{8.171}$$

The following approximations can be compared with these results.

The five-point formula (8.159) yields the following six equations:

$$\left.\begin{aligned}
\varphi_1 \quad -\tfrac{1}{4}\varphi_2 -\tfrac{1}{4}\varphi_3 \qquad\qquad\qquad\qquad &= 25 \\
-\tfrac{1}{2}\varphi_1 \;+ \varphi_2 \qquad -\tfrac{1}{4}\varphi_4 \qquad\qquad\qquad &= 25 \\
-\tfrac{1}{4}\varphi_1 \qquad\quad + \varphi_3 -\tfrac{1}{4}\varphi_4 -\tfrac{1}{4}\varphi_5 \qquad &= 0 \\
-\tfrac{1}{4}\varphi_2 -\tfrac{1}{2}\varphi_3 + \varphi_4 \qquad\quad -\tfrac{1}{4}\varphi_6 &= 0 \\
-\tfrac{1}{4}\varphi_3 \qquad\quad + \varphi_5 -\tfrac{1}{4}\varphi_6 &= 0 \\
-\tfrac{1}{4}\varphi_4 -\tfrac{1}{2}\varphi_5 + \varphi_6 &= 0
\end{aligned}\right\} . \tag{8.172}$$

Solving it directly yields

$$\left.\begin{aligned}
\varphi_1 &= \frac{300}{7} = 42.85, \qquad \varphi_2 = \frac{1475}{28} = 52.67, \\
\varphi_3 &= \frac{75}{4} = 18.75, \qquad \varphi_4 = 25, \\
\varphi_5 &= \frac{50}{7} = 7.14, \qquad \varphi_6 = \frac{275}{28} = 9.82 .
\end{aligned}\right\} \tag{8.173}$$

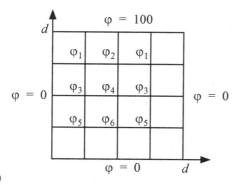

Fig. 8.9

Of course, these values deviate from the exact ones, but the maximum error is just under 5%.

One can also apply the nine-point formula (8.169). This requires to use the potentials at the corners of the square. At the discontinuities (*i.e.* at $x = 0$, $y = d$ and at $x = d$, $y = d$), one has to use the average, *i.e.* $\varphi = 50$. This gives the following equations:

$$
\left.
\begin{aligned}
\varphi_1 \quad -\frac{1}{5}\varphi_2 -\frac{1}{5}\varphi_3 -\frac{1}{20}\varphi_4 \qquad\qquad\qquad\quad &= \frac{55}{2} \\[4pt]
-\frac{2}{5}\varphi_1 \quad +\varphi_2 -\frac{1}{10}\varphi_3 -\frac{1}{5}\varphi_4 \qquad\qquad\qquad &= 30 \\[4pt]
-\frac{1}{5}\varphi_1 -\frac{1}{20}\varphi_2 \quad +\varphi_3 -\frac{1}{5}\varphi_4 -\frac{1}{5}\varphi_5 -\frac{1}{20}\varphi_6 &= 0 \\[4pt]
-\frac{1}{10}\varphi_1 -\frac{1}{5}\varphi_2 -\frac{2}{5}\varphi_3 \quad +\varphi_4 -\frac{1}{10}\varphi_5 -\frac{1}{5}\varphi_6 &= 0 \\[4pt]
-\frac{1}{5}\varphi_3 -\frac{1}{20}\varphi_4 \quad +\varphi_5 -\frac{1}{5}\varphi_6 &= 0 \\[4pt]
-\frac{1}{10}\varphi_3 -\frac{1}{5}\varphi_4 -\frac{2}{5}\varphi_5 \quad +\varphi_6 &= 0
\end{aligned}
\right\}
\qquad (8.174)
$$

and its solutions

$$
\left.
\begin{aligned}
\varphi_1 &= \frac{25.159}{92} = 43.2065, & \varphi_2 &= \frac{25.1095}{506} = 54.1007, \\[6pt]
\varphi_3 &= \frac{200}{11} = 18.1818, & \varphi_4 &= 25, \\[6pt]
\varphi_5 &= \frac{625}{92} = 6.7934, & \varphi_6 &= \frac{25.193}{506} = 9.5355 .
\end{aligned}
\right\}
\qquad (8.175)
$$

This coincides astonishingly well with the exact solutions eq. (8.170) and (8.171), respectively. The maximal error is at about 0.1%, and thereby about 50 times smaller than when using the five-point formula. The exact value for the potential $\varphi_4 = 25$ is obtained in both cases.

The so-called Gaussian elimination is oftentimes used for the direct solution of the systems of equations, whereby the coefficient matrix is transformed into a triangular matrix by applying elementary row operations. This results in an easily solvable system of equations. Another method is the so-called LR decomposition ("left-right decomposition"). In this case, the matrix is transformed into a product of two triangular matrices, which also allows to conveniently solve the system of equations.

Oftentimes the system is solved iteratively. For this purpose, one uses estimations for the potentials at the gridpoints. Then, by means of the respective formulas, one calculates new values etc., *i.e.* from the potentials of the *n*-th iteration steps ($\varphi_{i,j}^{(n)}$) one calculates the values of the *(n+1)*-th step ($\varphi_{i,j}^{(n+1)}$) as follows:

$$\varphi_{i,j}^{(n+1)} = \frac{1}{4}(\varphi_{i+1,j}^{(n)} + \varphi_{i-1,j}^{(n)} + \varphi_{i,j+1}^{(n)} + \varphi_{i,j-1}^{(n)}) . \tag{8.176}$$

This is the so-called *Jacobi method*. Convergence of the iteration is accelerated in case of the *Gauss-Seidel method*. Here, in step *(n+1)*, not only the old values *n* are used, but already the new values of the step *(n+1)* are used as much as they are available. Further acceleration of the convergence can be achieved through the *relaxation method*, which shall only be mentioned here, but not further discussed.

Of course, the iteration converges better, if the initial estimates are more accurate. At least for the current example, it is easy to find very useful estimates by using the five-point formula or – what amounts to the same – using the mean value theorem. Consider initially only an inner gridpoint in the center, then based on the boundary values one estimates a value of $\varphi_1^{(0)} = 25$. Now one narrows the grid according to Fig. 8.9. From $\varphi_4^{(0)}$ and the boundary values, the corners in particular, one obtains an estimate for $\varphi_1^{(0)}$ and $\varphi_5^{(0)}$, namely $\varphi_1^{(0)} = 175/4 \approx 44$ and $\varphi_5^{(0)} = 25/4 \approx 6$. This allows one to estimate the remaining potentials, $\varphi_2^{(0)} = 53$, $\varphi_3^{(0)} = 19$, $\varphi_6^{(0)} = 9$. Now, based on these values one iterates by use of eq. (8.176). It remains left for the reader to convince himself, that this converges towards the approximations given in (8.173), but not towards the exact value. The Gauss-Seidel method accelerates the convergence. In contrast to the Jacobi method, the symmetry of the values for the potential of the present problem is initially lost, even though this will, of course, finally converge towards the symmetric approximation solution.

These proceedings may be modified in many ways. The grid spacing for the various coordinate directions may be chosen with different values. The lattice does not need to be uniform, *i.e.* the grid may be variable inside a region. This leads to the method of local grid refinement, if inside a particular region an extremely fine discretization is needed in order to achieve a required accuracy.

Of course, the method of finite differences is applicable to all sorts of differential equations. Space or time dependent problems (*for example* diffusion equations or wave equations), generally require one to also discretize the time.

Depending on the strategy, there are two different types of difference equations. In the first case, the so-called *explicit methods*, all quantities of a given "time plane" are calculated from the immediately preceding time plane. In the second case, the *implicit methods*, the equations of one time plane contain the unknowns of the subsequent time plane. This distinction is of importance because the *per se* simpler explicit methods exhibit the disadvantage that they may be unstable, *i.e.* that errors may grow and the numerical results become useless. This only seemingly insignificant difference shall be illustrated by means of the **example** of the diffusion equation. We may discretize the equation

$$\frac{\partial^2 u}{\partial x^2} = A\frac{\partial u}{\partial t} \tag{8.177}$$

in the form

$$\frac{u_{i-1,k} - 2u_{i,k} + u_{1+i,k}}{h^2} = A\frac{u_{i,k+1} - u_{i,k}}{\Delta t} \tag{8.178}$$

but also in the form

$$\frac{u_{i-1,k+1} - 2u_{i,k+1} + u_{1+i,k+1}}{h^2} = A\frac{u_{i,k+1} - u_{i,k}}{\Delta t}. \tag{8.179}$$

In case of the explicit formulation (8.178), all $u_{i,k}$ can be directly calculated from the values $u_{i,0}$, generally all $u_{i,k+1}$ from $u_{i,k}$ etc. In contrast, the implicit formulation (8.179) requires more computing effort. The so-called *semi-implicit* methods constitute a compromise between the two methodologies whereby the two relations (8.178) and (8.179) are mixed by the weight factors α and $1 - \alpha$ ("*Euler factor*").

As an **example**, Fig. 8.10 shows the current density field inside a meandering thin-film resistor, which was calculated by means of the finite difference method. This is a two-dimensional, mixed boundary value problem ($\varphi = \varphi_1$ and $\varphi = \varphi_2$ at the two contacts on top and bottom, $\partial\varphi/\partial n = 0$ at the other boundaries). For this example, the domain decomposition method was additionally used. This allows one to reduce such a problem to finding the solutions in subregions, where the boundary conditions at the additional boundaries need to be initially estimated. The problem is then solved iteratively. More details on this method can be found in Bader [20]. In the present case, the area was partitioned entirely in rectangles.

8.7 Finite Elements Method

The method of Finite Elements has quickly gained significance. Although the effort in any particular application may be substantial, the method is in principle based on a simple and elegant idea. It is very flexible and thereby applicable to many types of problems. Oftentimes, it is superior to other methods, although this may not be the case for every type of problem. Corresponding to its significance,

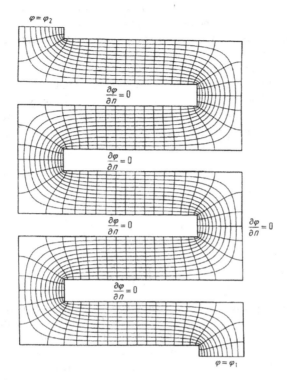

Fig. 8.10

the available literature covering finite elements is rather extensive. As examples, the books [17, 19, 21 through 27] shall be mentioned.

For this method, the region of definition of the unknown function or functions is partitioned into more or less arbitrarily shaped domains – just the finite elements, partial lines, partial surfaces, partial volumes, depending on the dimension of the region that needs discretization. Every finite element is associated with an approximate solution which is non vanishing only inside of that element. The approximate solution consists of a set of linearly independent basis functions (the so-called form functions) and a corresponding number of initially undetermined parameters. These parameters represent the value of the function itself, which it assumes at certain points of the finite element, the so-called *nodal points*. These nodal points, thereby, play a similar role as the gridpoints did in case of the finite difference method. The not so insignificant difference is based on the fact that the approximate solution assumed in the finite element method, approximates the unknown function at all points, not only at the nodal points. On the other hand, the method of finite difference may be regarded as a special case of the method of finite elements. Furthermore, even in case of the finite difference method, it is possible by proper interpolation, to assign values to all points The entire region has

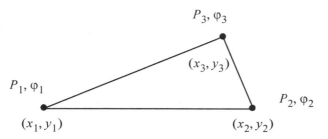

Fig. 8.11

to be filled with finite elements in such a way that every node at the boundary of a finite element coincides with the node of its neighbor elements, whereby the value of the functions have to match there as well. Together with the boundary conditions, one obtains a system of equations that allows to determine the value of the function at the nodal points. Thereby, one starts either with the method of the weighted residuals, usually in the form of the Galerkin method or – given the variational integral exists – one starts from the Rayleigh-Ritz method, which is equivalent to the Galerkin method.

Before providing more in depths details, we will illustrate what we have learned so far by a two-dimensional example. A particularly simple one-dimensional **example** that highlighted some of the important steps was already provided in Sect. 8.4.5, equations (8.128) and following. In the two-dimensional case, triangles can be used as finite elements, which need to fill the entire area. Fig. 8.11 shows one of these triangles with the corners P_1, P_2, P_3 and the corresponding values $\varphi_1, \varphi_2, \varphi_3$. The corners constitute the nodal points. Inside the triangle, we attempt to approximate the unknown function by a function of the form

$$\varphi = a + bx + cy .\tag{8.180}$$

Then, for every corner or nodal point, it has to be

$$\varphi_i = a + bx_i + cy_i , \qquad i = 1, 2, 3 .$$

a, b, c can be calculated from these three equations:

$$a = \frac{1}{D}\begin{vmatrix} \varphi_1 & x_1 & y_1 \\ \varphi_2 & x_2 & y_2 \\ \varphi_3 & x_3 & y_3 \end{vmatrix}, \quad b = \frac{1}{D}\begin{vmatrix} 1 & \varphi_1 & y_1 \\ 1 & \varphi_2 & y_2 \\ 1 & \varphi_3 & y_3 \end{vmatrix}, \quad c = \frac{1}{D}\begin{vmatrix} 1 & x_1 & \varphi_1 \\ 1 & x_2 & \varphi_2 \\ 1 & x_3 & \varphi_3 \end{vmatrix}, \tag{8.181}$$

where D is the coefficient determinant.

$$D = \begin{vmatrix} 1 & x_1 & y_1 \\ 1 & x_2 & y_2 \\ 1 & x_3 & y_3 \end{vmatrix} .$$

(8.182)

If one lets

$$f_1(x, y) = \frac{1}{D}[(x_2 y_3 - y_2 x_3) + (y_2 - y_3)x + (x_3 - x_2)y] ,$$

$$f_2(x, y) = \frac{1}{D}[(x_3 y_1 - y_3 x_1) + (y_3 - y_1)x + (x_1 - x_3)y] ,$$ (8.183)

$$f_3(x, y) = \frac{1}{D}[(x_1 y_2 - y_1 x_2) + (y_1 - y_2)x + (x_2 - x_1)y] ,$$

then for these so-called *form functions* we have

$$\boxed{f_i(x_k, y_k) = \delta_{ik}}$$ (8.184)

and the Ansatz (8.180) takes the form

$$\boxed{\varphi = \sum_{i=1}^{3} \varphi_i f_i(x, y)} .$$ (8.185)

Together with eq. (8.184) we obtain, as necessary:

$$\varphi_k = \sum_{i=1}^{3} \varphi_i f_i(x_k, y_k) = \sum_{i=1}^{3} \varphi_i \delta_{ik} = \varphi_k .$$ (8.186)

This means that eq. (8.185) represents the Ansatz within the observed finite element, in form of a linear combination of the form function and the coefficients represent the values of the function at the nodal points. These functions are non-vanishing only inside this particular finite element. The Ansatz for the problem overall, is finally obtained by superposition of all trial functions of all elements. The form function can also be interpreted as being so-called triangular coordinates. Every point in a triangle can be determined by the triangle coordinates ξ_1, ξ_2, ξ_3. As indicated in Fig. 8.12 , these are defined in such a way that the side opposite of P_i is associated with $\xi_i = 0$ and the line parallel to it that cuts through P_i is the line $\xi_i = 1$. On the parallel lines inbetween, ξ_i takes on distance proportional values. Of course, two of these three triangular coordinates are sufficient to uniquely characterize a point. The relation between the triangular coordinates and Cartesian coordinates is given by the coordinates of the corner points. We have

$$\left. \begin{array}{l} x = \xi_1 x_1 + \xi_2 x_2 + \xi_3 x_3 \\ y = \xi_1 y_1 + \xi_2 y_2 + \xi_3 y_{x3} \\ 1 = \xi_1 + \xi_2 + \xi_3 \end{array} \right\} .$$ (8.187)

From this, one obtains *for example*, for the point P_1 with $\xi_1 = 1$, $\xi_2 = 0$, $\xi_3 = 0$ just $x = x_1$, $y = y_1$, etc. Since the relation has to also be linear, it is

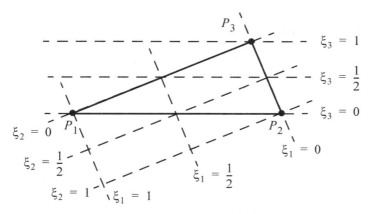

Fig. 8.12

Fig. 8.13

thereby proven. If one now calculates the triangular coordinates ξ_1, ξ_2, ξ_3 corresponding to a point x, y, then one precisely obtains the form functions

$$\xi_i(x,y) = f_i(x,y) \ . \tag{8.188}$$

The form functions thereby also represent a local coordinate system on the corresponding triangle. This knowledge is useful when calculating many of the integrals that are necessary when applying, *for example*, the Galerkin or the Rayleigh-Ritz method.

The described triangles, together with the linear form functions (8.183) and the related Ansatz (8.185) provide only a simple example. Practical applications use two- and three-dimensional finite elements of different types, with a varying number of nodal points and oftentimes much more complicated form functions of higher order. The details may become very voluminous and may be expressed in a complicated form, which of course, does not change the basically simple and elegant principle.

To simplify, the procedure shall be illustrated by means of the one-dimensional **example** which we have already mentioned in Sect. 8.4.5. Consider the region $x_0 \leq x \leq x_n$ and partition it as shown in Fig. 8.13 in n finite elements of length

$$h = \frac{x_n - x_0}{n} \ , \tag{8.189}$$

with the nodal points $x_i (i = 0, \dots n)$, where

$$x_{i-1} \le x \le x_i, \qquad x_i - x_{i-1} = h \ . \tag{8.190}$$

One proceeds in a strictly formal fashion, in order to highlight the analogy to the case of triangular finite elements. For the i-th finite element, let

$$\varphi = a + bx, \qquad x_{i-1} \le x \le x_i \ , \tag{8.191}$$

where

$$\varphi_{i-1} = a + bx_{i-1}$$

$$\varphi_i = a + bx_i \ . \tag{8.192}$$

Therefore

$$a = \frac{\begin{vmatrix} \varphi_{i-1} & x_{i-1} \\ \varphi_i & x_i \end{vmatrix}}{\begin{vmatrix} 1 & x_{i-1} \\ 1 & x_i \end{vmatrix}} = \frac{x_i \varphi_{i-1} - x_{i-1}\varphi_i}{h}, \ b = \frac{\begin{vmatrix} 1 & \varphi_{i-1} \\ 1 & \varphi_i \end{vmatrix}}{\begin{vmatrix} 1 & x_{i-1} \\ 1 & x_i \end{vmatrix}} = \frac{\varphi_i - \varphi_{i-1}}{h}$$

$$\tag{8.193}$$

and

$$\varphi = f_{i1}\varphi_{i-1} + f_{i2}\varphi_i$$

with

$$f_{i1} = \frac{x_i - x}{h}, \quad f_{i2} = \frac{x - x_{i-1}}{h} \ , \tag{8.194}$$

where obviously

$$\left. \begin{array}{lll} f_{i1}(x_{i-1}) = 1, & f_{i2}(x_{i-1}) = 0, & f_{i1} + f_{i2} = 1 \\ f_{i1}(x_i) = 0, & f_{i2}(x_i) = 1, & f_{i1}x_{i-1} + f_{i2}x_i = x \end{array} \right\} \ . \tag{8.195}$$

These relations are analogous to eqs.(8.180) through (8.188), which we have discussed for the triangles. The form functions (8.194) represents local coordinates in the i-th element, just as the triangle coordinates do for the triangular finite element. Next, one finds an approximation for the Poisson equation

$$\nabla^2\varphi = -g(x), \qquad x_0 \le x \le x_n \tag{8.196}$$

with the boundary conditions

$$\varphi(x_0) = \varphi_0, \quad \varphi(x_n) = \varphi_n \ . \tag{8.197}$$

This requires to minimize the integral

$$I = \int_a^b \{[\varphi'(x)]^2 - 2\varphi(x)g(x)\} \, dx \ . \tag{8.198}$$

The part stemming from the i-th finite element is

$$I = \int_{x_{i-1}}^{x_i} \left\{ \left[-\frac{\varphi_{i-1}}{h} + \frac{\varphi_i}{h} \right]^2 - 2f_{i1}g(x)\varphi_{i-1} - 2f_{i2}g(x)\varphi_i \right\} dx$$

$$= \frac{1}{h}(\varphi_{i-1}^2 - 2\varphi_{i-1}\varphi_i + \varphi_i^2) - 2G_{i1}\varphi_{i-1} - 2G_{i2}\varphi_i \qquad (8.199)$$

where

$$G_{i1,2} = \int_{x_{i-1}}^{x_i} f_{i1,2}(x)g(x)dx \ . \qquad (8.200)$$

Minimization of this, results in the following parts:

$$\left. \begin{array}{l} \dfrac{\partial I}{\partial \varphi_{i-1}} = 2\left(\dfrac{\varphi_{i-1}}{h} - \dfrac{\varphi_i}{h} - G_{i1} \right) \\[3mm] \dfrac{\partial I}{\partial \varphi_i} = 2\left(-\dfrac{\varphi_{i-1}}{h} + \dfrac{\varphi_i}{h} - G_{i2} \right) \end{array} \right\} \ . \qquad (8.201)$$

The corresponding coefficient matrix is called *element matrix*. Collecting all contributions of all finite elements gives the overall matrix of the linear system of equations which, ultimately will need to be solved.

Now, we choose $x_0 = 0$, $x_n = 1$ and $g(x) = x$. The exact solution in this case is

$$\varphi = \varphi_0 + (\varphi_n - \varphi_0)x + \frac{x - x^3}{6} \ . \qquad (8.202)$$

For the present case, one has

$$x_i = \frac{i}{n}, \quad h = \frac{1}{n} \qquad (8.203)$$

and

$$G_{i1} = \frac{3i-2}{6n^2}, \quad G_{i2} = \frac{3i-1}{6n^2} \ . \qquad (8.204)$$

Ultimately, one obtains the following equations for $i = 1, ..., n-1$:

$$\frac{\partial I}{\partial \varphi_i} = 2\left(-\frac{\varphi_{i-1}}{h} + \frac{2\varphi_i}{h} - \frac{\varphi_{i+1}}{h} - G_{i+1,1} - G_{i2} \right) = 0 \ . \qquad (8.205)$$

Eq. (8.204) yields

$$G_{i+1,1} + G_{i2} = \frac{3i+3-2+3i-1}{6n^2} = \frac{i}{n^2} \qquad (8.206)$$

and the equation system (8.205) takes on the form

$$-\varphi_{i-1} + 2\varphi_i - \varphi_{i+1} = \frac{i}{n^3} \ . \qquad (8.207)$$

Its solution is

$$\varphi_i = \varphi_0 + (\varphi_n - \varphi_0)\frac{i}{n} + \frac{n^2 i - i^3}{6n^3} \ . \tag{8.208}$$

At the nodal points, this coincides with the exact solution. The form function is used to interpolate in between the nodes. Comparison with eq. (8.162) shows that for the present case, the finite difference method would have yielded the same difference equations (8.207).

Despite its simple nature, this example exhibits the steps for applying finite elements.

a) Partitioning the region into finite elements.

b) For every finite element, approximate the function as a linear combination of the form functions, where the values of the functions at the nodal points are expressed as coefficients.

c) Determine the element matrix by evaluating the occurring integrals (which may require numerical methods).

d) The overall matrix is compiled from the element matrices. This step is potentially complicated. Sound indexing is paramount, in particular, in case of multi-dimensional problems and finite elements with many nodes.

e) Finally, the system of equations needs to be solved. The same methods can be used as already partly addressed in Sect. 8.6, presenting finite differences.

Despite its flexibility, finite elements (and also finite differences) are not suitable for problems involving infinite regions that include the external Dirichlet or Neumann boundary value problems. Oftentimes one uses finite regions, which has the disadvantage of introducing errors that are difficult to estimate because the boundary conditions on those artificially introduced boundaries are unknown and can only be estimated. An option is to tackle the problem by introducing so-called infinite elements. Two different approaches were taken so far in this respect. On one hand, one uses exponentially decaying form functions whereby a suitable exponent has to be chosen. On the other hand, one transforms the infinite elements by a suitable transformation onto finite elements. So far, it remains an open question whether the use of infinite elements leads to satisfactory results. In any case, problems involving infinite regions can be solved by means of the boundary element method, which we will discuss in the next section. For this method, the distinction between finite or infinite regions is insignificant. For many problems, to couple both methods is beneficial, where the finite elements are used inside a region and compatible boundary elements are then used on the outside boundary of this arbitrary region.

Fig. 8.14 presents the results of a diffusion problem that was solved with finite elements. It represents a rotationally symmetric magnetic field $B_z(r, \tau)$. Shown is the initial field and the fields for the dimensionless times $\tau = 0, 1 \ ; \ 0, 2$ $0, 3 \ ; \ 0, 4 \ ; \ 0, 5$. Comparison with the known analytic solution of the problem shows that the relative error of the presented numerical solution is $< 2.38 \cdot 10^{-4}$.

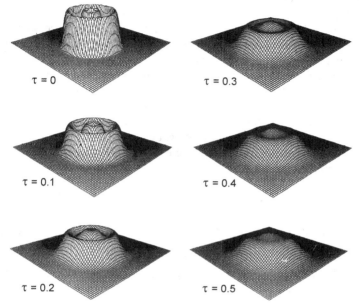

$\tau = 0$ $\tau = 0.3$

$\tau = 0.1$ $\tau = 0.4$

$\tau = 0.2$ $\tau = 0.5$

Fig. 8.14

8.8   Method of Boundary Elements

The Method of Boundary Elements is the most recent of the important numerical methods; a method, particularly interesting for electromagnetic field theory and virtually taylor made for many field theoretical problems. The reason is, that it emerges directly from the integral equations of field theory when discretizing them. It is valuable in both cases, as an independent method, as well as a supplement to the finite element method. An in-depth descriptions can be found in Brebbia or Brebbia, Telles, and Wrobel [28, 29].

Starting with the integral equations presented in Sect. 8.2 (or their corresponding integral equations for magnetostatic or time dependent problems), we will initially study only the surfaces (boundaries) of the observed regions. Solving these integral equations, initially only for the boundaries, yields the quantities by which the field of the finite or infinite region (this does not make any difference here) can be calculated, as described in Sect. 8.2. Sometimes a distinction between direct or indirect methods is made. The direct methods are based on eqs. (8.1) and (8.2), (8.6), and (8.7) or (in the trivial one-dimensional case) (8.16) and (8.17), while the indirect methods are based on equations of the type (8.11) through (8.14).

Discretization is achieved by partitioning the boundaries in surface elements or (in the two dimensional case) into line elements, which may be chosen in many different ways. The unknowns (that is φ or $\partial\varphi/\partial n$ in case of the direct methods, and surface charge σ or area density τ of the dipole moments in case of the indirect methods) are expressed on the boundary elements by trial functions with different form functions. In the simplest case, they are assumed constant on the boundary element. In the two-dimensional case, for instance, one may proceed as illustrated in Fig. 8.15 The boundary elements are here line elements on the boundary curve. The values at the center of each element are assumed to be constant, *i.e.* there is only one node per element (the so-called "constant element").

With $C = \pi$, and *for example* for the Laplace equation (*i.e.* when $\rho = 0$), from eq. (8.7), the potential on the boundary elements i ($i = 1 \ldots n$) yields

$$\pi\varphi_i = - \sum_{k=1}^{n} \left(\frac{\partial\varphi}{\partial n}\right)_k \int_{S_k} \ln|\mathbf{r}_i - \mathbf{r}_k| \, ds_k' + \sum_{k=1}^{n} \varphi_k \int_{S_k} \frac{\partial}{\partial n'} \ln|\mathbf{r}_i - \mathbf{r}_k| \, ds' . \qquad (8.209)$$

The result of evaluating these integrals (which each has to be calculated on the boundary elements referenced by S_k) is a linear system of equations of the form:

$$\sum_{k=1}^{n} A_{ik}\left(\frac{\partial\varphi}{\partial n}\right)_k + \sum_{k=1}^{n} B_{ik}\varphi_k = 0, \qquad i = 1, \ldots, n , \qquad (8.210)$$

or in matrix form

$$A\left(\frac{\partial\varphi}{\partial n}\right) + B\varphi = 0 , \qquad (8.211)$$

where φ and $\partial\varphi/\partial n$ represent the column vectors of the n values φ_k and $(\partial\varphi/\partial n)_k$, respectively and

$$A = (A_{ik}), \quad B = (B_{ik}) . \qquad (8.212)$$

The term $\pi\varphi_i$ on the left side of eq. (8.209) is included in B_{ii}. The system of equations (8.211) is initially homogeneous with respect to the $2n$ quantities φ_k and $(\partial\varphi/\partial n)_k$. In any case, half of these quantities are given (*for example*, in case of the mixed boundary value problem n_1 of the values φ_k and n_2 of the values $(\partial\varphi/\partial n)_k$, where $n = n_1 + n_2$). The remaining n quantities are unknown and need to be determined by solving this system of equations. Once the unknowns are determined, the potential may be calculated by means of eq. (8.6) at specific gridpoints within the entire region.

Fig. 8.15

Similarly, one might start from eqs. (8.11) and (8.12) and write them in the following form:

$$\varphi = \sum_{k=1}^{n} C_{ik}\sigma_k = \boldsymbol{C\sigma} \left. \begin{array}{c} \\ \\ \\ \end{array} \right| ,$$

$$\frac{\partial\varphi}{\partial n} = \sum_{k=1}^{n} D_{ik}\sigma_k = \boldsymbol{D\sigma} \qquad\qquad (8.213)$$

where σ represents the column vector of the n values σ_k and

$$\boldsymbol{C} = (C_{ik}), \quad \boldsymbol{D} = (D_{ik}) . \qquad\qquad (8.214)$$

Again, n of the $2n$ quantities φ_i and $(\partial\varphi/\partial n)_i$ are given. From the $2n$ equations, one selects the corresponding n equations and solves for the values σ_k. This allows to calculated φ at any point of the entire region.

The boundary element method is similar to the finite element method in the sense that at the boundary elements, trial functions similar to those used for finite elements are chosen by suitable form functions. The difference is that the weight functions are different. Indeed, the Boundary Element Method can be regarded as a special case of the method of weighted residuals, where the fundamental solution of the potential theory plays the role of the weight functions. More on this can be found in Brebbia, Telles, and Wrobel [29]. Thus, the various methods are somewhat interrelated. One distinguishes hereby the different formulations of the problem to be solved. For instance, in case of the Laplace equation, the residual

$$R = \int_V \psi\nabla^2\varphi\,d\tau$$

of the unknown function φ and the weight function ψ occurs. Integrating by parts once, by means of Green's integral theorem, one obtains – except for boundary integrals – the so-called *weak formulation* of the problem with the integral

$$\int_V \nabla\psi \bullet \nabla\varphi\,d\tau .$$

and a second integration by parts gives the so-called *inverse formulation* with the integral

$$\int \varphi\nabla^2\psi\,d\tau .$$

One should not read too much into this terminology. Nevertheless, it is remarkable and sometimes useful to realize that at each of these steps, the trial function φ has one derivative less and the weight functions requires one derivative more. Besides, as examples in applying these formulations, one can recognize the method of the variational integrals in the weak formulation, and in the inverse formulation, that of the Boundary Value Method. Other weight functions are possible in case of all three formulations. For instance, if in case of the inverse formulation, the basis and weight functions are identical, then one obtains the so-called *Trefftz method*, which shall not be covered here.

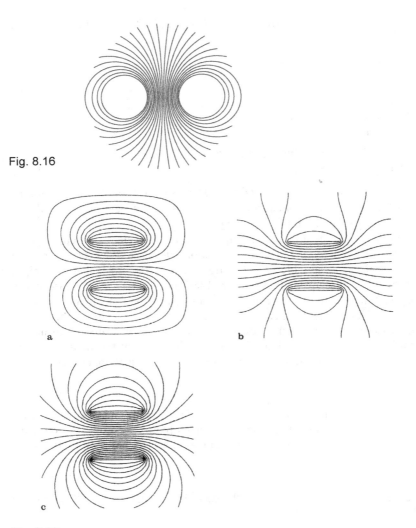

Fig. 8.16

Fig. 8.17

Fig. 8.16 shows the equipotential lines of two parallel conducting cylinders of potential $\pm \varphi_o$, calculated with the Direct Boundary Value Method. One obtains the expected result, namely the circles of Apollonius, which was already calculated in Sections 2.6.3 and 3.12. This represents a plane problem which was solved by using eqs. (8.6) and (8.7).

Fig. 8.17 and Fig. 8.18 present fields which were calculated by coupling of finite elements and boundary elements. Fig. 8.17 demonstrates the advantage of this approach when applied to problems in finite regions versus the use of solely using finite elements. If only finite elements are used, then the problem is solved by approximation in a region that has to be chosen not too small but finite, and on

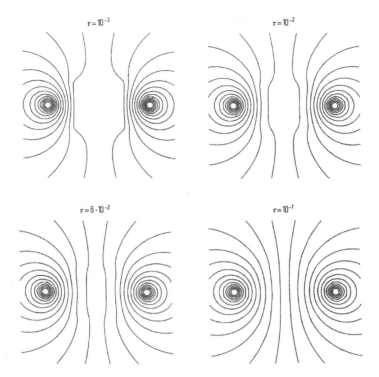

Fig. 8.18

its boundaries one has to, more or less arbitrarily, choose (roughly estimated) boundary conditions. The equipotential lines in Fig. 8.17a and Fig. 8.17b were calculated by only applying finite elements. Fig. 8.17a used $\varphi = 0$ and Fig. 8.17b used $\partial\varphi/\partial n = 0$ on the boundary. In contrast, Fig. 8.17c was obtained by coupling of finite elements and boundary elements, and it obviously delivers the much better results. The fields of Fig. 8.18 were also calculated by means of the coupled method. It represents an eddy current problem. There are two infinitely long wires, carrying a time independent constant current whose magnetic field penetrates an also infinitely long non-magnetic, conducting cylinder of square cross section. Fig. 8.18 shows the magnetic field shortly after suddenly turning on the current, i.e., at the dimensionless time $\tau = 10^{-3}$. The remaining illustrations show the field at $\tau = 10^{-2}$, $\tau = 6 \cdot 10^{-2}$, and $\tau = 10^{-1}$. Already for $\tau = 10^{-1}$, the final state is almost reached. The plots of the fields for times greater than $\tau = 10^{-1}$ can hardly be distinguished. The two problems of Fig. 8.17 and Fig. 8.18 were calculated by use of the so-called direct method while considering eqs. (8.6) and (8.7). Of course, when coupling, the boundary elements on the boundary, have to be chosen such that they are always compatible with those sides of the finite elements which form the boundary and in particular, that they agree at the nodal points.

8.9   Method of Image Charges

The Method of Image Charges is oftentimes suited to solve field theoretical problems. This is initially motivated by a number of specific problems, which may even be solved exactly by use of the image charge method, *for example*, the problem of a sphere within the field of a point charge or within a uniform field (sects. 2.6.1 and 2.6.2), the problem of a conducting cylinder within the field of a uniform line charge (Sect. 2.6.3), or that of a point charge in a dielectric half-space (Sect. 2.11.2). Analogous procedures exist as well for the case of stationary electric current fields (*e.g.*, Sect 4.5), in magnetostatics (*e.g.* Sect. 5.9), and for time dependent problems (*e.g.* Sect. 6.5.3).

For electrostatics, the perhaps most general statement on this subject is contained in Kirchhoff's theorem eq. (3.57), which is important in many respects. It states, among other things, that the fields inside an arbitrarily shaped region, generated by an arbitrary charge distribution located outside this region, can also be generated by placing suitable surface charges and dipole layers on the surface of this region and vice versa. The consequence is that boundary value problems can be treated as if the fields within the region of interest were generated by appropriately distributed charges outside of the region. These charges may occur in arbitrary configurations and form dipoles or more generally multipoles (as *for example*, for the case of a conducting sphere inside an electric field where two image charges occur, which form an ideal dipole). One should not interpret the term image charges to narrowly. It may represent arbitrary distributions of point charges, line charges, volume charges, or even an arbitrary distribution of multipoles.

Usually, one determines the assumed image charges by satisfying the given boundary conditions at selected points of the surface, *i.e.*, one uses the Collocation Method, perhaps in its overdetermined form. Therefore, the Method of the Weighted Residuals is ultimately where the potentials or the fields of the image charges represent the basis functions. Of course the Collocation Method could be replaced by other methods, like that of the least squares. However, the Collocation Method has the big advantage that it avoids potentially tedious integrations. If one attaches all image charges to the surface in form of surface charges, then the boundary element method emerges. This shows that the distinction between the methods is somewhat fuzzy.

The main problem of the various variants of image charge methods is that there is no clear methodological approach for their application. The user needs experience and good intuition for the peculiarities of the particular problem in order to determine the kind, and location of the image charge configuration. Once this is done, the remaining task is simple, particularly when using the collocation method. Consider n image charges, then the Ansatz, *for example* when using potentials, is

$$\varphi = \sum_{k=1}^{n} \varphi_k(\mathbf{r}, \mathbf{r}_k) M_k \; . \tag{8.215}$$

$\mathbf{r}_k$ is the location where the k-th image charge is located, $\varphi_k(\mathbf{r}, \mathbf{r}_k)$ is the potential of the unit charge (unit multipole), and M_k is the charge (multipole moment). If one has Dirichlet boundary condition

$$\varphi = f(\mathbf{r}_A) \tag{8.216}$$

to be fulfilled on the surface A, then one needs to choose (at least) n collocation points $\mathbf{r}_{Ai}$, $(i = 1, ..., n)$. This results in the system of equations

$$\sum_{k=1}^{n} \varphi_k(\mathbf{r}_{Ai}, \mathbf{r}_k) M_k = f(\mathbf{r}_{Ai}), \qquad i = 1, ..., n \tag{8.217}$$

or with

$$\varphi_k(\mathbf{r}_{Ai}, \mathbf{r}_k) = a_{ik}, \qquad f(\mathbf{r}_{Ai}) = f_i$$

$$\mathbf{A} = (a_{ik}), \qquad \mathbf{M} = \begin{pmatrix} M_1 \\ ... \\ M_n \end{pmatrix}, \qquad \mathbf{f} = \begin{pmatrix} f_1 \\ ... \\ f_n \end{pmatrix} \Biggr\}, \tag{8.218}$$

$$\mathbf{A}\mathbf{M} = \mathbf{f} \; . \tag{8.219}$$

The values of M_k have to be calculated from this system of equations. This approximation may not satisfy the given requirements. There are a number of ways to improve the solution. For instance, the number of image charges could be increased. Particularly, repeating the calculation with changed image charge locations may significantly improve the result. Another option is to not only consider the coefficients M_k to be variable, but also the locations r_k and then determine them by means of the collocation method. However, the big disadvantage of this approach is that the system of equations becomes non-linear.

Overall, it is questionable whether the image charge method will be of great significance in the future. Although there are a number of problems for which it is suitable, in general however, in particular the boundary element methods are theoretically better founded and represent a methodologically clearer path for the solution of such problems.

8.10 Monte-Carlo Method

The Monte-Carlo method is particularly suited for stochastic problems, which we will not discuss here. But it may as well be used to solve deterministic problems, *for example* to calculate definite integrals, for the solution of extrema problems, to solve linear systems of equations, etc. In the context of electromagnetic field theory, the Monte-Carlo method can be used to approach boundary value problems involving the Laplace and Poisson equation, as well as, boundary and initial value

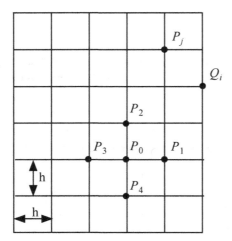

Fig. 8.19

problems of the diffusion equation. Although these problems are of deterministic nature, they are also problems whose solutions coincide with certain expectation values of suitable stochastic processes. An overview on Monte-Carlo methods give the books of Hengartner and Theodorescu or Buslenko and Schreider [30, 31].

Here, we will initially analyze the Dirichlet boundary value problem of the Laplace equation. To simplify, we choose a rectangular area as illustrated in Fig. 8.19. The area is discretized by introducing little squares where the length of the sides is h. Internal points are referenced by P and boundary points by Q. Let a symmetric random walk start at point P_0, as it was described in Sect. 8.5. For the first step, each of the neighboring points is reached by the probability $1/4$. This process is repeated until the boundary point Q_i is reached. This is where the random walk ends, i.e., the particle is absorbed. Let $W(P_j, Q_i) = W_{ji}$ be the probability for a random walk that starts at P_j and ends at Q_i. Then the relation for the points P_0 through P_4 drawn in Fig. 8.19 becomes

$$W(P_0, Q_i) = \frac{1}{4}\{W(P_1, Q_i) + W(P_2, Q_i) + W(P_3, Q_i) + W(P_4, Q_i)\} . \quad (8.220)$$

Furthermore, it must of course be:

$$W(Q_i, Q_j) = \delta_{ij}, \qquad W(Q_i, P_j) = 0 . \quad (8.221)$$

The reason is that a particle is absorbed at a boundary point Q_i. Now we define the quantity

$$F(P_j) = \sum_{i=1}^{s} W(P_j, Q_i)\varphi(Q_i) , \quad (8.222)$$

where s represents the total number of boundary points. Then, it follows from (8.22) that

$$F(P_0) = \frac{1}{4}\{F(P_1) + F(P_2) + F(P_3) + F(P_4)\} \qquad (8.223)$$

and

$$F(Q_j) = \sum_{i=1}^{s} W(Q_j, Q_i)\varphi(Q_i) = \sum_{i=1}^{s} \delta_{ji}\varphi(Q_i) = \varphi(Q_j) . \qquad (8.224)$$

These two equations establish the relation between the random walk and the boundary value problem where F has to be identified with the potential. The result is identical to the one obtained by applying finite differences. Eq. (8.223) corresponds to the five-point formula (8.159) while (8.224) represents the corresponding boundary conditions. On the other hand, one sees that the quantity F, the potential, can be interpreted in a statistical way, that is, we could say it is a result of rolling a dice. Let us start many random walks at every internal point, *for example* at P_j. They terminate with different probabilities at the different boundary points Q_i. which, in case of the Monte-Carlo method, are determined by a computer almost experimentally by simulation of the random walk process. By eq. (8.222), the potential at the point P_j is nothing else than the average

$$\boxed{\varphi(P_j) = \sum_{i=1}^{s} W(P_j, Q_i)\varphi(Q_i)} , \qquad (8.225)$$

obtained by using these probabilities. We may phrase this in a slightly different manner. We start many random walks at P_j and average the potentials of all boundary points that were thereby reached. The i-th random walk ends at Q_i with φ_i. Then

$$\boxed{\varphi(P_j) = \lim_{n \to \infty} \frac{1}{n} \sum_{i=1}^{n} \varphi_i} . \qquad (8.226)$$

The Poison equation

$$\nabla^2\varphi = -g(\mathbf{r}) \qquad (8.227)$$

may be treated in a similar manner. Here, we shall only provide the result:

$$\boxed{\varphi(P_j) = \sum_{i=1}^{s} W(P_j, Q_i)\varphi(Q_i) + \sum_{k=1}^{r} W(P_j, P_k)\bar{g}(P_k) + \bar{g}(P_j)} . \qquad (8.228)$$

r represents the number of all internal points, $W(P_j, P_k)$ is the probability that a particle starting at point P_j passes through the point P_k, and

$$\bar{g} = \frac{h^2}{4}g(\mathbf{r}) . \qquad (8.229)$$

The values $\bar{g}$ existing at the points passed, are averaged by the corresponding probabilities and are responsible for the second sum in eq. (8.228). This sum also contains the addend $W(P_j, P_j)\bar{g}(P_j)$. This term is necessary because the particle may pass through its starting point again. The additional term $\bar{g}(P_j)$ can be interpreted as the product of the probability 1 with $\bar{g}(P_j)$. The fact that the particle originates at point P_j implies that it certainly passes this point. This should not be considered again in $W(P_j, P_j)$. One might cancel the term $\bar{g}(P_j)$, which would require to replace $W(P_j, P_j)$ by $W(P_j, P_j) + 1$. If the potential vanishes on the boundary, then

$$\varphi(P_j) = \sum_{k=1}^{r} W(P_j, P_k)\bar{g}(P_k) + \bar{g}(P_j)$$. \hfill (8.230)

In this case, we may write

$$\varphi(P_j) = \lim_{n \to \infty} \frac{1}{n} \sum_{i=1}^{n} \bar{g}_i$$, \hfill (8.231)

that is, one averages the $\bar{g}$ over all internal points passed during very many random walks, including the starting point. In contrast to eq. (8.226) where the index i characterized subsequent random walks, here it addresses all passed internal points.

The results are related to the integral representations of the corresponding Green's functions. Comparison of eq. (8.225) and (3.94) reveals that $W(P_j, Q_i)$ – with the exception of constant factors – is the matrix that results from discretization of $-\partial G / \partial n$. Eq. (8.230) shows that $W(P_j, Q_k) + \delta_{ik}$ is just the discretized from of G in eq. (3.93). The Monte-Carlo method serves here, for the most part, in approximating the Green's functions by matrixes whose elements can be regarded as probabilities, and which can be either calculated or determined experimentally, namely by simulation of the random walk by a computer.

The solution of the diffusion equation is obtained in similar fashion. It shall not be discussed here. The interested reader is referred to the literature [30].

Two simple **examples** shall serve as illustration. The Dirichlet boundary value problem of the Laplace equation for the area given in Fig. 8.6, and the shown discretization with only three internal points shall be presented first. The potential at the top boundary shall be given as $\varphi = 100$ (dimensionless). For the remainder of the boundary it shall be $\varphi = 0$. From eq. (8.225) and with the probabilities (8.155) we obtain

$$\varphi_1 = \varphi_3 = \left(\frac{15}{56} + \frac{4}{56} + \frac{1}{56} \right) 100 = \frac{2000}{56} = \frac{250}{7}$$

$$\varphi_2 = \left(\frac{4}{56} + \frac{16}{56} + \frac{4}{56} \right) 100 = \frac{2400}{56} = \frac{300}{7}$$.

One can easily verify that using finite differences yields the same result.

Our second **example** is the one-dimensional problem shown in Fig. 8.5 where

$$\nabla^2 \varphi = x^2, \quad 0 \le x \le 1$$

$$\varphi(0) = \varphi_0, \quad \varphi(1) = \varphi_4$$.

For this one-dimensional case, one writes

a

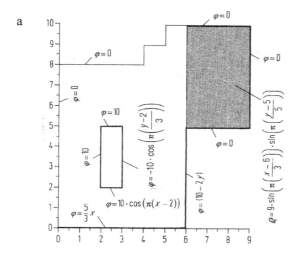

b

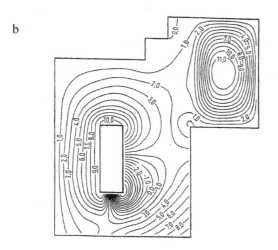

Fig. 8.20

$$h = \frac{1}{4}, \quad \bar{g} = g\frac{h^2}{2} = \frac{g}{32} = -\frac{x^2}{32}.$$

Eq.(8.228), together with the probabilities (8.154) give

$$\varphi_1 = \frac{3}{4}\varphi_0 + \frac{1}{4}\varphi_4 - \left(\frac{1}{2} \cdot \frac{1}{32 \cdot 16} + 1 \cdot \frac{4}{32 \cdot 16} + \frac{1}{2} \cdot \frac{9}{32 \cdot 16}\right) - \frac{1}{32 \cdot 16}$$

$$\varphi_2 = \frac{1}{2}\varphi_0 + \frac{1}{2}\varphi_4 - \left(1 \cdot \frac{1}{32 \cdot 16} + 1 \cdot \frac{4}{32 \cdot 16} + 1 \cdot \frac{9}{32 \cdot 16}\right) - \frac{4}{32 \cdot 16}$$

$$\varphi_3 = \frac{1}{4}\varphi_0 + \frac{3}{4}\varphi_4 - \left(\frac{1}{2} \cdot \frac{1}{32 \cdot 16} + 1 \cdot \frac{4}{32 \cdot 16} + \frac{1}{2} \cdot \frac{9}{32 \cdot 16}\right) - \frac{9}{32 \cdot 16}$$

or

$$\varphi_1 = \frac{3}{4}\varphi_0 + \frac{1}{4}\varphi_4 - \frac{5}{4^4}$$

$$\varphi_2 = \frac{1}{2}\varphi_0 + \frac{1}{2}\varphi_4 - \frac{9}{4^4}$$

$$\varphi_3 = \frac{1}{4}\varphi_0 + \frac{3}{4}\varphi_4 - \frac{9}{4^4} \ .$$

Here again, the finite differences yield the same result. Of course, when applying the Monte-Carlo method to an actual problem, then the necessary probabilities are not calculated but determined by means of random numbers, fed into a computer simulation. This requires many random walks to determine the necessary probabilities with sufficient reliability. The Monte-Carlo method is attractive but the necessary effort is significant because the probabilities need to be determined for every internal point.

Fig. 8.20 shows a plane Dirichlet boundary value problem that was calculated by means of the Monte-Carlo method (Fig. 8.20a: problem definition; Fig. 8.20b equipotential lines).

A   Appendices

A.1   Electromagnetic Field Theory and Photon Rest Mass

A.1.1   Introduction

Maxwell's equations constitute a significant foundation for science and technology. To continuously discuss them and question if, in light of new discoveries, these equations require modification or if they remain entirely accurate and thus valid, is rather natural. This leads to a more thorough understanding of the preconditions and peculiarities of such familiar theories, which often times are taken as too self-evident. Moreover, these discussions emphasize that electromagnetic Field theory is not an isolated body of knowledge, but closely intertwined with all of physical science.

At first sight, it may sound bizarre to relate the question of the exact validity of Coulomb's law of electrostatics, to the question on whether the rest mass of light quanta identically vanishes or not. This Appendix shall be dedicated to this issue, and its implications for electromagnetic field theory.

We start with the theorem of conservation of energy from classical mechanics. It states that the total energy W of a particle is constant:

$$W = \frac{1}{2}mv^2 + U = \frac{p^2}{2m} + U = \text{const.} \tag{A.1.1}$$

$(1/2)mv^2 = p^2/2m$ is the kinetic, U the potential energy of the particle which travels in a "conservative" force field, where m is its mass, v its velocity, and p its momentum. In Quantum Mechanics, physical quantities are replaced by operators. This transforms the energy theorem into Schrödinger's equation.

$$W \Rightarrow i\hbar\frac{\partial}{\partial t} : \qquad\qquad \text{operator of the total energy} \tag{A.1.2}$$

$$\mathbf{p} \Rightarrow -i\hbar\nabla : \qquad\qquad \text{operator of the momentum} \tag{A.1.3}$$

$$\frac{p^2}{2m} \Rightarrow -\frac{\hbar^2\nabla^2}{2m} = -\frac{\hbar^2}{2m}\Delta : \qquad \text{operator of the kinetic energy} \tag{A.1.4}$$

thus

$$i\hbar\frac{\partial}{\partial t} = -\frac{\hbar^2\nabla^2}{2m} + U \tag{A.1.5}$$

Eq. (A.1.5) constitutes the energy theorem in operator form. Applying it onto a function ψ gives the Schrödinger equation

$$i\hbar\frac{\partial\psi}{\partial t} = -\frac{\hbar^2}{2m}\nabla^2\psi + U\psi . \tag{A.1.6}$$

In case of fast ("relativistic") particles, when the rest mass is m_0, one has:

G. Lehner, *Electromagnetic Field Theory for Engineers and Physicists*,
DOI 10.1007/978-3-540-76306-2, © Springer-Verlag Berlin Heidelberg 2010

$$W^2 = c^2 p^2 + m_0^2 c^4 \ . \tag{A.1.7}$$

The associated wave equation, we could call it the relativistic Schrödinger equation, is the so-called "Klein-Gordon equation". It is obtained from (A.1.7) in the same manner as one finds the Schrödinger equation (A.1.6):

$$\left(i\hbar\frac{\partial}{\partial t}\right)^2 = c^2(i\hbar\nabla)^2 + m_0^2 c^4$$

or

$$-\hbar^2\frac{\partial^2\psi}{\partial t^2} = -c^2\hbar^2\nabla^2\psi + m_0^2 c^4\psi$$

$$\boxed{\nabla^2\psi - \frac{1}{c^2}\frac{\partial^2\psi}{\partial t^2} - \left(\frac{m_0 c}{\hbar}\right)^2\psi = 0} \ . \tag{A.1.8}$$

For $m_0 = 0$ (e.g., for photons when making the usual assumption that their rest mass vanishes) this becomes

$$\nabla^2\psi - \frac{1}{c^2}\frac{\partial^2\psi}{\partial t^2} = 0 \ . \tag{A.1.9}$$

One thereby obtains the wave equation known from electromagnetic field theory. It has to be regarded as a special case of the Klein-Gordon equation (A.1.8), which conversely, is the generalization of the wave equation for particles whose rest mass does not vanish. There is a restriction one has to mention, namely that this is valid only for particles with integer number spin (Bosons), but not for those with half integer spin (Fermions). Another equation, also traceable to the Klein-Gordon equation applies to Fermions. This is the Dirac equation, which will not be discussed here.

Using the Compton wave length λ_c and introducing the abbreviation:

$$\kappa = \frac{m_0 c}{\hbar} = \frac{2\pi}{\lambda_c} \ , \tag{A.1.10}$$

lets one write the time-independent Klein-Gordon equation in the following way:

$$\nabla^2\psi - \kappa^2\psi = 0 \ . \tag{A.1.11}$$

Its simplest spherically symmetric solution is the so-called Yukawa potential.

$$\psi = \frac{C}{r}e^{-\kappa r} \tag{A.1.12}$$

This can easily be proven by substituting the radial part of the Laplace operator

$$\nabla^2 = \frac{1}{r^2}\frac{\partial}{\partial r}r^2\frac{\partial}{\partial r} \ . \tag{A.1.13}$$

Yukawa introduced this potential, named in his honor, into the theory of nuclear forces. From their short reach, he predicted the existence of a nuclear field whose quanta possess a rest mass of $m_0 \approx 200 m_e$ (m_e is the rest mass of electrons). He

a)   b)

Fig. A.1.1

called those particles Mesons, which was a brilliant prediction. The particles he referred to are now know as π-Mesons (after an initial confusion with the μ-Mesons).

All this applies in principle to photons as well. Should it turn out that these have a rest mass that differs from zero (even by an arbitrarily small amount), then the potential of an electric point charge Q would not be the Coulomb potential

$$\varphi = \frac{Q}{4\pi\varepsilon_0 r} \ , \tag{A.1.14}$$

but the Yukawa potential

$$\varphi = \frac{Q}{4\pi\varepsilon_0 r} e^{-\kappa r} \ . \tag{A.1.15}$$

Consequently, Maxwell's equations would need to change in a significant way.

The Yukawa potential is not related to the formally similar potential, known in classical field theory as the Debye-Hückel potential:

$$\varphi = \frac{Q}{4\pi\varepsilon_0 r} e^{-r/d} \tag{A.1.16}$$

This potential is a result of classical field theory. The exponential decay is not related to the charge Q itself, but results from volume charges of the opposite sign, which are distributed in a spherically symmetric manner around Q and thereby shield the field of Q more and more as the distance to Q increases (hence the term shielded Coulomb potential, Sect. 2.3.2).

The Coulomb potential yields

$$\nabla \bullet \mathbf{D} = \rho \ . \tag{A.1.17}$$

The Yukawa potential for photons with $m_0 \neq 0$, $i.e.$, for $\kappa \neq 0$ yields

$$\nabla \bullet \mathbf{D} = \rho - \varepsilon_0 \kappa^2 \varphi \ . \tag{A.1.18}$$

Fig. A.1.1a illustrates the Coulomb field, while Fig. A.1.1b shows the Yukawa field. All field lines extend to infinity in case of the Coulomb field. In case of the Yukawa field, the number of field lines diminishes as the distance to the charge increases, even though there are no charges at those end points (the Debye-Hückel field would look the same way, but the field lines would end at charges).

If this were so, then Maxwell's equation

$$\nabla\times\mathbf{H} = \mathbf{g} + \frac{\partial \mathbf{D}}{\partial t} \qquad\qquad\qquad (A.1.19)$$

required modification because otherwise, this would be in conflict with the charge conservation. We introduce the vector potential $\mathbf{A}$, which by (7.183) allows to express

$$\mathbf{B} = \nabla\times\mathbf{A} \ , \qquad\qquad\qquad (A.1.20)$$

and apply the Lorentz gauge (7.186)

$$\nabla\bullet\mathbf{A} + \mu_0\varepsilon_0 \frac{\partial \varphi}{\partial t} = 0 \ , \qquad\qquad (A.1.21)$$

then, instead of (A.1.19), we now obtain

$$\nabla\times\mathbf{H} = \mathbf{g} + \frac{\partial \mathbf{D}}{\partial t} - \frac{\kappa^2}{\mu_0}\mathbf{A} \ . \qquad\qquad (A.1.22)$$

One can easily see that this satisfies the charge conservation. Taking the divergence of (A.1.22) gives

$$\nabla\bullet(\nabla\times\mathbf{H}) = \nabla\bullet\mathbf{g} + \frac{\partial}{\partial t}\nabla\bullet\mathbf{D} - \frac{\kappa^2}{\mu_0}\nabla\bullet\mathbf{A} = 0$$

and now applying (A.1.18), together with (A.1.21) gives

$$\nabla\bullet\mathbf{g} + \frac{\partial}{\partial t}(\rho - \varepsilon_0\kappa^2\varphi) - \frac{\kappa^2}{\mu_0}\left(-\mu_0\varepsilon_0\frac{\partial\varphi}{\partial t}\right) = 0,$$

that is,

$$\nabla\bullet\mathbf{g} + \frac{\partial\rho}{\partial t} = 0 \ .$$

The other of Maxwell's equations remain unchanged. We now have the following system of equations, which is called the *Proca-equations*:

$$\nabla\times\mathbf{E} = -\frac{\partial\mathbf{B}}{\partial t} \qquad\qquad\qquad (A.1.23)$$

$$\nabla\times\mathbf{H} = \mathbf{g} + \frac{\partial\mathbf{D}}{\partial t} - \frac{\kappa^2}{\mu_0}\mathbf{A} \qquad\qquad (A.1.24)$$

$$\nabla\bullet\mathbf{D} = \rho - \varepsilon_0\kappa^2\varphi \qquad\qquad\qquad (A.1.25)$$

$$\nabla\bullet\mathbf{B} = 0 \qquad\qquad\qquad\qquad (A.1.26)$$

Of course, for $\kappa = 0$ this results again in Maxwell's equations.

First, a remarkable fact is that the Proca equations, besides the usual fields $\mathbf{E}$, $\mathbf{D}$, $\mathbf{B}$, $\mathbf{H}$, the volume charge density ρ, and the current density $\mathbf{g}$, also contain the potentials $\mathbf{A}$ and φ. These potentials actually are part of the theory (for finite photon rest mass) and not auxiliary fields, introduced later to simplify the mathematical calculations. They are real fields that can not be eliminated. On a side note: Ultimately, they are real also in the context of Maxwell's theory, as the experiment of Bohm-Aharonov reveals (see Appendix A.3).

A consequence of eqs. (A.1.23) through (A.1.26) is that the potentials occur in Poynting's theorem, which one obtains in its generalized form from these eqs.:

$$\nabla \bullet \left(\mathbf{E} \times \mathbf{H} + \frac{\kappa^2}{\mu_0} \varphi \mathbf{A} \right) + \frac{\partial}{\partial t} \left[\frac{B^2}{2\mu_0} + \frac{\varepsilon_0 E^2}{2} + \kappa^2 \left(\frac{\varepsilon_0 \varphi^2}{2} + \frac{A^2}{2\mu_0} \right) \right] = -\mathbf{E} \bullet \mathbf{g} \ .$$

(A.1.27)

Using

$$\mathbf{S} = \mathbf{E} \times \mathbf{H} + \frac{\kappa^2}{\mu_0} \varphi \mathbf{A}$$

(A.1.28)

and the energy density

$$w = \frac{B^2}{2\mu_0} + \frac{\varepsilon_0 E^2}{2} + \kappa^2 \left(\frac{\varepsilon_0 \varphi^2}{2} + \frac{A^2}{2\mu_0} \right)$$

(A.1.29)

gives

$$\nabla \bullet \mathbf{S} + \frac{\partial}{\partial t} w = -\mathbf{E} \bullet \mathbf{g} \ .$$

(A.1.30)

Both, the Poynting vector $\mathbf{S}$, as well as the energy density, contain additional terms that include the potentials φ and $\mathbf{A}$. And again, for $\kappa = 0$, we obtain the classical results, which we have discussed in Sect. 2.14.

Substituting the relations

$$\mathbf{E} = -\nabla \varphi - \frac{\partial \mathbf{A}}{\partial t}$$

(A.1.31)

and

$$\mathbf{B} = \nabla \times \mathbf{A}$$

(A.1.32)

into the Proca equations yield the inhomogeneous wave equations in the following form.

$$\nabla^2 \varphi - \frac{1}{c^2} \frac{\partial^2 \varphi}{\partial t^2} - \kappa^2 \varphi = -\frac{\rho}{\varepsilon_0}$$

(A.1.33)

$$\nabla^2 \mathbf{A} - \frac{1}{c^2} \frac{\partial^2 \mathbf{A}}{\partial t^2} - \kappa^2 \mathbf{A} = -\mu_0 \mathbf{g}$$

(A.1.34)

They differ from the equations of the classical theory by the additional terms $-\kappa^2 \varphi$ and $-\kappa^2 \mathbf{A}$. Then, for the static case we have for φ

$$\nabla^2 \varphi - \kappa^2 \varphi = -\frac{\rho}{\varepsilon_0} \ ,$$

(A.1.35)

from which for a point charge

$$\rho = Q\delta(\mathbf{r}) , \tag{A.1.36}$$

we find the Yukawa potential

$$\varphi = \frac{Q}{4\pi\varepsilon_0 r} e^{-\kappa r} , \tag{A.1.37}$$

from which we started. For an arbitrary distribution of volume charges $\rho(\mathbf{r})$ we find

$$\varphi = \int \frac{\rho(\mathbf{r}')}{4\pi\varepsilon_0|\mathbf{r}-\mathbf{r}'|} e^{-\kappa|\mathbf{r}-\mathbf{r}'|} d\tau' . \tag{A.1.38}$$

Compare this to the classical result (2.20)

$$\varphi = \int \frac{\rho(\mathbf{r}')}{4\pi\varepsilon_0|\mathbf{r}-\mathbf{r}'|} d\tau' , \tag{A.1.39}$$

which agrees with the Yukawa potential for the special case of $\kappa = 0$.

All this has peculiar ramifications for field theory, which can be demonstrated via several simple examples. Furthermore, comparison of these theoretical consequences with experimental results allows to gain further insight into the nature of light quanta.

A.1.2   Examples

A.1.2.1 Uniformly Charged Spherical Surface

Consider a simple electrostatic problem, the field of a uniformly charged spherical surface. The solution shall be just given, not derived. That this is the correct solution can easily be verified. The fields are purely radial (Fig. A.1.2) and the electric field is

$$E_{ir} = \frac{\kappa r_0 \sigma_0}{\varepsilon_0} e^{-\kappa r_0} \left[\frac{\kappa r \cosh(\kappa r) - \sinh(\kappa r)}{(\kappa r)^2} \right] , \tag{A.1.40}$$

$$E_{er} = \frac{\kappa r_0 \sigma_0}{\varepsilon_0} e^{-\kappa r} \sinh(\kappa r_0) \left[\frac{1 + \kappa r}{(\kappa r)^2} \right] \tag{A.1.41}$$

and the potentials are

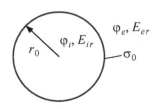

Fig. A.1.2

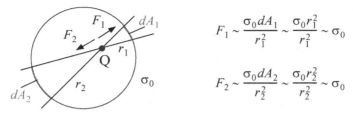

$$F_1 \sim \frac{\sigma_0 dA_1}{r_1^2} \sim \frac{\sigma_0 r_1^2}{r_1^2} \sim \sigma_0$$

$$F_2 \sim \frac{\sigma_0 dA_2}{r_2^2} \sim \frac{\sigma_0 r_2^2}{r_2^2} \sim \sigma_0$$

Fig. A.1.3

$$\varphi_i = \frac{r_0 \sigma_0}{\varepsilon_0} \cdot \frac{e^{-\kappa r_0} \sinh(\kappa r)}{\kappa r}, \tag{A.1.42}$$

$$\varphi_e = \frac{r_0 \sigma_0}{\varepsilon_0} \cdot \frac{e^{-\kappa r} \sinh(\kappa r_0)}{\kappa r}. \tag{A.1.43}$$

What is extraordinary here, is that φ_i is not constant and therefore $E_i \neq 0$. Essentially, this is evident: It is a unique characteristic of the Coulomb field, *i.e.*, of the quadratic distance relation, that there are no forces inside of a uniformly charged spherical surface (this is also true for gravity). Any other field law does not result in vanishing forces. Fig. A.1.3 makes this obvious. Search for a field inside of a uniformly charged spherical shell $E_i \neq 0$ is therefore one of the oldest methods to verify Coulomb's law. From today's perspective, we can interpret these experiments as an effort to measure the rest mass of light quanta.

Now, let's consider another electrostatic problem, the field and the capacitance of an ideal plate capacitor.

A.1.2.2 The Plane Capacitor and its Capacitance

According to Fig. A.1.4, we distinguish 5 regions, indicated by the indices 1 through 5. The regions outside the capacitor are 1 and 5. The conducting plates are in the regions 2 and 4. The region 3 is inside, between the plates. The potentials are

$$\varphi_1 = -\frac{V}{2} e^{\kappa(z+b)} \tag{A.1.44}$$

$$\varphi_2 = -\frac{V}{2} \tag{A.1.45}$$

$$\varphi_3 = +\frac{V}{2} \frac{\sinh[\kappa z]}{\sinh[\kappa a]} \tag{A.1.46}$$

$$\varphi_4 = +\frac{V}{2} \tag{A.1.47}$$

$$\varphi_5 = +\frac{V}{2} e^{-\kappa(z-b)}, \tag{A.1.48}$$

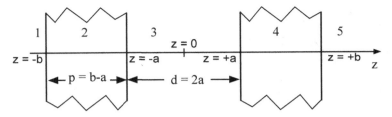

Fig. A.1.4

where V is the voltage on the capacitor. One can easily verify that (A.1.11) and all required boundary conditions are satisfied. The result is peculiar in several respects.

a) φ_1 and φ_5 are location dependent, *i.e.*, there are non-vanishing electric fields in regions 1 and 5.

b)   There is no field inside the plates and the potential is constant. This is chiefly the reason for the need of volume charges, which result from (A.1.25):

$$\rho = \rho_0 = \varepsilon_0 \kappa^2 \varphi = \pm \frac{\varepsilon_0 \kappa^2 V}{2} .$$ (A.1.49)

c)   The gradient of the potential has discontinuities at $z = \pm a$ and $z = \pm b$ [, *i.e.*, the electric field is discontinuous, which means that there are surface charges on both surfaces, the inner and outer surface of the plate. They are:

$$\sigma_{\pm a} = \pm \frac{\varepsilon_0 \kappa V}{2} \coth \frac{\kappa d}{2} ,$$ (A.1.50)

$$\sigma_{\pm b} = \pm \frac{\varepsilon_0 \kappa V}{2} .$$ (A.1.51)

From this, we find the capacitance to be

$$C = \frac{Q}{V} = \frac{A \left[\frac{\varepsilon_0 \kappa V}{2} + \frac{\varepsilon_0 \kappa V}{2} \coth \frac{\kappa d}{2} + d \frac{\varepsilon_0 \kappa^2 V}{2} \right]}{V}$$

$$= C_0 \frac{\kappa d}{2} \left[1 + \kappa d + \coth \frac{\kappa d}{2} \right] ,$$ (A.1.52)

where

$$C_0 = \frac{\varepsilon_0 A}{d} ,$$ (A.1.53)

is the classical capacitance and A the area of the capacitor plates. For $\kappa \to 0$, $\coth \kappa d/2 \to 1/(\kappa d/2)$ and therefore $C \to C_0$. One obtains the same result when calculating the total energy in all 5 regions according to eq. (A.1.29) and then adding

$$\frac{1}{2}CV^2 = W_1 + W_2 + W_3 + W_4 + W_5 = 2W_1 + 2W_2 + W_3 \;, \tag{A.1.54}$$

because of symmetry

$$W_1 = W_5 \tag{A.1.55}$$

and

$$W_2 = W_4 \;. \tag{A.1.56}$$

This simple example makes it obvious, that one must part with the familiar pictures.

A.1.2.3 The Ideal Electric Dipole

As a further example, consider the field of an electric point dipole, located at the origin and pointing in the positive z-direction. Then

$$\varphi = \frac{p\cos\theta}{4\pi\varepsilon_0 r^2}(1 + \kappa r)e^{-\kappa r} \tag{A.1.57}$$

and the related field is

$$E_r = \frac{p\cos\theta}{4\pi\varepsilon_0 r^3}(2 + 2\kappa r + \kappa^2 r^2)e^{-\kappa r}\;, \tag{A.1.58}$$

$$E_\theta = \frac{p\sin\theta}{4\pi\varepsilon_0 r^3}(1 + \kappa r)e^{-\kappa r}\;, \tag{A.1.59}$$

$$E_\varphi = 0\;. \tag{A.1.60}$$

Of course, $\kappa = 0$ yields the classical result (Sect. 2.5, eq. (2.60)):

$$\varphi = \frac{p\cos\theta}{4\pi\varepsilon_0 r^2}\;. \tag{A.1.61}$$

A.1.2.4 The Ideal Magnetic Dipole

Consider a magnetic dipole m at the origin and oriented in the positive z-direction, then we get for the vector potential (in spherical coordinates)

$$\mathbf{A} = \langle 0, 0, A_\varphi\rangle\;, \tag{A.1.62}$$

where

$$A_\varphi = \frac{m\sin\theta}{4\pi r^2}(1 + \kappa r)e^{-\kappa r} \tag{A.1.63}$$

and for the magnetic field

$$B_r = \frac{m\cos\theta}{4\pi r^3}(2 + 2\kappa r)e^{-\kappa r}\;, \tag{A.1.64}$$

$$B_\theta = \frac{m\sin\theta}{4\pi r^3}(1 + \kappa r + \kappa^2 r^2)e^{-\kappa r}, \tag{A.1.65}$$

$$B_\varphi = 0 . \tag{A.1.66}$$

In the classical situation ($\kappa = 0$), it is permissible to calculate the magnetic dipole field outside of the origin from the gradient of a scalar magnetic potential, eq. (5.55)

$$\psi = \frac{m\cos\theta}{4\pi\mu_0 r^2} . \tag{A.1.67}$$

It has the same form as the classical potential of the electric dipole (A.1.61). From this results a magnetic dipole field that has the same form as the electric dipole field. Formally, a distinction between them is not possible. Not so for $\kappa \neq 0$. Both dipole fields, (A.1.58) through (A.1.60) and (A.1.64) through (A.1.66) exhibit different forms. This is necessary, since because of (A.1.24), the magnetic dipole field is not irrotational, even in the static case:

$$\nabla\times\mathbf{H} = -\frac{\kappa^2}{\mu_0}\mathbf{A} . \tag{A.1.68}$$

Therefore, to obtain it from a scalar potential is impossible. Eq. (A.1.68) applies to all magnetostatic fields in current free regions.

The magnetic dipole field can be expressed as superposition

$$\mathbf{B} = \mathbf{B}_1 + \mathbf{B}_2 , \tag{A.1.69}$$

where

$$B_{1r} = \frac{m}{4\pi} \cdot \frac{2\cos\theta}{r^3}\left(1 + \kappa r + \frac{\kappa^2 r^2}{3}\right)e^{-\kappa r} , \tag{A.1.70}$$

$$B_{1\theta} = \frac{m}{4\pi} \cdot \frac{\sin\theta}{r^3}\left(1 + \kappa r + \frac{\kappa^2 r^2}{3}\right)e^{-\kappa r} , \tag{A.1.71}$$

$$B_{1\varphi} = 0 \tag{A.1.72}$$

and

$$B_{2r} = -\frac{m}{4\pi} \cdot \frac{\cos\theta}{r^3} \cdot \frac{2}{3}\kappa^2 r^2 e^{-\kappa r} , \tag{A.1.73}$$

$$B_{2\theta} = +\frac{m}{4\pi} \cdot \frac{\sin\theta}{r^3} \cdot \frac{2}{3}\kappa^2 r^2 e^{-\kappa r} , \tag{A.1.74}$$

$$B_{2\varphi} = 0 . \tag{A.1.75}$$

The part of the field $\mathbf{B}_1$ at the surface of a sphere with a fixed radius r exhibits the characteristics of a classical dipole field, where the dipole moment appears changed by a factor $(1 + \kappa r + \kappa^2 r^2/3)e^{-\kappa r}$. The additional field $\mathbf{B}_2$ has the same magnitude everywhere on the surface of the sphere but has only a component parallel to the axis.

$$B_{2z} = -\frac{m}{4\pi r^3} \cdot \frac{2}{3}\kappa^2 r^2 e^{-\kappa r} . \tag{A.1.76}$$

This fact is one to which we will return later.

A.1.2.5 Plane Waves

Interesting are also questions related to wave propagation. Here also, there are significant discrepancies to the classical theory. We restrict ourselves to plane waves in the infinite homogeneous space without currents or charges. For this case, (A.1.33) and (A.1.34) read

$$\nabla^2\varphi - \frac{1}{c^2}\frac{\partial^2\varphi}{\partial t^2} - \kappa^2\varphi = 0 , \tag{A.1.77}$$

$$\nabla^2\mathbf{A} - \frac{1}{c^2}\frac{\partial^2\mathbf{A}}{\partial t^2} - \kappa^2\mathbf{A} = 0 . \tag{A.1.78}$$

For plane waves travelling in the positive z-direction we have

$$\varphi = \varphi_0 e^{i(\omega t - kz)} , \tag{A.1.79}$$

$$\mathbf{A} = \mathbf{A}_0 e^{i(\omega t - kz)} . \tag{A.1.80}$$

Substituting these in the wave equations (A.1.77) and (A.1.78) yields the dispersion relation

$$(-ik)^2 - \frac{1}{c^2}(i\omega)^2 - \kappa^2 = 0$$

or

$$\frac{\omega^2}{c^2} = k^2 + \kappa^2 . \tag{A.1.81}$$

From a strictly formal point of view, it has the same form as *e.g.* that of a plasma wave

$$\frac{\omega^2}{c^2} = k^2 + \frac{\omega_p^2}{c^2} , \qquad \left(\omega_p^2 = \frac{ne^2}{\varepsilon_0 m_e}\right) . \tag{A.1.82}$$

However, the reason is of an entirely different nature. Because of the Lorentz gauge (A.1.21) the potential is

$$\varphi = \frac{\omega k A_{0z}}{k^2 + \kappa^2} e^{i(\omega t - kz)} \tag{A.1.83}$$

and one obtains the following fields:

$$B_x = +ik A_{0y} e^{i(\omega t - kz)} , \tag{A.1.84}$$

$$B_y = -ik A_{0x} e^{i(\omega t - kz)} , \tag{A.1.85}$$

$$B_z = 0 \; , \tag{A.1.86}$$

and

$$E_x = -ikA_{0x}e^{i(\omega t - kz)} \; , \tag{A.1.87}$$

$$E_y = -ikA_{0y}e^{i(\omega t - kz)} \; , \tag{A.1.88}$$

$$E_z = -i\frac{\omega\kappa^2}{k^2 + \kappa^2}A_{0z}e^{i(\omega t - kz)} \; . \tag{A.1.89}$$

This is a remarkable result. We obtain three (instead of the classically two) independent solutions.

a) If only $A_{0x} \neq 0$: We obtain one linearly polarized TEM wave (with E_x and B_y) as in the classical theory.

b) If only $A_{0y} \neq 0$: We obtain a second linearly polarized TEM wave (with E_y and B_x) as in the classical theory.

c) If only $A_{0z} \neq 0$: There is only an electric field in propagation direction but no magnetic field at all. This is a longitudinal wave, which does not exist in this form within the classical theory. The classical theory allows for longitudinal waves only if there are volume charges. Classically we have

$$\nabla \bullet \mathbf{D} = \rho$$

and with $\rho = 0$ we obtain

$$\nabla \bullet \mathbf{D} = 0 \; .$$

For a plane wave

$$\mathbf{D} = \mathbf{D}_0 e^{i(\omega t - \mathbf{k} \bullet \mathbf{r})}$$

and also

$$\nabla \bullet \mathbf{D} = -i\mathbf{k} \bullet \mathbf{D} = 0 \; ,$$

i.e., $\mathbf{k}$ and $\mathbf{D}$ are perpendicular to each other. Likewise, because of $\nabla \bullet \mathbf{B} = 0$, for the magnetic field of a plane wave we have

$$\mathbf{k} \bullet \mathbf{B} = 0 \; .$$

However, because of (A.1.25) this is no longer the case if $\kappa \neq 0$. Now, for $\rho = 0$ we have

$$\nabla \bullet \mathbf{D} = -\varepsilon_0 \kappa^2 \varphi \; ,$$

which allows longitudinal waves to exist.

Furthermore, the dispersion relation (A.1.81) bears remarkable consequences. The phase velocity becomes

$$v_{ph} = \frac{\omega}{k} = \frac{\omega}{\sqrt{\dfrac{\omega^2}{c^2} - \kappa^2}} = \frac{c}{\sqrt{1 - \dfrac{\kappa^2 c^2}{\omega^2}}} \geq c \; . \tag{A.1.90}$$

Conversely, the group velocity is

$$v_G = \frac{d\omega}{dk} = \frac{1}{\dfrac{dk}{d\omega}} = c\sqrt{1 - \frac{\kappa^2 c^2}{\omega^2}} \le c \ . \tag{A.1.91}$$

Furthermore

$$v_G \cdot v_{ph} = c^2 \ . \tag{A.1.92}$$

Contrary to the classical case, we now have dispersion, even for plane waves in lossless, uniform media (*e.g.*, even in a vacuum), and furthermore, it is no longer $v_G = v_{ph}$. Another important point is that the frequency may no longer be arbitrarily small. We obtain the lowest possible frequency from the limit $k \to 0$ (cutoff frequency)

$$\omega_G = c\kappa = \frac{m_0 c^2}{\hbar} \tag{A.1.93}$$

to which we will return. The relations are illustrated in (Fig. A.1.5). After

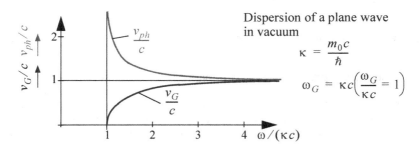

Fig. A.1.5

multiplication by $\hbar^2 c^2$ and using (A.1.10), the dispersion relation (A.1.81) can be written in the form

$$\hbar^2 \omega^2 = c^2 (\hbar^2 k^2 + m_0^2 c^2) \ . \tag{A.1.94}$$

Now, the energy of a light quantum is

$$W = \hbar\omega \tag{A.1.95}$$

and its momentum is

$$\mathbf{p} = \hbar\mathbf{k} \tag{A.1.96}$$

thus, we obtain

$$W^2 = c^2 p^2 + m_0^2 c^4 \ , \tag{A.1.97}$$

which is nothing else than the relativistic energy of a particle with a momentum of p, which was our starting point (eq. (A.1.7)).

The wave equation is basically nothing else than the energy theorem, which is the reason why the dispersion relation stemming from the wave equation leads back to the energy theorem. This is also true in the classical case, $\kappa = 0$. In this case we have

$$\omega = ck \, ,$$

$$\hbar\omega = c\hbar k \, ,$$

$$W = cp \, .$$

We do not intend to increase the number of examples any further. It has been shown that on the basis of the current theory, many familiar concepts from the classical field theory are no longer valid. It remains the question, what the actual rest mass of light quanta really is and whether or not we have to give up Maxwell's equations in favor of the Proca equations. We have to start by stating that based on our current knowledge, we can not definitely answer this question. However, all measurements taken so far, and all interpretations of the known electromagnetic phenomena do not suggest that the light quanta's rest mass m_0 is anything but zero. Every arbitrarily accurate measurement is limited by its own precision in its ability to make statements on upper limits of the rest mass m_0. We will conclude by elaborating some more on this issue.

A.1.3   Measurements and Conclusions

A.1.3.1 Magnetic Fields of Earth and Jupiter

Known from measurements of the Earth's magnetic field (including satellite data) is that, if at all, this field deviates only very little from that of a classical dipole field. It is safe to say, that at the equator, the additional field of $\mathbf{B}_2$ given by (A.1.76) is much smaller than the field $\mathbf{B}_1$ in (A.1.70) through (A.1.72), at least by a factor of $4 \cdot 10^{-3}$. This means that

$$B_2 = \frac{m}{4\pi r^3} \cdot \frac{2\kappa^2 r^2}{3} e^{-\kappa r} \leq 4 \cdot 10^{-3} B_1 \left(\theta = \frac{\pi}{2}\right)$$

$$\approx 4 \cdot 10^{-3} \frac{m}{4\pi r^3} e^{-\kappa r} \quad .$$

Therefore,

$$\frac{2}{3}\kappa^2 r^2 \leq 4 \cdot 10^{-3}$$

and

$$\kappa = \frac{m_0 c}{\hbar} \leq \frac{\sqrt{6 \cdot 10^{-3}}}{r}$$

or

$$m_0 \leq \frac{\sqrt{6 \cdot 10^{-3}}\hbar}{rc} \; ,$$

where r is the earth's radius and with the values

$$\hbar = \frac{h}{2\pi} = \frac{1}{2\pi} 6.6 \cdot 10^{-34} \text{Js} \; ,$$

$$r = 6.4 \cdot 10^6 \text{m} \; ,$$

$$c = 3 \cdot 10^8 \frac{\text{m}}{\text{s}} \; ,$$

this gives

$$m_0 \leq 4.2 \cdot 10^{-51} \text{kg} \tag{A.1.98}$$

that is, should the mass m_0 differ from zero, it may only be as large as this rather tiny value.

An even smaller limit can be found analyzing satellite data from the magnetic field of Jupiter [32], namely

$$m_0 \leq 8 \cdot 10^{-52} \text{kg} \; . \tag{A.1.99}$$

A.1.3.2 Schumann-Resonances

According to (A.1.93), the frequency of electromagnetic, plane waves may not become arbitrarily small but its lower limit depends on m_0. Of course, low frequency goes with long wave length, which prohibits laboratory experiments to answer this question.

However, the earth with the lower limit of the ionosphere constitutes a rather large resonant cavity whose resonant frequencies are know as Schumann-resonances. The lowest resonance is at about 8 Hz. Suppose (A.1.93) were applicable unchanged to resonant cavities (which is not really precise because the dispersion relation for a resonant cavity depends on parameters describing its geometric shape), then we obtain.

$$\omega_G = c\kappa = \frac{m_0 c^2}{\hbar} \leq 2\pi \cdot 8$$

and

$$m_0 \leq \frac{8h}{c^2} \approx 6 \cdot 10^{-50} \text{kg} \tag{A.1.100}$$

A more accurate calculation, taking into account the mentioned geometry factor of the system [33] yields a slightly larger upper limit:

$$m_0 \leq 4 \cdot 10^{-49} \text{kg} \; . \tag{A.1.101}$$

A summary of various values and methods to determine this limit is listed in Fig. A.1.6, which is based on a similar figure by Goldhaber and Nieto [32].

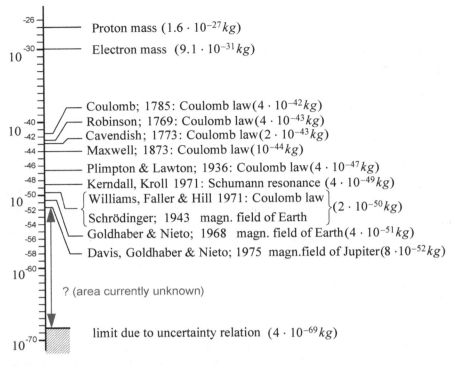

Fig. A.1.6

A.1.3.3 Fundamental Limit -- The Uncertainty Relation

The question of the rest mass of light quanta requires a certain degree of caution. Strictly speaking, the question whether the mass is exactly zero or not is not so useful. To determine the period of a process requires to observe it for a time span of at least one period. A photon rest mass of zero correlates to a cutoff frequency of zero, and its observation requires an infinite amount of time, while the age of the universe is finite. Considering the apparent age of the universe provides the largest possible observation time and thus, at best, we can find

$$m_0 \approx \frac{hf}{c^2} = \frac{h}{c^2 \tau}$$

where τ is the age of the universe and $f = 1/\tau$. Therefore

$$m_0 c^2 \tau = h$$

or

$$W\tau = h \ . \tag{A.1.102}$$

This is basically Heisenberg's uncertainty relation. The statement is, that it makes no sense to measure the mass to an arbitrarily small limit. The smallest mass or mass defect which we may consider is

$$m_0 = \frac{h}{c^2 \tau} = 2.33 \cdot 10^{-68} \text{kg} \ , \tag{A.1.103}$$

whereby we have based this on $\tau = 10^{10}$a or $\tau = 3.15 \cdot 10^{17}$s. More exact measurements of the mass required measurement durations beyond the age of the universe. Therefore, whether m_0 is exactly zero or not, is not the right question to ask. Even though not from a mathematical perspective, nevertheless, a mass of 10^{-68}kg can physically be identified with zero or be at least indistinguishable from zero. However, Fig. A.1.6 reveals that between the current findings and the fundamental limit lies an unexplored area of 17 orders of magnitude. When basing the calculation on the uncertainty principle itself $\Delta W \cdot \Delta t = \hbar$, the limit becomes $m \geq 4 \cdot 10^{-69}$kg, compared to above, a value reduced by a factor of 2π.

The question whether or not it is necessary to modify Maxwell's equations remained unanswered. Nevertheless, one could verify, that within the limits of our current accuracy in taking measurements, we may rest assured that we can trust Maxwell's equations. There is no know phenomenon which Maxwell's equations would not be able to describe with sufficient accuracy.

A.2   Magnetic Monopoles and Maxwell's Equation

A.2.1   Introduction

When looking at Maxwell's equations (1.77), it is easy to see that they are entirely symmetric, as long as there are neither charges nor currents present. Adding charges or currents makes them unsymmetrical (1.72). The reason is the existence of electric charges and currents, but nonexistence of magnetic charges or currents, at least as far as our current knowledge goes.

Conversely, if such magnetic charges or currents exist, Maxwell's equations would be symmetric. When pondering this, the question whether there are no such magnetic charges or currents, is almost inevitable. How does one know that there are not any? The only answer is that so far, no one has discovered them. The symmetry issue is sufficient motivation for a quest for them. There are other arguments as well. They originate from Dirac and have enabled him to make hypothetical predictions of potentially existing magnetic monopoles as integer multiples of an elementary, magnetic charge (that is a fundamental magnetic quantum). Dirac's argument is of quantum mechanical nature and results from the quantization of the angular momentum. Dirac's magnetic quantum is related to the electric charge quantum

$$e = 1.6 \cdot 10^{-19} C .$$

The existence of magnetic charges would thereby also explain why the electric charge is quantized. Both, magnetic and electric charge would be quantized and their quantization would be a result of the quantization of the angular momentum (a purely mechanical quantity!). Indeed, dimensional analysis of the product of magnetic and electric charge reveals that it has the dimensions of angular momentum:

$$F = \frac{Q_e \cdot Q_e'}{4\pi\varepsilon_0 r^2} \quad \text{(Coulomb's law for electric charges)} ,$$

$$F = \frac{Q_m \cdot Q_m'}{4\pi\mu_0 r^2} \quad \text{(Coulomb's law for magnetic charges)} .$$

Therefore, analysis of the dimenions reveals

$$[Q_e \cdot Q_m] = [F\sqrt{\varepsilon_0\mu_0} r^2] = \left[kg \cdot \frac{m}{s^2} \cdot \frac{s}{m} \cdot m^2 \right]$$

$$= \left[kg \cdot \frac{m}{s} \cdot m \right] = [\text{angular momentum}] .$$

Suppose magnetic charges existed, then we might ask how Maxwell's equations be altered. One finds (as explained in Sect. 1.12) that

$$-\nabla \times \mathbf{E} = \mathbf{g}_m + \frac{\partial \mathbf{B}}{\partial t} , \qquad (A.2.1)$$

$$\nabla \times \mathbf{H} = \mathbf{g}_e + \frac{\partial \mathbf{D}}{\partial t} , \qquad (A.2.2)$$

$$\nabla \bullet \mathbf{D} = \rho_e , \qquad (A.2.3)$$

$$\nabla \bullet \mathbf{B} = \rho_m . \qquad (A.2.4)$$

Besides the electric charge density ρ_e, we now have the magnetic charge density ρ_m. $\mathbf{B}$ is no longer source free. Besides the electric current density $\mathbf{g}_e$, we now have the magnetic current density $\mathbf{g}_m$. The need for the occurrence of $\mathbf{g}_m$ in the law of induction (A.2.1) is a result of charge conservation, which we must require for both electric and magnetic charges. Taking the divergence of (A.2.1) and (A.2.2) gives

$$-\nabla \bullet (\nabla \times \mathbf{E}) = 0 = \nabla \bullet \mathbf{g}_m + \nabla \bullet \frac{\partial \mathbf{B}}{\partial t} ,$$

$$\nabla \bullet (\nabla \times \mathbf{H}) = 0 = \nabla \bullet \mathbf{g}_e + \nabla \bullet \frac{\partial \mathbf{D}}{\partial t} .$$

When applying (A.2.3) and (A.2.4), we obtain the continuity equations, which now expresses charge conservation for both types of charges.

$$\nabla \bullet \mathbf{g}_m + \frac{\partial \rho_m}{\partial t} = 0 \ , \tag{A.2.5}$$

$$\nabla \bullet \mathbf{g}_e + \frac{\partial \rho_e}{\partial t} = 0 \ . \tag{A.2.6}$$

However, the ultimate authority in such questions is not our desire for symmetry, but reality. In any case, an answer to the question on the existence of magnetic "monopoles" requires experimental verification. Science does not know any other avenue. Where do we stand here?

An answer requires first to refine our language, which is necessary in order to pose the question with sufficient accuracy and meaning. It will turn out that one has to be very precise in formulating the question, in order to avoid getting lost in chasing bogus problems.

A.2.2 Dual Transformations

One starts by realizing that in the generalized Maxwell equations as of (A.2.1) through (A.2.4), the electric and magnetic fields, charges, and currents are not uniquely defined. If we start from these equations and apply the following *dual transformation*:

$$
\begin{aligned}
\mathbf{E} &= \mathbf{E}'\cos\xi + \mathbf{H}'\sin\xi \cdot f \\
\mathbf{D} &= \mathbf{D}'\cos\xi + \mathbf{B}'\sin\xi \cdot f^{-1} \\
\mathbf{H} &= -\mathbf{E}'\sin\xi \cdot f^{-1} + \mathbf{H}'\cos\xi \\
\mathbf{B} &= -\mathbf{D}'\sin\xi \cdot f + \mathbf{B}'\cos\xi \\
\rho_e &= \rho_e'\cos\xi + \rho_m'\sin\xi \cdot f^{-1} \quad \text{(analogue for } Q_e = \int \rho_e d\tau) \\
\rho_m &= -\rho_e'\sin\xi \cdot f + \rho_m'\cos\xi \quad \text{(analogue for } Q_m = \int \rho_m d\tau) \\
\mathbf{g}_e &= \mathbf{g}_e'\cos\xi + \mathbf{g}_m'\sin\xi \cdot f^{-1} \\
\mathbf{g}_m &= -\mathbf{g}_e'\sin\xi \cdot f + \mathbf{g}_m'\cos\xi
\end{aligned} \tag{A.2.7}
$$

(ξ is an arbitrary, dimensionless parameter, f is a factor with the dimensions of a resistor), then new equations for the transformed quantities $\mathbf{E}'$, $\mathbf{D}'$, $\mathbf{B}'$, $\mathbf{H}'$, ρ', $\mathbf{g}'$ emerge:

$$-\nabla \times \mathbf{E}' = \mathbf{g}'_m + \frac{\partial}{\partial t}\mathbf{B}' \tag{A.2.8}$$

$$+\nabla \times \mathbf{H}' = \mathbf{g}'_e + \frac{\partial}{\partial t}\mathbf{D}' \tag{A.2.9}$$

$$\nabla \bullet \mathbf{D}' = \rho_e' \tag{A.2.10}$$

$$\nabla \cdot \mathbf{B'} = \rho_m' , \tag{A.2.11}$$

i.e., Maxwell's equations again. Consequently, there is a continuum of dual transformations, for which Maxwell's equations are *invariant*. Invariant are also all other important and observable quantities which can be constructed from the fields. These quantities include *for example*, energy density and Poynting vector:

$$\mathbf{D} \bullet \mathbf{E} + \mathbf{H} \bullet \mathbf{B} = \mathbf{E'} \bullet \mathbf{D'} + \mathbf{H'} \bullet \mathbf{B'} , \tag{A.2.12}$$

$$\mathbf{E} \times \mathbf{H} = \mathbf{E'} \times \mathbf{H'} , \tag{A.2.13}$$

as well as the relations for the generalized forces

$$\mathbf{F} = Q_e(\mathbf{E} + \mathbf{v} \times \mathbf{B}) + Q_m(\mathbf{H} - \mathbf{v} \times \mathbf{D}) , \tag{A.2.14}$$

$$\mathbf{F'} = Q_e(\mathbf{E'} + \mathbf{v} \times \mathbf{B'}) + Q_m(\mathbf{H'} - \mathbf{v} \times \mathbf{D'}) \tag{A.2.15}$$

where

$$\mathbf{F} = \mathbf{F'}. \tag{A.2.16}$$

Q_e and Q_m (also Q'_e and Q'_m) are electric and magnetic charges, respectively. They are volume integrals of the respective densities ρ_e and ρ_m (also ρ'_e and ρ'_m), which transform like the charges. The overall conclusion is that there is no measurement or experiment which might force us to adopt one system of quantities over another, and require it to be the only true system. The consequence is that it is not possible to absolutely, positively distinguish between electric and magnetic charges. No one can prevent us from, *for example*, also assigning a magnetic charge to an electron.

The transformation (A.2.7) may become more plausible if we consider some special cases of it.

a)   For $\xi = 0$, $\cos\xi = 1$, $\sin\xi = 0$ we find:

$$\left.\begin{matrix} \mathbf{E} = \mathbf{E'} \\ \mathbf{D} = \mathbf{D'} \\ \mathbf{H} = \mathbf{H'} \\ \mathbf{B} = \mathbf{B'} \end{matrix}\right\} ; \qquad \left.\begin{matrix} \rho_e = \rho_e' \\ \rho_m = \rho_m' \end{matrix}\right\} ; \qquad \left.\begin{matrix} \mathbf{g}_e = \mathbf{g}_e' \\ \mathbf{g}_m = \mathbf{g}_m' \end{matrix}\right\} . \tag{A.2.17}$$

This implies that nothing has happened. This is the identity transformation.

b)   For $\xi = \pm(\pi/2)$, $\cos\xi = 0$, $\sin\xi = \pm 1$ one finds

$$\left.\begin{matrix} \mathbf{E} = \pm\mathbf{H'} \cdot f \\ \mathbf{D} = \pm\mathbf{B'} \cdot f^{-1} \\ \mathbf{H} = \mp\mathbf{E'} \cdot f^{-1} \\ \mathbf{B} = \mp\mathbf{D'} \cdot f \end{matrix}\right\} ; \quad \left.\begin{matrix} \rho_e = \pm\rho_m' \cdot f^{-1} \\ \rho_m = \mp\rho_e' \cdot f \end{matrix}\right\} ; \quad \left.\begin{matrix} \mathbf{g}_e = \pm\mathbf{g}_m' \cdot f^{-1} \\ \mathbf{g}_m = \mp\mathbf{g}_e' \cdot f \end{matrix}\right\} . \tag{A.2.18}$$

This case transforms all electric quantities into magnetic ones, and vice versa. In exactly this sense is the field of an oscillating magnetic dipole, which we have found in eqs. (7.306) and (7.307), the field dual to the oscillating electric dipole of eqs. (7.264) and (7.265). Dual transformations are generally rather useful. They can be used, in conjunction with already obtained solutions to Maxwell's equations, to construct additional solutions by a suitable dual transformation.

c) For $\xi = \pi$, $\cos\xi = -1$, $\sin\xi = 0$ we find:

$$\left.\begin{aligned} \mathbf{E} &= -\mathbf{E}' \\ \mathbf{D} &= -\mathbf{D}' \\ \mathbf{H} &= -\mathbf{H}' \\ \mathbf{B} &= -\mathbf{B}' \end{aligned}\right\} \; ; \qquad \left.\begin{aligned} \rho_e &= -\rho_e' \\ \rho_m &= -\rho_m' \end{aligned}\right\} \; ; \qquad \left.\begin{aligned} \mathbf{g}_e &= -\mathbf{g}_e' \\ \mathbf{g}_m &= -\mathbf{g}_m' \end{aligned}\right\} \; . \qquad \text{(A.2.19)}$$

All quantities have now changed their sign. Obviously, one may, and this is apprehensible, refer to all positive charges as negative ones and vice versa, as long as we allow the fields to change their sign as well.

We conclude that Maxwell's equations do not allow for an absolute distinction between electric and magnetic charges.

To illustrate this more, we consider a hypothetical particle with an electric charge Q_e' as well as a magnetic charge Q_m'. Then we transform:

$$Q_e = Q_e'\cos\xi + Q_m'\sin\xi \cdot f^{-1} \qquad \text{(A.2.20)}$$

$$Q_m = -Q_e'\sin\xi \cdot f + Q_m'\cos\xi \; , \qquad \text{(A.2.21)}$$

and choose ξ such that $Q_m = 0$, i.e.,

$$\tan\xi = \frac{Q_m'}{Q_e' \cdot f} = \frac{\sin\xi}{\cos\xi} \; . \qquad \text{(A.2.22)}$$

This makes

$$Q_e \neq 0 , \qquad \text{(A.2.23)}$$

$$Q_m = 0 \; . \qquad \text{(A.2.24)}$$

We could have chosen

$$\tan\xi = \frac{\sin\xi}{\cos\xi} = -\frac{Q_e' \cdot f}{Q_m'} , \qquad \text{(A.2.25)}$$

which would have resulted in

$$Q_e = 0 , \qquad \text{(A.2.26)}$$

$$Q_m \neq 0 \; . \qquad \text{(A.2.27)}$$

Consequently, one has the choice to pick, *ad libitum*, for one and the same particle, either to possess a purely electric charge, or a purely magnetic charge, or concurrently having a magnetic and an electric charge. This allows us, without limitation, to regard all known particles as having both, electric and magnetic charge, while all of Maxwell's equations apply now in their full symmetric form.

The question arises, if the quest to determine whether magnetic charges exist or not, is really an insignificant pseudo question? It is only so, if we do not pose the question accurately enough.

So far, we have considered a single elementary particle, for which we were able to let $Q_m = 0$ by a suitable choice of a parameter. For multiple, different types of particles, this is concurrently possible only if

$$\frac{Q_m{}'}{Q_e{}'} = f \cdot \tan \xi$$

assumes the same value for all of them. If this is not the case, then the dual transformation does not enable us to let all magnetic charges vanish. In this case, there are fundamental magnetic charges which can not be removed by a transformation. This means that, essentially, the only relevant question is whether the ratio

$$\frac{Q_m{}'}{Q_e{}'}$$

has the same value for all elementary particles or not. This allows one to pose the question for the existence of magnetic monopoles accurately. Our starting point, which was the question of the symmetry of Maxwell's equations has now become almost insignificant. If one chooses so, one can make Maxwell's equations symmetric, regardless of whether there are essential magnetic charges or not.

This clarification is important. It is also an interesting example to illustrate how asking the appropriate questions in a precise manner can be tantamount to not get lost in useless discussions of irrelevant questions.

A.2.3   Properties of Magnetic Monopoles

Now, we return to the afore mentioned Dirac monopole. When studying the interaction between a particle with the charges $Q_e \neq 0$, $Q_m = 0$ and another one with the charges $Q_e = 0$, $Q_m \neq 0$, Dirac developed the hypothesis that it has to be:

$$Q_m Q_e = nh , \qquad\qquad\qquad\qquad (A.2.28)$$

where n is an integer. We have already encountered this product above and found that its dimension is that of angular momentum. From a quantum mechanical perspective, it appears as the appropriate assumption, that this quantity is quantized, just as the angular momentum itself. If

$$Q_e = e \qquad\qquad\qquad\qquad (A.2.29)$$

then it should be

$$Q_m = \frac{nh}{e} . \qquad\qquad\qquad\qquad (A.2.30)$$

This represents a relatively large charge in the following sense. Consider the force between two such charges

$$F_m = \frac{n^2 h^2}{e^2 4\pi\mu_0 r^2} \tag{A.2.31}$$

and compare it to the force between two electrons with the same distance

$$F_e = \frac{e^2}{4\pi\varepsilon_0 r^2} \, . \tag{A.2.32}$$

The ratio between the two forces is

$$\frac{F_m}{F_e} = \frac{n^2 h^2 \varepsilon_0}{e^4 \mu_0} = \frac{n^2}{(2\alpha)^2} = 4692 n^2 \, , \tag{A.2.33}$$

where

$$\alpha = \frac{e^2}{2h}\sqrt{\frac{\mu_0}{\varepsilon_0}} \approx \frac{1}{137} \tag{A.2.34}$$

is an important dimensionless, natural constant, it is the so-called Sommerfeld fine-structure constant. The charge of these Dirac monopoles is large in the sense that even for $n = 1$, the force between such monopoles is about 5000 times larger than the force between two electrons.

Instead, one can characterize the magnetic elementary quantum by the magnetic flux which it creates. This magnetic flux is, because of (A.2.4), equal to the charge, just as in the analogue case of the electric flux:

$$\phi = \frac{nh}{e} = n \cdot 4.135 \cdot 10^{-15} \text{Wb} \, . \tag{A.2.35}$$

The magnetic field B, created thereby in a distance of 1m, can also be used for characterization

$$B = \frac{Q_m}{4\pi\mu_0 r^2} = \frac{nh}{4\pi\mu_0 e r^2} = n \cdot 2.618 \cdot 10^{-10} \text{T} \quad (\text{for } r = 1\text{m}) \, . \tag{A.2.36}$$

A.2.4   The Search for Magnetic Monopoles

All these thoughts have triggered an avid search for these magnetic poles. The questions whether
a)   magnetic charges exist at all, and if
b)   potentially existing magnetic charges obey Dirac's hypothesis,

are still entirely open. Publications in the year 1975 which reported finding magnetic monopoles turned out to be indefensible.

Dirac's hypothesis is usually the basis for the search for these magnetic monopoles. This enables to calculate the effects for which to search. The experiments are of various kinds. Some use accelerators to potentially create

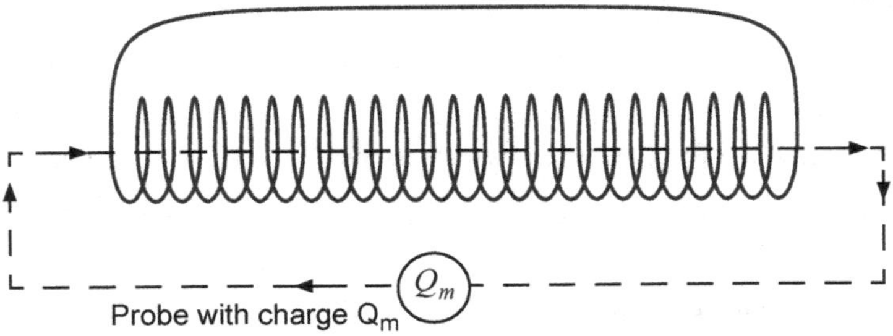

superconductive coil with N_s turns

Probe with charge Q_m
(It is moved N_p times along the indicated path.
e.g. N_p = 400)

Fig. A.2.1

magnetically charged particles, while others study cosmic radiation as a conceivable source. If magnetic charges can be created in probes or are already present there, then the process of detection is, in principle, achieved by pulling them out by means of magnetic fields, and possibly accelerate them with the magnetic field onto a detector.

Rocks from our Moon and from meteorites have been studied, since these were exposed to cosmic radiation for a long time and could have absorbed magnetic monopoles from this radiation. The magnetic monopoles can be detected in the probe when moving it. This creates a magnetic current, which by (A.2.1) creates the electric field

$$\nabla \times \mathbf{E} = -\mathbf{g}_m \, , \tag{A.2.37}$$

just as a magnetic field can be created by an electric current. This method is called the Alvarez method. In practice, a superconducting coil is used for this purpose. The electric field causes a current, which is then detected by the thereby created magnetic flux (Fig. A.2.1).

The relation for this case is (A.2.1) where $\mathbf{E} = 0$. One then has

$$\mathbf{g}_m = -\frac{\partial \mathbf{B}}{\partial t}$$

or

$$\int \mathbf{g}_m \bullet d\mathbf{A} = -\frac{\partial \phi}{\partial t}$$

$$|\phi| = \left| \int (\int \mathbf{g}_m \bullet d\mathbf{A}) dt \right| = \left| N_s N_p Q_m \right| \, , \tag{A.2.38}$$

that is, the magnetic flux induced in the coil is proportional to the magnetic charge Q_m, the number of turns of the coil N_s, and the number of passages of the probe through the coil N_p. In principle, this is a relatively simple method.

One might pose the question regarding the magnetic monopoles not only in conjunction with new types of particles, but also with respect to known particles. Then we may certainly assume that for electrons $Q_m = 0$ – which is obvious from the discussion on dual transformation. This determines all electric and magnetic quantities. Any furhter transformation becomes now impossible. Under these circumstances, the magnetic charges of other particles, *for example*, those of nucleons (protons or neutrons) may be different from zero. If this were the case, then we had to have a thereby caused magnetic field on the earth's surface. Since the field on the earth's surface is less than 1 Gauss, we can estimate that the upper limit of the magnetic charges of nucleons has to be

$$Q_m \leq 10^{-40}\,\text{Wb} \ .$$

This charge would be smaller by a factor $4 \cdot 10^{+25}$ than the smallest possible value allowed by Dirac's hypothesis. Conclusion is that either nucleons do not possess a magnetic charge or Dirac's hypothesis is wrong.

In closing, we note that the positive proof for the existence of magnetic particles has thus far not been achieved, so that this interesting question remains still open.

A.3   On the Significance of Electromagnetic Fields and Potentials (Bohm-Aharonov Effects)

A.3.1   Introduction

The classical field theory describes the force which an electric charge Q_1, located at $\mathbf{r}_1$ exerts on a charge Q_2, located at $\mathbf{r}_2$ by

$$\mathbf{F} = \frac{Q_1 Q_2 (\mathbf{r}_2 - \mathbf{r}_1)}{4\pi\varepsilon_0 |\mathbf{r}_2 - \mathbf{r}_1|^3} \ . \tag{A.3.1}$$

This is Coulomb's law, in which the force seems to represent an action at a distance. This is unsatisfactory and we therefore use a different formulation. Suppose that the charge creates a field $\mathbf{E}(\mathbf{r})$ in the entire space, and this field acts on other charges, where

$$\mathbf{F} = m\frac{d^2}{dt^2}\mathbf{r} = Q\mathbf{E}(\mathbf{r}) \ . \tag{A.3.2}$$

Thus, the action of a particle on another particle is described by the field that the former exhibits at the location of the latter. When also considering the magnetic force, the Lorentz force, we find:

$$\mathbf{F} = m\frac{d^2}{dt^2}\mathbf{r} = Q\mathbf{E}(\mathbf{r}) + Q\mathbf{v} \times \mathbf{B}(\mathbf{r}) \ , \tag{A.3.3}$$

where $\mathbf{B}(\mathbf{r})$ represents the magnetic induction at the location of the particle. Eq. (A.3.3) represents the equation of motion for an arbitrary particle in an arbitrary magnetic field, and thus, also describes the action of the field on the particle in the sense of classical mechanics in a complete manner. This describes a local interaction and not an action at a distance.

In quantum mechanics – for non-relativistic particles – Schrödinger's equation takes the place of the classical equation of motion. It results from the *Hamiltonian* of classical mechanics. For an arbitrary system, the Hamiltonian is a function of the canonical momentum and location coordinates p_k and q_k,

$$H = H(p_k, q_k) \ . \tag{A.3.4}$$

The Hamiltonian is usually referenced by the letter H and should not be confused with the magnetic field. In this terminology, the classical equation of motion is expressed by Hamilton's differential equations

$$\frac{dp_k}{dt} = -\frac{\partial H}{\partial q_k} \tag{A.3.5}$$

$$\frac{dq_k}{dt} = \frac{\partial H}{\partial p_k} \ . \tag{A.3.6}$$

The Hamiltonian for a particle of mass m, located in a force field with the potential $U(x_1, x_2, x_3)$ is

$$H = \frac{p^2}{2m} + U = \frac{p_1^2 + p_2^2 + p_3^2}{2m} + U(x_1, x_2, x_3) \ , \tag{A.3.7}$$

where the

$$p_i = m\dot{x}_i \tag{A.3.8}$$

are the components of the momentum. Then

$$\frac{dp_i}{dt} = -\frac{\partial U_i}{\partial x_i} \tag{A.3.9}$$

and

$$\frac{dx_i}{dt} = \frac{p_i}{m} \tag{A.3.10}$$

thus

$$m\frac{d^2}{dt^2}x_i = -\frac{\partial U}{\partial x_i} \ , \tag{A.3.11}$$

i.e. the classical (Newton's) equation of the motion.

For a particle in an electromagnetic field we have

$$H = \frac{(\mathbf{p} - Q\mathbf{A})^2}{2m} + Q\varphi \ , \tag{A.3.12}$$

where $\mathbf{A}$ and φ are the electromagnetic potentials which allow to calculate $\mathbf{E}$ and $\mathbf{B}$:

$$\mathbf{E} = -\nabla\varphi - \frac{\partial \mathbf{A}}{\partial t} \tag{A.3.13}$$

$$\mathbf{B} = \nabla \times \mathbf{A} \tag{A.3.14}$$

$\mathbf{p}$ is the canonical momentum,

$$\mathbf{p} = m\mathbf{v} + Q\mathbf{A} \ , \tag{A.3.15}$$

which should not be confused with the ordinary momentum $m\mathbf{v}$, into which it transforms in the limit $A = 0$. In this case, Hamilton's differential equations result in the afore mentioned equation of motion (A.3.3), which we will not prove here.

Replacing the physical quantities by operators, as we have done before in Sect. A.1, we obtain:

$$H \Rightarrow \hat{H} = i\hbar \frac{\partial}{\partial t} \tag{A.3.16}$$

$$\mathbf{p} \Rightarrow \hat{\mathbf{p}} = -i\hbar \nabla \ . \tag{A.3.17}$$

This allows to write Schrödinger's equation in the following form

$$i\hbar \frac{\partial \psi}{\partial t} = \hat{H}(-i\hbar \nabla, \mathbf{q})\psi \ . \tag{A.3.18}$$

In particular, for a particle in an electromagnetic field, we obtain from (A.3.12)

$$i\hbar \frac{\partial \psi}{\partial t} = \left[\frac{(-i\hbar \nabla - Q\mathbf{A})^2}{2m} + Q\varphi \right] \psi \ . \tag{A.3.19}$$

Particular care is advised when calculating the square because the momentum operator and the location operator, or the location dependent quantities, respectively (here $\mathbf{A} = \mathbf{A}(\mathbf{r})$), do not commute. Written more explicitly, we obtain:

$$i\hbar \frac{\partial \psi}{\partial t} = \left[\frac{-\hbar^2 \nabla^2 + i\hbar Q\mathbf{A} \bullet \nabla + i\hbar Q\nabla \bullet \mathbf{A} + Q^2 A^2}{2m} + Q\varphi \right] \psi \ . \tag{A.3.20}$$

A.3.2   The Role of Fields and Potentials

The equation of motion which applies to classical electromagnetic field theory is

$$\boxed{\mathbf{F} = m\frac{d^2\mathbf{r}}{dt^2} = Q\mathbf{E} + Q\mathbf{v} \times \mathbf{B}} \ . \tag{A.3.21}$$

Its counterpart in quantum mechanics is the Schrödinger equation

$$i\hbar \frac{d\psi}{dt} = \left[\frac{(-i\hbar \nabla - Q\mathbf{A})^2}{2m} + Q\varphi \right] \psi \ . \tag{A.3.22}$$

One equation contains the fields $\mathbf{E}$ and $\mathbf{B}$, while the other uses the potentials $\mathbf{A}$ and φ. Within the classical theory, the potentials were introduced formally as auxiliary quantities, whose task was to simplify the solution of Maxwell's equations, and they live up to their task. Two of the four of Maxwell's equations are automatically solved by (A.3.13) and (A.3.14), while the other two result in the inhomogeneous wave equations (7.187) and (7.188). It is easily possible to eliminate the fields from the equation of motion (A.3.21) and replace them with the potentials. The question arises, if it is also possible to eliminate the potentials in Schrödinger's equation in favor of the actual fields **E** and **B**. This is not possible, at least not without difficulties. If one insists to do so, the best place to start are the inhomogeneous wave equations (7.187) and (7.188). Their solutions are the retarded potentials (7.195) and (7.196). Furthermore, one may express ρ and **g** by **E** and **B**.

$$\rho = \varepsilon_0 \nabla \bullet \mathbf{E}$$

and

$$\mathbf{g} = \frac{1}{\mu_0} \nabla \times \mathbf{B} - \varepsilon_0 \frac{\partial \mathbf{E}}{\partial t} \ .$$

This allows to write the retarded potentials (7.195), (7.196) in the following form

$$\varphi(\mathbf{r}, t) = \int \frac{\nabla \bullet \mathbf{E}\left(\mathbf{r}', t - \frac{|\mathbf{r}-\mathbf{r}'|}{c}\right)}{4\pi|\mathbf{r}-\mathbf{r}'|} \, d\tau' \tag{A.3.23}$$

$$\mathbf{A}(\mathbf{r}, t) = \int \frac{\nabla \times \mathbf{B}\left(\mathbf{r}', t - \frac{|\mathbf{r}-\mathbf{r}'|}{c}\right) - \frac{1}{c^2}\frac{\partial}{\partial t}\mathbf{E}\left(\mathbf{r}', t - \frac{|\mathbf{r}-\mathbf{r}'|}{c}\right)}{4\pi|\mathbf{r}-\mathbf{r}'|} \, d\tau' \ . \tag{A.3.24}$$

It does not appear to be too beneficial to substitute these expressions into the Schrödinger equation. The thereby emerging formula would be very complicated, without carrying any particular benefit. Furthermore, they have the inconvenient property that the wave function $\psi(\mathbf{r})$ does not only depend on the fields $\mathbf{E}(\mathbf{r})$ and $\mathbf{B}(\mathbf{r})$ at their respective location, but also on the integrals of these fields in the entire space, that is, the advantage of the local interaction is lost. This would defeat the original intent for introducing these potentials in the classical theory. In contrast, the local interaction remains in place when we keep the potentials **A** and φ in the Schrödinger equation and regard them as real, not replaceable fields (in contrast to simply auxiliary quantities). Our subsequent discussion will show that this is also necessary based on different, even more fundamental reasons.

For our later use, we will analyze two special cases of Schrödinger's equation (A.3.22).

a)   The first case covers $\varphi = 0$

$$i\hbar\frac{\partial\psi}{\partial t} = \frac{(-i\hbar\nabla - Q\mathbf{A})^2}{2m}\psi \ . \tag{A.3.25}$$

If ψ_0 is a solution of the equation for $\mathbf{A} = 0$,

$$i\hbar\frac{\partial\psi_0}{\partial t} = \frac{(-i\hbar\nabla)^2}{2m}\psi_0 \tag{A.3.26}$$

then

$$\psi = \psi_0\exp\left[i\frac{Q}{\hbar}\int_{r_0}^{r}\mathbf{A}\bullet d\mathbf{s}\right] , \tag{A.3.27}$$

is a solution to (A.3.25) because

$$(-i\hbar\nabla - Q\mathbf{A})\psi_0 \exp\left[i\frac{Q}{\hbar}\int_{r_0}^{r}\mathbf{A}\bullet d\mathbf{s}\right] = \exp\left[i\frac{Q}{\hbar}\int_{r_0}^{r}\mathbf{A}\bullet d\mathbf{s}\right](-i\hbar\nabla)\psi_0$$

and

$$(-i\hbar\nabla - Q\mathbf{A})^2\psi_0 \exp\left[i\frac{Q}{\hbar}\int_{r_0}^{r}\mathbf{A}\bullet d\mathbf{s}\right] = \exp\left[i\frac{Q}{\hbar}\int_{r_0}^{r}\mathbf{A}\bullet d\mathbf{s}\right](-i\hbar\nabla)^2\psi_0 \ .$$

b) The second case covers $\mathbf{A} = 0$ and $\varphi = \varphi(t)$ (*i.e.* φ is independent of the location and thus, $\mathbf{E} = 0$):

$$i\hbar\frac{\partial\psi}{\partial t} = \left(-\frac{\hbar^2}{2m}\nabla^2 + Q\varphi\right)\psi \ . \tag{A.3.28}$$

If ψ_0 is a solution of equation (A.3.26), then a solution of eq. (A.3.28) is

$$\psi = \psi_0 \exp\left[-i\frac{Q}{\hbar}\int_{t_0}^{t}\varphi(t')dt'\right] \ . \tag{A.3.29}$$

The premise that $\varphi = \varphi(t)$ is satisfied *e.g.*, for a particle that moves inside a Faraday cage whose surface has a time dependent potential.

The potentials cause, in both cases, an additional phase factor with the phase shift of $(Q/\hbar)\int\mathbf{A}\bullet d\mathbf{r}$ and $(-Q/\hbar)\int\varphi(t)dt$, respectively.

A.3.3   The Ehrenfest Theorems

Despite the significant differences between classical mechanics and quantum mechanics, they do not mutually contradict themselves. Stated somewhat simplified, a result of quantum mechanics is that the averages of physical quantities behave like classical mechanics predicts. This is expressed in Ehrenfest's Theorems. Using the notation $\langle g \rangle$ to express the average of the physical quantity g, allows to obtain the following relations from Schrödinger's equation:

$$\frac{d}{dt}\langle\mathbf{r}\rangle = \frac{\langle\mathbf{p}\rangle}{m} \tag{A.3.30}$$

$$\frac{d}{dt}\langle\mathbf{p}\rangle = -\langle\,\nabla U(\mathbf{r})\,\rangle \tag{A.3.31}$$

or when combining the two

$$m\frac{d^2}{dt^2}\langle\mathbf{r}\rangle = -\langle\,\nabla U(\mathbf{r})\,\rangle \ . \tag{A.3.32}$$

This is the classical equation of motion, which occurs here as a consequence of Schrödinger's equation. Deviations from these averages are very improbable when dealing with macroscopic systems, which renders the difference between (A.3.32) and the classical results negligible. However, this is not true for microscopic systems.

Looking at the motion of a particle in an electromagnetic field, one obtains from Schrödinger's equation (A.3.32), after some tedious mathematics, Ehrenfest's theorem in the following form:

$$m\frac{d^2}{dt^2}\langle \mathbf{r} \rangle = \langle Q\mathbf{E} + \frac{Q}{2}(\mathbf{v} \times \mathbf{B} - \mathbf{B} \times \mathbf{v}) \rangle .$$ (A.3.33)

We conclude that quantum mechanics also permits to calculate the "mean" particle trajectory by means of $\mathbf{E}$ and $\mathbf{B}$, even in a form analogous to the classical equation of the motion. This applies only to the average and does not describe the exact motion of any particular particle in an electromagnetic field. The form of (A.3.33) may initially look peculiar, but results from the fact that velocity in quantum mechanics is an operator related with the momentum, what arises from (A.3.15) and (A.3.17). This operator does not commute with the location operator or any of the location dependent operators, *e.g.* $\mathbf{A} = \mathbf{A}(\mathbf{r})$ or $\mathbf{B} = \mathbf{B}(\mathbf{r})$. The classical expression is of course:

$$\frac{1}{2}(\mathbf{v} \times \mathbf{B} - \mathbf{B} \times \mathbf{v}) = \frac{1}{2}(\mathbf{v} \times \mathbf{B} + \mathbf{v} \times \mathbf{B}) = (\mathbf{v} \times \mathbf{B}) .$$ (A.3.34)

In quantum mechanics, these expressions may not be equated. However, one can see that in the limit of classical mechanics, Ehrenfest's theorem in the form (A.3.33), just yields the Lorentz force.

A.3.4   Magnetic Field and Vector Potential in an infinitely long Coil

Consider an infinitely long coil as shown in Fig. A.3.1. Its magnetic field can be expressed by the vector potential $\mathbf{A}$, where

$$\mathbf{B} = \nabla \times \mathbf{A} .$$ (A.3.35)

The magnetic flux through an arbitrary surface is:

$$\phi = \int_a \mathbf{B} \bullet da = \int_a \nabla \times \mathbf{A} \bullet da = \oint \mathbf{A} \bullet d\mathbf{s} .$$ (A.3.36)

$\mathbf{B}$ is gauge invariant, *i.e.*, is not affected by a transition from one vector potential to another of different gauge, where both vector potentials may only differ by the gradient of an arbitrary function. This makes the flux ϕ gauge invariant as well, as (A.3.36) makes obvious, because $\oint \mathbf{A} \bullet d\mathbf{s}$ can not change in this case ($\oint \nabla f \bullet d\mathbf{s} = 0$). Fig. A.3.1 shows the graph of $B_z(r)$ and $A_\varphi(r)$. Under the

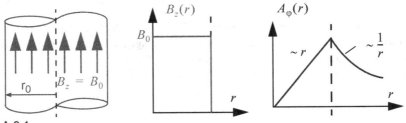

Fig. A.3.1

currently assumed gauge, **A** possesses only a φ-component $A_\varphi(r)$. Now, (see also Sect. 5.2.3):

$$B_z(r) = \begin{cases} B_0 & \text{for } r \leq r_0 \\ 0 & \text{for } r > r_0 \end{cases} \tag{A.3.37}$$

and

$$A_\varphi(r) = \begin{cases} \dfrac{r}{2} B_0 & \text{for } r \leq r_0 \\ \dfrac{r_0^2}{2r} B_0 & \text{for } r \geq r_0 \end{cases} \tag{A.3.38}$$

Interesting to realize is that the magnetic field outside of the coil vanishes, while the vector potential does not. This fact will be of our interest next. We will discuss the question whether the vector potential outside of the coil might influence, in any way, charged particles whose behavior is described by Schrödinger's equation. For this purpose, we will study experiments where electron beams interfere at a double slit, as was described by Bohm and Aharonov [34].

A.3.5 Interference of Electron Beams at Double Slit

A double slit as illustrated in Fig. A.3.2 shall be used for interference experiments with electron beams. Starting without a coil and its magnetic field behind the slit, we obtain an interference pattern on the screen with intensity maxima and minima due to the impinging electron rays. The quantities (a, d, L, r_1, r_2, x), defined in Fig. A.3.2 allow to calculate the difference in the geometric path length.

$$a = r_2 - r_1 = \sqrt{L^2 + \left(x + \frac{d}{2}\right)^2} - \sqrt{L^2 + \left(x - \frac{d}{2}\right)^2} . \tag{A.3.39}$$

If $x \ll L$, then

$$a \approx \frac{xd}{L} . \tag{A.3.40}$$

This corresponds to a phase difference of

$$\alpha = \frac{2\pi}{\lambda} \cdot \frac{xd}{L} , \tag{A.3.41}$$

if λ denotes the wave length of matter associated with the electrons, which is a result of the de Broglie relation

$$\lambda = \frac{h}{p} , \tag{A.3.42}$$

where p represents the momentum of the electron.

Repeating this experiment with the coil and its magnetic field in place, the electrons from each of the interfering rays travel within the range of the vector potential created by the coil. The coil shall be assumed to be ideal, *i.e.* has no fringing fields. The ray of electrons shall not be able to penetrate the inside of the coil. Expressed differently, the wave function of the electrons and the magnetic induction of the field inside the coil shall not overlap. This results in additional phase differences between the rays along the paths C_1 and C_2, respectively:

$$\beta_1 = \frac{Q}{\hbar}\int_{C_1} \mathbf{A} \bullet d\mathbf{s} \tag{A.3.43}$$

and

$$\beta_2 = \frac{Q}{\hbar}\int_{C_2} \mathbf{A} \bullet d\mathbf{s} . \tag{A.3.44}$$

Material for the interference is the phase difference

$$\beta = \beta_1 - \beta_2 = \frac{Q}{\hbar}\oint \mathbf{A} \bullet d\mathbf{s} = \frac{Q}{\hbar}\phi , \tag{A.3.45}$$

where ϕ is the flux contained inside the coil. This result is very peculiar. As long as we only regard the phase difference β, then only the overall flux is relevant, not the specific spatial distribution of the magnetic field which produces this flux. The flux causes a shift of the maxima and minima of the interference picture on the screen by the distance

$$\Delta x = \frac{L\lambda}{2\pi d}\cdot\beta = \frac{L\lambda}{2\pi d}\cdot\frac{Q}{\hbar}\phi = \frac{L\lambda Q\phi}{hd} , \tag{A.3.46}$$

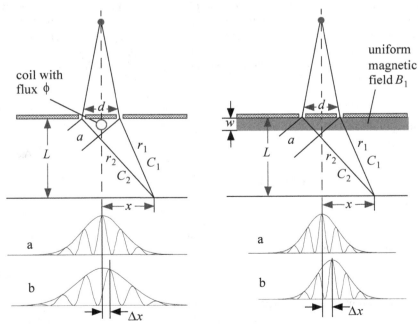

Fig. A.3.2   Fig. A.3.3

which results from (A.3.41) when replacing α by β and x by Δx. Fig. A.3.2 shows the interference pattern picture a) without and b) with magnetic field.

Precise measurements show [35], that although the maxima and minima shift by Δx as expressed by (A.3.36) and is depicted in Fig. A.3.2, the envelope remains unchanged, however. This is the result predicted by Bohm and Aharonov and meanwhile also verified experimentally, most recently and most clearly by [36]. This effect is not explainable by the **B** field alone. From the perspective of classical mechanics, the coil should not influence those passing rays (at least under the condition that the **B** field of the coil does not overlap with the wave function ψ, which is not easy to achieve experimentally and gave rise to many controversies on the validity of the Bohm-Aharonov effect).

Interesting is also to consider a variation of the experiment of Fig. A.3.2. This is described by Fig. A.3.3. Here, the coil and its magnetic field is replaced by a region, carrying a uniform field (perpendicular to the paper plane). The width of this region is w. Of course, without magnetic field, the result is just as before. With the magnetic field, we obtain a flux, which is what is relevant here. The flux is approximately (*i.e.*, if $x \ll L$)

$$\phi = B_1 wd \tag{A.3.47}$$

and the thereby caused shift of the maxima and minima of the interference pattern is then with (A.3.36)

$$\Delta x = \frac{L\lambda Q}{hd} B_1 wd = \frac{L\lambda Q B_1 w}{h} . \tag{A.3.48}$$

Fig. A.3.3 illustrates, just as Fig. A.3.2, the interference pattern a) without and b) with a magnetic field. However, in contrast to the case of Fig. A.3.2, now the entire interference pattern, including the envelope, shifts by Δx in x-direction. This is plausible and fits well with the classical understanding. For $x \ll L$, the Lorentz force causes

$$\Delta p_x = QvB_1 \tau = QvB_1 \frac{w}{v} = QB_1 w , \tag{A.3.49}$$

where τ represents the time which it takes for a particle to pass through the region of the uniform magnetic field. If v is its velocity, then

$$\tau = \frac{w}{v} . \tag{A.3.50}$$

Furthermore,

$$\frac{\Delta x}{L} \approx \frac{\Delta p_x}{p} = \frac{QB_1 w}{\frac{h}{\lambda}} = \frac{\lambda Q B_1 w}{h} . \tag{A.3.51}$$

After multiplication by L, this gives exactly the previous result, eq. (A.3.48), which we have derived there in a rather different way. All electrons are deflected by exactly the same angle ($\approx \Delta x/L$), *i.e.*, the entire interference pattern is shifted by

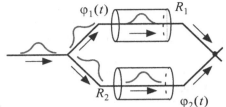

Fig. A.3.4

the distance Δx. Regarding the sign, we note that for electrons ($Q < 0$) and for magnetic fields which point out of the paper plane, the shift is to the right.

The difference between the two variants of the experiment (namely one where the coil that has no outside magnetic field, and the other where a uniform magnetic field is applied), is rather peculiar. Nevertheless, on the basis of Ehrenfest's theorems, on one hand, and the phase difference caused by the vector potential on the other, it should be conceptually clear. In order to determine the root cause of the maxima and minima, one knows that this only depends on the phase difference, and this means it solely depends on the enclosed flux ($\oint \mathbf{A} \cdot d\mathbf{s}$). The average of a particle trajectory is determined by the Lorentz force, and this means the magnetic induction, whereby the particular form of the Lorentz force given in eq. (A.3.33) has to be considered. All this is true from a quantum mechanical perspective as well. Consequently, the center of mass of the particle trajectory remains unchanged in the case of the coil without fringe field, and therefore, the envelope remains unchanged as well. In contrast in the case of the uniform field, the electrons are deflected both, classically and quantum mechanically by the same angle. The averages shift accordingly. Therefore, in this case, both the envelope and the interference pattern are shifted by the same distance Δx. This corresponds to a shift of the center of mass (of the average) by exactly this distance Δx. With this, we have gained a clear and easily understandable picture of the process.

An entirely different, nevertheless related experiment (which also goes back to the mentioned work of Bohm and Aharonov) is sketched in Fig. A.3.4.

Each of the two rays travels through a shielded cavity in form of a tube R_1 and R_2, respectively. Then the potentials $\varphi_1(t)$ and $\varphi_2(t)$ are applied to the pipes during the time when the wave packets travel through a tube but are not too close to the ends (where there may be fringe fields). This creates a phase difference

$$\beta_1 = -\frac{Q}{\hbar}\int \varphi_1(t)dt \tag{A.3.52}$$

and

$$\beta_2 = -\frac{Q}{\hbar}\int \varphi_2(t)dt \tag{A.3.53}$$

If $\beta_1 \neq \beta_2$, then an interference pattern with maxima and minima corresponding to the difference

$$\beta = \beta_1 - \beta_2 \tag{A.3.54}$$

is visible at a screen, where as before, eq. (A.3.46) applies. This is as peculiar as the previous experiment, involving the magnetic field of a coil. Even though inside the pipes we have

$$\mathbf{E} = -\nabla\varphi = 0, \tag{A.3.55}$$

the potential still influences the electrons which pass through it. Similarly to the previous explanation, it is important to realize that the electric field is not sufficient to describe the interaction. True also in this case is that the envelope of the interference pattern (*i.e.*, the "center of mass" of the incident electrons) does not shift if $\mathbf{E} = 0$. Merely the position of the maxima and minima is affected by the potentials $\varphi_1(t)$ and $\varphi_2(t)$. Conversely, when electrons pass through non-vanishing electric fields, then with shift of the center of mass of the rays, the entire pattern, including envelope shifts. Both obey the proposition of Ehrenfest's theorems.

A.3.6   Conclusions

We have found that quantum mechanics provides a new perspective on the reality of the fields $\mathbf{E}$ and $\mathbf{B}$, on one side, and the potentials $\mathbf{A}$ and φ on the other. Maxwell's equations are thereby unaffected. They can be accepted as correct even in light of quantum mechanics. However, the interaction of charged particles with electromagnetic fields exhibits effects which can not be explained classically by the equation of the motion, and the fields of Maxwell's equations can not sufficiently describe these effects. Particularly, the potentials $\mathbf{A}$ and φ are indispensable and one needs to consider them as separate fields. It would be easier to dispense of the classical fields $\mathbf{E}$ and $\mathbf{B}$ because the potentials are, anyway, a means within the classical theory to eliminate $\mathbf{E}$ and $\mathbf{B}$, while the reverse in quantum mechanics is not possible, at least not without major difficulties.

A.4   Liénard-Wiechert Potentials

A very interesting, special case of the retarded potentials as of eqs. (7.195) and (7.196) results, if we consider a charged particle travelling on an arbitrarily prescribed trajectory $\mathbf{r}_0(t)$. This case is described by:

$$\rho = Q\delta[\mathbf{r} - \mathbf{r}_0(t)] \tag{A.4.1}$$

$$\mathbf{g} = Q\dot{\mathbf{r}}_0\delta[\mathbf{r} - \mathbf{r}_0(t)]$$

$$= Q\mathbf{v}_0\delta[\mathbf{r} - \mathbf{r}_0(t)] . \tag{A.4.2}$$

Q represents the charge of the particle and

$$\dot{\mathbf{r}}_0(t) = \frac{d\mathbf{r}_0(t)}{dt} = \mathbf{v}_0(t) \tag{A.4.3}$$

is its velocity. The potentials are

$$\varphi(\mathbf{r}, t) = \frac{Q}{4\pi\varepsilon_0} \int \frac{\delta\left[\mathbf{r}' - \mathbf{r}_0\left(t - \frac{|\mathbf{r} - \mathbf{r}'|}{c}\right)\right]}{|\mathbf{r} - \mathbf{r}'|} d\tau' \tag{A.4.4}$$

$$\mathbf{A}(\mathbf{r}, t) = \frac{Q\mu_0}{4\pi} \int \frac{\mathbf{v}_0\left(t - \frac{|\mathbf{r} - \mathbf{r}'|}{c}\right)\delta\left[\mathbf{r}' - \mathbf{r}_0\left(t - \frac{|\mathbf{r} - \mathbf{r}'|}{c}\right)\right]}{|\mathbf{r} - \mathbf{r}'|} d\tau' \ . \tag{A.4.5}$$

Evaluating these integrals is difficult, despite the delta function in there. Its argument could be a complicated function of $\mathbf{r}'$, but has to vanish. In general we have

$$\int F(\mathbf{r}')\delta[\mathbf{f}(\mathbf{r}')]d\tau' = \int F(x', y', z')\delta[\mathbf{f}(x', y', z')]dx' dy' dz'$$

$$= \int F(x', y', z')\delta[\mathbf{f}(x', y', z')]\frac{1}{D}df_x df_y df_z \ , \tag{A.4.6}$$

where D is the functional determinant

$$D = \begin{vmatrix} \dfrac{\partial f_x}{\partial x'} & \dfrac{\partial f_x}{\partial y'} & \dfrac{\partial f_x}{\partial z'} \\[2mm] \dfrac{\partial f_y}{\partial x'} & \dfrac{\partial f_y}{\partial y'} & \dfrac{\partial f_y}{\partial z'} \\[2mm] \dfrac{\partial f_z}{\partial x'} & \dfrac{\partial f_z}{\partial y'} & \dfrac{\partial f_z}{\partial z'} \end{vmatrix} \ . \tag{A.4.7}$$

For the present case, this determinant is

$$D = 1 - \frac{\mathbf{v}_0\left(t - \frac{|\mathbf{r} - \mathbf{r}'|}{c}\right) \bullet (\mathbf{r} - \mathbf{r}')}{c|\mathbf{r} - \mathbf{r}'|} \ . \tag{A.4.8}$$

Using this, we obtain the so-called Liénard-Wiechert Potentials

$$\varphi(\mathbf{r}, t) = \frac{Q}{4\pi\varepsilon_0|\mathbf{r} - \mathbf{r}'|\left[1 - \dfrac{\mathbf{v}_0\left(t - \frac{|\mathbf{r} - \mathbf{r}'|}{c}\right) \bullet (\mathbf{r} - \mathbf{r}')}{c|\mathbf{r} - \mathbf{r}'|}\right]} \tag{A.4.9}$$

and

$$\mathbf{A}(\mathbf{r}, t) = \varepsilon_0\mu_0\mathbf{v}_0\varphi(\mathbf{r}, t) = \frac{\mathbf{v}_0}{c^2}\varphi(\mathbf{r}, t) \ . \tag{A.4.10}$$

$\mathbf{r}'$ is defined by the requirement that the argument of the δ-functions in eqs. (A.4.4) and (A.4.5) has to vanish, that is, it has to be

$$\mathbf{r}' = \mathbf{r}_0\left(t - \frac{|\mathbf{r} - \mathbf{r}'|}{c}\right) \ , \tag{A.4.11}$$

which means that $\mathbf{r}'$ is a function of $\mathbf{r}$ and t, and depending on the given trajectory might be difficult to determine. $\mathbf{r}'$ defines the location where the particle was at the retarded time.

A relatively simple, special case is that of a particle which moves with constant speed.

$$\mathbf{r}_0 = \mathbf{v}_0 t \ . \tag{A.4.12}$$

For this case, after some calculations which we skip, one obtains

$$\varphi(\mathbf{r}, t) = \frac{Qc}{4\pi\varepsilon_0 \sqrt{(c^2 t - \mathbf{v}_0 \bullet \mathbf{r})^2 + (c^2 - v_0^2)(r^2 - c^2 t^2)}} \ , \tag{A.4.13}$$

$$\mathbf{A} = \frac{\mathbf{v}_0 \varphi}{c^2} \ . \tag{A.4.14}$$

Of course, for the case $\mathbf{v}_0 = 0$, the potential has to be

$$\varphi = \frac{Q}{4\pi\varepsilon_0 r} \ , \tag{A.4.15}$$

as it is indeed. Without restricting generality, one may assume that $\mathbf{v}_0 = \langle v_0, 0, 0 \rangle$, which results in

$$\varphi = \frac{Qc}{4\pi\varepsilon_0 \sqrt{(c^2 t - v_0 x)^2 + (c^2 - v_0^2)(x^2 + y^2 + z^2 - c^2 t^2)}} \tag{A.4.16}$$

$$A_x = \frac{v_0 \varphi}{c^2}, \qquad A_y = 0, \qquad A_z = 0 \tag{A.4.17}$$

and for the electric field

$$\left. \begin{array}{l} E_x = -\dfrac{\partial \varphi}{\partial x} - \dfrac{\partial A_x}{\partial t} = \dfrac{Qc}{4\pi\varepsilon_0} \cdot \dfrac{(c^2 - v_0^2)(x - v_0 t)}{\sqrt{(c^2 t - v_0 x)^2 + (c^2 - v_0^2)(x^2 + y^2 + z^2 - c^2 t^2)}^3} \\[4mm] E_y = -\dfrac{\partial \varphi}{\partial y} = \dfrac{Qc}{4\pi\varepsilon_0} \cdot \dfrac{(c^2 - v_0^2)y}{\sqrt{(c^2 t - v_0 x)^2 + (c^2 - v_0^2)(x^2 + y^2 + z^2 - c^2 t^2)}^3} \\[4mm] E_z = -\dfrac{\partial \varphi}{\partial z} = \dfrac{Qc}{4\pi\varepsilon_0} \cdot \dfrac{(c^2 - v_0^2)z}{\sqrt{(c^2 t - v_0 x)^2 + (c^2 - v_0^2)(x^2 + y^2 + z^2 - c^2 t^2)}^3} \end{array} \right\} \tag{A.4.18}$$

The ratio is therefore

$$E_x : E_y : E_z = (x - v_0 t) : y : z \ , \tag{A.4.19}$$

that is, the force lines emerge as straight lines from the location $\langle v_0 t, 0, 0 \rangle$, namely the point where the particle is located at that particular moment. Conversely, the field is not spherically symmetric. It depends on the angle α, which the force lines enclose with the x-axis at any particular location of the particle (Fig. A.4.1).

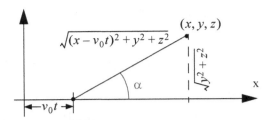

Fig. A.4.1

Using the relation

$$\sin^2\alpha = \frac{y^2 + z^2}{(x - v_0 t)^2 + y^2 + z^2}$$

allows to re-write eq. (A.4.18) in the following way:

$$\mathbf{E} = \frac{Q}{4\pi\varepsilon_0} \cdot \frac{(\mathbf{r} - \mathbf{v}_0 t)}{|\mathbf{r} - \mathbf{v}_0 t|^3} \cdot \frac{\left(1 - \frac{v_0^2}{c^2}\right)}{\sqrt{1 - \frac{v_0^2}{c^2}\sin^2\alpha}^3} \quad . \tag{A.4.20}$$

This form makes it obvious that the field lines emerge as straight lines from the location of the charge. Its magnitude is

$$E = \frac{Q}{4\pi\varepsilon_0} \cdot \frac{1}{|\mathbf{r} - \mathbf{v}_0 t|^2} \cdot \frac{\left(1 - \frac{v_0^2}{c^2}\right)}{\sqrt{1 - \frac{v_0^2}{c^2}\sin^2\alpha}^3} \quad , \tag{A.4.21}$$

and it clearly reveals the dependency on the angle. The field is minimal for $\alpha = 0$

$$E_{min} = \frac{Q}{4\pi\varepsilon_0} \cdot \frac{1}{|\mathbf{r} - \mathbf{v}_0 t|^2} \cdot \left(1 - \frac{v_0^2}{c^2}\right) , \tag{A.4.22}$$

and its maximum is for $\alpha = \pi/2$

$$E_{max} = \frac{Q}{4\pi\varepsilon_0} \cdot \frac{1}{|\mathbf{r} - \mathbf{v}_0 t|^2} \cdot \frac{1}{\sqrt{1 - \frac{v_0^2}{c^2}}} \quad . \tag{A.4.23}$$

Let's pick a specific value for v_0/c, e.g.: $v_0/c = 0.6$, then

$$\frac{E_{min}}{E_{max}} = \sqrt{1 - \frac{v_0^2}{c^2}}^3 \approx 0.5 \quad .$$

This field, that depends on the angle was already mentioned in Ch. 1 (Sect. 1.10).

We will discuss another interesting **example**: A particle which oscillates harmonically about the origin which is described by

$$\mathbf{r}_0 = \langle 0, 0, d\sin\omega t \rangle , \tag{A.4.24}$$

$$\mathbf{v}_0 = \langle 0, 0, \omega d\cos\omega t \rangle . \tag{A.4.25}$$

If we let d be very small, then the first order approximation gives

$$|\mathbf{r} - \mathbf{r}'| \approx r\left(1 - \frac{zz'}{r^2}\right) ,$$

$$z' \approx d\sin\left[\omega\left(t - \frac{r}{c}\right)\right] ,$$

and

$$\varphi \approx \frac{Q}{4\pi\varepsilon_0 r} \cdot \frac{1}{\left(1 - \dfrac{zz'}{r^2}\right)\left(1 - \dfrac{z\dot z'}{cr}\right)} \approx \frac{Q}{4\pi\varepsilon_0 r} \cdot \left(1 + \frac{zz'}{r^2} + \frac{z\dot z'}{cr}\right)$$

that is

$$\varphi \approx \frac{Q}{4\pi\varepsilon_0 r} + \frac{Qd\cos\theta}{4\pi\varepsilon_0} \cdot \left\{ \frac{\sin\left[\omega\left(t - \frac{r}{c}\right)\right]}{r^2} + \frac{\omega\cos\left[\omega\left(t - \frac{r}{c}\right)\right]}{cr} \right\} , \tag{A.4.26}$$

$$A_x = A_y = 0, \quad A_z \approx \frac{Q}{4\pi\varepsilon_0 r} \cdot \frac{\dot z'}{c^2} \approx \frac{Qd\omega\cos\left[\omega\left(t - \frac{r}{c}\right)\right]}{4\pi\varepsilon_0 r c^2} . \tag{A.4.27}$$

Adding a charge $-Q$ at rest at the origin gives the overall potential (using $Qd = p_0$)

$$\varphi \approx \frac{p_0\cos\theta}{4\pi\varepsilon_0} \cdot \left\{ \frac{1}{r^2}\sin\left[\omega\left(t - \frac{r}{c}\right)\right] + \frac{\omega}{cr}\cos\left[\omega\left(t - \frac{r}{c}\right)\right] \right\} \tag{A.4.28}$$

$$A_x = A_y = 0, \quad A_z = \frac{\mu_0 p_0 \omega}{4\pi r}\cos\left[\omega\left(t - \frac{r}{c}\right)\right] . \tag{A.4.29}$$

These represent the equations for the retarded potentials of the Hertz dipole, eqs. (7.266) and (7.267), while then the potential was expressed in spherical coordinates. This result is not surprising. The positive charge, oscillating around a negative charge at rest represents, in this context, an oscillating dipole. For the purposes of the radiation, the negative charge at rest is immaterial. The oscillating particle creates the same radiation as the oscillating dipole does. The difference in the potential is merely by the potential of the point charge at rest $Q/4\pi\varepsilon_0 r$, which is unrelated to the radiation.

A.5   The Helmholtz Theorem

A.5.1   Derivation and Interpretation

Helmholtz's theorem, simplified somewhat, states that an arbitrary vector field can uniquely be expressed by the totality of its sources and vortices. The easiest way to conceptualize why this has to be so is by means of a hydrodynamic model. Consider a finite or infinite volume, within which there is a fluid, initially at rest. One may add sources, sinks, and generate vortices. These, in conjunction with the boundary conditions at the surface, will uniquely determine the resulting flux field.

This theorem summarizes many of the properties which we had used in previous Sections. It also sheds an interesting light on Maxwell's equations as such. Their task is to accurately describe two vector fields. In light of Helmholtz's theorem, this is best achieved by expressing all of their sources and vortices. This is exactly what Maxwell's equations achieve, and they do this in a very simple and elegant manner. The fields are thereby not independent of each other. They are correlated by the fact that the time derivative of each one represents the curl of the other field.

Consider a vector field $\mathbf{W}$ in a finite or infinite volume V with the surface a. Given are its sources and vortices

$$\nabla \bullet \mathbf{W} = \rho(\mathbf{r}) \tag{A.5.1}$$

$$\nabla \times \mathbf{W} = \mathbf{g}(\mathbf{r}) . \tag{A.5.2}$$

$\rho(\mathbf{r})$ and $\mathbf{g}(\mathbf{r})$ are arbitrary densities of sources and vortices. In the electrostatic case it would be $\mathbf{W} = \mathbf{D}$, $\mathbf{g} = 0$, and ρ the charge density. In the magnetostatic case it would be $\mathbf{W} = \mathbf{H}$, $\rho = 0$, and $\mathbf{g}$ the current density. A further assumption shall be that there are no sources nor vortices at infinity (otherwise, these needed separate consideration). This allows to express $\mathbf{W}$ in the following manner.

$$\mathbf{W} = -\nabla \left[\int_V \frac{\rho(\mathbf{r}')}{4\pi|\mathbf{r} - \mathbf{r}'|} d\tau' - \oint_a \frac{\mathbf{W}(\mathbf{r}') \bullet d\mathbf{a}}{4\pi|\mathbf{r} - \mathbf{r}'|} \right]$$
$$+ \nabla \times \left[\int_V \frac{\mathbf{g}(\mathbf{r}')}{4\pi|\mathbf{r} - \mathbf{r}'|} d\tau' + \oint_a \frac{\mathbf{W}(\mathbf{r}') \times d\mathbf{a}}{4\pi|\mathbf{r} - \mathbf{r}'|} \right] . \tag{A.5.3}$$

Introducing the abbreviation ϕ and $\mathbf{A}$ for the expression in the brackets, we obtain

$$\mathbf{W} = -\nabla\phi + \nabla \times \mathbf{A} . \tag{A.5.4}$$

This is Helmholtz's theorem. Its relation to many results in field theory is apparent. The proof is easy. It starts from eq. (3.53)

$$\mathbf{W}(\mathbf{r}) = \int_V \mathbf{W}(\mathbf{r}')\delta(\mathbf{r} - \mathbf{r}')d\tau' , \tag{A.5.5}$$

and in conjunction with eq. (3.56)

$$\delta(\mathbf{r} - \mathbf{r}') = -\frac{1}{4\pi}\nabla_r^2 \frac{1}{|\mathbf{r} - \mathbf{r}'|} \tag{A.5.6}$$

this gives

$$\mathbf{W(r)} = \int_V \mathbf{W(r')}\left(-\frac{1}{4\pi}\nabla_r^2\frac{1}{|\mathbf{r}-\mathbf{r'}|}\right)d\tau' = -\nabla_r^2\int\frac{\mathbf{W(r')}}{4\pi|\mathbf{r}-\mathbf{r'}|}d\tau' .$$

(A.5.7)

Applying the vector identity eq. (5.11)

$$\nabla\times(\nabla\times\mathbf{A}) = \nabla(\nabla\bullet\mathbf{A}) - \nabla^2\mathbf{A}$$

(A.5.8)

gives

$$\mathbf{W(r)} = \nabla_r\times\left[\nabla_r\times\int_V\frac{\mathbf{W(r')}}{4\pi|\mathbf{r}-\mathbf{r'}|}d\tau'\right] - \nabla_r\left[\nabla_r\bullet\int_V\frac{\mathbf{W(r')}}{4\pi|\mathbf{r}-\mathbf{r'}|}d\tau'\right] .$$

(A.5.9)

Therefore

$$\phi = \nabla_r\bullet\int_V\frac{\mathbf{W(r')}}{4\pi|\mathbf{r}-\mathbf{r'}|}d\tau' = -\int_V\frac{\mathbf{W(r')}}{4\pi}\nabla_{r'}\frac{1}{|\mathbf{r}-\mathbf{r'}|}d\tau'$$

$$= -\int_V\nabla_{r'}\bullet\frac{\mathbf{W(r')}}{4\pi|\mathbf{r}-\mathbf{r'}|}d\tau' + \int_V\frac{\nabla_{r'}\mathbf{W(r')}}{4\pi|\mathbf{r}-\mathbf{r'}|}d\tau'$$

$$\boxed{\phi = \int_V\frac{\rho(\mathbf{r'})}{4\pi|\mathbf{r}-\mathbf{r'}|}d\tau' - \oint_a\frac{\mathbf{W(r')}\bullet d\mathbf{a'}}{4\pi|\mathbf{r}-\mathbf{r'}|}}$$

(A.5.10)

and

$$\mathbf{A} = \nabla_r\times\int_V\frac{\mathbf{W(r')}}{4\pi|\mathbf{r}-\mathbf{r'}|}d\tau' = -\int_V\frac{\mathbf{W(r')}}{4\pi}\times\nabla_r\frac{1}{|\mathbf{r}-\mathbf{r'}|}d\tau'$$

$$= +\int_V\frac{\mathbf{W(r')}}{4\pi}\times\nabla_{r'}\frac{1}{|\mathbf{r}-\mathbf{r'}|}d\tau'$$

$$= -\int_V\nabla_{r'}\times\left(\frac{\mathbf{W(r')}}{4\pi|\mathbf{r}-\mathbf{r'}|}\right)d\tau' + \int_V\frac{\nabla_{r'}\times\mathbf{W(r')}}{4\pi|\mathbf{r}-\mathbf{r'}|}d\tau' .$$

Now applying Gauss' integral theorem in the form of eq. (5.68) gives

$$\boxed{\mathbf{A} = \int_V\frac{\mathbf{g(r')}}{4\pi|\mathbf{r}-\mathbf{r'}|}d\tau' + \oint_a\frac{\mathbf{W(r')}\times d\mathbf{a'}}{4\pi|\mathbf{r}-\mathbf{r'}|}} .$$

(A.5.11)

If we take an infinitely large volume where the location of the sources and vortices are finite, then all surface integrals vanish. Conversely, for a finite volume, the surface integrals have to be considered as well. They also have significance from a plausibility perspective and could have been introduced, even without the above provided formal prove. We will demonstrate this through an example by means of the field given in Fig. A.5.1 The field is uniform inside a cylinder and vanishes outside. Obviously, it must have sources and vortices. Sources and sinks are generally at locations where the normal component of the field is discontinuous, and it is rotational at those locations where the tangential field component is discontinuous. The surface density of the sources is

$$\sigma(\mathbf{r}) = -\mathbf{W}\bullet\mathbf{n}$$

(A.5.12)

while that of the vortices is

$$\mathbf{k(r)} = \mathbf{W}\times\mathbf{n} ,$$

(A.5.13)

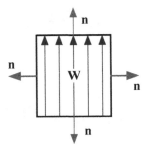

Fig. A.5.1

where **n** is the unit vector normal to the surface (pointing outward). This means, the field in Fig. A.5.1 has sources at the bottom, sinks (negative sources) at the top, while the vortices are on the sides of the cylinder. If one now considers a volume that is larger than the cylinder itself, then all those surface sources and surface vortices are inside this volume and consequently part of the volume integral, where they appear in the form of δ-functions, which transforms the volume integrals into surface integrals. We will return to this point in detail in the form of an example.

There is a relationship between the Helmholtz theorem and the previously proven theorem (3.57). An irrotational field, that is otherwise arbitrary, can be expressed in the form of a scalar potential. Then with the current notation one writes

$$\mathbf{W} = -\nabla\phi = -\nabla\left[\int_V \frac{\rho(\mathbf{r}')d\tau'}{4\pi|\mathbf{r}-\mathbf{r}'|} - \oint_a \frac{\mathbf{W}(\mathbf{r}')\bullet d\mathbf{a}'}{4\pi|\mathbf{r}-\mathbf{r}'|} - \oint_a \frac{\phi(\mathbf{r}')}{4\pi}\cdot\frac{\partial}{\partial n'}\frac{1}{|\mathbf{r}-\mathbf{r}'|}da'\right].$$

(A.5.14)

Using Helmholtz's theorem with $\mathbf{g}(\mathbf{r}) = 0$ gives the same field

$$\mathbf{W} = -\nabla\left[\int_V \frac{\rho(\mathbf{r}')d\tau'}{4\pi|\mathbf{r}-\mathbf{r}'|} - \oint_a \frac{\mathbf{W}(\mathbf{r}')\bullet d\mathbf{a}'}{4\pi|\mathbf{r}-\mathbf{r}'|}\right] + \nabla\times\left[\oint_a \frac{\mathbf{W}(\mathbf{r}')\times d\mathbf{a}'}{4\pi|\mathbf{r}-\mathbf{r}'|}\right].$$

(A.5.15)

Although the field is uniquely defined by its sources and vortices, still there may be several ways to express it. The field consists of three parts. The first two are the same in both representations. The third part can be expressed by a scalar or a vector potential. We have found in Sect. 3.4.7 that this third part of the scalar potential is that of a dipole layer. Again, as before in Sect 5.3, one encounters the equivalence of eddy ring and dipole layer. This means that we can imagine the field to be created by the surface sources and surface vortices. This equivalence is not as astonishing as it may seem at first. The vortices are nothing else than the discontinuity of the tangential components. The boundary conditions (2.117) reveal that dipole layers, if they are non-uniform, also cause such discontinuities.

A.5.2   Examples

A.5.2.1 Uniform Field inside a Sphere

Consider the field $\mathbf{W}$ given in Fig. A.5.2, which is uniform inside the sphere and vanishes everywhere outside. There are no sources nor vortices inside. Therefore, the field can be calculated solely from the surface integrals. Using

$$\sigma = -W\cos\theta \tag{A.5.16}$$

and

$$k_\varphi = W\sin\theta , \tag{A.5.17}$$

together with the Helmholtz theorem gives

$$\phi(r, \theta) = -\frac{W}{4\pi}\int_a \frac{\cos\theta_0}{|\mathbf{r} - \mathbf{r}_0|}da_0 = -\frac{W}{4\pi}\oint_a \frac{P_1^0(\cos\theta_0)}{|\mathbf{r} - \mathbf{r}_0|}da_0 \tag{A.5.18}$$

and

$$A_r = A_\theta = 0$$

$$A_\varphi(r, \theta) = \frac{W}{4\pi}\oint_a \frac{\cos(\varphi - \varphi_0)\sin\theta_0}{|\mathbf{r} - \mathbf{r}_0|}da_0 = \frac{W}{4\pi}\oint_a \frac{P_1^1(\cos\theta_0)\cos(\varphi - \varphi_0)}{|\mathbf{r} - \mathbf{r}_0|}da_0 \tag{A.5.19}$$

where

$$|\mathbf{r} - \mathbf{r}_0| = \sqrt{r^2 + r_0^2 + 2rr_0[\sin\theta\sin\theta_0\cos(\varphi - \varphi_0) + \cos\theta\cos\theta_0]} . \tag{A.5.20}$$

Note that, as was mentioned frequently, calculation of A_φ should be based on Cartesian coordinates. This results, as before in eq. (5.44), in the additional factor $\cos(\varphi - \varphi_0)$ in the integrand. Both integrals are not elementary integrals, however, the series expansion of the inverse distance in spherical coordinates as Legendre polynomials allows for evaluation in an elegant way. The general integral is

$$J = \oint \frac{P_{n'}^{m'}(\cos\theta_0)\cos[m'(\varphi - \varphi_0)]}{|\mathbf{r} - \mathbf{r}_0|}da_0$$

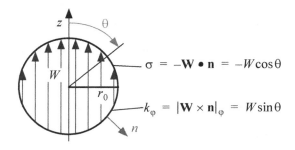

Fig. A.5.2

$$J = \oint (P_n^{m'}(\cos\theta_0)\cos[m'(\varphi - \varphi_0)]) \sum_{n=0}^{\infty} \sum_{m=0}^{n} \frac{(2-\delta_{0m})}{r_0} \left\{ \begin{array}{c} \left(\dfrac{r}{r_0}\right)^n \\ \left(\dfrac{r_0}{r}\right)^{n+1} \end{array} \right.$$

$$\cdot \frac{(n-m)!}{(n+m)!} P_n^m(\cos\theta_0) P_n^m(\cos\theta)\cos[m(\varphi - \varphi_0)]da_0 \; . \qquad \text{(A.5.21)}$$

Using the orthogonality relation (3.300) gives

$$J = \frac{4\pi r_0}{2n'+1} \left\{ \begin{array}{c} \left(\dfrac{r}{r_0}\right)^{n'} \\ \left(\dfrac{r_0}{r}\right)^{n'+1} \end{array} \right\} P_n^{m'}(\cos\theta) \; . \qquad \text{(A.5.22)}$$

The potential for the cases $r < r_0$ (ϕ_i) and $r > r_0$ (ϕ_a) becomes

$$\left. \begin{array}{l} \phi_i = -\dfrac{W}{3} r\cos\theta = -\dfrac{W}{3} z \\[2mm] \phi_a = -\dfrac{W}{3}\dfrac{r_0^3}{r^2}\cos\theta \end{array} \right\} \qquad \text{(A.5.23)}$$

and the vector potential is

$$\left. \begin{array}{l} A_{\varphi i} = +\dfrac{W}{3} r\sin\theta \\[2mm] A_{\varphi a} = +\dfrac{W}{3}\dfrac{r_0^3}{r^2}\sin\theta \end{array} \right\} \; . \qquad \text{(A.5.24)}$$

The corresponding field inside is
$$\mathbf{W}_i = -\nabla\phi_i + \nabla \times \mathbf{A}_i \; .$$

It possesses only a z-component.

$$W_{iz} = \frac{1}{3}W + \frac{2}{3}W = W \; , \qquad \text{(A.5.25)}$$

which is created by $1/3$ from sources and $2/3$ from vortices. On the outside, there are only dipole fields which cancel mutually and therefore
$$\mathbf{W}_a = 0 \; . \qquad \text{(A.5.26)}$$

All this should not be surprising. From previous Sections we know, that surface charge densities proportional to $\cos\theta$ and current densities at the surface in the azimuthal direction which are proportional to $\sin\theta$, cause uniform fields inside and dipole fields outside. One corresponds to the electric field of a uniformly polarized sphere, the other to the magnetic field of a uniformly magnetized sphere.

By (A.5.14), one can also express the entire field **W** by means of a scalar potential alone. In this case, the above vector potential has to be replaced by

$$\phi_3 = -\frac{1}{4\pi}\oint_a \phi(\mathbf{r}_0)\frac{\partial}{\partial r_0}\frac{1}{|\mathbf{r}-\mathbf{r}_0|}da_0 \ . \tag{A.5.27}$$

In order to be able to calculate this potential, one has to know the potential at the surface, which does by no means imply that one can arbitrarily fix this potential. This would lead to over determination of the problem, as we have discussed in Sect. 3.4. The purely scalar potential inside, that corresponds to the field of Fig. A.5.2, has to be

$$\phi_{gi} = -Wz = -Wr\cos\theta \ . \tag{A.5.28}$$

This gives

$$\phi_3 = +\frac{Wr_0}{4\pi}\oint_a P_1^0(\cos\theta_0)\frac{\partial}{\partial r_0}\frac{1}{|\mathbf{r}-\mathbf{r}_0|}da_0 \ .$$

With eq. (A.5.22), one can re-write this

$$\phi_3 = \frac{Wr_0}{4\pi}\cdot\frac{\partial}{\partial r_0}\left[\frac{1}{r_0}\left\{\frac{\left(\frac{r}{r_0}\right)^1}{\left(\frac{r_0}{r}\right)^2}\right\}\right]\cdot\frac{4\pi r_0^2}{3}\cos\theta \ ,$$

i.e.,

$$\left.\begin{array}{l}\phi_{3i} = -\dfrac{2}{3}Wr\cos\theta = -\dfrac{2}{3}Wz\\[2ex]\phi_{3a} = +\dfrac{1}{3}W\dfrac{r_0^3}{r^2}\cos\theta\end{array}\right\} \ . \tag{A.5.29}$$

Combining this with the potential eq. (A.5.23) yields

$$\left.\begin{array}{l}\phi_{gi} = -Wr\cos\theta = -Wz\\[1.5ex]\phi_{ga} = 0\end{array}\right\} \ , \tag{A.5.30}$$

which was to be proven.

The just presented example gave us a "hands on experience", illustrating that, and how, a field, created by surface vortices can, ad libitum, be expressed by either the corresponding vector potential or by the potential of the equivalent dipole layer. Vortices, (*i.e.* discontinuities of the tangential field component) occur only, if the dipole layer is inhomogeneous or if the potential under the integral eq. (A.5.27) is not constant. We know from Sect. 3.4 that the surface density of the dipole moment is

$$\tau = -\varepsilon_0\phi(\mathbf{r}_0) \ . \tag{A.5.31}$$

If the boundary coincides with an equipotential surface, then the dipole layer is uniform and there are no vortices, *i.e.*, the fields must vanish. We will illustrate this. Consider the very simple field depicted in Fig. A.5.3, which vanishes both, inside and outside the sphere. The potential shall be constant on the surface of the sphere:

$$\phi(\mathbf{r}_0) = C \ . \tag{A.5.32}$$

This corresponds to the case when in (A.5.14) only the third term is non-vanishing. Then one gets for the potential

$$\phi = -\frac{C}{4\pi}\oint_a \frac{\partial}{\partial r_0}\frac{1}{|\mathbf{r}-\mathbf{r}_0|} da_0 \ .$$

Using (A.5.22) and because of $P_0^0 = 1$, this results in

$$\phi = -\frac{C}{4\pi}\cdot\frac{\partial}{\partial r_0}\left[\frac{1}{r_0}\left\{\left(\frac{r_0}{r}\right)\right\}\right]\cdot 4\pi r_0^2 P_0^0 \ ,$$

or

$$\left.\begin{aligned}\phi_i &= C = -\frac{\tau}{\varepsilon_0}\\ \phi_a &= 0\end{aligned}\right\} \ . \tag{A.5.33}$$

This again, confirms one of our previous results, namely that of Sect. 2.5.3, eqs. (2.72) and (2.73) in particular. The potentials are constant and the fields vanish, just like we had assumed. Nevertheless, inside the dipole layer, there exists an infinitely strong field, which creates exactly the potential difference C.

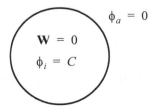

Fig. A.5.3

A.5.2.2 Point Charge inside a Conducting Hollow Sphere

Here, we will revisit the problem of a sphere which we have repeatedly discussed and solved, in Sect. 2.6.1 via the method of images, and in Sect. 3.8.2.3 by the separation of variables. The sphere shall have a radius of r_s. The potential on the surface shall be $\varphi = 0$ and the charge shall be on the z-axis at $z = r_0$. Then by (3.336)

$$
\phi = \frac{Q}{4\pi\varepsilon_0} \sum_{n=0}^{\infty} \left[\frac{1}{r_0} \left\{ \begin{matrix} \left(\frac{r}{r_0}\right)^n \\ \left(\frac{r_0}{r}\right)^{n+1} \end{matrix} \right\} - \frac{r_0^n r^n}{r_s^{2n+1}} \right] P_n^0(\cos\theta) \ , \tag{A.5.34}
$$

and the radial field at the sphere's surface is by (3.340)

$$
E_r = \frac{Q}{4\pi\varepsilon_0} \sum_{n=0}^{\infty} [2n+1] \frac{r_0^n}{r_s^{n+2}} P_n^0(\cos\theta) \ . \tag{A.5.35}
$$

The surface of the sphere constitutes an equipotential surface, that is, there are no tangential components (vortices).

The Helmholtz theorem under these conditions gives

$$
\phi = \int_V \frac{\nabla \bullet \mathbf{E}(\mathbf{r}')d\tau'}{4\pi|\mathbf{r}-\mathbf{r}'|} - \oint_a \frac{E_r(\mathbf{r}')da'}{4\pi|\mathbf{r}-\mathbf{r}'|} \ . \tag{A.5.36}
$$

The volume integral results with

$$
\nabla \bullet \mathbf{E}(\mathbf{r}') = \frac{Q}{\varepsilon_0}\delta(\mathbf{r}-\mathbf{r}') \tag{A.5.37}
$$

in the first part of the potential obtained in eq. (A.5.34). On the other hand, the surface integral, together with (A.5.35) provides the other part of that potential, which can be shown by means of eq. (A.5.22). This second part is nothing else than the potential of the image charge. However, in the context of this reflection, notice that this does not constitute a method to solve the problem. After all, we had to use the initially unknown field at the surface. Rather, the purpose here was to illustrate the content of the Helmholtz theorem by means of an example.

It shall also be noted that only compatible quantities may be substituted into Helmholtz theorem. In conjunction with the point charge Q, it would be possible to consider fields different from the one given in (A.5.35), which would constitute a different problem. In any case, the field prescribed at the surface has to possess a flux which fits to the totality of all sources. Consider a field with radial components at the surface of the following form:

$$
E_r = \sum_{n=0}^{\infty} A_n P_n^0(\cos\theta) \ , \tag{A.5.38}
$$

then the total flux created by this field $\mathbf{E}$ at the surface is

$$
\oint E_r(\mathbf{r}')da' = A_0 4\pi r_s^2 \ . \tag{A.5.39}
$$

If the total charge inside the volume is Q, then it has to be

$$A_0 4\pi r_s^2 = \frac{Q}{\varepsilon_0} \, , \tag{A.5.40}$$

which determines the first term of the series expansion (A.5.38) and agrees with (A.5.35).

A.6   Maxwell's Equations and Relativity

A.6.1   Galilean and Lorentz Transformation

Classical physics as coined by Newton, is based on an understanding which appears self-evident to us, namely that time and space are absolute realities. The result of this understanding is the supposition that the laws of physics are invariant under the so-called *Galilean Transformation*. Consider a physical process in two different reference frames, we shall call one Σ and the other Σ', where Σ' moves with a constant velocity $\mathbf{v}$ relative to the rest frame Σ. Both are *inertial systems*, *i.e.*, there shall be no inertial forces present. We will not discuss the important implications that come along with this restriction. The relation between the location vector $\mathbf{r}$ and the time t in system Σ on one hand, and the location vector $\mathbf{r}'$ and the time t' in system Σ' is as follows

$$\left. \begin{array}{l} \mathbf{r}' = \mathbf{r} - \mathbf{v}t \\ t' = t \end{array} \right\} \, . \tag{A.6.1}$$

The coordinate axes of the two systems shall be parallel to each other and they shall coincide for $t = 0$. The two equations (A.6.1) constitute the so-called *Galilean Transformation*. One of its consequences is the familiar addition of velocities. If a point particle moves with the velocity $\dot{\mathbf{r}}$ in system Σ, then its velocity in the other system (Σ') is $\dot{\mathbf{r}}' = \dot{\mathbf{r}} - \mathbf{v}$ and thus

$$\dot{\mathbf{r}} = \dot{\mathbf{r}}' + \mathbf{v} \, . \tag{A.6.2}$$

We had assumed, and this is important here, that the velocity is constant. The acceleration is therefore the same in both reference frames

$$\ddot{\mathbf{r}}' = \ddot{\mathbf{r}} \, . \tag{A.6.3}$$

Consequently, if there is a force $\mathbf{F}$, then the equation of the motion is the same in both frames

$$\mathbf{F} = m\ddot{\mathbf{r}} = m\ddot{\mathbf{r}}' \, , \tag{A.6.4}$$

having assumed that the mass is the same in both reference frames (which will not be the case once we consider relativistic effects). Though simplified and abbreviated, this is precisely what is meant when saying, that the *laws of classical physics are invariant under the Galilean Transformation*.

After having theoretically derived the speed of the propagation of electromagnetic waves in vacuum and identified it as the speed of light in vacuum,

$$c = \frac{1}{\sqrt{\varepsilon_0 \mu_0}} \,,$$

(A.6.5)

and then had also proven experimentally that waves really existed (Heinrich Hertz, 1887/88), the question became inevitable, for which reference frame this statement applied, *i.e.*, in which reference frame does light actually propagate with that speed. From a classical perspective, there is only one reference frame with this property. The speed of light according to eq. (A.6.2) is different for all other reference frames. Initially, it seemed quite natural, yes downright self-evident, to suppose that this unique reference frame (where the light had the given velocity) was the surrounding absolute space in which the fixed stars were at rest (not the earth, however). (From our current knowledge of space, with its vastness and the incredible dynamic of the behavior of these fixed stars within their spiral nebulas etc., this was a very naive conception.) In order for waves to propagate in this absolute space, it was assumed to be filled with a suitable medium, which was called *ether*. This *ether* was to play the same role for light waves as, *for example*, gas plays for the propagation of sound waves. Since the earth moves in the *ether*, the speed of light as seen from the earth should be different, exactly by its speed through the *ether*, which means that eq. (A.6.2) should be applicable. By means of this equation, it should be possible to determine the speed of the earth within the *ether* (and thereby its speed relative to the absolute space). These chains of reasoning triggered an intensive discussion and hectic experimental activity. Many experiments were performed with the intention to confirm the just described understanding (the famous Michelson interference experiment was one of them), but all failed. The final result shall be presented now without going into further details of these discussions. The result, which on one hand, is simple, but on the other, vigorously objects our immediate conception, was first formulated by Einstein (1905) in this form. *It says that the speed of light in vacuum has the same value in any uniformly moving reference frame (in all inertial systems), regardless of their speed* v, *relative to each other.* This forms the basis for the *theory of special relativity* (not to confuse with the later formulated *theory on general relativity*, which is not subject of this discussion). Therefore, *Maxwell's equations are not Galilei invariant.*

The theory of relativity that emerged from this understanding was mandated by the experimental results and inevitable. It was the compulsory result of the experimental experience, *i.e.*, it would have emerged, even without Einstein. Nevertheless, it rightly is connected to his name because he was the first to face the necessary and inevitable consequences, which required to radically part with our conception, and he formulated the results of the research of his time, and furthermore, because he has done this in an ingenious and clear manner.

The theory of relativity has caused discussions, even beyond the realm of Physics. Many of these discussions were more or less useful, frequently even bizarre and senseless, sometimes initiated by people who did not even understand it. Many of the debates, then and now, are distracted by the term relativity, which in a naive way is taken out of context and the theory of relativity is rejected because of some alleged philosophical reasons by which it is not permissible that everything is relative. On the other hand, people misuse the theory and claim that certain statements are relative, *i.e.*, not absolute, even though they are entirely unrelated to the theory. Feynman has commented on this in an ironic way in his famous textbook [37, Section 16-1, "Relativity and the Philosophers"]. *On the other hand, let's note for the record, that Einstein's theory is ultimately a very clear, even simple theory, which replaced the theory that preceded it historically (this preceding theory could be called Galilean theory of relativity). Furthermore, it shall be noted that Einstein's theory, by no means contradicts its predecessor but generalizes it because it includes it as a limit.*

The fundamental statement of the theory of relativity is that the quantity

$$c^2 = \frac{x^2 + y^2 + z^2}{t^2}$$

has the same value in every inertial frame, or that

$$\left. \begin{array}{l} x^2 + y^2 + z^2 - c^2 t^2 = \sum_{i=1}^{4} x_i^2 \\ \langle x, y, z, ict \rangle = \langle x_1, x_2, x_3, x_4 \rangle = \langle x_i \rangle \end{array} \right\}$$

(A.6.6)

is invariant. The mathematical transformation which ensures this, is the so-called *Lorentz transformation*. In contrast to the Galilean Transformation, it also includes the time. Time is now also dependent on the reference frame. To derive this transformation is not difficult. It only requires the *orthogonal transformation* of the coordinates of an n-dimensional space (here four-dimensional), which is well known in mathematics.

To illustrate the content of eq. (A.6.6), consider an arbitrary point $\mathbf{r}_0$ in the laboratory frame Σ, from which a spherical wave emerges at time $t = t_0$. In the reference frame Σ', this wave emerges from the point $\mathbf{r}_0'$ at the time $t' = t_0'$. By suitable translation in space and time, it is always possible to achieve a situation where $\mathbf{r}_0 = 0$, $\mathbf{r}_0' = 0$, $t_0 = 0$, and $t_0' = 0$. This does not impose a limit on its generality.

A.6.2   Lorentz Transformation as an Orthogonal Transformation

Consider an n-dimensional Euclidean space where two Cartesian coordinate systems coexist, but are rotated against each other. A particular location is represented in one system by $\langle x_i \rangle$ and in the other by $\langle x_i' \rangle$, where

$$x_i' = \sum_{k=1}^{n} L_{ik} x_k \qquad (i, k = 1, \ldots n) .$$

(A.6.7)

This transformation must possess the property

$$\sum_{i=1}^{n} x_i'^2 = \sum_{i=1}^{n} x_i^2 = \text{invariant} , \tag{A.6.8}$$

that is, the distance must be invariant. This is the case if the matrix operator

$$L = (L_{ik}) \tag{A.6.9}$$

is *orthogonal*. Its properties can easily be derived

$$\sum_{i=1}^{n} x_i'^2 = \sum_{i=1}^{n} \left(\sum_{k=1}^{n} L_{ik} x_k \right) \left(\sum_{l=1}^{n} L_{il} x_l \right) = \sum_{k=1}^{n} x_k^2 = \sum_{k=1}^{n} \sum_{l=1}^{n} \delta_{kl} x_k x_l .$$

It is therefore

$$\sum_{i=1}^{n} L_{ik} L_{il} = \delta_{kl} . \tag{A.6.10}$$

Of course we have

$$L^{-1} L = \hat{1} ,$$

i.e., unity and therefore

$$\sum_{i=1}^{n} L_{ki}^{-1} L_{il} = \delta_{kl} . \tag{A.6.11}$$

Comparing eqs. (A.6.10) and (A.6.11) yields

$$\boxed{L_{ki}^{-1} = L_{ik} , \qquad L^{-1} = L^T} . \tag{A.6.12}$$

This is exactly what represents the property of the orthogonal transformation operator, namely that its inverse operator L^{-1} is equal to its transposed operator L^T (where $L_{ki}^T = L_{ik}$). Obviously, one also has

$$LL^{-1} = \hat{1} ,$$

$$\sum_{i=1}^{n} L_{ki} L_{il}^{-1} = \sum_{i=1}^{n} L_{ki} L_{li} = \delta_{kl} . \tag{A.6.13}$$

The two eqs. (A.6.10) and (A.6.13) express that both, the column vectors as well as the row vectors of an orthogonal operator are unit vectors which are orthogonal to each other.

With this mathematical preparation, we now turn our attention to the Lorentz transformation. For that purpose, as before in eq. (A.6.6), we will use the four-dimensional vector

$$\langle x, y, z, ict \rangle = \langle x_1, x_2, x_3, x_4 \rangle = \langle x_i \rangle , \tag{A.6.14}$$

where now, the four-dimensional distance $\sum_{i=1}^{4} x_i^2$ has to be invariant. Formally, this

is the same problem as that of a four-dimensional Euclidean space. The difference is merely that one of the coordinates ($x_4 = ict$) and its corresponding matrix elements are now complex valued. This is therefore also called *pseudo-Euclidean four-dimensional space-time*, which is also called *Minkowski space*.

Initially, we shall assume that the relative velocity **v** has only an x_1 component v_1:

$$\mathbf{v} = \langle v_1, 0, 0 \rangle \tag{A.6.15}$$

This suggests to suppose that $y = x_2$ and $z = x_3$ remain unchanged in this case, while only $x = x_1$ and $ict = x_4$ will be changed by the transformation. This assumption is not necessary but simplifies matters in the following discussion. The transformation may then be written in the following form:

$$\left.\begin{array}{lll}
x_1' &=& L_{11}x_1 & +L_{14}x_4 \\
x_2' &=& x_2 \\
x_3' &=& x_3 \\
x_4' &=& L_{41}x_1 & +L_{44}x_4
\end{array}\right\} \tag{A.6.16}$$

When $x_1 = v_1 t$, then the origin of the moving reference frame Σ' is at $x' = 0$, which allows one to write

$$x_1' = f(v_1)(x_1 - v_1 t) = f(v_1)x_1 - \frac{v_1 f(v_1)}{ic}ict = f(v_1)x_1 - \frac{v_1 f(v_1)}{ic}x_4 \ .$$

Thus

$$L_{14} = -\frac{v_1}{ic}L_{11} = i\frac{v_1}{c}L_{11} \ . \tag{A.6.17}$$

Because of the orthogonality of L, one has

$$L_{11}^2 + L_{14}^2 = 1 \tag{A.6.18}$$

$$L_{41}^2 + L_{44}^2 = 1 \tag{A.6.19}$$

$$L_{11}L_{41} + L_{14}L_{44} = 0 \ . \tag{A.6.20}$$

Using eqs. (A.6.17) and (A.6.18) gives

$$L_{11} = \frac{1}{\sqrt{1 - \frac{v_1^2}{c^2}}} \ , \qquad L_{14} = \frac{i\frac{v_1}{c}}{\sqrt{1 - \frac{v_1^2}{c^2}}} \ . \tag{A.6.21}$$

Furthermore, from this and eq. (A.6.20) follows

$$L_{41} = -\frac{L_{14}}{L_{11}}L_{44} = -\frac{iv_1}{c}L_{44} \ ,$$

and by (A.6.19) this yields

$$L_{44} = \frac{1}{\sqrt{1 - \frac{v_1^2}{c^2}}} , \qquad L_{41} = -\frac{i\frac{v_1}{c}}{\sqrt{1 - \frac{v_1^2}{c^2}}} . \qquad \text{(A.6.22)}$$

Introducing the frequently used abbreviation

$$\beta = \frac{v}{c} = \frac{v_1}{c} , \qquad \text{(A.6.23)}$$

finally, the Lorentz transformation emerges

$$\boxed{\begin{aligned} x_1' &= \frac{x_1 + i\beta x_4}{\sqrt{1 - \beta^2}} \\ x_2' &= x_2 \\ x_3' &= x_3 \\ x_4' &= \frac{-i\beta x_1 + x_4}{\sqrt{1 - \beta^2}} \end{aligned}} \qquad \text{(A.6.24)}$$

or expressed with t and t'

$$\boxed{\begin{aligned} x_1' &= \frac{x_1 - v_1 t}{\sqrt{1 - \beta^2}} \\ x_2' &= x_2 \\ x_3' &= x_3 \\ t' &= \frac{-\frac{v_1}{c^2} x_1 + t}{\sqrt{1 - \beta^2}} \end{aligned}} \qquad \text{(A.6.25)}$$

One can easily verify that all requirements are met and, in particular, eq. (A.6.6) is invariant.

The case of an arbitrary velocity $\mathbf{v} = \langle v_1, v_2, v_3 \rangle$ can be reduced to the just derived result. For this purpose, decompose $(\mathbf{r} - \mathbf{v}t)$ into components parallel and perpendicular to $\mathbf{v}$,

$$(\mathbf{r} - \mathbf{v}t)_{\parallel} = [(\mathbf{r} - \mathbf{v}t) \bullet \mathbf{v}] \frac{\mathbf{v}}{v^2}$$

$$(\mathbf{r} - \mathbf{v}t)_{\perp} = (\mathbf{r} - \mathbf{v}t) - [(\mathbf{r} - \mathbf{v}t) \bullet \mathbf{v}] \frac{\mathbf{v}}{v^2}$$

and obtain

$$\mathbf{r'} = \frac{(\mathbf{r}-\mathbf{v}t)_{\parallel}}{\sqrt{1-\beta^2}} + (\mathbf{r}-\mathbf{v}t)_{\perp} = \mathbf{r} + \frac{(\mathbf{r}\bullet\mathbf{v})\mathbf{v}}{v^2}\left[\frac{1}{\sqrt{1-\beta^2}}-1\right] - \frac{\mathbf{v}t}{\sqrt{1-\beta^2}}$$

$$t' = \frac{-\frac{1}{c^2}(\mathbf{r}\bullet\mathbf{v})+t}{\sqrt{1-\beta^2}}$$

.(A.6.26)

Of course, letting $\mathbf{v} = \langle v_1, 0, 0\rangle$ gives the result of our previous transformation (A.6.25). Also easy to verify is that (A.6.26), indeed, is invariant as required, *i.e.*

$$r'^2 - c^2 t'^2 = r^2 - c^2 t^2 .$$

The L operator (when using $\beta = v/c$) is obtained from (A.6.26) and in the following form:

$$L = \begin{bmatrix} 1+\frac{v_1^2}{v^2}\left[\frac{1}{\sqrt{1-\beta^2}}-1\right] & \frac{v_1 v_2}{v^2}\left[\frac{1}{\sqrt{1-\beta^2}}-1\right] & \frac{v_1 v_3}{v^2}\left[\frac{1}{\sqrt{1-\beta^2}}-1\right] & \frac{iv_1}{c\sqrt{1-\beta^2}} \\ \frac{v_1 v_2}{v^2}\left[\frac{1}{\sqrt{1-\beta^2}}-1\right] & 1+\frac{v_2^2}{v^2}\left[\frac{1}{\sqrt{1-\beta^2}}-1\right] & \frac{v_2 v_3}{v^2}\left[\frac{1}{\sqrt{1-\beta^2}}-1\right] & \frac{iv_2}{c\sqrt{1-\beta^2}} \\ \frac{v_1 v_3}{v^2}\left[\frac{1}{\sqrt{1-\beta^2}}-1\right] & \frac{v_2 v_3}{v^2}\left[\frac{1}{\sqrt{1-\beta^2}}-1\right] & 1+\frac{v_3^2}{v^2}\left[\frac{1}{\sqrt{1-\beta^2}}-1\right] & \frac{iv_3}{c\sqrt{1-\beta^2}} \\ \frac{-iv_1}{c\sqrt{1-\beta^2}} & \frac{-iv_2}{c\sqrt{1-\beta^2}} & \frac{-iv_3}{c\sqrt{1-\beta^2}} & \frac{1}{\sqrt{1-\beta^2}} \end{bmatrix}.$$

(A.6.27)

Its *inverse transformation is obtained simply by replacing v by -v.* It becomes immediately obvious that the so obtained inverse matrix is $L^{-1} = L^T$. Letting $\beta \to 0$ results in the Galilean Transformation. This means that the theory of relativity includes classical physics in this limit and thereby generalizes it.

We have, thereby, gained the Lorentz transformation, first in its simple form of eqs. (A.6.24) or (A.6.25) for $\mathbf{v} = \langle v_1, 0, 0\rangle$ and then in its more complicated form of eq. (A.6.27) for $\mathbf{v} = \langle v_1, v_2, v_3\rangle$. Next, we want to consider some of its fundamental consequences. For the most part, we will use the simpler form, which is sufficient for many purposes.

A.6.3   Some Consequences of the Lorentz Transformation

A.6.3.1 Lorentz Contraction

Consider a rod, oriented parallel to the x_1 and the x_1' axis, respectively. Its length in system Σ is $l = x_{1e} - x_{1a}$. What is its length $l' = x_{1e}' - x_{1a}'$ in system Σ'? The task requires caution. One has to ensure that the coordinate points of both ends in

Σ' are determined at the same time t'. This needs to be recognized because simultaneousness in Σ does not imply simultaneousness in Σ'. Therefore, one requires that $t_{1e}' = t_{1a}'$, and obtain with (A.6.25)

$$t_a - \frac{v_1}{c^2}x_{1a} = t_e - \frac{v_1}{c^2}x_{1e}, \qquad t_e - t_a = \frac{v_1}{c^2}(x_{1e} - x_{1a}) \ .$$

Furthermore

$$l' = x_{1e}' - x_{1a}' = \frac{x_{1e} - x_{1a} - v_1(t_e - t_a)}{\sqrt{1-\beta^2}} = \frac{x_{1e} - x_{1a} - \frac{v_1^2}{c^2}(x_{1e} - x_{1a})}{\sqrt{1-\beta^2}},$$

$$\boxed{l' = (x_{1e} - x_{1a})\sqrt{1-\beta^2} = l\sqrt{1-\beta^2}} \ . \tag{A.6.28}$$

From the moving system Σ', one "observes" the rod shortened by the factor $\sqrt{1-\beta^2}$. This phenomenon is the so-called *Lorentz-contraction*. The word observes in the previous sentence was used in quotation marks because caution is of order when interpreting this result, in order to avoid misconception – which has frequently occurred.

The Lorentz-contraction causes, *for example*, that a sphere in motion for an observer at rest changes to a flattened ellipsoid. However, it does not mean that a visually observed sphere is seen by an observer as a flattened ellipsoid, nor that a photographic picture would show a flattened ellipsoid, as it was frequently stated in the past. This mistake was cleared only relatively late by Penrose [38] and Terrell [39]. Since the speed of light is finite, visual or photographic observation of an object gives a distorted picture, because it shows parts with different distances to the observer at different times. In case of a large distance to the object (or small solid angle) the object appears in its natural shape and length, but rotated. If the object is a sphere, then the observer sees a rotated sphere, but not an ellipsoid. If the object is a rod, then the observer sees a rod with its natural length but rotated. The angle is such that the rod's projection onto the direction of the motion represents the Lorentz-contraction. We pass on a detailed discussion here. Those details are found in the mentioned publications [38, 39]. It shall be emphasized that there is no doubt on the reality of the Lorentz contraction. This is a real and experimentally verified effect. The described facts simply mean that a visual observation is not a suitable method to experimentally verify the Lorentz contraction. The fact that a visually observed object, precisely because of the Lorentz contraction remains visible in its natural shape is very remarkable. The reason is that the distortion is compensated for by the different optical paths. In this sense, the theory of relativity restores the "vividness" of objects in motion, which were otherwise lost due to the finite speed of light.

A.6.3.2 Time Dilatation

Let an object (*e.g.* a clock) be located at position x_1, which emits signals at different times with $\Delta t = t_2 - t_1$. Then for system Σ' one has

$$\Delta t' = t_2' - t_1' = \frac{\left(t_2 - \frac{v_1}{c^2}x_1\right) - \left(t_1 - \frac{v_1}{c^2}x_1\right)}{\sqrt{1-\beta^2}} = \frac{t_2 - t_1}{\sqrt{1-\beta^2}} = \frac{\Delta t}{\sqrt{1-\beta^2}}$$

$$\boxed{\Delta t' = \frac{\Delta t}{\sqrt{1-\beta^2}}}. \tag{A.6.29}$$

From the perspective of the moving system, the temporal distance appears expanded by the factor $1/\sqrt{1-\beta^2}$. Therefore, a clock runs slower and slower as it moves faster and faster. This is the so-called *time dilatation*. This can be proven experimentally, *e.g.*, by means of very fast instable particles (radioactive decay), where its half-life relative to an observer at rest is elongated versus its half-live in the rest frame. There are so-called μ-mesons contained in high altitude radiation. They are instable and very clearly show the effect of time dilatation. Time dilatation is also the root cause of the so-called *twin paradox*.

A.6.3.3 Relativistic Addition of Velocities

Let a particle in the Σ' system have the velocity $\mathbf{u'} = \langle u_1', u_2', u_3' \rangle$. What is its velocity in the Σ frame? First, using $\beta = v/c = v_1/c$, gives

$$u_1 = \frac{dx_1}{dt} = \frac{\frac{dx_1'}{dt'} + v_1}{\sqrt{1-\beta^2}} \cdot \frac{dt'}{dt}$$

and

$$\frac{dt'}{dt} = \frac{1 - \frac{v_1 u_1}{c^2}}{\sqrt{1-\beta^2}} = \frac{\sqrt{1-\beta^2}}{1 + \frac{v_1 u_1'}{c^2}}. \tag{A.6.30}$$

Therefore

$$\boxed{u_1 = \frac{u_1' + v_1}{1 + \frac{v_1 u_1'}{c^2}}, \qquad u_1' = \frac{u_1 - v_1}{1 - \frac{v_1 u_1'}{c^2}}}. \tag{A.6.31}$$

Furthermore

$$u_2 = \frac{dx_2}{dt} = \frac{dx_2'}{dt'} \cdot \frac{dt'}{dt} = u_2' \frac{dt'}{dt}, \qquad u_3 = u_3' \frac{dt'}{dt},$$

and with eq. (A.6.30) we obtain

$$u_2 = u_2' \cdot \frac{1 - \dfrac{v_1 u_1}{c^2}}{\sqrt{1-\beta^2}} = u_2' \cdot \frac{\sqrt{1-\beta^2}}{1 + \dfrac{v_1 u_1'}{c^2}}$$

$$u_3 = u_3' \cdot \frac{1 - \dfrac{v_1 u_1}{c^2}}{\sqrt{1-\beta^2}} = u_3' \cdot \frac{\sqrt{1-\beta^2}}{1 + \dfrac{v_1 u_1'}{c^2}}$$

(A.6.32)

Eqs. (A.6.31) and (A.6.32) represent the relativistic theorem for addition of velocities, which for small velocities ($v \ll c$) reduces to the classical vector addition ($u_1 = u_1' + v_1, u_2 = u_2', u_3 = u_3'$).

Consider a few special cases: For $\mathbf{u}' = \langle c, 0, 0 \rangle$ one has also $\mathbf{u} = \langle c, 0, 0 \rangle$.

If $\quad \mathbf{u}' = \langle 0, c, 0 \rangle,\quad$ then $\quad \mathbf{u} = \langle v_1, c\sqrt{1 - \dfrac{v_1^2}{c^2}}, 0 \rangle,\quad$ where

$u^2 = v_1^2 + c^2(1 - v_1^2/c^2) = c^2$ and thus $u = c$. Next, more general, let

$\mathbf{u}' = \langle c\cos\alpha', c\sin\alpha', 0 \rangle$

then

$$\mathbf{u} = \langle \frac{c\cos\alpha' + v_1}{1 + \dfrac{v_1\cos\alpha'}{c}}, \frac{c\sin\alpha'\sqrt{1 - \dfrac{v_1^2}{c^2}}}{1 + \dfrac{v_1\cos\alpha'}{c}}, 0 \rangle = \langle c\cos\alpha, c\sin\alpha, 0 \rangle, \qquad (A.6.33)$$

where

$$\cos\alpha = \frac{\cos\alpha' + \dfrac{v_1}{c}}{1 + \dfrac{v_1\cos\alpha'}{c}} \qquad \sin\alpha = \frac{\sin\alpha'\sqrt{1 - \dfrac{v_1^2}{c^2}}}{1 + \dfrac{v_1\cos\alpha'}{c}}. \qquad (A.6.34)$$

Whereby

$\cos^2\alpha + \sin^2\alpha = \cos^2\alpha' + \sin^2\alpha' = 1$

and

$u = u' = c$. (A.6.35)

This means that an object (*for example*. light), moving with the speed of light, has the same speed in every reference frame. This is, of course, necessary and was the starting point of our reflections (Sect. 6.1). Nevertheless, the direction of the motion (that is the propagation direction of *e.g.* light) depends on the reference frame. This is the so-called *aberration* (of light), which is described by the relations (A.6.34).

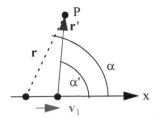

Fig. A.6.1

For $v_1 = c$, the velocity $\mathbf{u} = \langle c, 0, 0 \rangle$ is independent of $\mathbf{u}'$. This also is an immediate and plausible consequence of the invariance of the speed of light.

A.6.3.4 Aberration and Doppler Effect

Consider a moving, point-like light source from which a spherical wave emerges (Fig. A.6.1). At point P, an observer in the rest frame Σ sees it at the angle α, while an observer moving with the light source (reference frame Σ') sees the source at the angle α'. The quantities measured then in Σ have the phase factor

$$\exp\left[i\left(\omega t - \frac{\omega}{c}x_1 \cos\alpha - \frac{\omega}{c}x_2 \sin\alpha\right)\right]$$

and in Σ', the phase factor is

$$\exp\left[i\left(\omega' t' - \frac{\omega'}{c}x_1' \cos\alpha' - \frac{\omega'}{c}x_2' \sin\alpha'\right)\right] .$$

These have to be compatible with the Lorentz transformation and mutually transform into each other. This requires that

$$\omega' t' - \frac{\omega'}{c}x_1' \cos\alpha' - \frac{\omega'}{c}x_2' \sin\alpha'$$

$$= \frac{\omega'\left(t - \frac{v_1 x_1}{c^2}\right)}{\sqrt{1-\beta^2}} - \frac{\omega'(x_1 - v_1 t)}{c\sqrt{1-\beta^2}}\cos\alpha' - \frac{\omega'}{c}x_2 \sin\alpha'$$

$$= \omega t - \frac{\omega}{c}x_1 \cos\alpha - \frac{\omega}{c}x_2 \sin\alpha .$$

By comparing coefficients of t we find

$$\boxed{\omega = \omega'\frac{1 + \frac{v_1}{c}\cos\alpha'}{\sqrt{1-\beta^2}}} \tag{A.6.36}$$

that is, the angular frequency ω depends on the reference system.

This represents the so-called *relativistic Doppler effect*. Comparing coefficients of x_1 and x_2, while also considering (A.6.36), results in eq. (A.6.34), *i.e.* the aberration of light.

These results amount to the phase $\mathbf{k} \bullet \mathbf{r} - \omega t$ being relativistically invariant, to which we will return in Sect. A.6.5.2.

A.6.4 Lorentz Transformation of Maxwell's equations

The experiments mentioned in Sect. A.6.1 to determine the speed of the earth within the absolute space (*ether*) had to fail, because electromagnetic waves in vacuum propagate in all inertial systems with the same speed, the vacuum-speed of light c. Ultimately, the reason for this is that the Maxwell equations are not Galilei invariant, but Lorentz invariant, which was only revealed by, and understood in light of Einstein's theory of relativity. Today, we are able to realize that Maxwell's equations were, and still have to be, the starting point for relativity and, even though it was not apparent when Maxwell stated them, his equations always contained relativity. This is the reason why it is so important to show that the Maxwell equations are indeed Lorentz invariant, that is, that these equations apply unchanged in all inertial reference frames, as long as the field quantities (*i.e.* the components of $\mathbf{E}$ and $\mathbf{B}$) are transformed in a suitable manner.

To show this, we consider the two reference frames Σ and Σ'. In Σ we have the coordinates $\mathbf{r}$, t and the fields $\mathbf{E}(\mathbf{r},t)$, $\mathbf{B}(\mathbf{r},t)$, in Σ' we have the coordinates $\mathbf{r}'$, t' and the fields $\mathbf{E}'(\mathbf{r}',t')$, $\mathbf{B}'(\mathbf{r}',t')$. We limit our examination to free space. Later we will apply a generally applicable formalism which allows in an elegant way to treat the general Maxwell equations. For the frame Σ, Maxwell's equations apply in the following form

$$
\left.
\begin{aligned}
\nabla \times \mathbf{B} &= \frac{1}{c^2}\frac{\partial \mathbf{E}}{\partial t} \qquad \frac{1}{c^2} = \varepsilon_0 \mu_0 \\[2mm]
\nabla \times \mathbf{E} &= -\frac{\partial \mathbf{B}}{\partial t} \\[2mm]
\nabla \bullet \mathbf{B} &= 0 \\[2mm]
\nabla \bullet \mathbf{E} &= 0
\end{aligned}
\right\} . \tag{A.6.37}
$$

The transformation of $\mathbf{r}$, t by the Lorentz transformation to $\mathbf{r}'$, t' is not difficult but tedious and therefore skipped here (this calculation can be found *e.g.* in Simonyi [40]) and results again in Maxwell's equations,

$$
\left.
\begin{aligned}
\nabla' \times \mathbf{B}' &= \frac{1}{c^2}\frac{\partial \mathbf{E}'}{\partial t'} \\[2mm]
\nabla' \times \mathbf{E}' &= -\frac{\partial \mathbf{B}'}{\partial t'} \\[2mm]
\nabla' \bullet \mathbf{B}' &= 0 \\[2mm]
\nabla' \bullet \mathbf{E}' &= 0
\end{aligned}
\right\} . \tag{A.6.38}
$$

The transformation of the field components is as follows:

$$E_x{}' = E_x, \qquad E_y{}' = \frac{E_y - v_1 B_z}{\sqrt{1-\beta^2}}, \qquad E_y{}' = \frac{E_z + v_1 B_y}{\sqrt{1-\beta^2}} \Bigg\}$$

$$B_x{}' = B_x, \qquad B_y{}' = \frac{B_y + \dfrac{v_1}{c^2}E_z}{\sqrt{1-\beta^2}}, \qquad B_y{}' = \frac{B_z - \dfrac{v_1}{c^2}E_z}{\sqrt{1-\beta^2}} \Bigg\} \tag{A.6.39}$$

When using $\mathbf{v} = \langle v_1, 0, 0 \rangle$, this can be written in a more elegant form

$$\mathbf{E}_\parallel{}' = \mathbf{E}_\parallel, \qquad \mathbf{E}_\perp{}' = \frac{\mathbf{E}_\perp + \mathbf{v} \times \mathbf{B}}{\sqrt{1-\beta^2}} \Bigg\}$$

$$\mathbf{B}_\parallel{}' = \mathbf{B}_\parallel, \qquad \mathbf{B}_\perp{}' = \frac{\mathbf{B}_\perp - \dfrac{1}{c^2}\mathbf{v} \times \mathbf{E}}{\sqrt{1-\beta^2}} \Bigg\} \tag{A.6.40}$$

Only the field components perpendicular to the velocity $\mathbf{v}$ change, while the parallel ones remain unchanged. This is just the reverse from what was found for the components of the location vector. This is no accident, as we shall see more clearly. The nabla symbols for curl $\nabla' \times$ and divergence $\nabla' \bullet$ have a prime as a reminder that $\mathbf{E}'$ and $\mathbf{B}'$ are functions of $\mathbf{r}'$, t', and therefore the derivative in eqs. (A.6.38) is with respect to $\mathbf{r}' = \langle x_1', x_2', x_3' \rangle$. Conversely, eqs. (A.6.40) are independent of any particular coordinate system and apply for arbitrary velocities $\mathbf{v} = \langle v_1, v_2, v_2 \rangle$.

The force on a particle in system Σ is $\mathbf{f}$, and $\mathbf{f}'$ in system Σ'

$$\begin{aligned} \mathbf{f} &= Q(\mathbf{E} + \mathbf{u} \times \mathbf{B}) \\ \mathbf{f}' &= Q(\mathbf{E}' + \mathbf{u}' \times \mathbf{B}') \end{aligned} \Bigg\} \tag{A.6.41}$$

This means that the forces on a particle have the same form in every reference frame, *i.e.*, they are calculated within their respective reference frame from the electric and magnetic fields in the same way. Nevertheless $\mathbf{f} \neq \mathbf{f}'$. The equations for the transformation of $\mathbf{f}$ will be provided in Sect. A.6.5.2, eq. (A.6.72), which incidentally, can be used to derive the equations for the transformation of the electromagnetic fields. The charge Q is scalar, *i.e.*, is not transformed, which means that $Q = Q'$. Remarkable and easy to verify is also that the two following quantities are Lorentz invariant:

$$B'^2 - \frac{1}{c^2}E'^2 = B^2 - \frac{1}{c^2}E^2, \qquad \mathbf{E}' \bullet \mathbf{B}' = \mathbf{E} \bullet \mathbf{B}. \tag{A.6.42}$$

The reader shall be reminded of Sect. 6.1.2 and particularly eq. (6.6). Comparison with A.6.40 reveals that there, if we consider that $\mathbf{E}$ is the electric field in the moving frame, we had applied the non-relativistic approximation, *i.e.* the approximation for velocities $v_1 \ll c$, $\sqrt{1-\beta^2} \to 1$.

Furthermore, the reader shall be reminded of Appendix A.4. There, we have introduced to the Liénard-Wiechert potentials and calculated the problem of a uniformly moving charge as a specific example. Within the moving frame, the particle exhibits exclusively an electric Coulomb field and no magnetic field. The just provided transformation allows one to obtain the electric and the magnetic field, as experienced by the observer at rest. Carrying out the transformation yields precisely the field obtained in Appendix A.4, which demonstrates the fact that Maxwell's equations automatically provide the relativistically correct result. This result is limited to the special case of constant velocities because the Lorentz transformation is valid only for inertial systems, but not for accelerated reference frames. We will return in Sect. A.6.6.4 to the discussion of a charge of uniform velocity.

We will carry out the transformation of the fields in a more systematic and more elegant manner. Thereby, it will be revealed that the field components are simply components of an anti-symmetric four-dimensional tensor of rank two (which possesses just six independent components). From this follows immediately how to transform the field components. Moreover, this tensor allows one to express Maxwell's equations in a particularly compact and elegant form, which furthermore, makes its Lorentz invariance instantly plausible. This requires to occupy ourselves with the four-dimensional vectors of the Minkowski space (the so-called *4-vectors*) and their corresponding *Four-tensors*, *i.e.*, the tensor products of the 4-vectors.

A.6.5   4-Vectors and 4-Tensors

A.6.5.1 Definitions

When changing the reference frame of a 4-vector

$$\langle \mathbf{r}, ict \rangle = \langle x_1, x_2, x_3, x_4 \rangle \ , \tag{A.6.43}$$

where the fourth component is $x_4 = ict$, then the Lorentz transform

$$L = (L_{ik}) \tag{A.6.44}$$

is applied and it holds that

$$x_i' = \sum_{k=1}^{4} L_{ik} x_k \ . \tag{A.6.45}$$

Generally, only vectors that transform in this manner are called 4-vectors

$$a_i' = \sum_{k=1}^{4} L_{ik} a_k \ . \tag{A.6.46}$$

Quantities which transform like multiple products of the components of 4-vectors are called tensors of the respective rank. The expressions

$$b'_{ik} = \sum_{l=1}^{4} \sum_{m=1}^{4} L_{il} L_{km} b_{lm} , \tag{A.6.47}$$

$$c'_{ikl} = \sum_{m=1}^{4} \sum_{n=1}^{4} \sum_{p=1}^{4} L_{im} L_{kn} L_{lp} c_{mnp} , \tag{A.6.48}$$

represent a tensor of rank 2, rank 3, etc. In this sense, a 4-vector is a tensor of rank 1. A scalar quantity is a tensor of rank zero, for which

$$d = d' . \tag{A.6.49}$$

The scalar product (corresponding to the summation over a common index, which is also referred to as rank reduction) reduces the rank of the tensor by one each time. The scalar product of two 4-vectors $\sum_{i=1}^{4} a_i b_i$ represents an invariant scalar quantity. The scalar product of a 4-vector and a 4-tensor of rank two $\sum_{i=1}^{4} a_i b_{ik}$ yields a 4-vector, etc.

The coefficients of the Lorentz transform for $\mathbf{v} = \langle v_1, 0, 0 \rangle$ are given by (A.6.24) and by (A.6.27) for $\mathbf{v} = \langle v_1, v_2, v_2 \rangle$.

A.6.5.2 Some Important 4-Vectors

An important vector in the three-dimensional space is the ∇-vector. In the four-dimensional space it becomes a 4-vector:

$$\boxed{\nabla = \left\langle \frac{\partial}{\partial x_1}, \frac{\partial}{\partial x_2}, \frac{\partial}{\partial x_3}, \frac{\partial}{\partial x_4} \right\rangle = \left\langle \frac{\partial}{\partial x_1}, \frac{\partial}{\partial x_2}, \frac{\partial}{\partial x_3}, \frac{1}{ic} \frac{\partial}{\partial t} \right\rangle} , \tag{A.6.50}$$

Now, we explore the continuity equation (1.58) in this context

$$\nabla \bullet g + \frac{\partial \rho}{\partial t} = \frac{\partial}{\partial x_1} g_{x1} + \frac{\partial}{\partial x_2} g_{x2} + \frac{\partial}{\partial x_3} g_{x3} + \frac{1}{ic} \frac{\partial}{\partial t} (ic\rho) = 0 ,$$

which is nothing else than the four-dimensional divergence of the 4-vector of the current density or short *4-current density*

$$\boxed{\langle \mathbf{g}, ic\rho \rangle = \langle g_{x1}, g_{x2}, g_{x3}, ic\rho \rangle} . \tag{A.6.51}$$

It is a very remarkable and consequential result that the three components of the current density, together with $ic\rho$ (that is the volume charge density, for the most part) form a 4-vector. That this is true, can be shown on one hand, but is also a result of the fact that the divergence of the 4-vector vanishes, *i.e.* it is invariant.

The Lorentz gauge (eq. (7.186)) represents a four-dimensional divergence

$$\nabla \bullet \mathbf{A} + \mu_0 \varepsilon_0 \frac{\partial \varphi}{\partial t} = \nabla \bullet \mathbf{A} + \frac{1}{c^2} \frac{\partial \varphi}{\partial t}$$

$$= \frac{\partial}{\partial x_1} A_{x1} + \frac{\partial}{\partial x_2} A_{x2} + \frac{\partial}{\partial x_3} A_{x3} + \frac{1}{ic} \frac{\partial}{\partial t} \left(\frac{ic\varphi}{c^2} \right) = 0 . \tag{A.6.52}$$

This brings to light the fact, that the vector potential, together with $A_{x4} = (i\varphi)/c$ as its fourth component forms a *4-potential*,

$$\langle \mathbf{A}, i\frac{\varphi}{c} \rangle = \langle A_{x1}, A_{x2}, A_{x3}, i\frac{\varphi}{c} \rangle .$$

(A.6.53)

The scalar product of the four-dimensional ∇-vector with itself yields the four-dimensional analogue to the Laplacian

$$\nabla \bullet \nabla = \frac{\partial^2}{\partial x_1^2} + \frac{\partial^2}{\partial x_2^2} + \frac{\partial^2}{\partial x_3^2} - \frac{1}{c^2}\frac{\partial^2}{\partial t^2} = \nabla^2 - \frac{1}{c^2}\frac{\partial^2}{\partial t^2} = \Box .$$

(A.6.54)

This scalar operator is the d'Alembertian and frequently represented by the square symbol $\Box$. This allows to combine the two inhomogeneous wave equations (7.187) and (7.188) into a single equation:

$$\Box \langle \mathbf{A}, \frac{i\varphi}{c} \rangle = -\mu_0 \langle \mathbf{g}, ic\rho \rangle .$$

(A.6.55)

This represents the *four-dimensional inhomogeneous wave equation* for the 4-potential with the 4-current density as its inhomogeneity. It illustrates in a simple way that here we have Lorentz-invariant equations.

The Lorentz gauge resulted in a four-dimensional divergence, which means it is Lorentz invariant. This poses an advantage over other gauges which are not Lorentz invariant.

Another important 4-vector, particularly for mechanical applications, is the *4-momentum*. Consider a particle in the reference frame Σ with the rest mass m_0 and the velocity $\mathbf{u} = \langle u_1, u_2, u_3 \rangle$. $\mathbf{u}$ is used to distinguish the particle velocity from the relative velocity $\mathbf{v}$ of the two reference frames Σ and Σ'. We assign the momentum vector

$$\mathbf{p} = \frac{m_0 u}{\sqrt{1 - \frac{u^2}{c^2}}}$$

(A.6.56)

to the particle in system Σ. For the reference frame Σ', the vector shall be

$$\mathbf{p}' = \frac{m_0 u'}{\sqrt{1 - \frac{u'^2}{c^2}}} .$$

(A.6.57)

The relation between $\mathbf{u}$ and $\mathbf{u}'$ is given by eqs. (A.6.31) and (A.6.32), which allow to derive the interesting and oftentimes useful relation

$$\frac{1}{\sqrt{1 - \frac{u'^2}{c^2}}} = \frac{1}{\sqrt{1 - \frac{u^2}{c^2}}} \cdot \frac{1 - \frac{u_1 v_1}{c^2}}{\sqrt{1 - \frac{v_1^2}{c^2}}} .$$

(A.6.58)

Then reapplying eqs. (A.6.31) and (A.6.32) gives

$$p_1' = \frac{m_0 u_1'}{\sqrt{1 - \frac{u'^2}{c^2}}} = \frac{m_0}{\sqrt{1 - \frac{u^2}{c^2}}} \cdot \frac{1 - \frac{u_1 v_1}{c^2}}{\sqrt{1 - \frac{v_1^2}{c^2}}} \cdot \frac{u_1 - v_1}{1 - \frac{u_1 v_1}{c^2}} = \frac{p_1 - \frac{m_0 v_1}{\sqrt{1 - \frac{u^2}{c^2}}}}{\sqrt{1 - \frac{v_1^2}{c^2}}}$$

$$p_2' = \frac{m_0 u_2'}{\sqrt{1 - \frac{u'^2}{c^2}}} = \frac{m_0}{\sqrt{1 - \frac{u^2}{c^2}}} \cdot \frac{1 - \frac{u_1 v_1}{c^2}}{\sqrt{1 - \frac{v_1^2}{c^2}}} u_2 \frac{\sqrt{1 - \frac{v_1^2}{c^2}}}{1 - \frac{u_1 v_1}{c^2}} = \frac{m_0 u_2}{\sqrt{1 - \frac{u^2}{c^2}}} = p_2$$

$$p_3' = p_3$$

$$\left. \right\} \quad . \text{(A.6.59)}$$

We continue by introducing the masses m and m', as well as the energies W and W' in the frames Σ and Σ', respectively.

$$\boxed{m = \frac{m_0}{\sqrt{1 - \frac{u^2}{c^2}}} = \frac{W}{c^2} \, , \quad m' = \frac{m_0}{\sqrt{1 - \frac{u'^2}{c^2}}} = \frac{W'}{c^2} \, ,} \qquad \text{(A.6.60)}$$

with the relation

$$m' = \frac{W'}{c^2} = \frac{m_0}{\sqrt{1 - \frac{u'^2}{c^2}}} = \frac{m_0}{\sqrt{1 - \frac{u^2}{c^2}}} \cdot \frac{1 - \frac{u_1 v_1}{c^2}}{\sqrt{1 - \frac{v_1^2}{c^2}}} = \frac{-p_1 \frac{v_1}{c^2} + \frac{W}{c^2}}{\sqrt{1 - \frac{v_1^2}{c^2}}} \, . \qquad \text{(A.6.61)}$$

Comparison of eqs. (A.6.59) and (A.6.61) with (A.6.24) reveals that the vector

$$\boxed{\langle \mathbf{p}, \frac{iW}{c} \rangle = m(\mathbf{u}, ic) = \frac{m_0}{\sqrt{1 - \frac{u^2}{c^2}}} \langle \mathbf{u}, ic \rangle} \qquad \text{(A.6.62)}$$

is a 4-vector. We call it the 4-momentum. The quantity m is the velocity dependent mass, which depends on the reference frame, along with the speed $\mathbf{u}$ of the particle. W is the particle's energy, which also depends on the reference frame. *Eq. (A.6.60) contains the famous Einstein relation between mass and energy ($W = mc^2$) and the known relation between m and m_0.* Now, the background and the reasons for these relations becomes apparent, which rest in the fact that the vector (A.6.62) is a 4-vector. For very small velocities $\mathbf{u}$, its three spatial components give the classical momentum $m_0 \mathbf{u}$, which justifies the name 4-momentum.

The rest mass m_0 is invariant, the just defined velocity dependent mass is not, however. Dividing the 4-momentum by m_0 gives another 4-vector, the so-called 4-velocity

$$\boxed{\frac{1}{\sqrt{1 - \dfrac{u^2}{c^2}}} \langle \mathbf{u}, ic \rangle} \ . \tag{A.6.63}$$

Conversely, dividing the 4-momentum by m gives another vector $(\mathbf{u}, ic)$, but this is not a 4-vector. The velocity $\mathbf{u}$ can not be regarded as part of a 4-vector. Eqs. (A.6.31) and (A.6.32) also reveal that $\mathbf{u}$ transforms <u>rather</u> differently than the spatial part of the 4-vector. Only the factor $1/(\sqrt{1 - u^2/c^2})$ creates a 4-vector. The absolute value of the 4-momentum is invariant, of course,

$$p^2 - \frac{W^2}{c^2} = \text{invariant} \ .$$

$W = m_0 c^2$ for $p = 0$ and we obtain

$$p^2 - \frac{W^2}{c^2} = -m_0^2 c^2 \ .$$

Rearranging reveals the important relation which we have used before, eq. (A.1.7):

$$\boxed{W^2 = m_0^2 c^4 + p^2 c^2} \ . \tag{A.6.64}$$

When specifically considering electromagnetic radiation, one finds that it also possesses momentum and energy, where both are proportional to each other. Expressing both for a light quantum (which is not necessary to do), we get

$$\mathbf{p} = \hbar k \ , \qquad \mathbf{E} = \hbar \omega \tag{A.6.65}$$

and the 4-vector of the momentum becomes

$$\hbar \langle \mathbf{k}, \frac{i\omega}{c} \rangle \ . \tag{A.6.66}$$

Consequently,

$$\langle \mathbf{k}, \frac{i\omega}{c} \rangle \tag{A.6.67}$$

is a 4-vector whose fourth component is, essentially, the angular frequency. We might call it the 4-vector of the wave number whose scalar product is invariant with the 4-vector of the location.

$$\langle \mathbf{k}, \frac{i\omega}{c} \rangle \bullet (\mathbf{r}, ict) = \mathbf{k} \bullet \mathbf{r} - \omega t = \text{invariant} \ . \tag{A.6.68}$$

This invariance is nothing else than the phase, whose invariance we have already noted in A.6.3.4. Furthermore:

$$\langle \mathbf{k}, \frac{i\omega}{c} \rangle^2 = k^2 - \frac{\omega^2}{c^2} = \text{invariant} = 0 \ . \tag{A.6.69}$$

We obtain the dispersion relation, which is of course invariant as well.

Another 4-vector is the 4-force

$$F = \frac{1}{\sqrt{1 - \dfrac{u^2}{c^2}}} \left\langle \mathbf{f}, \frac{i\,\mathrm{d}W}{c\,\mathrm{d}t} \right\rangle .$$

(A.6.70)

W is the particle energy, introduced with eq (A.6.60) and $\mathbf{f}$ the 3-force. With this, we have the relativistic equation of the motion

$$\mathbf{f} = \frac{\mathrm{d}}{\mathrm{d}t} \frac{m_0 \mathbf{u}}{\sqrt{1 - \dfrac{u^2}{c^2}}} .$$

(A.6.71)

Note that $\mathbf{f}$ does not represent the spatial part of the 4-force and does not transform like that. Rather, it results in (see *e.g.* [45], p 296ff):

$$f_x' = f_x - \frac{v u_y'}{c^2 \sqrt{1 - \beta^2}} f_y - \frac{v u_z'}{c^2 \sqrt{1 - \beta^2}} f_z = f_x - \frac{v u_y}{c^2 \left(1 - \dfrac{v u_x}{c^2}\right)} f_y - \frac{v u_z}{c^2 \left(1 - \dfrac{v u_x}{c^2}\right)} f_z$$

$$f_y' = \frac{1 + \dfrac{v u_x'}{c^2}}{\sqrt{1 - \beta^2}} f_y = \frac{\sqrt{1 - \beta^2}}{1 - \dfrac{v u_x}{c^2}} f_y \;; \qquad f_z' = \frac{1 + \dfrac{v u_x'}{c^2}}{\sqrt{1 - \beta^2}} f_z = \frac{\sqrt{1 - \beta^2}}{1 - \dfrac{v u_x}{c^2}} f_z$$

(A.6.72)

This transformation allows for a very remarkable rearrangement

$$\mathbf{f}' = \begin{bmatrix} f_x' \\ f_y' \\ f_z' \end{bmatrix} = \begin{bmatrix} f_x \\ f_y/(\sqrt{1 - \beta^2}) \\ f_z/(\sqrt{1 - \beta^2}) \end{bmatrix} + \begin{bmatrix} u_x' \\ u_y' \\ u_z' \end{bmatrix} \times \begin{bmatrix} 0 \\ f_z \\ -f_y \end{bmatrix} \frac{v}{c^2 \sqrt{1 - \beta^2}} .$$

(A.6.73)

Suppose *e.g.* that $\mathbf{f} = Q\mathbf{E}$, then the transformed force $\mathbf{f}'$ contains the Lorentz force in the form of the customary vector product $Q\mathbf{u} \times \mathbf{B}$. However, any other arbitrary force results in a force analogous to the Lorentz force.

Finally, the 4-acceleration shall be mentioned as well:

$$\frac{1}{\sqrt{1 - \dfrac{u^2}{c^2}}} \cdot \frac{\mathrm{d}}{\mathrm{d}t} \frac{\langle \mathbf{u}, ic \rangle}{\sqrt{1 - \dfrac{u^2}{c^2}}} .$$

(A.6.74)

Here again, the 3-acceleration $\mathbf{b} = \dfrac{\mathrm{d}\mathbf{u}}{\mathrm{d}t}$ does not constitute the spatial part of the 4-acceleration. The transformation of $\mathbf{b}$ yields (see *e.g.*, [45], p 296)

$$b_x' = \left(\frac{1 + \dfrac{vu_x'}{c^2}}{\sqrt{1-\beta^2}}\right)^3 b_x = \left(\frac{\sqrt{1-\beta^2}}{1 + \dfrac{vu_x}{c^2}}\right)^3 b_x$$

$$b_y' = \left(\frac{1 + \dfrac{vu_x'}{c^2}}{\sqrt{1-\beta^2}}\right)^2 \left[b_y + \frac{vu_y'}{c^2\sqrt{1-\beta^2}}b_x\right] = \left(\frac{\sqrt{1-\beta^2}}{1 - \dfrac{vu_x}{c^2}}\right)^2 \left[b_y + \frac{vu_y}{c^2\left(1 - \dfrac{vu_x}{c^2}\right)}b_x\right]$$

$$b_z' = \left(\frac{1 + \dfrac{vu_x'}{c^2}}{\sqrt{1-\beta^2}}\right)^2 \left[b_z + \frac{vu_z'}{c^2\sqrt{1-\beta^2}}b_x\right] = \left(\frac{\sqrt{1-\beta^2}}{1 - \dfrac{vu_x}{c^2}}\right)^2 \left[b_z + \frac{vu_z}{c^2\left(1 - \dfrac{vu_x}{c^2}\right)}b_x\right]$$

$$\tag{A.6.75}$$

These equations for the transformation of **b** contain a portion that is in the form of a vector product, similarly to eqs. (A.6.73) for **f** before.

Clearly, for $c \to \infty$, it is both, $\mathbf{f}' = \mathbf{f}$ and $\mathbf{b}' = \mathbf{b}$.

If the force **f** can be expressed by a potential, *i.e.*, can be expressed by the potential energy U, then

$$-\nabla U = \mathbf{f} \tag{A.6.76}$$

and the equation of the motion takes the following form

$$-\nabla U = \frac{d}{dt}\frac{m_0 \mathbf{u}}{\sqrt{1 - \dfrac{u^2}{c^2}}} . \tag{A.6.77}$$

Multiplying by $\mathbf{u} = \dfrac{d\mathbf{x}}{dt}$ gives

$$-\nabla U \bullet \frac{d\mathbf{x}}{dt} = -\frac{dU}{dt} = \mathbf{u} \bullet \frac{d}{dt}\frac{m_0 \mathbf{u}}{\sqrt{1 - \dfrac{u^2}{c^2}}} .$$

We also have

$$\mathbf{u} \bullet \frac{d}{dt} \frac{m_0\mathbf{u}}{\sqrt{1-\frac{u^2}{c^2}}} = \frac{d}{dt} \frac{m_0 u^2}{\sqrt{1-\frac{u^2}{c^2}}} - \frac{m_0\mathbf{u}}{\sqrt{1-\frac{u^2}{c^2}}} \bullet \frac{d\mathbf{u}}{dt} = \frac{d}{dt} \frac{m_0 u^2}{\sqrt{1-\frac{u^2}{c^2}}} + \frac{d}{dt}\left(m_0 c^2 \sqrt{1-\frac{u^2}{c^2}}\right)$$

$$= \frac{d}{dt}\left[\frac{m_0 u^2}{\sqrt{1-\frac{u^2}{c^2}}} + \frac{m_0 c^2\left(1-\frac{u^2}{c^2}\right)}{\sqrt{1-\frac{u^2}{c^2}}}\right] = \frac{d}{dt}\left(\frac{m_0 c^2}{\sqrt{1-\frac{u^2}{c^2}}}\right)$$

and

$$\boxed{\frac{d}{dt}(U+mc^2) = 0}\;, \tag{A.6.78}$$

that is, $U+mc^2$, *the sum of potential and kinetic energy, including rest energy of the particle $m_0 c^2$ is a conserved quantity.* This represents a relativistic generalization of the related theorem of classical mechanics, which results when the velocities approach zero:

$$\frac{d}{dt}(U+mc^2) = \frac{d}{dt}\left[U+\frac{m_0 c^2}{\sqrt{1-\frac{u^2}{c^2}}}\right] = \frac{d}{dt}\left[U+m_0 c^2\left(1+\frac{1}{2}\frac{u^2}{c^2}+\ldots\right)\right]$$

$$\approx \frac{d}{dt}\left[U+m_0 c^2 + \frac{1}{2}m_0 u^2\right] = 0\;. \tag{A.6.79}$$

A.6.5.3 Field Tensor F

The 4-potential allows to calculate all components of **B** and **E**.

$$B_{x1} = \left[\frac{\partial A_{x3}}{\partial x_2} - \frac{\partial A_{x2}}{\partial x_3}\right] \qquad E_{x1} = -\frac{\partial \varphi}{\partial x_1} - \frac{\partial A_{x1}}{\partial t} = ic\left[\frac{\partial A_{x4}}{\partial x_1} - \frac{\partial A_{x1}}{\partial x_4}\right]$$

$$B_{x2} = \left[\frac{\partial A_{x1}}{\partial x_3} - \frac{\partial A_{x3}}{\partial x_1}\right] ; \quad E_{x2} = -\frac{\partial \varphi}{\partial x_2} - \frac{\partial A_{x2}}{\partial t} = ic\left[\frac{\partial A_{x4}}{\partial x_2} - \frac{\partial A_{x2}}{\partial x_4}\right] .$$

$$B_{x3} = \left[\frac{\partial A_{x2}}{\partial x_1} - \frac{\partial A_{x1}}{\partial x_2}\right] \qquad E_{x3} = -\frac{\partial \varphi}{\partial x_3} - \frac{\partial A_{x3}}{\partial t} = ic\left[\frac{\partial A_{x4}}{\partial x_3} - \frac{\partial A_{x3}}{\partial x_4}\right]$$

$$\tag{A.6.80}$$

All these quantities are formally calculated like the components of the curl of a vector. Naturally, in the three-dimensional space, there are only three components, which can also be expressed by a three-dimensional vector (simply the curl) even though curl does not constitute an ordinary vector. For the four-dimensional space, there are six such components, which sometimes are called a "6-vector". In a sense, this represents the curl of a vector in the four-dimensional space. Nevertheless, in mathematical terms, this is a tensor of rank 2, a so-called *field tensor* with the components

$$F_{ik} = \left[\frac{\partial A_{x_k}}{\partial x_i} - \frac{\partial A_{x_i}}{\partial x_k} \right] . \tag{A.6.81}$$

Each of the two terms in this summation, individually embodies a tensor of rank 2 in the sense of (A.6.47), *i.e.*, it constitutes a tensor product of the four-dimensional ∇-vector and the 4-potential. The field tensor is anti-symmetric and its components have the following property:

$$F_{ik} = -F_{ki} , \qquad F_{ii} = 0 , \tag{A.6.82}$$

namely, it has six independent components, which by (A.6.80), essentially, are just the components of **E** and **B**

$$F = \begin{bmatrix} 0 & +B_{x3} & -B_{x2} & -\frac{i}{c}E_{x1} \\ -B_{x3} & 0 & +B_{x1} & -\frac{i}{c}E_{x2} \\ +B_{x2} & -B_{x1} & 0 & -\frac{i}{c}E_{x3} \\ \frac{i}{c}E_{x1} & \frac{i}{c}E_{x2} & \frac{i}{c}E_{x3} & 0 \end{bmatrix} . \tag{A.6.83}$$

It is obvious, even without further calculation, how the field components transform, namely as of (A.6.47). For the simple case of $\mathbf{v} = \langle v_1, 0, 0 \rangle$, we obtain the same result as eqs. (A.6.39) and (A.6.40), respectively, which can easily be verified by substitution.

Strictly speaking, the components of the curl of a three-dimensional vector (as well as the vector product of two vectors) are also the components of an anti-symmetric tensor of rank 2.

Next we take Maxwell's equations

$$\left. \begin{aligned} \nabla \times \mathbf{B} &= \mu_0 \mathbf{g} + \frac{1}{c^2} \frac{\partial \mathbf{E}}{\partial t} \\[2mm] \nabla \bullet \mathbf{E} &= \frac{\rho}{\varepsilon_0} \\[2mm] \nabla \times \mathbf{E} &= -\frac{\partial \mathbf{B}}{\partial t} \\[2mm] \nabla \bullet \mathbf{B} &= 0 \end{aligned} \right\} \tag{A.6.84}$$

and reformulate them by means of F. When substituting the respective tensor components of (A.6.83) into (A.6.84), then from the third and the fourth equation we obtain

$$\frac{\partial F_{34}}{\partial x_2} + \frac{\partial F_{42}}{\partial x_3} + \frac{\partial F_{23}}{\partial x_4} = 0$$

$$\frac{\partial F_{41}}{\partial x_3} + \frac{\partial F_{13}}{\partial x_4} + \frac{\partial F_{34}}{\partial x_1} = 0$$

$$\frac{\partial F_{12}}{\partial x_4} + \frac{\partial F_{24}}{\partial x_1} + \frac{\partial F_{41}}{\partial x_2} = 0$$

$$\frac{\partial F_{23}}{\partial x_1} + \frac{\partial F_{31}}{\partial x_2} + \frac{\partial F_{12}}{\partial x_3} = 0 \ .$$

These four equations can be combined into a single equation

$$\boxed{\frac{\partial F_{kl}}{\partial x_i} + \frac{\partial F_{li}}{\partial x_k} + \frac{\partial F_{ik}}{\partial x_l} = 0}$$

(A.6.85)

where (i, k, l) represent the number sequences $(1, 2, 3)$, $(2, 3, 4)$, $(3, 4, 1)$, or $(4, 1, 2)$, (that is, they are three different numbers out of the range $1 .. 4$ in cyclical arrangement). The first two equations of (A.6.84) yield

$$\boxed{\sum_{k=1}^{4} \frac{\partial}{\partial x_k} F_{ik} = \mu_0 \langle \mathbf{g}, ic\rho \rangle}$$

(A.6.86)

The components of the first equation are obtained for $i = 1, 2, 3$, while for $i = 4$ one gets the components of the second equation. As for eq. (A.6.85), this represents a tensor of rank 3. Eq. (A.6.86) is a vector equation. The term on the left originates from the field tensor by taking the divergence (namely, by the scalar product with the four-dimensional ∇-vector). This constitutes a reduction of the tensor's rank by one and thus creates a vector out of the field tensor (rank 2). The vector on the right side is essentially the 4-current density.

Both, eq. (A.6.85) and (A.6.86) provide a representation of Maxwell's equations that is very elegant, and immediately reveals their Lorentz invariance. However, this form is also very abstract and due to its unfamiliarity, it has lost the conceptual clarity of the customary form of Maxwell's equations. Both forms are entirely equivalent. One is advised to employ both, depending on the type of problem to be solved.

The purpose of this Appendix is not to generalize and re-write the entire field theory in a relativistic form. This has been done by various authors which are listed for further reference [40 - 51]. Here, we will limit ourselves to a few simple problems with the intent to further clarify the terminology and their interdependence.

A.6.6 Examples

A.6.6.1 Surface Charges and their Fields

Consider two infinite, parallel planes carrying the uniform surface charge $\pm\sigma$ (Fig. A.6.2). This causes a uniform field between the planes with the magnitude $E_{x_2} = \sigma/\varepsilon_0$. Now we observe the situation from within a reference frame Σ', which moves with the constant velocity $\mathbf{v}$ in the direction parallel to the x_1-axis. The question is now, what field E_{x_2}' is observed from within the moving frame?

First, we note that *the entire charge inside a volume is relativistically invariant*. This is a well verified, a very important, and a seemingly trivial result. Nevertheless, charge densities are not invariant, which can also be seen from the fact that it occurs as the fourth component of a 4-vector. Applying the Lorentz transformation (A.6.24) to (A.6.51) for $\mathbf{g} = 0$ gives

$$\rho' = \frac{\rho}{\sqrt{1 - \dfrac{v^2}{c^2}}}, \qquad g_{x_1} = \frac{-\rho v}{\sqrt{1 - \dfrac{v^2}{c^2}}}. \qquad (A.6.87)$$

Focusing on the surface elements of the charged plane (Fig. A.6.2), we find that the charge of such a surface element is invariant. From the moving system's perspective, a surface element is shortened in the direction of the motion due to the Lorentz contraction by a factor $\sqrt{1 - v^2/c^2}$. The surface charge density is therefore increased by the same factor

$$\sigma' = \frac{\sigma}{\sqrt{1 - \dfrac{v^2}{c^2}}}, \qquad (A.6.88)$$

which can easily be found by taking the limit of (A.6.87). Both equations, (A.6.87) and (A.6.88) can therefore be apprehended conceptually as a consequence of the Lorentz contraction. Naturally, along with increasing σ' goes an increase of E_{x_2}' by the same factor

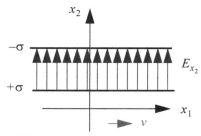

Fig. A.6.2

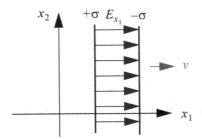

Fig. A.6.3

$$E'_{x_2} = \frac{E_{x_2}}{\sqrt{1 - \dfrac{v^2}{c^2}}} , \tag{A.6.89}$$

which conforms with the transformation equations (A.6.39) and (A.6.40), as $\mathbf{B} = 0$. From the perspective of system Σ', the moving surface charges correspond to surface current densities. Calculating their magnetic field $\mathbf{B'}$ yields

$$B'_{x_3} = -\frac{v E_{x_2}}{c^2 \sqrt{1 - \beta^2}} = -\frac{v\sigma}{\varepsilon_0 c^2 \sqrt{1 - \beta^2}} = -\frac{\mu_0 v\sigma}{\sqrt{1 - \beta^2}} . \tag{A.6.90}$$

which agrees with eqs. (A.6.39) and (A.6.40). Of course, all fields vanish in the outside space.

Next, we repeat the analysis for the situation depicted in Fig. A.6.3. Now the charged surfaces are perpendicular to $\mathbf{v}$, and therefore

$$\sigma' = \sigma \tag{A.6.91}$$

and

$$E'_{x_1} = E_{x_1} , \tag{A.6.92}$$

whereby eq. (A.6.92) also agrees with (A.6.39) and (A.6.40).

Furthermore, (A.6.87) applies as well. Nonetheless, the applicable equation for σ' is (A.6.91) but not (A.6.88). The reason is that the surface charge is obtained as the limit of a layer with finite extend.

$$\left. \begin{aligned} &\sigma = \rho d (\rho \to \infty, d \to 0); && \rho' = \frac{\rho}{\sqrt{1 - \beta^2}} \\ &d' = d\sqrt{1 - \beta^2} ; && \rho'd' = \rho d , && \sigma' = \sigma \end{aligned} \right\} . \tag{A.6.93}$$

The moving charges cause currents in the Σ' frame, but for symmetry reasons, these do not cause a magnetic field. Similarly, the equations of the transformation do not yield a magnetic field because of $\mathbf{E} \| \mathbf{v}$. Instead of the fields, one might transform the 4-potential. While in the Σ frame the vector potential is $\mathbf{A} = 0$, in the Σ' frame we have $A \neq 0$:

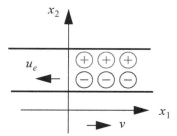

Fig. A.6.4

$$\mathbf{A}' = \langle A_{x_1}'(x_1'), 0, 0 \rangle; \qquad \mathbf{B}' = \nabla' \times \mathbf{A}' = 0 \ . \tag{A.6.94}$$

We have realized already in Sect. (A.6.4) that the component of the location vector which is parallel to the relative velocity $\mathbf{u}$ changes, while the component perpendicular to it remains unchanged, and that this behavior is reversed for the field components.

A.6.6.2 Currents and Volume Charges

If the 4-vector of the current density in the rest frame Σ is $\langle g_{x_1}, 0, 0, ic\rho \rangle$, then for the reference frame Σ' we have

$$\left.\begin{aligned}
g_{x_1}' &= \frac{1}{\sqrt{1 - \dfrac{v^2}{c^2}}}(g_{x_1} - v\rho), \qquad g_{x_2}' = 0, \qquad g_{x_3}' = 0 \\[2mm]
ic\rho' &= \frac{1}{\sqrt{1 - \dfrac{v^2}{c^2}}}\left(-\frac{iv g_{x_1}}{c} + ic\rho\right)
\end{aligned}\right\} \ . \tag{A.6.95}$$

Even if $\rho = 0$, it still is $\rho' \neq 0$, namely

$$\rho' = -\frac{v g_{x_1}}{c^2 \sqrt{1 - \dfrac{v^2}{c^2}}} \ . \tag{A.6.96}$$

This might come as a surprise at first, but can be explained quite well on a conceptual level. As an **example**, Fig. A.6.4 depicts a metallic conductor. From the perspective of the rest frame Σ, it contains ions and electrons with the volume charge densities $+\rho_0$ and $-\rho_0$. The electrons travel in negative x_1-direction with the velocity u_e. The total charge density is $\rho_0 - \rho_0 = 0$. The question is now what is the perspective of an observer in a reference frame that moves with velocity v in the positive x_1-direction? First, the volume charge density of the electrons in their own reference frame (within which they are at rest) is

$$\rho_{0e} = -\rho_0 \sqrt{1 - \frac{u_e^2}{c^2}} \; .$$

The velocity of the observer relative to the rest system of the electrons is given by eq. (A.6.31) as

$$\frac{v + u_e}{1 + \frac{vu_e}{c^2}} \; .$$

Therefore, in the reference frame Σ' of the moving observer we have for the electrons

$$\rho_e' = \rho_{0e} \frac{1}{\sqrt{1 - \frac{1}{c^2}\left(\frac{v + u_e}{1 + \frac{vu_e}{c^2}}\right)^2}} = -\frac{\rho_0 \sqrt{1 - \frac{u_e^2}{c^2}}}{\sqrt{1 - \frac{1}{c^2}\left(\frac{v + u_e}{1 + \frac{vu_e}{c^2}}\right)^2}}$$

$$= -\frac{\rho_0 \sqrt{1 - \frac{u_e^2}{c^2}}\left(1 + \frac{vu_e}{c^2}\right)}{\sqrt{1 - \frac{v^2}{c^2} - \frac{u_e^2}{c^2} + \frac{v^2 u_e^2}{c^4}}} = -\frac{\rho_0 \sqrt{1 - \frac{u_e^2}{c^2}}\left(1 + \frac{vu_e}{c^2}\right)}{\sqrt{1 - \frac{u_e^2}{c^2}} \sqrt{1 - \frac{v^2}{c^2}}}$$

$$\rho_e' = -\rho_0 \frac{1 + \frac{vu_e}{c^2}}{\sqrt{1 - \frac{v^2}{c^2}}} \; .$$

The volume charge density for the ions is

$$\rho_i' = +\rho_0 \frac{1}{\sqrt{1 - \frac{v^2}{c^2}}} \; .$$

This makes the total charge density in the moving system Σ'

$$\rho' = \rho_i' + \rho_e' = \frac{\rho_0}{\sqrt{1 - \frac{v^2}{c^2}}}\left(1 - 1 - \frac{vu_e}{c^2}\right) = -\frac{\rho_0 vu_e}{c^2 \sqrt{1 - \frac{v^2}{c^2}}} = -\frac{v g_{x1}}{c^2 \sqrt{1 - \frac{v^2}{c^2}}} \; ,$$

which agrees with (A.6.96). For g_{x1} one finds:

$$g'_{x\,1} = -\rho_e' \frac{v+u_e}{1+\dfrac{vu_e}{c^2}} - \rho_i' v = \rho_0 \frac{1+\dfrac{vu_e}{c^2}}{\sqrt{1-\dfrac{v^2}{c^2}}} \cdot \frac{v+u_e}{1+\dfrac{vu_e}{c^2}} - \frac{\rho_0 v}{\sqrt{1-\dfrac{v^2}{c^2}}}$$

$$g'_{x\,1} = \frac{\rho_0}{\sqrt{1-\dfrac{v^2}{c^2}}}(v+u_e-v) = \frac{\rho_0 u_e}{\sqrt{1-\dfrac{v^2}{c^2}}} = \frac{g_{x1}}{\sqrt{1-\dfrac{v^2}{c^2}}} \,, \tag{A.6.97}$$

which agrees with (A.6.95) for $\rho = 0$. Of course, the magnitude of the current density 4-vector is invariant, $i.e.$, for $\rho = 0$ we have

$$g_{x1}^2 = g'^2_{x\,1} - c^2\rho'^2 \,. \tag{A.6.98}$$

A.6.6.3 Force of a Current on a moving Charge

If, instead, we replace the observer who moves with velocity $\mathbf{v}$, by a charged particle, then a force is exerted on this particle. In the system where the conductor is at rest, this is caused by its magnetic field in the form of the Lorentz force. Now, there is no electric force because of $\rho = 0$. Conversely, there exists no Lorentz force in the reference frame where the charged particle is at rest. Instead, now there exist an electric field which is created by the volume charges and which exerts a force on the particle. This electric field is purely radial

$$E_r' = \frac{r_0^2 \pi \rho'}{2\pi\varepsilon_0 r} \,. \tag{A.6.99}$$

r_0 is the radius of the conductor and $r_0^2\pi$ its cross-sectional area. Using ρ' from eq. (A.6.96) gives

$$E_r' = \frac{-r_0^2\pi}{2\pi\varepsilon_0 r} \cdot \frac{vg_{x1}}{c^2\sqrt{1-\dfrac{v^2}{c^2}}} = \frac{-vI}{2\pi\varepsilon_0 r \dfrac{1}{\varepsilon_0\mu_0}\sqrt{1-\dfrac{v^2}{c^2}}} = \frac{-\mu_0 vI}{2\pi r\sqrt{1-\dfrac{v^2}{c^2}}} \,. \tag{A.6.100}$$

I represents the current in the rest sytem of the conductor. This results in

$$\frac{\mu_0 I}{2\pi r} = B_\varphi \,, \tag{A.6.101}$$

which is nothing else than the azimuthal component of the conductor's magnetic field in the rest system. Therefore

$$E_r' = \frac{-vB_\varphi}{\sqrt{1-\dfrac{v^2}{c^2}}} = \frac{(\mathbf{v}\times\mathbf{B})_{r'}}{\sqrt{1-\dfrac{v^2}{c^2}}} \tag{A.6.102}$$

that is, we obtain exactly the same field as we would get when transforming the fields according to eqs. (A.6.39) and (A.6.40). The transformed magnetic field is explained by the current density when transformed according to (A.6.97).

These examples illustrated that the seemingly peculiar phenomena of the theory of relativity are not as implausible as they appear at first. Based only on the assumption of the Lorentz contraction and the relativistic addition of velocities, we were able to derive everything else in a conceptually clear manner. This demonstrates that the Lorentz force is, basically, also an electrical force which is caused by relativistic effects. Fundamentally, the magnetic field is not necessary. Nevertheless, our example has shown that the introduction of the magnetic field is useful and description of the relations is thereby tremendously simplified. The calculation without the magnetic field required to first find the force on the particle in its rest frame and subsequently transform this force into the given reference frame.

A.6.6.4 Field of a Uniformly Moving Point Charge

We have presented the Liénard-Wiechert Potentials in Sect. A.4, and as a special case thereof, those of a uniformly moving point charge.

Oftentimes, applying relativity simplifies the process to solve electromagnetic problems tremendously. For this purpose, one first solves the problem in a reference frame which makes the problem particularly simple. Then, that solution is transformed into the given reference frame. Eqs. (A.4.16) and (A.4.17) of Section A.4 represent the potentials for the field of a charge moving parallel to the x-axis with velocity v_0.

The potentials in the moving reference frame Σ' are

$$\mathbf{A}' = 0 , \qquad \varphi' = \frac{Q}{4\pi\varepsilon_0\sqrt{x'^2 + y'^2 + z'^2}} . \tag{A.6.103}$$

The 4-potential in Σ' is therefore

$$\langle 0, 0, 0, \frac{iQ}{4\pi\varepsilon_0 c\sqrt{x'^2 + y'^2 + z'^2}} \rangle .$$

Using the Lorentz operator L as of (A.6.27) gives for the 4-potential in Σ

$$
\begin{bmatrix} A_x \\ A_y \\ A_z \\ \dfrac{i\varphi}{c} \end{bmatrix} = L^{-1} \begin{bmatrix} 0 \\ 0 \\ 0 \\ \dfrac{i\varphi'}{c} \end{bmatrix} = \begin{bmatrix} \dfrac{1}{\sqrt{1-\beta^2}} & 0 & 0 & \dfrac{-iv_0}{c\sqrt{1-\beta^2}} \\ 0 & 1 & 0 & 0 \\ 0 & 0 & 1 & 0 \\ \dfrac{iv_0}{c\sqrt{1-\beta^2}} & 0 & 0 & \dfrac{1}{\sqrt{1-\beta^2}} \end{bmatrix} \begin{bmatrix} 0 \\ 0 \\ 0 \\ \dfrac{i\varphi'}{c} \end{bmatrix},
$$

i.e.,

$$
A_x = \frac{v_0 \varphi'}{c^2 \sqrt{1-\beta^2}}, \qquad A_y = A_z = 0, \qquad \varphi = \frac{\varphi'}{\sqrt{1-\beta^2}}
$$

$$
A_x = \frac{v_0 \varphi}{c^2}, \qquad \varphi = \frac{Q}{4\pi\varepsilon_0 \sqrt{1-\beta^2}\sqrt{x'^2 + y'^2 + z'^2}}. \tag{A.6.104}
$$

When replacing x', y', and z' by x, y, and z according to eq. (A.6.25), then after some simple calculations, one obtains just the potentials of eqs. (A.4.16) and (A.4.17), as is has to be. This way of solving the problem is indeed much simpler than it was before. We learn that it can be beneficial to determine which choice of a particular reference frame simplifies the solution of the problem. Furthermore, this proceeding fosters a deeper understanding of a problem and its solution.

A.6.7 Final Remarks

It was not the objective of this Appendix to present a detailed treatment of the entire theory of relativity. Nevertheless, because it has emerged from the electromagnetic field theory as a mandatory consequence, and conversely, the electromagnetic field theory can only be understood completely in light of the theory of relativity, this Appendix's intention was to foster a deeper understanding. It became clear at many occasions throughout the book, that relativity emerges frequently and its understanding is necessary to comprehend the topic at hand.

In closing, Sommerfeld shall be quoted. At the beginning of the Section on the four-dimensional formulation of Maxwell's equations, he states: "I wish to create the impression in my audience that the true mathematical form of those constructs unfolds only now, just like a mountain view when the fog breaks open" [40, p 197]. A little later: "From the point of view of Maxwell's equations, the theory of relativity is self evident. Simply from the form of Maxwell's equations, a mathematician, whose eyes were trained by Klein in the Erlanger program, should have been able to see and read out its transformation group, including all of its kinematical and optical consequences" [40, p 200]. These statements do by no means limit Einstein's extraordinary accomplishments. However, there can be no doubt, that if relativity were wrong, along with relativity, Maxwell's equations had to be disposed of as well. Expressed differently, the convincing proofs for the validity of Maxwell's equations by numerous, different experiments, carried out independently from each other, provide an equally convincing testimony for the validity of the theory of relativity.

Bibliography

[1]   Purcell, E. M.: The fields of moving charges. In: Berkeley physics Course. Vol. 2, New York: McGraw-Hill, 1965
[2]   Moon, P.; Spencer, D. E.: Field Theory Handbook. 2nd ed. Berlin: Springer, 1988 und: Field Theory for Engineers. Princeton, Toronto, London: Van Nostrand, 1961
[3]   Ryshik, I. M.; Gradstein, I. S.: Summen-, Produkt- und Integraltafeln. Berlin: VEB Deutscher Verlag der Wissenschaften, 1957
[4]   Erdelyi, A.; Magnus, W.; Oberhettinger, F.; Tricomi, F. G.: Higher Transcendental Functions, New York, Toronto, London: McGraw-Hill, Vol. I 1953, Vol. 11 1953, Vol. III 1955
[5]   same authors: Tables of Integral Transforms. New York, Toronto, London: McGraw-Hill, 2 volumes, 1954
[6]   Smirnow, W. 1.: Lehrgang der Höheren Mathematik. Berlin: VEB Deutscher Verlag der Wissenschaften, Teil I 1967, Teil 11 1966, Teil 111, 1 und 111, 2 1967, Teil IV, 1966, Teil V 1967
[7]   Watson, G. N.: A Treatise on the Theory of Bessel Functions. Cambridge: University Press, 1958
[8]   Stratton, J. A.: Electromagnetic Theory. New York, London: McGraw-Hill, 1941
[9]   Morse, P. M.; Feshbach, H.: Methods of Theoretical Physics. Part I, 11. New York: McGraw-Hill, 1953
[10]   Petrovskij, I. G.: Vorlesungen über die Theorie der Integralgleichungen. Würzburg: Physica-Verlag, 1953
[11]   Sternberg, W.: Potentialtheorie 1 Die Elemente der Potentialtheorie, Potentialtheorie 11 Die Randwertaufgaben der Potentialtheorie, 2 Bände, Sammlung Güschen. Berlin, Leipzig: Walter de Gruyter, 1925, 1926 C123 Kellogg, 0 . D.: Foundations of Potential Theory. Berlin: Julius Springer, 1929
[13]   Günther, N. M.: Die Potentialtheorie und ihre Anwendungen auf Grundaufgaben der mathematischen Physik. Leipzig: B. G. Teubner, 1957
[14]   Walter, W.: Einführung in die Potentialtheorie. Mannheim, Wien, Zürich: Bibliographisches Institut, 1971
[15]   Martensen, E.: Potentialtheorie: Stuttgart, B. G. Teubner, 1968
[16]   Sigl, R: Einführung in die Potentialtheorie, 2. Auflage. Karlsruhe: Wichmann, 1989
[17]   Davies, A. J.: The Finite Element Method-A First Approach. Oxford: Clarendon Press, 1986
[18]   Zurmühl, R.; Falk, S.: Matrizen und ihre Anwendung, Band 1, Grundlagen, 6. Auflage. Berlin, Heidelberg, New York: Springer, 1992
[19]   Marsal, D.: Finite Differenzen und Elemente. Berlin etc.: Springer Verlag, 1989
[20]   Bader, G.: Domain Decomposition Methoden für gemischte elliptische Randwertprobleme. Archiv für Elektrotechnik 145 (74) 1990
[21]   Zienkiewicz, 0. C.: The Finite Element Method, Fourth Edition. 2 Bände, London etc.: McGraw-Hill, 1989
[22]   Silvester, P. P.; Ferrari, R. L.: Finite elements for engineers, Second Edition. Cambridge: Cambridge University Press, 1990
[23]   Dhatt, G.; Touzot, G.: The Finite Element Method Displayed. Chichester, New York etc.: John Wiley & Sons, 1984
[24]   Strang, G.; Fix, G. J.: An Analysis of the finiteelement method. New Jersey: Prentice-Hall, 1973
[25]   Bathe, K.-J.: Finite-Elemente Methoden. Berlin etc.: Springer-Verlag, 1990
[26]   Schwarz, H. R.: Methode der finiten Elemente. Stuttgart: B. G. Teubner, 1984
[27]   Kämmel, G.; Franeck, H.; Recke, H.-G.: Einführung in die Methode der finiten Elemente. 2. Auflage, München, Wien: Carl Hanser, 1990
[28]   Brebbia, C. A.: The Boundary Element Method for Engineers. London, Plymouth: Pentech Press, 1984
[29]   Brebbia, C. A.; Telles, J. C. F.; Wrobel, L. C.: Boundary Element Techniques. Berlin etc.: Springer Verlag, 1984
[30]   Hengartner, W.; Theodorescu, R.: Einführung in die Monte-Carlo-Methode. München, Wien: Carl Hanser, 1978
[31]   Buslenko, N. P.; Schreider, J. A.: Die Monte-Carlo-Methode und ihre Verwirklichung mit elektronischen Digitalrechnern. Leipzig: B. G. Teubner, 1964
[32]   Goldhaber, A. S.; Nieto, M. M: The mass of the photon. Scientific American, 234, Nir. 5 (Mai 1976) 86

[33]   Kroll, N. M.: Concentric spherical cavities and limits on the photon rest mass. Phys. Rev. Letters 27 (1971) 340

[34]   Aharonov, Y.; Bohm, D.: Significance of electromagnetic potentials in the quantum theory. Phys. Rev. 115 (1959) 485

[35]   Olariu, S.; Popescu, I. 1.: The quantum effects of electromagnetic fluxes. Rev. Mod. Phys., 57 (1985) 339

[36]   Tonomura, A.; Noboyuki, 0 . ; Matsuda, T.; Kawasaki, T.; Endo, J.; Yano, S.; Yamada, H.: Evidence for Aharonov-Bohm effect with maanetic field completely shielded from electron wave. Phvs. - . Rev. Letters 56 (1986) 729

[37]   Feynman, R.P.; Leighton, R.B.; Sands, M.: Vorlesungen über Physik. München, Wien: R. Oldenbourg, Band I, Teil 1, 1974

[38]   Penrose, R.: The apparent shape of a relativistically moving sphere. Proc. Cambridge Phil. Soc. 55 (1959)

[39]   Terrell, J.: Invisibility of the Lorentz Contraction. Phys. Rev. 116 (1959) 1041

[40]   Sommerfeld A.: Vorlesungen über Theoretische Physik, Band 111, Elektrodynamik (revidiert von F. Bopp und J. Meixner, Nachdruck der 4. durchgesehenen Auflage). Thun, Frankfurt/M.: Verlag Harn Deutsch, 1988

[41]   Simonyi, K.: Theoretische Elektrotechnik. Leipzig, Berlin, Heidelberg: Johann Ambrosius Barth, Edition Deutscher Verlag der Wissenschaften, 10. Auflage 1993, S. 921 ff.

[42]   French, A.P.: Die spezielle Relativitätstheorie, MIT Einführungskurs Physik. Braunschweig: Fr. Vieweg & Sohn, 1971

[43]   Rosser, W.G.V: Introductory Special Relativity. London, New York, Philadelphia: Taylor & Francis, 1991

[44]   Mould, R. A.: Basic Relativity. New York, Berlin, Heidelberg: Springer-Verlag, 1994

[45]   Melcher, H.: Relativitätstheone in elementarer Darstellung mit Aufgaben und Lösungen. Berlin: VEB Deutscher Verlag der Wissenschaften, 1974

[46]   Papapetrou, A.: Spezielle Relativitätstheorie. Berlin: VEB Deutscher Verlag der Wissenschaften, 5. Auflage 1975

[47]   Resnick, R.: Einführung in die Spezielle Relativitätstheorie. Stuttgart: Kleti-Verlag, 1976

[48]   Schröder, U.E.: Spezielle Relativitätstheone. Thun, Frankfurt/M.: Verlag Harn Deutsch, 2. Auflage 1987

[49]   Rindler, W.: Introduction to Special Relativity. Oxford: Clarendon Press, 1991

[50]   Ruder, H.; Ruder, M.: Die Spezielle Relativitätstheorie. Braunschweig, Wiesbaden: Friedr. Vieweg & Sohn, 1993

[51]   Van Bladel, J.: Relativity and Engineering. Berlin, Heidelberg, New York, Tokyo: Springer-Verlag, 1984

Generally Recommended Reading

Jackson, J. D.: Classical electrodynamics. Third ed. New York, London, Sydney, Toronto: John Wiley & Sons, 1999

Feynman, R. P.; Leighton, R. B.; Sands, M.: Vorlesungen über Physik. München, Wien: R.Oldenbourg, Band I1 Teil 1, 1973, Band I1 Teil 2, 1974

Simonyi, K.: Theoretische Elektrotechnik. Leipzig, Berlin, Heidelberg: Johann Ambrosius Barth, Edition Deutscher Verlag der Wissenschaften, 10. Aufl. 1993

Stratton, J. A.: Electromagnetic Theory. New York, London: McGraw-Hill, 1941

Smythe, W. R.: Static and Dynamic Electricity. New York: McGraw-Hill, 1968

Durand, E.: Electrostatique Tome I, Les Distributions. Paris: Masson et Cie, 1964

Durand, E.: Electrostatique Tome 11, Problemes Generaux Conducteurs. Paris: Masson et Cie, 1966

Durand, E.: Electrostatique Tome 111, Methodes de Calcul Dielectriques. Paris: Masson et Cie, 1966

Durand, E.: Magnétostatique. Pans: Masson et Cie, 1968

Index